I0044399

Bioenergy: Emerging Trends

Bioenergy: Emerging Trends

Edited by Alice Wheeler

New York

Published by Syrawood Publishing House,
750 Third Avenue, 9th Floor,
New York, NY 10017, USA
www.syrawoodpublishinghouse.com

Bioenergy: Emerging Trends
Edited by Alice Wheeler

International Standard Book Number: 978-1-64740-120-7 (Hardback)

Cataloging-in-Publication Data

Bioenergy : emerging trends / edited by Alice Wheeler.
p. cm.
Includes bibliographical references and index.
ISBN 978-1-64740-120-7
1. Biomass energy. 2. Bioenergetics. 3. Energy crops. I. Wheeler, Alice.
TP339 .B56 2022
662.88--dc23

TABLE OF CONTENTS

PREFACE

Bioenergy is the energy harnessed from materials that are derived from biological sources. The primary source of bioenergy is biomass, which is an organic material that stores sunlight in the form of chemical energy. Biomass can be utilized for the production of heat and energy, and can be used as a raw material in industrial processes. Biomass fuel may include animal waste, wood and wood waste, sugarcane, crop residues, and various other by-products of agriculture, farming and animal husbandry. Crops such as corn, soybeans, willow, sorghum, jatropha, etc. are specifically grown for the production of biofuel. Biomass can also be converted into transportation fuels like biodiesel and ethanol, as well as usable forms of energy like biogas. Unlike fossil and nuclear fuels, biomass is a renewable energy source that is based on the carbon cycle. It can contribute to waste management, provide fuel security and slow down climate change. This book contains some path-breaking studies and emerging trends in bioenergy production and utilization. The topics included herein are of utmost significance and bound to provide incredible insights to readers. This book is appropriate for students and experts seeking detailed information about the current and future perspectives of renewable energy.

After months of intensive research and writing, this book is the end result of all who devoted their time and efforts in the initiation and progress of this book. It will surely be a source of reference in enhancing the required knowledge of the new developments in the area. During the course of developing this book, certain measures such as accuracy, authenticity and research focused analytical studies were given preference in order to produce a comprehensive book in the area of study.

This book would not have been possible without the efforts of the authors and the publisher. I extend my sincere thanks to them. Secondly, I express my gratitude to my family and well-wishers. And most importantly, I thank my students for constantly expressing their willingness and curiosity in enhancing their knowledge in the field, which encourages me to take up further research projects for the advancement of the area.

Editor

1

Soil carbon and belowground carbon balance of a short-rotation coppice: assessments from three different approaches

GONZALO BERHONGARAY, MELANIE S. VERLINDEN, LAURA S. BROECKX, IVAN A. JANSSENS and REINHART CEULEMANS

Department of Biology, Research Centre of Excellence on Plant and Vegetation Ecology, University of Antwerp, Universiteitsplein 1, B-2610 Wilrijk, Belgium

Abstract

Uncertainty in soil carbon (C) fluxes across different land-use transitions is an issue that needs to be addressed for the further deployment of perennial bioenergy crops. A large-scale short-rotation coppice (SRC) site with poplar (*Populus*) and willow (*Salix*) was established to examine the land-use transitions of arable and pasture to bioenergy. Soil C pools, output fluxes of soil CO_2, CH_4, dissolved organic carbon (DOC) and volatile organic compounds, as well as input fluxes from litter fall and from roots, were measured over a 4-year period, along with environmental parameters. Three approaches were used to estimate changes in the soil C. The largest C pool in the soil was the soil organic carbon (SOC) pool and increased after four years of SRC from 10.9 to 13.9 kg C m^{-2}. The belowground woody biomass (coarse roots) represented the second largest C pool, followed by the fine roots (Fr). The annual leaf fall represented the largest C input to the soil, followed by weeds and Fr. After the first harvest, we observed a very large C input into the soil from high Fr mortality. The weed inputs decreased as trees grew older and bigger. Soil respiration averaged 568.9 g C m^{-2} yr^{-1}. Leaching of DOC increased over the three years from 7.9 to 14.5 g C m^{-2}. The pool-based approach indicated an increase of 3360 g C m^{-2} in the SOC pool over the 4-year period, which was high when compared with the −27 g C m^{-2} estimated by the flux-based approach and the −956 g C m^{-2} of the combined eddy-covariance + biometric approach. High uncertainties were associated to the pool-based approach. Our results suggest using the C flux approach for the assessment of the short-/medium-term SOC balance at our site, while SOC pool changes can only be used for long-term C balance assessments.

Keywords: bioenergy, carbon fluxes, carbon pools, land-use change, poplar, *Populus* sp., second-generation biofuels, soil organic carbon

Introduction

The cultivation of soils with arable crops produces a net carbon (C) flux from the soil to the atmosphere, contributing to the increased greenhouse effect (Le Quéré, *et al.* 2013) and also reducing soil fertility and water quality (Lal, 2004). Afforestation, on the other hand, has been highly recommended to restore C stocks in the soil (Smith *et al.*, 1997). Worldwide studies tried to estimate the ability of the soil to sequester C back from the atmosphere (Jones *et al.*, 2005; Liang *et al.*, 2005; Batjes, 2008; Schulp *et al.*, 2008) and to mitigate the (anthropogenic and agricultural) emissions of CO_2 into the atmosphere.

Short-rotation coppice (SRC) cultures are defined as high-density plantations of fast-growing trees for rotations from 2 to 8 years. At the end of each rotation, the trees are harvested at the base, resulting in the regeneration of new shoots from the remaining stump and the roots. Due to their fast growth and high yield, poplars (*Populus*) and willows (*Salix*) are the most widely used tree species in SRC. Wood chips from SRC biomass can be burned, gasified or co-fired with coal to produce renewable electricity and/or renewable heat. This type of bioenergy from lignocellulosic feedstock is called second-generation bioenergy. Therefore, second-generation bioenergy crops, such as SRC, are considered interesting management options both to sequester C in European croplands, and to partially replace the consumption of fossil fuels (Smith, 2004). However, the cultivation of SRC is more comparable with an arable crop cultivation than with afforestation, despite the woody nature of the planted poplars or willows. Although SRC has been applied since the early 1970s (Hansen *et al.*,

Correspondence: Reinhart Ceulemans
e-mail: reinhart.ceulemans@uantwerpen.be

1979) as an alternative bioenergy source with a high potential, its capacity to sequester C remains unclear (Walter *et al.*, 2015).

Currently, bioenergy is the most significant renewable energy, contributing to almost 80% of the renewable energy supply (IEA, 2014). The European Union is dedicated to increase the amount of renewable energy used to 20% of the total energy consumption by 2020, while simultaneously reducing C emissions by 20% by 2020 (EU 2009). However, there is still a lack of quantitative information on the changes in soil organic C (SOC) for a land-use change (LUC) to second-generation bioenergy crops with respect to historical land covers (i.e. arable, grass) and current land management practices (genotypes, planting density, harvest) (Harris *et al.*, 2015). Moreover, comprehensive studies on SOC dynamics and greenhouse gas emissions under SRC are limited; subsoil processes and C losses through leaching remain largely unknown (Agosti *et al.*, 2015). A large-scale operational SRC plantation in Belgium, that has been intensively studied within the POPFULL project, provided the opportunity to examine the soil C balance of SRC under the prevailing conditions. The overall aims of this study were to quantify the C balance of the soil of a recently LUC to SRC, and to evaluate the potential of SRC for soil C sequestration. The specific objectives were (i) to close the carbon balance of the belowground compartment of an SRC plantation and (ii) to compare three methodological approaches. As the closure of the carbon balance is a complex and difficult task, we used three different methodologies to reach the primary objective.

Materials and methods

Experimental site

The POPFULL project aimed at the full-system analysis of bioenergy production from an SRC of poplars and involved both an experimental approach at a representative field site and a modelling part (POPFULL project; http://uahost.uantwerpen.be/popfull/). The experimental field site is located in Lochristi, Belgium (51°06′N, 03°51′E), and consists of a high-density plantation of large monospecific and monogenotypic blocks of both poplar (*Populus* spp.) and willow (*Salix* spp.). Lochristi is located 11 km from Ghent in the province of East-Flanders at an altitude of 6.25 m above sea level with a flat topography. The long-term average annual temperature at the site is 9.5 °C, and the average annual precipitation is 726 mm (Royal Meteorological Institute of Belgium). The region is pedologically described as a sandy region and has poor natural drainage; the soil type according the World Reference Base (WRB) is Anthrosol (FAO, 2015). The total area of the site is 18.4 ha. The two former land-use types of the site were (i) agriculture, consisting of cropland (ryegrass, wheat, potatoes, beets, and most recently monoculture corn with regular nitrogen (N) fertilization at a rate of 200–300 kg ha^{-1} yr^{-1} as liquid animal manure and chemical fertilizers) and (ii) extensively grazed pasture. For more information on the site and the planting scheme, see Broeckx *et al.* (2012) and the Supporting Information (SI).

A detailed soil analysis was carried out in March 2010, prior to planting. The analysis characterized the soil type as a sandy texture. In the upper soil layer, C and N concentrations were significantly lower in cropland as compared with pasture ($P < 0.05$) and decreased exponentially with depth in both former land-use types. Table 1 presents a detailed analysis of nutrients and soil variables for both land-use types (see also Broeckx *et al.*, 2012).

After soil preparation by intensive ploughing (40–70 cm depth), tilling and a pre-emergent herbicide treatment, a total of 14.5 ha were planted between 7 and 10 April 2010 with 25-cm-long dormant and unrooted cuttings from 12 poplar and three willow genotypes in monogenotypic blocks in a double-row planting scheme with a commercial leek planter (Broeckx *et al.*, 2012). The distance between the narrow rows was 75 cm and that between the wide rows was 150 cm. The distance between trees within a row was 110 cm, yielding an overall density of 8000 trees per ha. The total length of individual rows ranged from 45 m up to more than 325 m. Manual and chemical weed control was applied during the first and the second years. Neither fertilization nor irrigation was applied during the entire lifetime of the plantation thus far. Two small portions of the field remained untouched with pasture as unplanted control plots for the determination of soil changes. The plantation was managed in two-year rotations. After the first rotation (two growing seasons), the plantation was harvested on 2–3 February 2012 using commercially available SRC harvesters (Berhongaray *et al.*, 2013b). From there on, trees continued to grow as a coppice culture with multiple shoots for two more years in the second rotation; the second harvest took place on 18–20 February 2014 (Vanbeveren *et al.*, 2015). Eddy-covariance techniques were used to monitor the main greenhouse gases (CO_2, N_2O, CH_4) and recorded continuously from June 2010 till present (Zenone *et al.*, 2016). Environmental variables as soil temperature, air temperature, relative humidity, wind speed and precipitations were also monitored. All sensors for these measurements were placed in the immediate proximity of the eddy-covariance mast (see Data S1).

By the reason of the high labour intensity, and to limit the variability caused by different species and genotypes, only two poplar genotypes were assessed for the soil C and belowground C balance: that is Koster (*Populus deltoides* Marsh × *P. nigra* L.) and Skado (*P. trichocarpa* Hook. × *P. maximowiczii* Henry). Both genotypes were chosen because they are genetically and phenotypically contrasting, and they represented the range of productivity values for the entire plantation (see Broeckx *et al.*, 2012 for more details on the productivity of the genotypes).

Carbon pools

Soil organic matter. The C content in the soil organic matter (SOM), known as the soil organic C (SOC), was assessed before

Table 1 Soil bulk density (kg dm^{-3}), carbon concentrations (%) and carbon content (kg m^{-2}) in the soil organic matter (SOM) at different depths before the planting (2010) and after four years (2014) of short-rotation coppice culture. As no differences were detected between genotypes (Skado and Koster), data were pooled. The means are presented for both previous land-use types, and for both narrow and wide rows. Values from narrow and wide rows were averaged taking into account the proportional area they occupied per m^{-2}

	2010						2014									
	Cropland			Pasture			Cropland					Pasture				
							Narrow		Wide		Average	Narrow		Wide		Average
Depth	BD	C		BD	C		BD	C	BD	C	C	BD	C	BD	C	C
cm	kg dm^{-3}	%	kg m^{-2}	kg dm^{-3}	%	kg m^{-2}	kg dm^{-3}	%	kg dm^{-3}	%	kg m^{-2}	kg dm^{-3}	%	kg dm^{-3}	%	kg m^{-2}
0–15	1.48	1.52	3.37	1.28	1.94	3.73	1.47	1.62	1.49	1.58	3.55	1.47	1.58	1.49	1.65	3.62
15–30	1.44	1.41	3.05	1.42	1.36	2.90	1.12	1.51	1.53	1.49	3.13	1.12	1.90	1.53	1.89	3.96
30–45	1.45	1.06	2.32	1.44	1.18	2.54	1.79	1.43	1.72	1.27	3.45	1.79	1.43	1.72	1.48	3.82
45–60	1.47	0.75	1.66	1.42	1.08	2.30	1.73	1.06	1.78	1.03	2.75	1.73	1.63	1.78	1.22	3.59
60–75	1.57	0.56	1.32	1.46	0.63	1.39										
75–90	1.55	0.34	0.79	1.57	0.36	0.85										

BD, bulk density; C, carbon.

plantation establishment (March 2010) and after the second rotation (March 2014). A random sampling was performed at 110 locations in March 2010, of which 60 locations matched with the distribution of the two studied genotypes Skado and Koster, and eight locations with the control pasture. These 68 locations were revisited, and the soil was re-sampled in March 2014. Half of the 60 locations at the plantation were sampled in each former land-use type, and within each land-use type half (i.e. 15) in each of the two row spacings. In March 2010, the soil was sampled up to a depth of 90 cm, while the repeated sampling in March 2014 was only up to 60 cm depth as a previous analysis only showed roots until 60 cm depth (Berhongaray *et al.*, 2015). In both campaigns, independent samples were taken every 15 cm on each of the 68 locations using a 2.5 cm diameter and 15 cm length corer (Eijkelkamp Agrisearch equipment, Arnhem, the Netherlands). Bulk density (BD) samples were taken independently using soil BD corer of 5 cm length and with a 5 cm diameter. Carbon mass fractions were determined in three replicates per sample (see below under section *Chemical analysis of soil and biomass samples*). From the C mass fractions and the BD, the C pool per 15 cm depth interval was calculated, and cumulated over 90 cm for the 2010 samples and over 60 cm for the 2014 samples. SOC data were transformed to equivalent soil mass to account for differences in BD between the soil conditions (i.e. previous land-use type and row spacing). The estimations of SOC at equivalent soil mass were performed for masses of 200, 400, 650 and 900 kg m^{-2} using spline functions as previously described (Berhongaray *et al.*, 2013a). For the spline functions, the soil mass was used as the independent variable and SOC as the dependent variable. Interpolations were made by adding or by removing a portion of the soil to reach the desired soil mass assuming that transitions between soil layers were smooth and continuous.

Stumps, coarse and medium-sized roots. Root biomass was determined by excavation of the root system immediately after the two harvests. In February 2012, five trees of different stem diameters (from 20 mm to 60 mm diameter at 22 cm above the soil) were selected within both genotypes (Koster and Skado) for each of both former land-use types. In February 2014, only four trees per genotype and per land-use type were excavated. In both excavation campaigns, the remaining stumps and roots were excavated over an area of 1.1 m × 1.125 m (planting distance in the rows x sum of half inter-row distances). All roots within this area were collected, assuming that roots from adjacent trees compensated for roots of the selected tree growing outside the sampled area (Levillain *et al.*, 2011; Razakamanarivo *et al.*, 2012). The excavation depth was limited to 60 cm, as very few roots were observed under 60 cm (Berhongaray *et al.*, 2015). Coarse roots (Cr; Ø > 5 mm) and medium-sized roots (Mr; Ø = 2–5 mm) were sampled; total dry biomass (DM) of these roots (Cr) and of the remaining 15-cm-high stump (Stu) was determined after oven drying at 70 °C. As neither a significant effect was found for genotype nor for former land-use type, all data were pooled. Belowground woody biomass and stump biomass were plotted against basal area, and an allometric regression was fitted. Estimations of the average belowground woody biomass and of the stump biomass pool were made from the diameter inventory of each sampling year, that is from winter 2012 (January 2012) and from winter 2014 (January 2014) as explained in Berhongaray *et al.* (2015). Dried root wood material was grated for C analyses. An average of the C mass fractions was used for calculating the belowground woody C pool.

Fine roots. The fine root (Fr, Ø < 2 mm) biomass pool was annually estimated using the soil core methodology. Intact soil samples were taken using an 8 cm diameter × 15 cm deep hand-driven corer (Eijkelkamp Agrisearch equipment) at the end of each growing season, that is: in winter 2011 (December 2010–February 2011), winter 2012 (December 2011–February 2012), winter 2013 (December 2012), and in winter 2014 (December 2013–January 2014). Winter samples were taken only from the first 15 cm. Samples from different depths were collected during two campaigns in summer, that is: August 2011 and August 2012. In August 2011, sampling was performed in six different soil layers (0–15 cm, 15–30 cm, 30–45 cm, 45–60 cm, 60–75 cm and 75–90 cm, whereas in August 2012 four different soil layers (0–15 cm, 15–30 cm, 30–45 cm, 45–60 cm) were sampled. After each sampling campaign, samples were transported to the laboratory and stored in a freezer until processed. All roots were picked from the sample by hand while (i) separating out weed roots from poplar roots, (ii) sorting poplar roots in dead and living roots, and (iii) sorting poplar roots according to diameter classes (<2 mm and >2 mm). The roots were sorted by visual inspection as previously described (Berhongaray *et al.*, 2013c). The sorting of dead (necromass) and living (biomass) roots was based on the darker colour and the poorer cohesion between the cortex and the periderm of the dead roots (Janssens *et al.*, 1999). Following washing, fine poplar roots were oven dried at 70 °C for one to four days to determine the standing (fine) root biomass per soil surface area and expressed in g DM m^{-2}. More details on root collection and on data processing can be found in Berhongaray *et al.* (2013c,d).

Carbon fluxes

Belowground inputs: fine root productivity and root C input. Sequential soil coring was used to determine Fr mass and Fr production for the second growing season of the first rotation (i.e. 2011) and the first growing season of the second rotation (i.e. 2012). From February 2011 to November 2012, the upper 15 cm of soil layer was sampled every 2–3 weeks (except for the winter when the sampling intensity was decreased) using an 8 cm diameter × 15 cm deep hand-driven corer (Eijkelkamp Agrisearch equipment). During 2011, 20 samples were collected at every sampling campaign for each genotype. During 2012, the number of samples was different at each sampling date, following the expected intrinsic variability of the Fr biomass based on the experience of the previous year (i.e. 2011). Based on our previously described methodology (Berhongaray *et al.*, 2013d), the number of samples in 2012 varied from 12 in winter to 20 in summer. At each sampling campaign in 2011 and in 2012, half of the samples were collected in the narrow and half in the wide rows, randomly distributed over the planted area within the former pasture land-use type. The samples were transported to the laboratory and stored in a freezer until processed. Once in the laboratory, fine roots were processed as described previously (Berhongaray *et al.*, 2013d). Twenty-one Fr weight of one sample core picked for *x* min (i.e. 5–20 min) was converted into total Fr mass in the sample (i.e. after 60-min picking duration) using Richard's equation (Berhongaray *et al.*, 2013d) and expressed in g DM m^{-2}. Subsamples of dried roots were ground for C and N analysis.

For 2011 and 2012 (second growing season of the first rotation and first growing season of the second rotation), root production (*P*) was calculated using the 'decision matrix' approach (Fairley & Alexander, 1985). All differences in biomass and necromass were taken into account during the calculation, assuming that the living and dead pools of roots were continuously changing. This approach was better than using the significant differences between root mass of consecutive sampling dates, especially in the case of frequent sampling (Brunner *et al.*, 2013), as in our sampling campaign. To calculate annual root production, all productivity values from sampling periods were summed from the beginning until the end of the year. Root productivity calculations and the comparison of different methods were previously described in more detail (Berhongaray *et al.*, 2013c).

By the reason of time restrictions, Fr production was estimated with the in-growth core technique in the second growing season of the second rotation (2013). This method provided reliable estimates for *P* with much less labour time. In December 2012, ten 2.2-mm mesh bags (10 cm diameter × 0.40 m depth) were installed for each genotype, so 20 in total. Each mesh bag was refilled with root-free original soil obtained from the root biomass assessment. Root-free soils were stored in plastic bags, and care was taken to refill the holes with soil with exactly the same stratification. The in-growth cores were harvested after one year in December 2013. The in-growth cores were divided into two samples from 0 to 15 cm depth and from 15 to 30 cm depth, and the separated samples were stored in plastic bags until processed. Consequently, only the first 30 cm of the in-growth cores was used to make it comparable to the 15-cm increment soil coring approach, and the bottom 10 cm of the in-growth cores (from 30 to 40 cm) were discarded. The samples were processed in the same way as the samples from the soil coring approach. The *P* was estimated from the quantity of total root mass produced (biomass and necromass) in the considered period of time and expressed in g DM m^{-2} yr^{-1}. For periods in which Fr production was not measured, interpolation and extrapolation methods were used. For example, to calculate *P* for 2010, we used the ratio of *P* and Fr biomass from 2011 and the Fr biomass from 2010.

The turnover rate is widely used to estimate root-derived C inputs to the soil. An assumption of this root turnover approach is that annual Fr production equals fine root mortality on an annual basis. However, the approach of the turnover rate is only valid in steady-state systems, as, for example mature forests, but not in actively growing systems such as our SRC poplar plantation. In mature forests, the amount of roots produced is the same as those that die at the end of the growing season; they represent the C inputs. In a growing system, part of the productivity is used to form the growing standing biomass. We used the following approach to estimate C inputs from roots (I_{root}) that consider the increments in root biomass:

$$I_{root} = (P - \Delta Br) * C\% \quad (1)$$

where P is the root productivity in g DM m^{-2} yr^{-1}; ΔBr is the difference between root biomass at the end and at the beginning of the growing season in g DM m^{-2} yr^{-1}; and C% is the fraction of C (g C g DM^{-1}). In our study, this methodology only applied to the fine roots. As a result of the absence of mortality of medium-sized and coarse roots, productivity of these last mentioned root classes was estimated using ΔB, and the C input was equal to zero.

From the in-growth technique, we obtained evidence for an identical vertical distribution of Fr and root P, that is the proportion of P at one specific soil depth was similar to the proportion of Fr at the same depth. For years 2011 and 2012, the C inputs from Fr were extrapolated up to 60 cm depth using the measured P from the first 15 cm and the vertical distribution of Fr in each year (see above for the fine root depth-distribution measurements).

Aboveground inputs

Leaf fall. Leaf litter was collected each year (2010, 2011, 2012 and 2013) during the period of leaf fall from early September to December in two plots of 5 × 6 trees for each genotype within each former land-use type (n = 8). In each plot, three perforated litter traps (*i.e.* plastic litter baskets) of 57 cm × 39 cm were placed on the ground along a diagonal transect between the rows covering the wide and the narrow inter-row spacings. Every two weeks the litter traps of each plot were emptied and leaf dry biomass was determined after oven drying at 70 °C for one week. The collected leaf biomass was cumulated over time to obtain the yearly leaf C input (I_{leaves}).

Weeds and grasses. Before the soil was ploughed in March 2010, the former pasture land was covered by grasses. To account for the C input from these grasses, the aboveground biomass from grasses was harvested in five randomly distributed plots of a contiguous pasture land. Aboveground biomass from weeds was measured after they reached the maximum standing biomass (after flowering) only in two growing seasons, that is: August 2011 and August 2013. In 2011, six randomly distributed plots of 1 m^2 were harvested under each genotype and from the previous pasture land area. In 2013, four plots of 1 m^2 were harvested under each genotype and previous land-use type combination, that is 16 plots in total. In each plot, the weeds were cut at ground level and put in paper bags. The collected weed biomass was oven dried for 10 days at 70 °C and the DM expressed in g DM m^{-2}. For years for which we did not measure the weed biomass, we estimated it using the root biomass quantified on these years and the root:shoot ratios from the measured years. The weeds died annually, and the total (weed) biomass was considered as an input to the soil. We estimated the aboveground annual C input from the weeds (I_{weed}) using the C mass fraction reported in Fortunel *et al.* (2009).

Harvest losses. Harvest losses were estimated from samples collected at the field site after both harvests, that is early March 2012 and mid-March 2014. Two different harvest techniques were used and compared during each harvest, that is two mechanical harvesters in February 2012 (Berhongaray *et al.*, 2013b) and a mechanical *vs.* a manual harvesting in February 2014 (Vanbeveren *et al.*, 2015). To estimate the harvest losses, harvested woody debris and woody biomass material were collected from the soil surface on four areas of 1 m^2 within the land area harvested by each harvesting technique for the two genotypes. The collected biomass material and debris were transported to the laboratory and dried in a drying oven at 60–70 °C until constant weight. The harvest losses were expressed in g DM m^{-2}, and later expressed as C inputs ($I_{harvest}$) using the C mass fraction. More details can be found in Berhongaray *et al.* (2013b) and Vanbeveren *et al.* (2015).

Carbon outputs

Soil CO_2 efflux. Soil CO_2 efflux was continuously monitored using an automated soil CO_2 flux system (LI-8100; LI-COR Biosciences, Lincoln, NE, USA) from December 2010 to January 2012 and from May 2012 to January 2014. Sixteen long-term chambers operating as closed systems were connected to an infrared gas analyser through a multiplexer (LI-8150; LI-COR Biosciences). The 16 chambers were spatially distributed over the plantation. Soil CO_2 efflux was extrapolated for the periods without measurements by a neural network analysis (using MATLAB; 7.12.0, 2011; Mathworks, Natick, MA, USA) based on soil temperature, which was also continuously monitored throughout the year. Values of CO_2 efflux were integrated over time to obtain the cumulated CO_2 efflux. More details can be found in Verlinden *et al.* (2013a).

Partitioning of soil respiration. To calculate the SOC balance (see below under *Carbon balance*), we quantified the contribution of roots and SOM decomposition to the CO_2 emission from the soil. The soil CO_2 efflux (R_s) is the result of CO_2 release from two main sources: (i) microbial decomposition of SOM (heterotrophic respiration, R_h) and (ii) root-derived respiration (autotrophic respiration). We partitioned R_s based on the spatial and the temporal variations in root biomass, in soil temperature, in soil water content and in soil respiration, following the methodology described in Data S3, as follows:

$$R_s = R_h + R_m + R_{gr} \quad (2)$$

where R_m is the CO_2 from the maintenance of root biomass, this rate is assumed to be linearly related to the root biomass to be maintained; R_{gr} is the cost of the formation of new root structures and is assumed to be proportional to the growth rate of the roots; R_h is consequently assumed to be the C output from the SOM pool. The results were annualized and expressed in g C m^{-2} yr^{-1}.

Dissolved organic carbon. For the analysis of dissolved organic carbon (DOC) in the soil, 10 groundwater samples were collected monthly from August 2011 until July 2013 from six PVC water tubes (length x diameter: 2 m × 5 cm) distributed randomly under the two genotypes. Water samples were collected using a 2-m plastic tube connected to a glass bottle by applying a vacuum of 60 kPa. After collection, the samples were stored at 4 °C and sent to an external laboratory (SGS, Antwerp, Belgium) within 24 h. DOC concentrations were determined with a Shimadzu TOC VPH analyser (Shimadzu corp., Japan, 2001) with IR detection after thermal oxidation.

Leaching from the belowground system (see below for a description of the system) was estimated using DOC concentrations and the soil water balance. The soil water balance was calculated as the difference between the monthly cumulative precipitation minus the monthly evapotranspiration, considering positive values as water excess and leaching (Data S4.1). Precipitation was monitored from June 2010 onwards using a tipping-bucket rain gauge (model 3665R; Spectrum Technologies Inc., Plainfield, IL, USA) installed next to the eddy-covariance mast (see Data S1). A LI-7000 fast response gas analyser (LiCor) was used to continuously measure latent heat from air samples at the eddy-covariance mast from June 2010 onwards. Latent heat flux was converted into evapotranspiration using air temperature and latent heat of vaporization. The annual leaching of DOC was calculated by summing the monthly products of DOC concentrations and water excess. For months without DOC data, the average DOC concentration was used. The annual DOC leaching was also calculated using annual averages of DOC concentration, and annual precipitation and evapotranspiration.

Chemical analysis of soil and biomass samples

Soil samples as well as dried biomass from wood, leaves and roots were ground and analysed by dry combustion with an NC element analyser (NC-2100 element analyser; Carlo Erba Instruments, Milano Italy). Soil and plant mass were converted to C mass using the average C mass fraction and expressed in g C m^{-2}. The means from different row spacings were calculated separately, and then, the scaled-up averages were calculated taking into account the proportion of the land area that each row spacing occupied.

Carbon balance

The boundaries of the belowground system that we considered for our C balance were the top of the soil surface and a soil depth of 60 cm (Fig. 1). Three different approaches were used to quantify the changes in the SOC, that is (i) the pool change-based approach, (ii) the component flux-based approach and (iii) the combined eddy-covariance + biometric approach. The pool change-based approach was performed comparing initial and final SOC at equivalent soil mass. The SOC balance was calculated via the component flux-based approach as:

$$\Delta SOC = I_{leaves} + I_{roots} + I_{weeds} + I_{harvest} - R_h - DOC \quad (3)$$

where the C inputs from the different plant components were expressed in g C m^{-2}. A few minor components of possible C losses were not measured at the soil level and were thus not taken into account for the SOC balance, that is non-CO_2 losses as CO, CH_4, volatile organic compounds (VOCs) to the atmosphere and herbivory. Moreover, these components only represent a tiny, negligible portion of the soil C emissions (Asensio *et al.*, 2007; Görres *et al.*, 2015).

Finally, the SOC balance was calculated with the combined eddy-covariance + biometric approach as:

$$\Delta SOC = NECB - \Delta B \quad (4)$$

where NECB was the net ecosystem C balance representing the overall ecosystem C balance from all sources and sinks – the net ecosystem exchange (NEE, net CO_2 flux from the ecosystem to the atmosphere); the net CH_4 efflux; the net efflux; the net DOC leaching loss; and the net lateral transfer of C out of the ecosystem by processes such as anthropogenic transport or harvest; –ΔB is the change in the standing biomass (Stu + Cr + Mr + Fr). More details on this last approach were described in detail in Data S5.

Statistical analyses

Data were analysed with different linear models. A two-way analysis of variance (ANOVA) was performed using land-use type and genotype as fixed factors, also including their interactions. More complicated models considered climate, plant and soil variables. These were tested as covariates ($P \leq 0.05$) and included in the model when significant. In the case of a significant genotype effect, pairwise comparisons were performed using a Tukey's post hoc test ($P \leq 0.05$). Regression and correlation analyses were performed to search for relationships among variables, the significance of which was tested by an F test ($P \leq 0.05$).

Uncertainty analysis

The primary obstacles for applying the C balance approach were as follows: (i) the quantification of the annual fluxes of the inputs, the outputs and the changes in the C pools with a reasonable precision, and (ii) the accumulation of errors in the calculation of the C balance as a sum of many components, each with their own error. A combination of error propagation formulas and Bayesian methods as Monte Carlo simulations was used for the uncertainty analysis, following the *IPCC Good Practice Guidance and Uncertainty Management in National Greenhouse Gas Inventories* (IPCC, 2006). The methodology for uncertainty analysis has been explained in detail in Data S6.

Results

Carbon pools

As for nearly all terrestrial biomes, the largest C pool in the soil was situated in the SOM. The SOC pool in the first 60 cm of the soil before the planting was on average 10.9 kg C m^{-2} (109 Mg C ha^{-1}) vs. 13.9 kg C m^{-2} (139 Mg C ha^{-1}) after 4 years of SRC (Table 1). Changes in BD were also observed, especially in the wide rows. Before planting, the vertical distribution of

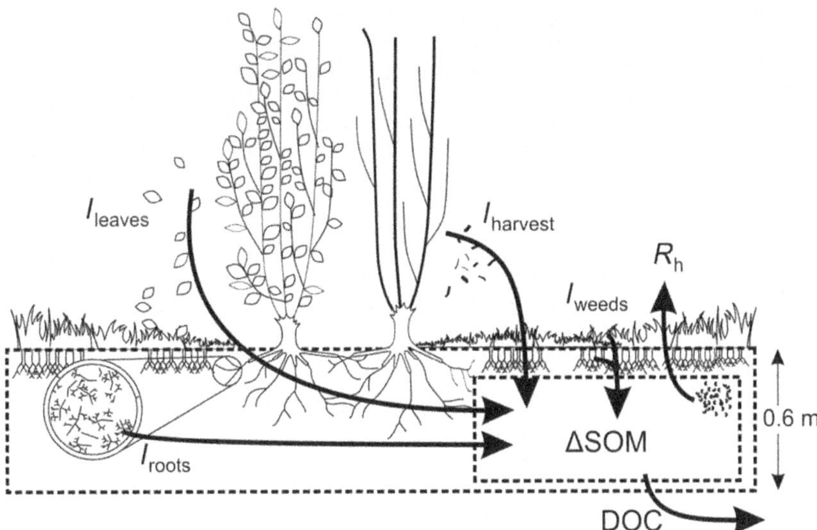

Fig. 1 Representation of the soil organic matter (SOM) carbon balance approach. The dashed lines around Δ-SOM indicate the boundaries that are being considered for the SOM C balance. C, carbon; C fluxes: I_{leaves}, leaf C input; I_{roots}, root C input; $I_{harvest}$, harvest loses C input; I_{weeds}, weed C inputs; R_h, heterotrophic respiration; DOC, dissolved organic C.

C differed between both land-use types. In the first layer (0–15 cm), the C% was higher in previous pasture, while in the second layer (15–30 cm), the C% was higher in previous cropland ($P < 0.05$). This vertical distribution was disrupted during the ploughing just before the planting of the SRC. Furthermore, in 2014 the C% was higher in the second layer of the previous pasture as compared to the previous cropland, indicating that the soil was ploughed upside down. The soil layer that was the top layer in 2010 was found in 2014 at a depth of approx. 30 cm. This higher C% is likely a combination of movement of the soil from intensive ploughing and the SRC cultivation for four years. After the conversion to SRC the C% showed a clear spatial distribution, with higher values in the narrow rows than in the wide rows ($P < 0.05$). No differences were found in the control pasture between March 2010 and March 2014 at any depth (see Data S8).

When SOC changes were analysed at the same soil mass (Table 2), we observed a loss of C in the top layer (0–200 kg m^{-2} to 0–15 cm) for the former pasture ($P = 0.05$). In the former cropland, only small, but not significant, C losses were found in the second layer (200–400 kg m^{-2} to 15–30 cm). However, after losses in the first layers, we observed an accumulation of C in the deeper layers for both land-use types. An overall sequestration of C was found in the entire soil profile (0–900 kg m^{-2} to 0–60 cm) with repeated soil samplings. At equivalent soil masses, the SOC pool in the 0–900 kg m^{-2} (0–60 cm) layer before the planting was on average 11.2 kg C m^{-2} and increased to 14.6 kg C m^{-2} after four years of SRC (Table 2). The higher SOC sequestration was evidenced in the previous pasture land.

The total accumulation of C in Cr, Mr and Stu after four years of SRC was smaller than the changes in SOC (Table 3). The annual change in C stored in the Cr averaged 18.4 g C m^{-2} yr^{-1}. This annual change in C was much larger in genotype Skado on the previous cropland, with 22.5 g C m^{-2} yr^{-1}, than in the other treatments, which averaged 17.0 g C m^{-2} yr^{-1} ($P < 0.05$). The higher Cr for 'Skado cropland' per unit of land area (*i.e.* m^{-2}) compared to 'Skado pasture' could be explained by the lower tree mortality that resulted in a higher plant density per area (Berhongaray *et al.*, 2015). The Mr biomass remained constant between both sampling campaigns, representing about 22% of the total root biomass. No differences were found in Fr between both genotypes ($P = 0.05$). In general, Fr biomass was lower in previous pasture than in previous cropland. Among the plant C pools belowground, the highest amount of C was stored – after four years of SRC – in the woody biomass (Cr and Stu), followed by the Fr and Mr.

Carbon inputs

The annual leaf fall represented the largest C input to the soil. The total amount of leaf fall increased with the age of the trees, from 2010 to 2013. This C input was exceeded only by the aboveground inputs from weeds in the former pasture land in 2011 and by the Fr in the year 2012, just after the first harvest. After the first harvest, we observed a very high Fr mortality that resulted

Table 2 Carbon in SOM at equivalent soil mass before planting (2010) and after four years (2014) of a short-rotation coppice culture for genotypes Skado and Koster. The difference between 2010 and 2014 (DELTA) is also given. SDs are provided in brackets, and significant differences (ANOVA, $P < 0.05$) between different land uses and years at the same equivalent soil mass are represented with different letters

		2010		2014		Δ	
	Soil mass kg m^{-2}	Cropland kg C m^{-2}	Pasture	Cropland	Pasture	Cropland	Pasture
Koster	0–200	3.16 (0.32)a	3.93 (0.29)b	3.30 (0.54)a	3.22 (0.91)a	0.13 (0.62)	−0.71 (0.96)
	200–400	3.10 (1.46)ab	2.05 (0.40)a	3.56 (0.55)bc	4.36 (1.88)c	0.46 (1.63)	2.31 (1.93)
	400–650	3.27 (1.01)a	2.82 (0.90)a	3.87 (1.99)a	3.84 (2.27)a	0.60 (2.25)	1.02 (2.45)
	650–900	2.13 (1.15)ab	1.80 (1.66)a	4.89 (2.24)c	3.93 (5.16)bc	2.76 (2.54)	2.13 (5.43)
	0–900	11.65 (2.23)a	10.61 (2.01)a	15.61 (3.09)b	15.37 (6.01)b	3.95 (3.81)	4.76 (6.33)
Skado	0–200	3.00 (0.59)a	3.79 (0.44)b	3.09 (0.94)a	3.22 (0.72)a	0.09 (1.12)	−0.56 (0.85)
	200–400	3.23 (0.58)a	2.90 (0.33)a	3.19 (1.07)a	4.09 (0.85)b	−0.04 (1.22)	1.18 (0.92)
	400–650	2.74 (1.35)a	3.45 (0.35)a	3.00 (1.38)a	3.79 (1.03)a	0.26 (1.95)	0.34 (1.09)
	650–900	1.15 (1.01)a	2.44 (0.95)b	2.20 (1.72)ab	4.85 (2.73)c	1.06 (2.00)	2.41 (2.89)
	0–900	10.12 (1.91)a	12.58 (1.16)a	11.49 (2.63)a	15.94 (3.13)b	1.37 (3.24)	3.37 (3.34)

SOM, soil organic matter.

Table 3 Overview of the belowground carbon pools in the short-rotation coppice culture: fine roots (Fr), medium-sized roots (Mr), coarse roots (Cr), stumps (Stu) and soil organic matter (SOM), before planting (winter 2010) and at the end of each growing season (winters 2011, 2012, 2013 and 2014). No differences were detected in Fr for genotypes (Skado and Koster) under both previous land-use types (cropland and pasture). Fr data were pooled, and the mean and SE (in brackets) are presented. For all other pools, significant differences (ANOVA, $P < 0.05$) were detected; the mean and the range given by the mean values of the combination of genotype*land-use type are presented. SE, standard error. See Berhongaray *et al.* (2015) for more information on the statistics

		Fr (0–1 mm)		Fr (1–2 mm)		Mr (2–5 mm)		Cr (>5 mm)		Stu		SOM	
	Depth	Mean g C m^{-2}	SE	Mean	SE	Mean	Range	Mean	Range	Mean	Range	Mean	Range
Winter 2010	0–15 cm	0.00	–	0.00	–	0.00	–	0.00	–	0.00	–	3473	3260–3700
	0–60 cm	0.00	–	0.00	–	0.00	–	0.00	–	0.00	–	10325	9570–11600
Winter 2011	0–15 cm	4.5	±1.48	1.2	±0.29	–	–	–	–	–		–	–
	0–60 cm	–	–	–	–	–	–	–	–	–	–	–	–
Winter 2012	0–15 cm	14.2	±0.77	7.1	±1.28	18.2	26–65	40.1	27–51	–	–	–	–
	0–60 cm	33.9	–	21.0	–	41.2	74–118	51.9	33–67	129.3	93 –156	–	–
Winter 2013	0–15 cm	10.4	±0.92	6.1	±0.81	–	–	–	–	–	–	–	–
	0–60 cm	24.8	–	12.0	–	–	–	–	–	–	–	–	–
Winter 2014	0–15 cm	22.6	±1.96	13.2	±4.03	19.6	32–68	34.8	32–43	–	–	3242	3170–3330
	0–60 cm	54.1	–	26.0	–	41.4	86–120	73.6	66–90	167.6	152–205	14046	11 000–15 260

in a large C input into the soil. During the early stages of land conversion from agriculture to the SRC, annual soil C inputs from weed roots far exceeded those from the poplar trees (Table 4). This was more evident in the former pasture land than in the previous cropland. The contribution of inputs from weed decreased as trees grew older and bigger, while the harvest losses increased. However, the C inputs to the soil after both harvests strongly depended on the operated harvesting machine. The losses during the harvesting reached up to 10.7% of the potential harvestable aboveground biomass (Berhongaray *et al.*, 2013b). On average, these C inputs due to the harvest losses were as high as the Fr C inputs.

Carbon losses

Over the three years of the measurements, R_s averaged across treatments was 568.9 g C m^{-2} yr^{-1}. For all treatments, R_s was higher in summer than in winter. R_s continuously increased from 2011 to 2013 in the former cropland, while in the previous pasture, it remained quite stable. Overall R_s was much higher in the previous pasture and under the genotype Skado. Narrow

Table 4 Inputs and outputs (release) of carbon (C) from/to the belowground soil system for both previous land-use types and both genotypes (Koster and Skado). Uncertainties were calculated using Monte Carlo simulations

	Year	Aboveground inputs						Belowground inputs				Output					
		Leaves	SD	Weeds	SD	Harv	SD	Weeds	SD	Fr	SD	Rh	SD	DOC	SD	Balance	SD
Cropland																	
Koster	2010	34	1.43	39	27.8	0	–	32	21.4	58	52.8	312	113.1	7.8	2.5	−156.3	119.8
	2011	72	0.76	54	31.4	45	25.7	175	181.0	73	55.2	314	110.4	9.4	2.0	95.2	164.2
	2012	115	9.74	14	10.1	0	–	85	71.1	125	42.4	285	110.2	12.8	2.8	41.6	137.3
	2013	151	7.82	16	9.7	40	27.8	50	43.2	74	44.0	282	103.6	14.6	3.1	34.5	122.2
Total		372	12.6	124	44.2	85	37.9	341	200.3	331	97.8	1193	218.7	45	5.3	15.0	281.8
Skado	2010	32	11.57	7	4.3	0	–	5	2.8	43	36.3	385	113.6	7.8	2.5	−305.6	118.7
	2011	122	31.05	38	20.4	183	101.6	121	111.3	46	35.9	325	84.1	9.4	2.0	175.4	136.3
	2012	161	46.71	6	4.1	0	–	67	69.7	108	55.3	354	108.3	12.8	2.8	−25.0	134.2
	2013	197	13.96	1	0.7	67	33.3	35	30.3	7	7.1	380	115.9	14.6	3.1	−86.8	123.4
Total		512	58.9	53	21.3	250	106.9	227	134.8	204	75.6	1444	212.4	45	5.3	−242.1	254.9
Pasture																	
Koster	2010	38	4.20	116	63.2	0	–	67	38.6	56	51.5	305	114.2	7.8	2.5	−36.4	144.4
	2011	127	1.93	255	110.4	17	10.6	123	72.2	94	92.0	307	110.5	9.4	2.0	300.2	142.3
	2012	66	32.94	35	20.7	0	–	133	95.0	116	76.5	303	86.1	12.8	2.8	34.0	156.7
	2013	168	20.19	13	9.1	15	9.4	17	17.3	66	50.7	296	99.8	14.6	3.1	−31.8	113.7
Total		399	38.9	418	129.2	32	14.1	340	126.6	332	139.7	1210	206.5	45	5.3	266.1	282.3
Skado	2010	48	15.00	91	52.6	0	–	52	35.9	69	59.4	492	114.4	7.8	2.5	−240.1	171.4
	2011	174	8.99	160	82.4	74	45.7	159	90.2	67	57.8	553	128.3	9.4	2.0	70.2	167.2
	2012	233	40.56	13	8.4	0	–	105	75.2	154	122.2	417	124.9	12.8	2.8	75.7	181.9
	2013	197	36.95	4	2.3	88	50.4	46	41.2	57	51.0	429	131.6	14.6	3.1	−51.8	154.7
Total		652	57.6	268	98.1	162	68.0	362	129.5	346	156.2	1892	249.9	45	5.3	−145.9	337.6

Harv, losses after harvest; Fr, fine roots; Rh, heterotrophic respiration; DOC, dissolved organic C.
All values are in g C m^{-2} yr^{-1}.

rows had higher R_s rates than the wide rows (Data S2). This was related to the higher root biomass in the narrow rows (Berhongaray *et al.*, 2013c). The variation in the monthly R_s was correlated both with soil temperature at 10 cm and with root biomass increment. This allowed to describe the relationship for soil respiration partitioning in root related (autotrophic; R_r) respiration and in R_h. On an annual basis, R_h accounted from 48 to 79% of the total annual R_s. It ranged from 79% to 95% in winter, and from to 41% to 83% depending on the model used (Fig. S2.4).

We observed a cumulative increase of DOC over the three years of study (2011, 2012 and 2013). The leaching of DOC calculated on a monthly basis increased exponentially from 7.9 (in 2010) to 9.3 (in 2011), 12.8 (in 2012) and 14.5 g C m^{-2} (in 2013). The DOC leaching calculated on an annual basis was a bit lower than on a monthly basis; this was because the calculated water balance on an annual basis was lower than the one calculated on a monthly basis. However, DOC leaching also exponentially increased over the years as presented in Table 4. There was no difference ($P = 0.05$) in DOC concentration between the former land-use types.

Carbon balance

Contradicting results were obtained by the pool change-based approach as compared to the flux-based and the combined eddy-covariance + biometric approaches (Fig. 2). The pool change-based approach resulted in an average SOC increase of 4360 g C m^{-2} for genotype Koster and 2370 g C m^{-2} for Skado. However, the flux-based approach resulted in a small increase of 140 g C m^{-2} for genotype Koster and a small decrease of −194 g C m^{-2} for Skado. The main C inputs to the soil resulted from the leaf litter fall, from annual weeds and fine roots (Table 4). The total C inputs over the four years ranged from a potential minimum of 730 g C m^{-2} to a potential maximum of 1530 g C m^{-2} depending on the genotype, on the previous land-use type and on the used harvesting machine. The main C flux released from the soil came from soil respiration; the leaching of DOC represented only a very minor proportion (<3%). The total C released from the soil ranged from 1193 g C m^{-2} to 1892 g C m^{-2} for the four years. If we added the C stored in the woody biomass pools, the belowground system resulted in a net gain of C after four years of SRC in both genotypes and both former land-use types. However, with the combined eddy-covariance + biometric approach – which integrated over different genotypes and land uses – net C losses of −956 g C m^{-2} were estimated over the four-year period.

Uncertainties

Uncertainties in the SOC balance were highest in the pool change-based approach, followed by the eddy-covariance + biometric and the flux-based approaches (Fig. 2). Although the estimations with the pool change-based approach were highly sensitive to the BD data, most of the uncertainty was on the C% data, contributing to 82% of the uncertainty. Among the seven variables included in the eddy-covariance + biometric approach, the harvested biomass (see Fpc in Data S5) was the most sensitive and contributed to 84% of the uncertainty. In the flux-based approach, the degree of uncertainty was not related to the size of the flux. In this last mentioned approach, most of the uncertainty was on the belowground fluxes (Table 4). Rh estimations contributed with 32% to the uncertainty, followed by the weeds and fine root inputs with 38% combined. While uncertainties from the aboveground inputs were relatively low, the uncertainty was reduced by on average 18% with the use of Monte Carlo simulations as compared to the simple error propagation formulas (data not shown).

Discussion

Belowground pool and fluxes, and SOM C balance

Our pool-based approach indicated an average increase of 3360 g C m^{-2} (or 33.6 Mg ha^{-1}) in the SOM pool, which is a large value when compared with the flux-based and the combined eddy-covariance + biometric approaches (Fig. 2). High accumulations of SOC were found deep in the soil (below 30 cm depth), but small gains of C were also measured in the top layer. The soil was ploughed upside down before planting, putting C-rich soil soil deep in the soil where SOC decomposition processes were reduced due to low temperature, low priming effects (low root exudates) and frequent soil saturation by the water table. On the other hand, initially deep low-C soil was placed in the top layer, where most C inputs occurred. This initial low-C soil had a high potential to form soil structures and physically protect new SOC from decomposition. These two mechanisms for the top and the deep SOC might explain the C increases in the soil. Other studies also showed that soil depth significantly influenced SOC change rates and should thus be considered in C emission accounting in SRC cultures (Qin *et al.*, 2016). The belowground woody biomass (Stu, Cr, Mr) represented the second largest C pool of the SRC. This long-term belowground biomass also contributed to enhance the C sequestration along the four-year sequence (Pacaldo *et al.*, 2014). The value observed for the belowground biomass C

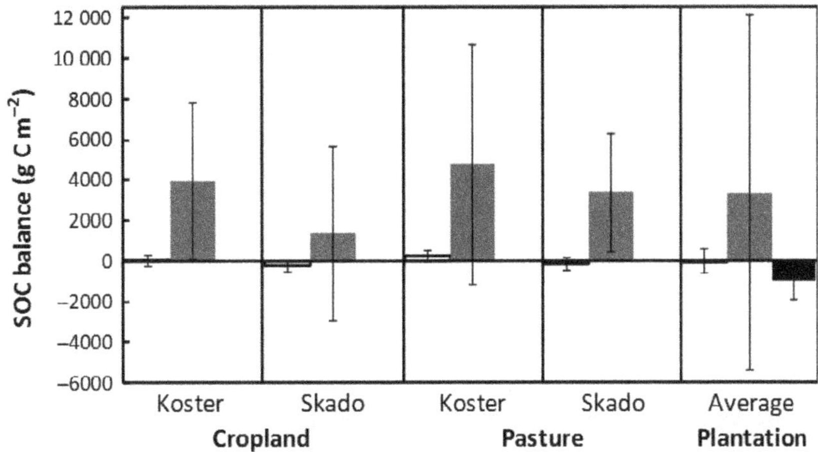

Fig. 2 Soil carbon balance using three different approaches where an increased SOC storage is displayed as positive, and a SOC loss is displayed as negative. The left bars represent the component flux-based approach (non-filled bars), the central bars represent the pool change-based approach (bars filled in grey), and the right-hand bars represent the eddy-covariance approach. Data from the two contrasting genotypes (i.e. Skado and Koster) and the two land uses (i.e. pasture and cropland) were averaged for the flux-based and the pool change-based methods. The combined eddy-covariance + biometric method represents the SOC change in the plantation, including the two land uses and multiple genotypes. Error bars indicate standard errors of the mean. SOC, soil organic carbon.

sequestration (240 g C m^{-2}) was much higher than the 90 g C m^{-2} reported for an SRC plantation in Canada (Arevalo *et al.*, 2011). This might be explained by the higher planting density at our site.

Although not all fluxes were continuously measured, especially in the former cropland, we were able to identify and quantify the main fluxes. Our estimates of the SOM C balance depended on the genotypes, on the weed control, and on the harvesting machines (Table 4). In the future selection of the appropriate management, the choice of the suitable genotype, the process of weeding and the efficiency of the harvesting process are all important for the SOC sequestration.

Effect of the previous land-use type

The flux-based C balance was lower in the previous cropland than in the former pasture land. This was explained by the higher C inputs in the former pasture. These higher inputs in the former pasture came in particular from leaf fall and from weeds. However, the C inputs were measured with less intensity (fewer locations and occasions) in cropland. This might have slightly altered the C balance in favour of the previous cropland. Nevertheless, the pool-based approach also showed higher accumulations of SOC in the former pasture land.

Changes of the total SOC pool as a result of land-use change from cropland and pasture to SRC in Central Europe were recently reported (Walter *et al.*, 2015) and ranged from −1.3 to 1.4 Mg C ha^{-1} yr^{-1} (converted from cropland) and from −0.6 Mg C ha^{-1} yr^{-1} to +0.1 Mg C ha^{-1} yr^{-1} (converted from pasture). Overall, there was no SOC change in the study of Walter *et al.* (2014) which is in line with results of a 20-year chronosequence for SRC plantations in the USA (Pacaldo *et al.*, 2013). These findings suggest that the C inputs from short-term components (as Fr, leaves, weeds) did not result in a SOC accumulation over time. In contrast, a chronosequence of SRC cultures in Canada showed that soils initially lost C while after two years soil C levels increased and reached the initial values in the seventh year (Arevalo *et al.*, 2011).

Effects of harvesting and of the presence of weeds. Our study showed that harvesting played an important role in the soil C balance. Overall, the inputs from harvest losses were as high as the Fr inputs. The C inputs from the harvest losses were higher in the former cropland, which can possibly be explained by the higher aboveground biomass productions (and yields) in the cropland. This demonstrated that the harvesting operation had an effect on the C balance of the system. However, the harvest losses have negative implications on the energy production; the more biomass that is left in the field, the less there is for energy generation. Litter fall is temporarily reduced in frequently harvested tree plantations (Jandl *et al.*, 2007); this reduces the C inputs and contributes to lower soil C stocks. The input of harvest losses into the soil may compensate for the smaller litter fall inputs. Additionally, we found an increased belowground input from Fr mortality after harvest. Apart from the changed C inputs, the harvest might have secondary effects. For example, harvesting changes the microclimate. Decomposition of forest floor C is temporarily stimulated after harvest, because the soil becomes warmer and possibly wetter due to the reduced evapotranspiration (Piene & Vancleve, 1978). Moreover, a harvested field is more exposed to wind and to erosion. Field studies in timber plantations showed that SOC decreased with increasing harvest intensity (Nave *et al.*, 2010). We found very high annual C inputs from weeds, especially in the first rotation. Annual grasses can offset the removal of C inputs in bioenergy crops by providing additional biomass and thus C input (Blanco-Canqui, 2013).

Soil CO_2 efflux. R_s constituted the largest flux to return belowground C to the atmosphere, and it represented the combined R_r and R_h. The R_s represented 55% of the total ecosystem respiration in our SRC (Verlinden *et al.*, 2013b), with roots representing about 47–79% of the total R_s (Data S2). The current study revealed a large R_s during the four years of SRC, ranging from 470 to 785 g C m^{-2} yr^{-1}. These values are within the range of R_s values of 740–970 g C m^{-2} yr^{-1} obtained in different willow SRC plantations in the USA under a similar planting scheme and comparable climatic conditions as our plantation. Other measurements of R_s on poplar SRC plantations were recorded over shorter time periods and are not comparable (Arevalo *et al.*, 2011).

In the former cropland, there was an increasing R_s throughout the years. This might contradict results from other SRCs where R_s remained rather constant over the years after agricultural lands were converted to SRC (Arevalo *et al.*, 2011). However, this increase was not observed in the previous pasture land. The higher R_s in the pasture compared with the cropland might be attributed to the higher initial SOC in previous pasture and to the higher root biomass and growth of genotype Skado (Verlinden *et al.*, 2015).

Doc. The annual DOC leaching increased exponentially throughout the years, and this was driven by the water balance. With regard to our DOC measurements, very similar annual estimates (7–13 g C m^{-2} yr^{-1}) were reported for forests in Belgium (Gielen *et al.*, 2011) and in Germany (Borken *et al.*, 2011). Moreover, for forests (Gielen *et al.*, 2011) as well as for agroecosystems (Brye *et al.*, 2001) the interannual variability of DOC fluxes is primarily driven by the water balance, in line with our observations.

Uncertainties

The uncertainties in the quantification of the C pools, of the C inputs and of the C outputs have different sources: (i) the measurement error involved in the data collection, and (ii) the prediction error, when a model was used for predictions. The measurement errors were due to the intrinsic variability of the measured variable and to errors in the measurements. Furthermore, in our SRC plantation spatial variability was generated by the double-row planting system, by the different genotypes and by the previous land-use types. The prediction errors resulted from (i) the model itself, through its error term and the uncertainty of its parameters and (ii) the uncertainty of the other variables used in the model (Molto *et al.*, 2013). The use of combined error propagation formulas and Monte Carlo simulations allowed a proper treatment of the uncertainties.

We were able to quantify the uncertainties in the estimation of the soil C balance, as well as the contribution by the various input variables to these uncertainties. Some variables showed a very low uncertainty (including a low standard deviation), as they were measured in a small area; the representativeness of these values was difficult to quantify. On the other hand, some input variables contributed only very little to the overall uncertainty; they represented a very low input value, but the uncertainty on these variables might be very large. Below, we review and discuss the uncertainties on the estimation of each input variable itself. This is relevant for future research to improve the estimates of key input variables.

Uncertainties in the flux-based approach

In general, soil characteristics are highly spatially variable over short distances. A high degree of uncertainty is created by the low capture of the spatial heterogeneity in the R_s estimations. The measurements of R_s were concentrated on a rather small area of the plantation because of various logistic reasons, as the restricted length of the instrument cables and the necessity of mains power supply (Verlinden *et al.*, 2013). Moreover, the proportion of R_h to R_s was high. The contribution of R_h has been estimated to be between 10 and 90% of R_s (Hanson *et al.*, 2000), with an average of 60%. Our models predicted the proportion of R_h within the range of previous studies, but close to the higher values. The other variables were measured over a larger area of the plantation and might have a lower spatial uncertainty. Uncertainties were also created by the upscaling models, by the calculation methods, etc. For example, the uncertainties associated with our estimations of the DOC leaching highly depended on the water balance estimation. Uncertainties in the estimations of Fr productivity were associated with the method used (Berhongaray *et al.*, 2013c), as well as with the R_s partitioning (Data S2). On the other hand, mycorrhizal inputs were not quantified at our plantation.

Mycorrhizal inputs are a dominant process for C input in poplar plantations (up to 68% of the total inputs), exceeding the input via leaf litter and fine root turnover (Godbold *et al.*, 2006). Aboveground inputs from weeds were also subject to a high uncertainty. This high uncertainty was created by the high spatial heterogeneity and the rather low sampling intensity and frequency. Due to time constraints and logistic management issues, aboveground weed biomass was measured with few replicates, that is only in two of the four years of the study, and only once in the previous pasture land area. The proper assessment of the uncertainties with the Monte Carlo simulation allowed the reduction of the uncertainty in the multivariable flux-based SOC balance.

Uncertainties in the pool-based approach

For the SOC determination, we captured the spatial heterogeneity. A strong determinant in the change of the SOC stock was the change in BD. The values of BD for March 2014 were too high for the soil below the 30 cm depth. Large uncertainties were associated to the BD estimations at these depths. However, the high BD values below the 30 cm depth were related to the high soil compaction measurements at the same depth (see Data S2), concluding that even with large uncertainties, the mean values were reasonable. If an average BD of 1.5 was used instead of 1.76, the SOC change would be half, which still represents a large positive SOC change. Moreover, the soil was a deep ploughed Anthrosol soil. Due to the deep ploughing, a high spatial heterogeneity of SOC has been induced. This is visible in Table 2 in the increasing standard deviation from 2010 to 2014. Also the more increased SOC pools in the subsoil layer as compared to the topsoil indicated a strong deep ploughing soil inversion effect. To reduce the uncertainty in the estimation of the pool change-based approach, we used a control pasture. No changes in SOC were found in the control grassland. This allowed to attribute all the SOC changes to the changes in the land use, and not to methodological or climatic reasons. Taken all considerations into account, the pool-based approach reflected an unrealistically high SOC change, limiting its use in the evaluation of short-term SOC changes.

Uncertainties in the combined eddy-covariance + biometric approach

The eddy-covariance measurements, including CO_2 and CH_4 fluxes, probably had a high degree of uncertainty due to the size of the footprint. Usually, the size of the footprint increases with changes in atmospheric stability from unstable (day) to stable (night) conditions, directly affecting the NEE estimations (Leclerc & Thurtell, 1990). DOCs and VOCs were measured with a smaller frequency or during shorter periods, but they only represented a tiny portion of the C balance and their impact on the uncertainty was small. The uncertainties on the harvested biomass were rather low. The total biomass yield of the site was recently quantified using three different methodologies with a very good agreement among them (Verlinden *et al.*, 2015).

The highly spatial variability together with the high resilience dynamics of the SOC stocks require long (>20 years) periods to quantify SOC changes using the pool-change approach (Guo & Gifford, 2002). On the other hand, medium-term (2–5 years) flux measurements can account for SOC changes, and provide better estimates about whether the soil pool or reservoir is functioning as a source or as a sink for C. Our results showed a small C increase in the belowground compartment of the SRC plantation. However, results from the entire life of an SRC (around 20 years) should be considered to substantiate the C storage potential of this type of bioenergy crop. Unfortunately, C sequestration in the soil is not permanent and it seems that forest soils are reaching an equilibrium (Janssens *et al.*, 2005). Compared to the reduced emissions of other GHG sources, which can continue indefinitely, C sequestration in the soil is therefore time-limited and finite. This limitation is explained by the sink saturation (Stewart *et al.*, 2007) and because further increases in forest areas are unlikely (Jandl *et al.*, 2007).

The additional heterogeneity made SOC pool assessments much more difficult than in common soils. So, the conclusions on the best choice of methods are limited to our site. Regardless of the soil C sequestration, the C fixation in bioenergy crops provides large benefits reducing CO_2 emissions which make bioenergy crops useful and beneficial. The results presented in this study are of high relevance for bioenergy crop models and for C stock estimations. Life cycle analysis studies of SRC for bioenergy will also benefit from this and similar soil C balance assessments. This information is also crucial for policymakers for the proper evaluation and further improvement of this renewable source of energy.

Acknowledgements

This work was supported by the European Research Council under the European Commission's Seventh Framework Programme (FP7/2007-2013) as ERC Advanced Grant agreement # 233366 (POPFULL), as well as by the Flemish Hercules Foundation as Infrastructure contract # ZW09-06. Further funding was provided by the Flemish Methusalem Programme and by the Research Council of the University of Antwerp. GB was supported by the Erasmus-Mundus External Cooperation, Consortium EADIC – Window Lot 16 financed by the European

Union Mobility Programme # 2009-1655/001-001. We gratefully acknowledge the excellent technical, logistic and field support of the entire POPFULL team, especially Nadine Calluy for laboratory analyses.

References

Agosti F, Gregory AS, Richter GM (2015) Carbon sequestration by perennial energy crops: is the jury still out? *Bioenergy Research*, **8**, 1057–1080.

Arevalo CBM, Bhatti JS, Chang SX, Sidders D (2011) Land use change effects on ecosystem carbon balance: from agricultural to hybrid poplar plantation. *Agriculture Ecosystems & Environment*, **141**, 342–349.

Asensio D, Peñuelas J, Filella I, Llusià J (2007) On-line screening of soil VOCs exchange responses to moisture, temperature and root presence. *Plant and Soil*, **291**, 249–261.

Batjes NH (2008) Mapping soil carbon stocks of Central Africa using SOTER. *Geoderma*, **146**, 58–65.

Berhongaray G, Alvarez R, De Paepe J, Caride C, Cantet R (2013a) Land use effects on soil carbon in the Argentine Pampas. *Geoderma*, **192**, 97–110.

Berhongaray G, El Kasmioui O, Ceulemans R (2013b) Comparative analysis of harvesting machines on an operational high-density short rotation woody crop (SRWC) culture: one-process versus two-process harvest operation. *Biomass and Bioenergy*, **58**, 333–342.

Berhongaray G, Janssens IA, King JS, Ceulemans R (2013c) Fine root biomass and turnover of two fast-growing poplar genotypes in a short-rotation coppice culture. *Plant and Soil*, **373**, 269–283.

Berhongaray G, King JS, Janssens IA, Ceulemans R (2013d) An optimized fine root sampling methodology balancing accuracy and time investment. *Plant and Soil*, **366**, 351–361.

Berhongaray G, Verlinden MS, Broeckx LS, Ceulemans R (2015) Changes in belowground biomass after coppice in two *Populus* genotypes. *Forest Ecology and Management*, **337**, 1–10.

Blanco-Canqui H (2013) Crop residue removal for bioenergy reduces soil carbon pools: how can we offset carbon losses? *Bioenergy Research*, **6**, 358–371.

Borken W, Ahrens B, Schulz C, Zimmermann L (2011) Site-to-site variability and temporal trends of DOC concentrations and fluxes in temperate forest soils. *Global Change Biology*, **17**, 2428–2443.

Broeckx LS, Verlinden MS, Ceulemans R (2012) Establishment and two-year growth of a bio-energy plantation with fast-growing *Populus* trees in Flanders (Belgium): effects of genotype and former land use. *Biomass and Bioenergy*, **42**, 151–163.

Brunner I, Bakker MR, Björk RG *et al.* (2013) Fine-root turnover rates of European forests revisited: an analysis of data from sequential coring and ingrowth cores. *Plant and Soil*, **362**, 357–372.

Brye KR, Norman JM, Bundy LG, Gower ST (2001) Nitrogen and carbon leaching in agroecosystems and their role in denitrification potential. *Journal of Environmental Quality*, **30**, 58–70.

EU (2009) *Directive 2009/28/EC of the European Parliament and of the Council of 23 April 2009 on the Promotion of the use of Energy From Renewable Sources and Amending and Subsequently Repealing Directives 2001/77/EC and 2003/30/EC*. Official Journal of the EUropean Union, EU, Brussels, Belgium.

Fairley RI, Alexander IJ (1985) Methods of calculating fine root production in forests. In: *Ecological Interactions in Soil* (ed. Fitter AH), pp. 37–42. Blackwell Science Inc, Oxford, UK.

FAO (2015) *World Reference Base (WRB) for Soil Resources 2014: International Soil Classification System for Naming Soils and Creating Legends for Soil Maps*. FAO, Rome, Italy.

Fortunel C, Garnier E, Joffre R *et al.* (2009) Leaf traits capture the effects of land use changes and climate on litter decomposability of grasslands across Europe. *Ecology*, **90**, 598–611.

Gielen B, Neirynck J, Luyssaert S, Janssens IA (2011) The importance of dissolved organic carbon fluxes for the carbon balance of a temperate Scots pine forest. *Agricultural and Forest Meteorology*, **151**, 270–278.

Godbold DL, Hoosbeek MR, Lukac M *et al.* (2006) Mycorrhizal hyphal turnover as a dominant process for carbon input into soil organic matter. *Plant and Soil*, **281**, 15–24.

Görres CM, Kammann C, Ceulemans R (2015) Soil greenhouse gas fluxes from a poplar bioenergy plantation: How long does former land use type matter? EGU Assembly 2015, Vienna, Austria. Vol. 17, 11675.

Guo LB, Gifford RM (2002) Soil carbon stocks and land use change: a meta analysis. *Global Change Biology*, **8**, 345–360.

Hansen EA, McNeel HA, Netzer DA *et al.* (1979) Short-rotation intensive culture practices for northern Wisconsin. In: Proceedings, 16th Annual Meeting, North American Poplar Council, Joint Meeting of the United States and the Canadian Chapters, pp. 47–63.

Hanson PJ, Edwards NT, Garten CT, Andrews JA (2000) Separating root and soil microbial contributions to soil respiration: a review of methods and observations. *Biogeochemistry*, **48**, 115–146.

Harris ZM, Spake R, Taylor G (2015) Land use change to bioenergy: a meta-analysis of soil carbon and GHG emissions. *Biomass and Bioenergy*, **82**, 27–39.

IEA (2014) *World Energy Outlook 2014*. International Energy Agency, Paris, France.

IPCC (2006) *IPCC Good Practice Guidance and Uncertainty Management in National Greenhouse Gas Inventories*. Institute for Global Environmental Strategies (IGES), Hayama, Japan.

Jandl R, Lindner M, Vesterdal L *et al.* (2007) How strongly can forest management influence soil carbon sequestration? *Geoderma*, **137**, 253–268.

Janssens IA, Sampson D A, Cermak J, Meiresonne L, Riguzzi F, Overloop S, Ceulemans R (1999) Above- and belowground phytomass and carbon storage in a Belgian Scots pine stand. *Annals of Forest Science*, **56**, 81–90.

Janssens IA, Freibauer A, Schlamadinger B *et al.* (2005) The carbon budget of terrestrial ecosystems at country-scale - a European case study. *Biogeosciences*, **2**, 15–26.

Jones RJA, Hiederer R, Rusco E, Montanarella L (2005) Estimating organic carbon in the soils of Europe for policy support. *European Journal of Soil Science*, **56**, 655–671.

Lal R (2004) Soil carbon sequestration impacts on global climate change and food security. *Science*, **304**, 1623–1627.

Le Quéré C, Andres RJ, Boden T *et al.* (2013) The global carbon budget 1959-2011. *Earth System Science Data*, **5**, 165–185.

Leclerc MY, Thurtell GW (1990) Footprint prediction of scalar fluxes using a Markovian analysis. *Boundary-Layer Meteorology*, **52**, 247–258.

Levillain J, Thongo-M'Bou A, Deleporte P, Saint-Andre L, Jourdan C (2011) Is the simple auger coring method reliable for below-ground standing biomass estimation in *Eucalyptus* forest plantations? *Annals of Botany*, **108**, 221–230.

Liang BC, Campbell CA, Mcconkey BG, Padbury G, Collas P (2005) An empirical model for-estimating carbon sequestration on the Canadian prairies. *Canadian Journal of Soil Science*, **85**, 549–556.

Molto Q, Rossi V, Blanc L (2013) Error propagation in biomass estimation in tropical forests. *Methods in Ecology and Evolution*, **4**, 175–183.

Nave LE, Vance ED, Swanston CW, Curtis PS (2010) Harvest impacts on soil carbon storage in temperate forests. *Forest Ecology and Management*, **259**, 857–866.

Pacaldo RS, Volk TA, Briggs RD (2013) No significant differences in soil organic carbon contents along a chronosequence of shrub willow biomass crop fields. *Biomass and Bioenergy*, **58**, 136–142.

Pacaldo R, Volk T, Briggs R (2014) Carbon sequestration in fine roots and foliage biomass offsets soil CO_2 effluxes along a 19-year chronosequence of shrub willow (*Salix x dasyclados*) biomass crops. *Bioenergy Research*, **7**, 1–8.

Piene H, Vancleve K (1978) Weight-loss of litter and cellulose bags in a thinned white spruce forest in Interior Alaska. *Canadian Journal of Forest Research*, **8**, 42–46.

Qin Z, Dunn JB, Kwon H, Mueller S, Wander MM (2016) Soil carbon sequestration and land use change associated with biofuel production: empirical evidence. *Global Change Biology Bioenergy*, **8**, 66–80.

Razakamanarivo RH, Razakavololona A, Razafindrakoto MA, Vieilledent G, Albrecht A (2012) Below-ground biomass production and allometric relationships of eucalyptus coppice plantation in the central highlands of Madagascar. *Biomass and Bioenergy*, **45**, 11–10.

Schulp CJE, Nabuurs GJ, Verburg PH (2008) Future carbon sequestration in Europe - Effects of land use change. *Agriculture, Ecosystems and Environment*, **127**, 251–264.

Smith P (2004) Carbon sequestration in croplands: the potential in Europe and the global context. *European Journal of Agronomy*, **20**, 229–236.

Smith P, Powlson DS, Glendining MJ, Smith JU (1997) Potential for carbon sequestration in European soils: Preliminary estimates for five scenarios using results from long-term experiments. *Global Change Biology*, **3**, 67–79.

Stewart CE, Paustian K, Conant RT, Plante AF, Six J (2007) Soil carbon saturation: concept, evidence and evaluation. *Biogeochemistry*, **86**, 19–31.

Vanbeveren SPP, Schweier J, Berhongaray G, Ceulemans R (2015) Operational short rotation woody crop plantations: manual or mechanised harvesting? *Biomass and Bioenergy*, **72**, 8–18.

Verlinden MS, Broeckx LS, Wei H, Ceulemans R (2013a) Soil CO_2 efflux in a bioenergy plantation with fast-growing *Populus* trees – influence of former land use, inter-row spacing and genotype. *Plant and Soil*, **369**, 631–644.

Verlinden MS, Broeckx LS, Zona D *et al.* (2013b) Net Ecosystem Production and carbon balance of an SRC poplar plantation during the first rotation. *Biomass and Bioenergy*, **56**, 412–422.

2

An integrated assessment of the potential of agricultural and forestry residues for energy production in China

JI GAO[1], AIPING ZHANG[1], SHU KEE LAM[2], XUESONG ZHANG[3,4], ALLISON M. THOMSON[5], ERDA LIN[1], KEJUN JIANG[6], LEON E. CLARKE[3], JAMES A. EDMONDS[3], PAGE G. KYLE[3], SHA YU[3], YUYU ZHOU[7] and SHENG ZHOU[8]

[1]*Institute of Environment and Sustainable Development in Agriculture, Chinese Academy of Agricultural Sciences, Beijing 100081, China,* [2]*Crop and Soil Sciences Section, Faculty of Veterinary and Agricultural Sciences, the University of Melbourne, Melbourne, Vic. 3010, Australia,* [3]*Joint Global Change Research Institute, Pacific Northwest National Laboratory and University of Maryland, College Park, MD 20740, USA,* [4]*Great Lakes Bioenergy Research Center, Michigan State University, East Lansing, MI 48824, USA,* [5]*Field to Market, The Alliance for Sustainable Agriculture, 777 N Capitol St. NE, Suite 803, Washington, DC 20002, USA,* [6]*Energy Research Institute (ERI), Beijing 100038, China,* [7]*Department of Geological & Atmospheric Sciences, Iowa State University, Ames, IA 50011, USA,* [8]*Institutes of Energy, Environment and Economy, Tsinghua University, Beijing 100084, China*

Abstract

Biomass has been widely recognized as an important energy source with high potential to reduce greenhouse gas emissions while minimizing environmental pollution. In this study, we employ the Global Change Assessment Model to estimate the potential of agricultural and forestry residue biomass for energy production in China. Potential availability of residue biomass as an energy source was analyzed for the 21st century under different climate policy scenarios. Currently, the amount of total annual residue biomass, averaged over 2003–2007, is around 15 519 PJ in China, consisting of 10 818 PJ from agriculture residues (70%) and 4701 PJ forestry residues (30%). We estimate that 12 693 PJ of the total biomass is available for energy production, with 66% derived from agricultural residue and 34% from forestry residue. Most of the available residue is from south central China (3347 PJ), east China (2862 PJ) and south-west China (2229 PJ), which combined exceeds 66% of the total national biomass. Under the reference scenario without carbon tax, the potential availability of residue biomass for energy production is projected to be 3380 PJ by 2050 and 4108 PJ by 2095, respectively. When carbon tax is imposed, biomass availability increases substantially. For the CCS 450 ppm scenario, availability of biomass increases to 9002 PJ (2050) and 11 524 PJ (2095), respectively. For the 450 ppm scenario without CCS, 9183 (2050) and 11 150 PJ (2095) residue biomass, respectively, is projected to be available. Moreover, the implementation of CCS will have a little impact on the supply of residue biomass after 2035. Our results suggest that residue biomass has the potential to be an important component in China's sustainable energy production portfolio. As a low carbon emission energy source, climate change policies that involve carbon tariff and CCS technology promote the use of residue biomass for energy production in a low carbon-constrained world.

Keywords: bioenergy, carbon tax, carbon capture and storage, climate policy, integrated assessment, residue biomass

Introduction

China's energy consumption has been soaring due to rapid increase in population and economic growth over the last decade. Its total energy consumption has increased from 44 022 PJ in 2001 to 110 055 PJ in 2013. Since 2011, China has been the largest energy consumer with oil and natural gas dependency rates of approximately 60% and 33%, respectively (Shi, 2013). The International Energy Agency (IEA) estimates that with an 80% oil dependency rate, China will overtake the United States to become the world's largest oil-demanding country by 2035 (IEA, 2010). Meanwhile, China has overtaken the United States as the world's largest carbon emitter since 2007 and is projected to account for half of the increase in global CO_2 emissions through 2035 (IEA, 2011). In December 2009, China's State Council announced that China will reduce its carbon intensity per unit of GDP by 40–45% by 2020, compared with 2005. Energy security, environmental health and greenhouse

Ji Gao and Aiping Zhang contributed equally to this work and should be considered co-first authors.

Correspondence: Erda Lin
e-mail: lined@ami.ac.cn; Kejun Jiang
e-mail: kjiang@eri.org.cn

gases (GHGs) mitigation have been a major impediment to China's sustainable development. Bioenergy is often regarded as an environmentally acceptable and more efficient alternative for energy production (IEA, 2007). China's abundance in biomass resources accentuates the potential of using biomass to promote its development in renewable energy in a carbon-constrained world.

Unlike fossil fuel, biomass energy generates low or even net-zero carbon emissions because CO_2 is recycled during the life cycle of using biomass for energy production. Therefore, temporary and permanent carbon storage based on biogenic sources is thought of as a key way to achieve low CO_2 concentrations and mitigate climate change (Guest *et al.*, 2013). Bioenergy with carbon capture and storage (Bio-CCS) can lead to negative carbon emissions (IEA, 2011). It could potentially have a 33% share of overall mitigation by the end of the century (Klein *et al.*, 2011) and is important to mitigating global warming.

Although large-scale production may incur negative impacts such as increasing food price, accelerating soil erosion and runoff, decreased farmland productivity, and loss of wildlife habitat and biodiversity (Pimentel, 1994; Cramer, 2007), biomass can be environmentally friendly and renewable when used in a sustainable and responsible manner (Gustavsson *et al.*, 2007). Most of the studies on biomass energy were focused on energy production potential, energy conversion technologies, and associated environmental, political and financial problems (Liao *et al.*, 2004; Elmore *et al.*, 2008; Sun *et al.*, 2011; Cui & Wu, 2012; Li *et al.*, 2012; Yu *et al.*, 2012; He *et al.*, 2013). However, previous studies have not examined the potential of biomass as a sustainable energy source in China with a global Integrated Assessment Modeling Framework.

In this study, we estimated the energy potential of agricultural and forestry residue biomass and quantity of residue retention, as well as their spatial distribution in China. We employed the Global Change Assessment Model (GCAM) to simulate the future potential of residue biomass from agricultural and forestry residues for energy production in China in response to global and national energy demand and climate change policies. Results obtained here would improve the understanding of how the development of residue biomass for energy production can help China achieve climate change mitigation goals and contribute to global mitigation efforts.

Materials and methods

Current availability of residue biomass

Determining potential availability of agricultural residues. The four main categories of residue biomass for energy production are agriculture, forestry, municipal solid wastes (MSW) and emerging energy crops. Agricultural residues refer to field (e.g., straw, stalks, stubble, leaves and seed pods) and processed (e.g., husks, seeds, bagasse, molasses and roots) residues from a variety of crops. Agricultural residues are used as fertilizer, forage, raw material for producing paper and generating energy for cooking and heating.

The total amount of agricultural residues was calculated using the estimated ratios of agricultural biomass residue to agricultural product in China (Bi *et al.*, 2008; Bi 2010; Table 1). Not all residue biomass was available due to residue retention and loss during transportation and storage, and subsequent processing. These factors were taken into consideration when estimating the maximum available supply of residue biomass. We used the collectable and usable coefficient (Table 1) of agricultural residues (the ratio of collectable and usable residues to the aboveground biomass of crop) to estimate the maximum available supply of biomass residue (Bi *et al.*, 2008). We also calculated the total potential energy supply by agricultural residues based on their heating values on a dry mass basis. For each crop, we also estimated a residue retention fraction (Table 1) as the amount of residue to be retained for erosion control and nutrient cycling. The yields of main agricultural products of different crops averaged over 2003–2007 are presented in Table 1 and used as the baseline crop yields data for GCAM, for which simulation starts from year 2005 through the end of the 21st century.

Determining the potential availability of forest residues. Forestry residues refer to wastes associated with the processing of forest products including logging residues, wood-processing residues and tending/thinning residues (Cai *et al.*, 2012). Logging residues originate from the harvesting operations and include stumps, roots, leaves, off-cuts, branches and sawdust. These residues are left on forestland. Wood-processing residues, or primary mill residues, are generated when processing roundwood at a sawmill, veneer mill, plywood mill or pulp mill. These residues include discarded logs, bark, sawdust and shavings (Liao *et al.*, 2004; Yuan, 2002). Tending/thinning residues are derived from the processing of tending and thinning of different forests and afforestation activities such as stumping, thinning and pruning. Forest residues are used for generating heat, electricity, liquid fuels and solid fuels (Tan *et al.*, 2010; MOA, 1998).

The total biomass production from logging and tending/thinning residues varies with forest type, location, and tree density and growth rate. The amount of forest residues was estimated by multiplying biomass yields by collectable coefficients of biomass (Table 2). Forests were divided into five categories according to the Forest Law of the People's Republic of China: timber stands, protection forest, economic forest, forest for special uses and firewood forest. In addition, residues from other kinds of forest were evaluated based on the number of trees, their productivity and collectability. In this category, we included sparse forest, shrubs, sipang forest and bamboo forest. Notably, orchards, urban greening forest and hedgerow may produce large amounts of biomass due to annual pruning, which might be a potential bioenergy source. Firewood was assumed to be entirely harvested and currently used for

Table 1 The theoretical maximum energy potential and energy potential availability of agricultural residues in 2003 and 2007

	Type of residue	Residue to product ratio (RPR)	Average Crop production (10^4 t) over (2003–2007)*	Sown area (10^3 ha)†	Water content (%)	Low heating value (KJ kg^{-1})	Collectable coefficient	Retention (t ha^{-1})	Maximum energy potential (PJ)	Energy potential availability (PJ)	Source
Rice	Straws	0.94	17761.68	28318.10	6.00[1]	14059.00	0.83	1.01	2354.78	1954.47	[1] Niu & Liu (1984)
–	Husks	0.21	17761.68	28318.10	9.00[1]	13067.00	0.95	0.00	487.39	463.02	
Wheat	Straws and stalks	1.30	9872.99	22749.90	13.50	14766.00	0.65	1.97	1895.20	1231.88	
Corn	Stalks	1.10	13787.72	26762.52	15.00	14356.00	0.90	0.57	2177.30	1959.57	
–	Cobs	0.21	13787.72	26762.52	9.70	14359.00	0.90	0.00	415.75	374.18	
Other grains	Straws and stalks	1.27	996.31	3917.91	11.35	14384.00	0.86	0.44	181.32	156.39	
Millet	Straws	1.40	173.67	897.19	13.50	14569.00	0.85	0.41	35.42	30.11	
Sorghum	Stalks	1.60	236.75	618.52	10.20	15105.00	0.90	0.61	57.22	51.50	
Barley	Straws and stalks	1.09	316.60	826.68	10.40[2]	13720.00	0.85	0.63	47.35	40.25	[2] Lv & Wang (1998)
Others	Straws and stalks	1.09	269.29	1575.52	11.30	14142.00	0.85	0.28	41.51	35.28	
Beans	Straws stalk leaves pod	1.60	2048.21	12505.55	5.10	14788.50	0.56	1.15	484.64	271.40	
Gram	Straws stalk pod	1.60	90.46	735.45	10.30	14615.00	0.56	0.87	21.15	11.85	
Small red bean	Straws stalk pod	1.60	32.98	219.59	10.30	14548.00	0.56	1.06	7.68	4.30	
Soy beans	Straws stalk leaves pod	1.60	1538.99	9310.13	10.30	15079.00	0.56	1.16	371.30	207.93	
Others	Straws stalk pod	1.60	385.79	2240.38	10.30	14912.00	0.56	1.21	92.05	51.55	
Tubers	Stem and leaves	0.77	3209.70	8924.16	11.80	14125.50	0.73	0.75	349.11	254.85	
Potato	Stem and leaves	0.96	1361.79	4528.09	11.30	13498.00	0.73	0.78	176.46	128.82	
Sweet potato	Stem and leaves	0.63	1847.91	4396.07	12.30	14753.00	0.73	0.72	171.75	125.38	
Cotton	Stalk and torus	5.00[3]	641.08	5521.40	15.00	14979.00	0.86	0.81	480.13	412.91	[3] Cui *et al.* (2008)
Oils crop	–	2.32	2832.62	14775.12	9.92	14775.12	0.78	0.98	971.04	757.41	
Peanut	Stalks	1.50	1360.35	4472.96	10.23[4]	15033.00	0.83	0.78	306.75	254.60	[4] Nan *et al.* (2008)
–	Peanut hull	0.28	1360.35	4472.96	7.80	15682.00	0.70	0.00	59.73	41.81	
Rapes	Stem, leaves, pod	2.87[5]	1183.85	6679.35	10.78[4]	14142.00	0.64	1.83	480.50	307.52	[5] Xie *et al.* (2011)
Sesame	Stem, leaves, pod	2.80	62.82	590.89	10.85[4]	15491.00	0.83	0.51	27.25	22.62	
Sunflower	Residues after Sunflower seed harvest	2.80	156.99	925.69	Air drying	15021.00	0.86	0.66	66.03	56.79	
Flax (linseed)	Hemps blade tips	2.01[5]	37.61	393.60	Air drying	15439.00	0.74	0.50	11.67	8.64	
Others (average)	Stem, leaves, pod	2.00	30.99	1712.63	Air drying	15491.00	0.85	0.05	9.60	8.16	

(continued)

Table 1 (continued)

	Type of residue	Residue to product ratio (RPR)	Average Crop production (10^4 t) over (2003–2007)*	Sown area (10^3 ha)†	Water content (%)	Low heating value (KJ kg^{-1})	Collectable coefficient	Retention (t ha^{-1})	Maximum energy potential (PJ)	Energy potential availability (PJ)	Source
Fiber crops[6]	Sticks, sheath, leaves, hempshell	2.86	93.01	310.10	Air drying	15491.00[7]	0.84	1.37	41.20	34.61	[6] Liu (1986); [7] Jing (2006)
Jute and hemp	Sticks sheath	1.90	9.11	33.62	Air drying	15491.00	0.87	0.67	2.68	2.33	
Flax (Linum)	Sticks sheath	1.10	50.75	123.86	Air drying	15491.00	0.82	0.81	8.65	7.09	
Cannabis sativa	Sticks sheath	3.00	4.69	15.53	Air drying	15491.00	0.86	1.27	2.18	1.87	
Ramie	Leaves hempshell	6.50	27.11	133.97	Air drying	15491.00	0.84	2.10	27.30	22.93	
Others (like jute and hemp)	–	1.90	1.36	3.11	Air drying	15491.00	0.85	1.24	0.40	0.34	
Sugar	Sugar bagasse	0.23	10262.47	1631.71	Air drying	15350	0.97	0.00	355.74	343.31	
–	Sugar cane stalk Sheath	0.10	10262.47	1631.71	Air drying	14902	0.76	1.53	152.93	115.80	
Sugarcane	Bagasse	0.24	9535.30	1421.16	Air drying	15491.00	0.97	0.48	354.51	343.87	
–	Cane stalk sheath	0.10	9535.30		Air drying	13816.00	0.70	–	131.74	92.22	
Sugar beet	Bagasse	0.04	727.17	210.55	13.50	13500.00[8]	0.90	0.14	3.93	3.53	[8] Wang & Liu (1984)
–	Stems leaves	0.10	727.17	210.55	Air drying	14235.00	0.75	–	10.35	7.76	
Tobacco	Stems leaves	1.60	243.95	1249.10	3.51	11300.00[9]	0.95	0.16	44.11	41.90	[9] Zhang *et al.* (2012)
Flue-cured Tobacco	Stems leaves	1.60	220.92	1136.71	3.51	11300.00[9]	0.95	0.16	39.94	37.94	
Others	Stems leaves	1.60	23.03	112.39	Air drying	11300.00[9]	0.95	0.16	4.16	3.96	
Vegetables and Melons	Vine stems shell	0.10	62453.90	19681.70	Air drying	13498.00[11]	0.50	1.59	843.00	421.50	
Total									10817.9	8419.02	

Residue to product ratio, water content, collectable coefficient and low heating value determined following Bi (2010) except [1] to [9].
Residue retention fraction means the minimum to return the field and determined by sown area, collectable coefficient, residue to product ratio (RPR) and crop production.
* and † come from NBS (2003–2007).

Table 2 The theoretical energy potential availability of forestry residues from 2004 to 2008

	Type of residue	Forest area (10^4 ha) (2004–2008)*	Product yield (kg ha^{-1})†	Collectable coefficient‡	Water content	Heating value (KJ kg^{-1})	Retention t ha^{-1}	Energy potential PJ yr^{-1}
Timber stands	Wood chips, sawdust, needle leaves, bark, branches, cone	6007.44	3750	0.50	Dry weight	18600.0	1.31	2095.09
Protection forest	Wood chips, sawdust, bark, branches	8194.68	3750	0.20	Dry weight	18600.0	0.53	1143.16
Forest for special uses	Wood chips, sawdust, bark, branches	1182.14	1875	0.10	Dry weight	18600.0	0.13	41.23
Firewood forest	Total train	174.73	3750	1.00	Dry weight	16747.0	0.00	109.73
Bamboo forest	Wood chips, sawdust, bark, branches	538.10	1875	0.10	Dry weight	17672.1	0.13	17.83
Economic forest	Wood chips, sawdust, bark, branches	2041.00	1875	0.10	Dry weight	18600.0	0.13	71.18
Sparse forest	Wood chips, sawdust, bark, branches	482.22	1875	0.50	Dry weight	18600.0	0.66	84.09
Shrubbery	Bark, branches	5365.34	938	0.50	Dry weight	18600.0	0.33	468.04
Sipang forest	Wood chips, sawdust, bark, branches	1121054.00 (10^4zhu)	2 (kg zhu^{-1})	0.50	Dry weight	18600.0	0.00	208.52
City greening forest; Hedgerow	Wood chips, sawdust, bark, branches	400.00[1]	1625	0.10	Dry weight	18600.0	0.11	12.09
Orchard	Fruitwood, pruning coconut shell, chestnut shell, walnut shell, etc.	996.66[2]	1875	0.10	Dry weight	18600.0	0.13	34.76
Mill	Lath, slab, woodshaving	10675.27	540 (kg m^{-3})	0.34	Dry weight	19500.0[3]	0.00	382.20
Waste wood products		2000.00[1]	250 (kg m^{-3})	0.34	Dry weight	19500.0[3]	0.00	33.15
Total								4701.06

*Comes from SFA (2009) except [1] Cai *et al.*, 2012 and [2] NBS (2004–2008), † and ‡ determined following MOA, 1998 and Lu, 1997; Heating value derives from MOA, 1998 except [3] Zhang *et al.*, 2008. Residue retention fraction determined by forest area, collectable coefficient, product yield based on assumption of the 30% availability of forest residues (MOA, 1998).

heating in rural areas. For wood-processing residues, the available amount was estimated based on the average annual production of roundwood in 2005 and 2009, which included net imported roundwood. These residues collectively accounted for ca. 34.4% of the total roundwood production (MOA, 1998). For forest residues, we also estimated (i) the maximum available supply of residue biomass based on the coefficients of collectable residues, (ii) the total potential energy supply from forestry residues according to their heating values (Table 2) and (iii) a residue retention value (Table 2).

While agricultural residues and forestry residues were the major focus of our study, MSW and energy crops were also considered and discussed. Note that the information provided here only reflects the gross amount of residues and energy potentials, which were derived based on the assumption that all the residues were economically exploitable and fully utilized.

The potential of residue biomass in the future

To better describe the interrelations between agriculture, food, bioenergy and climate change and understand the potential role of this energy resource in the future, the residue availability parameters particularly derived for China, as introduced in the above section, were incorporated into the Global Change Assessment Model (GCAM) to simulate future availability of residue biomass for bioenergy production in response to global mitigation policies.

The GCAM is a long-term partial equilibrium model with 32 energy/economy regions and 283 agro-ecological zones (AEZs). Besides, it also includes a reduced form carbon cycle and climate module and runs from 1990 to 2100 in 5-year time step. GCAM was designed to estimate the long-term changes in the global energy/economy, agriculture/land use and water use and further explore the interactions between sectors (Kim *et al.*, 2006). It will serve for understanding the potential ramifications of climate mitigation actions. GCAM has been used to investigate the potential roles of specific policy measures and different energy technologies such as bioenergy, CCS (carbon capture and storage), nuclear energy and other technologies used in different sectors Clarke *et al.*, 2007a; (Thomson *et al.*, 2011). We used the standard release of GCAM 3.0 with a thorough representation of bioenergy, agriculture and land use as described in (Wise *et al.*, 2009; Wise & Calvin 2011; Wise *et al.*, 2014). GCAM can model three types of commercial biomass energy including dedicated energy crops, municipal solid waste and residue biomass (Wise *et al.*, 2009; Luckow *et al.*, 2010; Kyle *et al.*, 2011). Biomass energy production from dedicated crops is mainly dependent on the availability and characteristics of land resources, technology options for production, competing land uses as well as bioenergy price in the context of energy markets. Potential energy production from residue biomass depends on crop production, harvest index and price of bioenergy. Potential production is also influenced by population and income. Carbon fluxes associated with terrestrial ecosystems were simulated in 15 different carbon pools (Wise *et al.*, 2009), which inform bioenergy production under a carbon-constrained world.

For this analysis, the GCAM was used to simulate future bioenergy production from residue biomass under a reference scenario and two policy scenarios with and without CCS that are targeted at 450 ppm atmospheric concentration of CO_2 by the end of the 21st century. The reference scenarios (Business as Usual) do not have greenhouse gas emissions constraints or taxes. For the policy scenario with carbon tax, we assumed that carbon emissions from the terrestrial ecosystems, fossil fuel and industrial sources are equally charged with a carbon price starting in 2020 and increasing at 5% per year through 2100. This scenario is noted as UCT (Universal Carbon Tax) (Edmonds *et al.*, 2008; Wise *et al.*, 2009). The carbon price pathway was set to limit atmospheric CO_2 concentration to 450 ppm. In the other policy scenario, bioenergy with CCS detailed in Clarke *et al.* (2007b) was also considered, which has been shown as an effective technology to greatly reduce CO_2 emissions for achieving low CO_2 concentration targets. The policy scenario without CCS would be of higher cost. We used different carbon price starting in 2020 at approximately 76 $\$\ t^{-1}\ C^{-1}$ (in 2005$) without CCS and 129 $\$\ t^{-1}\ C^{-1}$ (in 2005$) with CCS. Future crop productivity needs be considered for projecting the energy production from residue biomass in the future. In the reference scenario, change in crop yield was based on FAO projection until 2050 to ensure global food security (Briunsma, 2009). Consistent with the historical trend, we assumed yields increase at a slower growth rate in the developed countries, but a relatively high yield growth rate in the developing countries. For instance, in China, the crop yield increase rates are 0.83%,0.62% and 0.35% for 2020, 2035 and 2050, respectively (Kyle *et al.*, 2011). After 2050, the annual agriculture productivity changes converge to 0.25% for all crops and regions in the world. Global population growth pathway was inherited from United Nation's 2011 (Eom *et al.*, 2012). Chinese population and GDP growth was described in Jiang *et al.* (2009), peaking in around 2035 and decreasing thereafter due to population aging and low birth rate. We assumed that GDP increases at a fast growth rate in China before 2030, and changes to a lower growth rate close to other developed countries beyond that (Jiang *et al.*, 2009; Zhou *et al.*, 2013). In this study, we used the same social and macroeconomic drivers, including population, labor productivity and changes in crop productivity, for all the scenarios.

Results

Current availability of agricultural residues

The total amount of agricultural residues and available energy supply are about 10 818 PJ and 8419 PJ per year, respectively (Table 1). This energy supply is roughly 8% of the annual energy consumption (105 952 PJ) of China in 2012. The energy potential of rice residues (including rice husks) is the greatest (around 2418 PJ), followed by corn residues (including corn cobs) of 2334 PJ and wheat straws (1232 PJ). These three crop residues combined account for ca. 71% of the total potentially available energy supply. The total processing crop residues, including rice husks, corn cobs, sugar bagasse and peanut hull, account for approximately 1319 PJ and 1223 PJ, respectively, which dominantly represent about 12% and 15% of the total residue availability.

The potential availability of crop residues for energy production in China was also analyzed spatially in Table 3. South central China has the highest potential for crop residue-based energy production of ca. 2419 PJ, followed by east China with ca. 2198 PJ. North-east China has the lowest potential of ca. 648 PJ. Other districts, including south-west China, north-west China and north China, combined have the potential to provide crop residue-based energy production of more than 3156 PJ. In China, Henan, Shandong, Jiangsu, Guangxi, Sichuan, Hubei and Heilongjiang are the top seven provinces in terms of potential availability of agricultural residues, occupying 46.5% of the total availability. Rice residues are mainly available over the central south China, east China and south-west China, accounting for 86.7% (2098 PJ) of the national rice residue potential. Wheat residue is mainly available in north China, east China and south central China, which collectively account for about 80.5% of the total of 1232 PJ. Henan, Shandong and Anhui provinces in located in these three districts account for about 70.1% of the total wheat residue availability. The energy production potential of corn residues is distributed mainly over north-west China and north China, amounting to 51.4% of the total national availability. The availability of the two root crops, viz. sugarcane and sugar beet, is the highest in Guangxi, Yunnan and Guangdong of central south and south-west of China, accounting for 82.9% of the national energy production potential of 459 PJ (Table 3).

Current availability of forest residues

The total amount of forest residues for energy production was estimated at ca. 4274 PJ per year (Table 4) based on the data of the seventh National Forestry

Table 3 The spatial distribution of the theoretical energy potential available of agricultural residues in 2003 and 2007 at province level (PJ)

	Rice	Rice husks	Wheat	Corn	Corn cobs	Other grains	Beans	Tubers	Cotton	Oils crop	Peanut hull	Sugar bagasse	Sugar cane stalk sheath	Fiber crops	Tobacco	Vegetables and Melons	Total
Residue to product ratio (RPR)	0.94	0.21	1.3	1.1	0.21	1.27	1.6	0.77	5	2.32	0.28	0.23	0.1	2.86	1.6	0.1	
Heating value (KJ kg^{-1}) low	14 059	13 067	14 766	14 356	14 359	14 384	14788.5	14125.5	14 979	14775.1	15 682	15349.9	14 902	15 491	11 300	13 498	
Collectable coefficient	0.83	0.95	0.65	0.9	0.9	0.86	0.56	0.73	0.86	0.78	0.7	0.96504	0.7572	0.84	0.95	0.5	
Beijing	0.06	0.01	2.89	8.18	1.56	0.08	0.34	0.20	0.23	0.70	0.08	0.00	0.00	0.00	0.00	3.13	15.82
Tianjin	1.08	0.26	5.53	10.54	2.01	0.20	0.46	0.05	6.27	0.37	0.02	0.00	0.00	0.00	0.00	3.35	28.11
Hebei	5.47	1.30	139.89	176.11	33.63	8.96	6.87	7.86	41.08	39.68	4.17	1.31	0.44	0.22	0.14	45.45	474.78
Shanxi	0.10	0.02	28.51	86.16	16.45	9.97	5.66	5.11	7.20	6.09	0.09	0.35	0.12	0.01	0.12	6.53	155.97
Inner Mongolia	6.02	1.43	16.99	147.47	28.16	13.61	19.03	13.54	0.26	27.20	0.07	3.73	1.26	0.65	0.31	8.00	259.53
North China	**12.74**	**3.02**	**193.81**	**428.47**	**81.82**	**32.82**	**32.37**	**26.77**	**55.04**	**74.04**	**4.43**	**5.39**	**1.82**	**0.88**	**0.56**	**66.47**	**934.20**
Liaoning	46.24	10.95	0.78	156.39	29.86	14.59	6.53	4.01	0.19	10.49	1.10	0.14	0.05	0.01	0.49	15.08	265.95
Jilin	48.78	11.56	0.35	257.62	49.19	8.92	18.76	3.91	0.07	12.13	0.79	0.18	0.06	0.11	0.79	6.45	369.67
Heilongjiang	129.23	30.62	9.45	164.08	31.33	10.02	82.72	7.13	0.00	11.10	0.14	4.26	1.44	11.25	1.01	9.66	471.96
North-east China	**224.25**	**53.13**	**10.57**	**578.08**	**110.38**	**33.52**	**108.02**	**15.05**	**0.27**	**33.72**	**2.03**	**4.58**	**1.54**	**11.36**	**2.29**	**31.19**	**1107.58**
Shanghai	9.53	2.26	1.28	0.39	0.07	0.52	0.35	0.06	0.12	1.59	0.01	0.34	0.12	0.00	0.00	3.38	19.92
Jiangsu	183.18	43.40	97.33	27.94	5.33	13.69	11.56	4.55	23.43	52.16	1.57	0.67	0.23	0.17	0.01	26.35	484.67
Zhejiang	72.58	17.19	2.39	2.54	0.48	1.91	4.93	3.36	1.48	11.30	0.13	2.90	0.98	0.05	0.07	14.25	135.93
Anhui	136.36	32.30	109.58	38.77	7.40	7.55	15.68	8.51	21.97	64.78	2.31	0.81	0.27	1.29	0.44	15.11	453.40
Fujian	57.12	13.53	0.24	1.72	0.33	0.32	2.78	10.35	0.02	6.86	0.74	2.86	0.96	0.01	2.00	9.94	108.72
Jiangxi	180.95	42.87	0.31	0.85	0.16	0.18	3.40	4.32	6.31	20.94	1.08	2.56	0.86	0.45	0.42	8.63	273.06
Shandong	10.55	2.50	223.56	233.41	44.57	2.40	8.35	17.00	62.41	93.70	10.64	0.00	0.00	0.13	1.47	66.39	721.86
East China	**650.28**	**154.05**	**434.70**	**305.61**	**58.36**	**26.57**	**47.06**	**48.13**	**115.74**	**251.31**	**16.48**	**10.14**	**3.42**	**2.10**	**4.40**	**144.06**	**2197.57**
Henan	39.60	9.38	331.09	177.33	33.86	6.22	11.49	13.22	42.26	112.95	9.83	0.67	0.23	1.46	4.23	45.58	795.71
Hubei	160.70	38.07	30.32	27.02	5.16	2.63	7.69	10.19	28.40	74.34	1.78	1.31	0.44	1.96	1.48	21.57	406.12
Hunan	252.91	59.92	1.26	17.22	3.29	1.37	6.22	10.82	13.33	33.47	0.79	3.26	1.10	5.22	3.17	18.07	427.33
Guangdong	120.44	28.53	0.14	8.06	1.54	0.72	2.84	13.51	0.00	20.95	2.37	37.83	12.76	0.06	0.93	17.60	264.36
Guangxi	126.72	30.02	0.17	27.02	5.16	0.30	4.28	4.59	0.08	13.08	1.34	194.92	65.75	0.42	0.46	14.47	482.28
Hainan	14.78	3.50	0.00	0.83	0.16	0.09	0.24	2.76	0.00	2.15	0.24	11.86	4.00	0.04	0.00	2.57	42.84
Central south China	**715.15**	**169.42**	**362.98**	**257.47**	**49.16**	**11.32**	**32.76**	**55.10**	**84.08**	**256.95**	**16.35**	**249.85**	**84.27**	**9.15**	**10.27**	**119.86**	**2418.64**
Chongqing	52.04	12.33	8.72	31.34	5.98	1.36	4.69	20.00	0.01	9.75	0.24	0.37	0.13	0.44	1.46	5.90	148.54

(continued)

Table 3 (continued)

	Rice	Rice husks	Wheat	Corn	Corn cobs	Other grains	Beans	Tubers	Cotton	Oils crop	Peanut hull	Sugar bagasse	Sugar cane stalk sheath	Fiber crops	Tobacco	Vegetables and Melons	Total
Sichuan	159.64	37.82	54.02	79.91	15.26	9.57	15.16	34.83	1.49	57.91	1.70	4.36	1.47	2.34	2.85	18.79	480.17
Guizhou	50.25	11.90	7.92	48.09	9.18	1.70	4.97	16.05	0.04	20.20	0.20	2.28	0.77	0.06	5.74	5.69	175.66
Yunnan	68.76	16.29	13.40	64.00	12.22	7.59	10.55	13.69	0.01	7.81	0.14	51.93	17.52	3.02	12.79	6.78	294.13
Tibet	0.06	0.01	3.29	0.24	0.04	9.70	0.42	0.09	0.00	1.45	0.00	0.00	0.00	0.00	0.00	0.26	15.52
South-west China	**330.74**	**78.35**	**87.35**	**223.58**	**42.69**	**29.93**	**35.80**	**84.65**	**1.55**	**97.13**	**2.27**	**58.95**	**19.88**	**5.86**	**22.83**	**37.42**	**1114.03**
Shaanxi	8.71	2.06	48.82	62.25	11.89	2.54	4.99	5.56	5.04	11.40	0.22	0.08	0.03	0.04	0.97	6.59	159.08
Gansu	0.42	0.10	32.63	34.09	6.51	13.85	5.12	14.38	7.27	12.42	0.01	0.65	0.22	0.46	0.51	6.39	128.50
Qinghai	0.00	0.00	5.84	0.14	0.03	2.03	1.42	1.87	0.00	7.45	0.00	0.01	0.00	0.01	0.01	0.59	19.36
Ningxia	6.33	1.50	9.49	17.98	3.43	1.14	0.72	2.26	0.01	2.96	0.00	0.01	0.00	0.00	0.02	1.79	44.21
Xinjiang	5.85	1.38	45.68	51.90	9.91	3.27	3.14	1.07	143.97	9.97	0.02	13.66	4.61	4.75	0.03	7.14	296.42
North-west China	**21.31**	**5.05**	**142.46**	**166.36**	**31.77**	**22.82**	**15.40**	**25.14**	**156.27**	**44.20**	**0.25**	**14.41**	**4.86**	**5.26**	**1.54**	**22.50**	**647.58**
Total	1954.47	463.02	1231.88	1959.57	0.00	156.98	271.40	254.85	412.95	757.36	0.00	343.31	115.80	34.61	41.90	421.50	8419.59

Survey (2004–2008), when forest area reached 195 × 10^6 ha and covered 20.4% of China's land. The energy potential of timber stands is the greatest (around 2095 PJ), followed by protection forest of 1143 PJ. These four main forestry residues combined account for ca. 73.6% of the total potentially available energy supply. The total amount of residues from other forest types is about 778.48 PJ accounting for ca. 16.6% of the total potentially available energy supply. The total wood-processing residues account for approximately 415 PJ, which represents about 8.9% of the total residue availability. The residues of orchards, urban greening forest and hedgerow are of the lowest potential of ca. 46.9 PJ accounting 0.99% of the total forestry residues. The energy potential of logging (including tending/thinning residues) and mills residues was about 4286 PJ and 415 PJ, respectively, with north-east China, north China and south-west China having the largest amount of forest residues availability. The top five provinces in terms of forest residue availability are Yunnan, Heilongjiang, Inner Mongolia, Sichuan and Guangxi, which combined account for 41.4% of total logging residue.

Combined agricultural and forest residue availability

The total energy potential from all sources is about 12 693 PJ per year (Table 5), with agricultural residues contributing about 8419 PJ each year. The total energy potential of forest residues is ca. 4274 PJ each year (excluding city greening forest; hedgerow, mills and waste wood). Agricultural residues alone contribute more than 66% of the national energy potential of biomass residues. The spatial distribution of the potential availability of biomass residues for energy production is shown in Tables 5 and 6. The total residue availability was the highest in south central China (3347 PJ), followed by the east China and south-west China with 2862 and 2229 PJ, respectively. These three regions collectively account for over 66% of the national residue availability.

Future residue biomass availability under different scenarios

The total bioenergy potential from agricultural residues, forest residues and mills will reach 17 660, 21 710 and 21 980 PJ by 2050 under BAU, CCS450 and NOCCS450, respectively, and 17 320, 21 180 and 21 640 PJ by the end of the 21st century, as a result of an increase in food demand, agriculture productivity and crop price. The energy potential under the reference scenario is lower than that of the two policy scenarios (Table 6).

To project bioenergy production in the future, bioenergy price was calculated within the GCAM based on energy demand and competition with other energy

Table 4 The spatial distribution of the theoretical energy potential availability of forestry residues from 2004 to 2008 (PJ)

	Timber stands	Protection forest	Forest for special uses	Firewood forest	Bamboo forest	Economic forest	Sparse forest	Shrubbery	Sipang forest	Orchard	Total
Product yields (kg ha^{-1})	3750	3750	1875	3750	1875	1875	1875	938	2	1875	
Collectable coefficient	0.5	0.2	0.1	1.0	0.1	0.1	0.5	0.5	0.5	0.1	
Heating value KJ kg^{-1}	18 600	18 600	18 600	16 747	17672.1	18 600	18 600	18 600	18 600	18 600	
Beijing	0.88	4.13	0.12	0.08	0.00	0.57	0.04	3.32	0.36	0.27	9.76
Tianjin	0.24	0.61	0.02	0.00	0.00	0.13	0.03	0.17	0.24	0.13	1.55
Hebei	23.03	28.37	0.36	5.36	0.00	3.18	1.46	8.97	70.70	3.82	145.24
Shanxi	8.10	19.07	0.42	0.19	0.00	1.58	3.14	10.36	3.34	0.96	47.18
Inner Mongolia	148.02	154.20	5.28	0.00	0.00	0.69	11.84	61.32	1.61	0.17	383.14
North China	**180.27**	**206.38**	**6.20**	**5.63**	**0.00**	**6.16**	**16.50**	**84.14**	**76.24**	**5.36**	**586.87**
Liaoning	44.47	26.23	0.47	20.23	0.00	4.26	0.99	5.12	2.13	1.09	105.00
Jilin	128.10	44.41	1.39	0.81	0.00	0.31	2.15	1.38	0.76	0.25	179.57
Heilongjiang	186.15	169.93	5.52	1.60	0.00	0.50	2.84	0.56	1.95	0.14	369.18
North-east China	**358.72**	**240.57**	**7.38**	**22.64**	**0.00**	**5.08**	**5.98**	**7.07**	**4.83**	**1.49**	**653.75**
Shanghai	0.03	0.12	0.08	0.00	0.01	0.08	0.00	0.03	0.47	0.09	0.91
Jiangsu	18.40	2.38	0.16	0.08	0.12	1.03	0.06	0.10	8.37	0.64	31.35
Zhejiang	79.66	21.88	0.29	0.00	2.59	3.92	0.67	2.74	4.13	1.04	116.92
Anhui	63.40	10.93	0.25	2.18	1.07	1.98	1.23	3.06	7.22	0.36	91.68
Fujian	130.63	21.51	1.17	2.26	3.29	3.53	1.59	1.81	0.73	1.91	168.44
Jiangxi	127.03	50.18	1.21	6.04	2.82	4.20	0.78	1.73	1.95	1.04	196.97
Shandong	29.83	9.33	0.13	0.00	0.00	3.43	1.17	0.74	10.56	2.56	57.75
East China	**448.98**	**116.34**	**3.29**	**10.55**	**9.91**	**18.17**	**5.51**	**10.20**	**33.44**	**7.64**	**664.04**
Henan	43.97	19.22	0.60	1.51	0.07	1.78	1.12	5.36	20.88	1.43	95.94
Hubei	42.63	49.50	0.50	10.25	0.50	1.94	2.18	12.56	4.90	0.94	125.89
Hunan	152.33	36.58	0.87	1.61	2.08	5.52	2.40	12.38	3.36	1.50	218.63
Guangdong	152.75	25.97	1.79	2.11	1.35	4.55	2.43	5.81	0.92	3.44	201.12
Guangxi	195.87	28.35	1.16	5.43	0.99	6.87	2.68	21.50	2.41	2.99	268.24
Hainan	8.28	5.10	0.83	0.00	0.05	3.16	0.04	0.33	0.37	0.58	18.74
Central south China	**595.83**	**164.71**	**5.75**	**20.91**	**5.04**	**23.82**	**10.85**	**57.95**	**32.84**	**10.87**	**928.57**
Chongqing	18.13	16.22	0.46	0.40	0.41	0.67	2.55	8.90	13.76	0.63	62.14
Sichuan	116.68	103.77	2.86	3.05	1.61	3.47	8.73	63.49	22.67	1.61	327.93
Guizhou	37.75	35.07	0.83	9.25	0.44	1.68	3.29	10.92	3.51	0.40	103.14
Yunnan	241.81	81.67	5.52	22.31	0.30	5.81	9.03	32.90	6.16	0.82	406.34
Tibet	49.34	81.59	3.98	0.35	0.00	0.02	5.04	74.55	0.67	0.00	215.56
South-west China	**463.72**	**318.31**	**13.65**	**35.36**	**2.76**	**11.65**	**28.64**	**190.77**	**46.78**	**3.46**	**1115.10**
Shaanxi	45.49	53.18	1.15	14.05	0.12	4.05	5.02	16.04	2.97	2.86	144.94
Gansu	1.32	19.86	2.35	0.00	0.00	0.90	2.99	29.93	5.85	1.25	64.46
Qinghai	0.25	1.74	0.78	0.00	0.00	0.01	1.18	28.43	0.50	0.02	32.90
Ningxia	0.08	0.71	0.20	0.00	0.00	0.16	0.36	3.31	0.65	0.18	5.65
Xinjiang	0.42	21.35	0.49	0.60	0.00	1.20	7.06	40.20	4.41	1.62	77.35
North-west China	**47.57**	**96.85**	**4.96**	**14.65**	**0.12**	**6.31**	**16.61**	**117.91**	**14.38**	**5.93**	**325.30**
Total	2095.09	1143.16	41.23	109.73	17.83	71.18	84.09	468.04	208.52	34.76	4273.62

The theoretical energy potential availability of forestry residues did not include mill and waste products because of no available data at province level.

Table 5 The total theoretical energy potential availability of agriculture and forestry residues (PJ)

	Forestry residues	Agriculture residues	Total
Beijing	9.76	15.82	25.58
Tianjin	1.55	28.11	29.67
Hebei	145.24	474.78	620.02
Shanxi	47.18	155.97	203.15
Inner Mongolia	383.14	259.53	642.66
North China	**586.87**	**934.20**	**1521.07**
Liaoning	105.00	265.95	370.95
Jilin	179.57	369.67	549.24
Heilongjiang	369.18	471.96	841.14
North-east China	**653.75**	**1107.58**	**1761.33**
Shanghai	0.91	19.92	20.83
Jiangsu	31.35	484.67	516.02
Zhejiang	116.92	135.93	252.86
Anhui	91.68	453.40	545.09
Fujian	168.44	108.72	277.17
Jiangxi	196.97	273.06	470.03
Shandong	57.75	721.86	779.61
East China	**664.04**	**2197.57**	**2861.60**
Henan	95.94	795.71	891.65
Hubei	125.89	406.12	532.01
Hunan	218.63	427.33	645.96
Guangdong	201.12	264.36	465.48
Guangxi	268.24	482.28	750.52
Hainan	18.74	42.84	61.58
Central south China	**928.57**	**2418.64**	**3347.21**
Chongqing	62.14	148.54	210.68
Sichuan	327.93	480.17	808.10
Guizhou	103.14	175.66	278.80
Yunnan	406.34	294.13	700.47
Tibet	215.56	15.52	231.08
South-west China	**1115.10**	**1114.03**	**2229.12**
Shaanxi	144.94	159.08	304.02
Gansu	64.46	128.50	192.96
Qinghai	32.90	19.36	52.26
Ningxia	5.65	44.21	49.86
Xinjiang	77.35	296.42	373.78
North-west China	**325.30**	**647.58**	**972.88**
Total	4273.62	8419.59	12693.21

Table 6 The theoretical maximum energy potential under different scenario (EJ)

	2020	2035	2050	2065	2080	2095
BAU	16.70	17.46	17.66	17.77	17.63	17.32
CCS450	18.71	20.65	21.71	22.09	21.87	21.18
NOCCS450	19.16	21.08	21.98	22.45	22.33	21.64

sources. Figure 1 shows the bioenergy prices for BAU and two limitation concentration scenarios. Growth of bioenergy market prices over time is enhanced by carbon price in a perspective of economics. Note that the

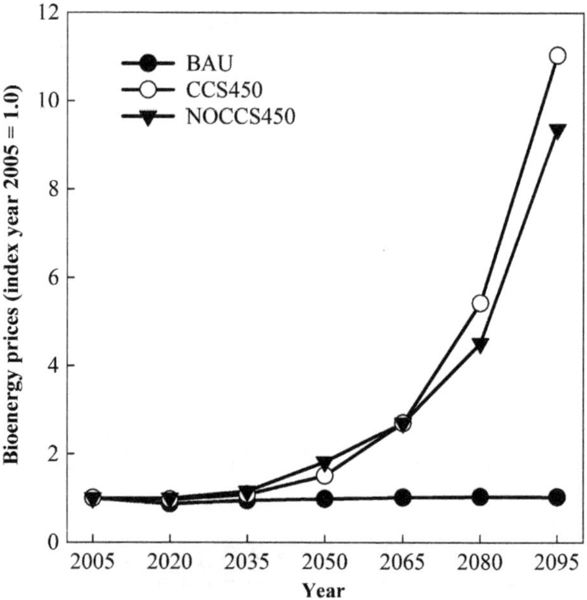

Fig. 1 Bioenergy prices along two alternative UCT CO_2 concentration target pathways (index year 2005 = 1.0). Growth of bioenergy market prices over time is enhanced by carbon price under climate policy scenarios in a perspective of economics.

carbon price was not added to the price of bioenergy based on the assumption of zero carbon emissions from bioenergy production (Wise *et al.*, 2009). Bioenergy price under the policy scenarios is much than that in the reference scenario after 2035. Among the two policy scenarios, bioenergy price under the NOCCS 450 ppm scenario has a competitive advantage compared to the CCS 450 ppm mitigation scenario after 2065, as evidenced by the trends of more bioenergy production from energy crop until the carbon prices are very high.

In the reference scenario without carbon tax, more and more residue biomass from agriculture and forestry becomes available along with increase in energy demand and energy prices and reaches a projected output of approximately 3380 PJ yr^{-1} by 2050 and 4108 PJ yr^{-1} by 2095 (Fig. 2). Under the UCT scenarios, the carbon price is charged for the emission of CO_2. This intensifies the demand and increases price of residue biomass for energy and further decreases the use of fossil fuels. The total bioenergy production from residue biomass is 9000 and 9180 PJ by 2050 under CCS 450 ppm and NOCCS 450 ppm, respectively, and 11 520 and 11 150 PJ by the end of the century, respectively (Fig. 2).

Figure 3 shows the carbon prices calculated within the GCAM that are required to drive a fundamental transformation of the global economy. To achieve the 450 ppm CO_2 concentration targets, the policy scenario without CCS will require higher carbon prices than the policy scenario with CCS, especially toward the end of

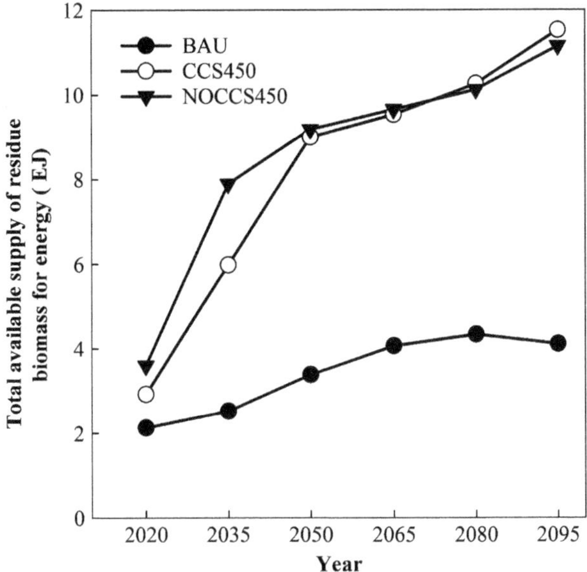

Fig. 2 The total available supply of residue biomass for energy under different scenarios from 2020 to 2095 (EJ). More and more residue biomass from agriculture and forestry becomes available along with increase in energy demand and energy prices, and the carbon price is charged under climate policy scenarios, which further intensifies the demand and increases price of residue biomass for energy.

the century. For example, the 2095 carbon price for the CCS 450 ppm scenario is 5004 \$ t^{-1} C^{-1}, which is much higher than the carbon price of 2955 \$ t^{-1} C^{-1} under NOCCS 450 ppm in 2095. However, the two policy scenarios do not differ substantially from each other in terms of supply of residue biomass after 2035 (Fig. 2).

Figure 4 shows the total bioenergy production (including residue biomass, energy crop and MSW) increases substantially over time under all three climate policy scenarios, but with higher bioenergy production under the CCS 450 ppm CO_2 scenario than that under the other two scenarios after 2050. The NOCCS 450 ppm CO_2 scenario projects more bioenergy production compared with the other two scenarios before 2050.

Figure 5 shows the proportion of bioenergy production from residue biomass over time in the total bioenergy production. In the reference scenario, bioenergy production from energy crops accounts for about 65% of the total bioenergy production after 2035, as a result of no CO_2 emission limitation. In the CCS 450 ppm CO_2 scenario, residue biomass meets nearly half all the bioenergy production in 2035, 53% by mid-century and 40% by the end of the century. In the NOCCS 450 ppm CO_2 scenario, residue biomass contributes ca. 60% of all the total bioenergy production in 2035, 55% by 2050 and

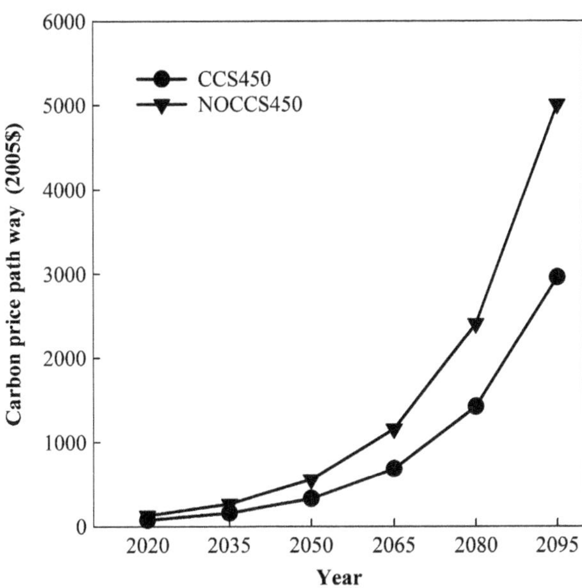

Fig. 3 Carbon price pathway under different climate policy scenario (2005\$). The carbon prices calculated within the GCAM that are required to drive a fundamental transformation of the global economy. The policy scenario without CCS will require higher carbon prices than the policy scenario with CCS, especially toward the end of the century for achieving the 450 ppm CO_2 concentration targets.

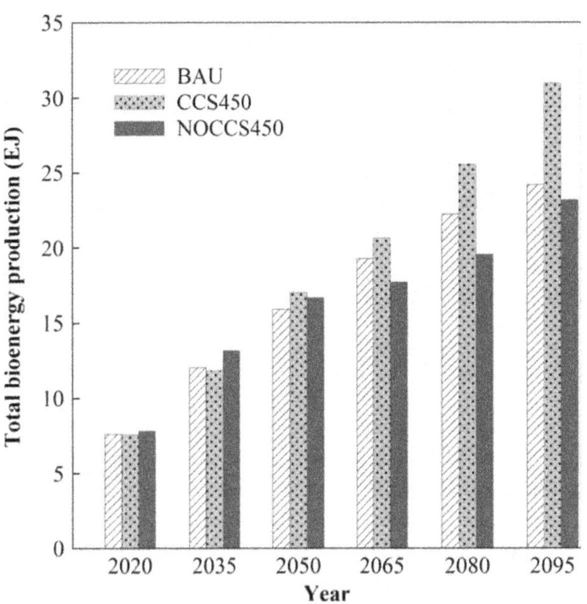

Fig. 4 The total bioenergy production (including residue biomass, energy crop and MSW) under different scenarios (EJ). In the future, the total bioenergy production substantially shows an increases, but with higher bioenergy production with the CCS scenario than that under two other scenarios after 2050, and the NOCCS scenario accounts for a great proportion in bioenergy production compared with two other scenarios before 2050.

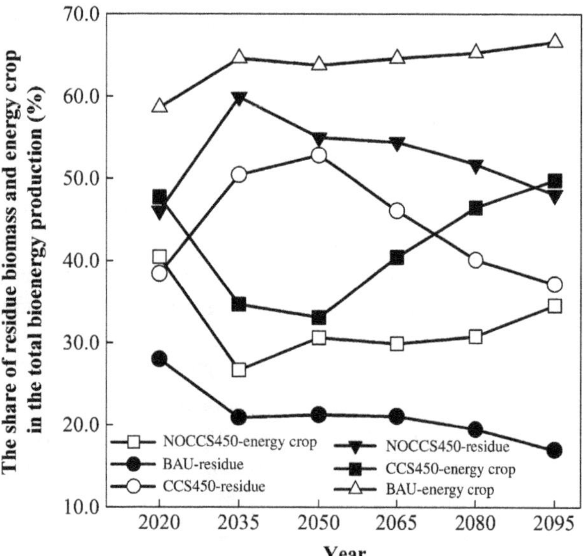

Fig. 5 The share of residue biomass and energy crop in the total bioenergy production under different scenarios. In the CCS scenario, residue biomass meets nearly half all the bioenergy production in 2035, 53% by mid-century and 40% by the end of the century. In the NOCCS scenario, residue biomass accounts for about 60% of all the total bioenergy production in 2035, 55% by 2050 and 48% by the 2095. Bioenergy production from energy crops contributes to about 65% of the total bioenergy production after 2035 in the BAU.

48% by the 2095. Total bioenergy production will contribute about 31% of the total energy production by 2050 and 35% by 2095. These results show that more biomass energy from residue biomass will be produced under climate policies without carbon tax and additional land and that trade-off between energy prices competitiveness, options of low carbon technology (CCS) and climate policy (carbon tax) is required for bioenergy production.

Discussion

We evaluated the energy potential of agricultural and forest residues in China and found that the total potential is about 12 693 PJ per year under current conditions. This is close to 10% of the total primary energy demand of China in 2013 (110 055 PJ). However, it is important to note these estimated values may be affected by the availability and reliability of data on crop species, harvest index, location, soil properties and seasonal variation (Liao *et al.*, 2004; Zhou *et al.*, 2011). The potential of agricultural residues as a bioenergy source is complicated by their numerous alternative uses including feeding, fodder, fertilizer, household fuels and industrial fuels. Currently, agricultural residues are mainly used for forage (24.5%), industry materials (3.9%), base material for edible mushrooms (2.3%), biogas (0.85%), direct field restoration (14.1–14.6%), direct combustion by farmers (24.9–30.7%), whereas the rest are lost during collection (15%), being discarded or directly burnt (12.3–20.5%) in the field (Bi *et al.*, 2008; Wang *et al.*, 2010). At present, the collectable and utilizable amount of agriculture residues as a bioenergy resource is estimated to be around only 23.9%. In addition, the actual availability is also limited by economic, social, environmental, institutional and policy incentives, logistical considerations, infrastructural and technological constraints, and availability of skilled personnel (Bi, 2010; Okello *et al.*, 2013).

Overall, most of residue biomass should be returned to the field for improving soil fertility through maintaining soil organic matter and soil structure. The reasonable residue incorporation rate of 3.0–4.5 t ha^{-1} has been reported to slightly increase soil organic carbon and crop yield for rice and wheat and 4.5–6.0 t ha^{-1} for corn in China. The amount of residue retention was 1 911 721 $\times$ 10^4 t, accounting for 22.7% of the total residue biomass in 2008. If the amount of residues returned directly to fields (92 $\times$ 10^6 t accounting for 10.9% of the total residue biomass in 2008) is also considered, the amount of residue retention represents one-third of the total residue biomass in 2008, when a residue retention ratio of 2.33 t ha^{-1} is used (Bi, 2010). Note that this residue retention ratio is lower than the desired residue retention ratio for maintaining sustainable agroecosystems, which is an important factor in collecting residue biomass for energy production.

Not all forest residues are harvestable. Some of them must be retained for maintaining nutrient levels and preventing soil erosion. This study identified that the logging and processing forest residues would potentially provide 4701 PJ of energy. Significant variation in the potential is observed, as influenced by numerous factors such as forest type, collectable fraction and geographical location. For example, the average yield of firewood forest in the southern mountain area is as high as 7.5 t ha^{-1}, but only 3.75 t ha^{-1} in the North Mountain area. The yield shrub forest is 0.75 t ha^{-1} over the country, with a collectable coefficient of 0.2 in the mountain area and 0.5 in the plains area (Yuan, 2002). Logging residues are usually located in remote regions, leading to difficulties for collecting and utilizing them. The amount of forest residues available used for renewable energy production is also affected by technical, ecological and environmental factors. In fact, the potential for renewable energy production from logging residues and wood-processing residues is estimated to be about 1286–1607 PJ and 228.5 PJ, respectively, accounting for 30–37.5% and 55% of logging residues and wood-processing residues in China (MOA, 2006).

In the future, climate policy is a key factor affecting the supply of residue biomass. Imposing carbon tax is projected to be an effective way to reduce CO_2 emissions and mitigate climate change (Wise *et al.*, 2009). Terrestrial carbon storage has been thought to be a low cost method to address the climate change. For example, soil carbon on croplands is a key component of terrestrial carbon storage. In China, croplands (over 130 M ha) contain 730 (329–1095) Tg C in the topsoil and are estimated to have sequestrated carbon at a rate of about 24.3 (11.0–36.5) Tg C yr^{-1} over the last 30 years (Yu *et al.*, 2012, 2013). Residue removal may lead to the loss of soil organic carbon, which can be minimized by improved management practices such as nitrogen fertilizer application, straw retention and incorporation and conservation tillage. These practices have been estimated to increase soil organic carbon from 38.5 Mg C ha^{-1} in 2010 to 56.9 Mg C ha^{-1} in 2050 on China's croplands (Yu *et al.*, 2013), which translate to \$2929 ha^{-1} in 2010 and \$18 711 ha^{-1} in 2050 with the carbon price under the CCS 450 ppm scenario. The carbon sequestration potential through optimal management is estimated to be approximately 2.39 Pg C over the next 40 years nationally (Smith *et al.*, 2007; Yu *et al.*, 2013). Nevertheless, the carbon sequestration potential in China after 2050 requires further evaluation under different future climate scenarios and management practices because soil carbon may reach a new equilibrium after 84 years of improved management practices and fertilizer amendment (Yan *et al.*, 2007; Yu *et al.*, 2013).

In this study, GCAM allows farmers to allocate the amount of residues to be retained on the field or to be removed for energy production, based on an economic assessment on carbon price, carbon stocks, the cost and benefit of the bioenergy (Wise *et al.*, 2009). Notably, imposing climate policies, such as carbon tax, can be difficult as it requires monitoring and evaluating terrestrial carbon emissions and stocks. Solutions to these barriers may require huge amounts of money to identify the landowners and transfer the decrease or increase of carbon stocks (Calvin *et al.*, 2014). One limitation of this study is that projected future utilization of residue biomass depends on a series of assumptions within GCAM including crop productivity, economic growth and land policy. Residue biomass production seems to be highly sensitive to future changes in crop productivity that may reduce land-use change emissions under the climate policy scenario (Wise *et al.*, 2009). In addition, the increasing demand for food to feed the rising world population may further limit residue availability (Gregg & Izaurralde, 2009; Gregg & Smith, 2010). As well, this will likely decrease unit mass collection cost and shift the supply curve accordingly. Dedicated energy crops will occupy more available agricultural land in order to achieve higher biomass yields while reducing production cost to compete with residue biomass by 2095. If crop yields increase only slightly or remain stable in the future, less residue biomass would be harvested because a higher proportion of the residue must be left in fields to maintain soil quality and reduce erosion. Management options that may increase residue removal rate include the practice of conservation tillage, better crop rotation and the introduction of catch crops. In addition, under a UCT regime, all carbon emissions to be taxed are simulated as the best policy to limit the CO_2 concentration. Forests with a higher below ground storage of carbon will be preferable due to their efficiency in limiting carbon emissions from land-use change. This implies that land polices limit the conversion from forests to bioenergy production and stress food production from agricultural lands (Calvin *et al.*, 2014).

In summary, China is the largest developing agricultural country in the world. Agricultural and forest residues in China have considerable potential to be available as a bioenergy source to provide ca. 10% of its total primary energy consumption in 2013. Accurate projection and successful utilization of residue biomass for energy production requires a comprehensive and multifactorial assessment. The integrated assessment results indicate that residue biomass for energy production could play an important role in mitigating the climate change. The production of bioenergy should be achieved in a sustainable way through optimal land management practices by conserving soil quality to enhance interactive economic, environmental and social purposes.

Acknowledgements

This work was supported by the Ministry of Science and Technology of the People's Republic of China (2013BAD11B03 and 2012CB955801) and the National Natural Science Foundation of China (71373142).

References

Bi YY (2010) Study of straw resources evaluation and utilization. Chinese Academy of Agriculture Science in China PHD Dissertation.

Bi YY, Wang DL, Gao CY, Wang YH (2008) *Straw Resources Evaluation and Utilization in China*. Chinese Agricultural Sciences and Technology, Beijing.

Briunsma J (2009) *The Resource Outlook to 2050: By How Much Do Land, Water, and Crop Yields Need to Increase by 2050? Expert Meeting on How to Feed the World in 2050*. Food and Agriculture Organization of the United Nations, Rome.

Cai F, Zhang L, Zhang CH (2012) Potential of woody biomass energy and its availability in China. *Journal of Beijing Forestry University (Social Sciences)*, **1**, 103–107.

Calvin K, Wise M, Kyle P, Patel P, Clarke L, Edmonds J (2014) Trade-offs of different land and bioenergy policies on the path to achieving climate targets. *Climatic Change*, **123**, 691–704.

Clarke L, Edmonds J, Jacoby H, Pitcher H, Reilly J, Richels R (2007a) *CCSP Synthesis and Assessment Product 2.1. Part A: Scenarios of Greenhouse Gas Emissions and Atmospheric Concentration*. U.S. Government Printing Office, Washington, DC.

Clarke L, Lurz J, Wise M, Edmonds J, Kim S, Smith S, Pitcher H (2007b) Model Documentation for the MiniCAM Climate Change Science Program Stabilization Scenarios: CCSP Product 2.1a. *PNNL Technical Report. PNNL-16735*.

Cramer J (ed.) (2007) *Testing Framework for Sustainable Biomass: Final Report from the Project Group "Sustainable Production of Biomass"*. Energy Transition's Interdepartmental Programme Management (IPM), The Netherlands.

Cui H, Wu R (2012) Feasibility analysis of biomass power generation in China. *Energy Procedia*, **16**, 45–52.

Cui M, Zhao LX, Tian YS *et al.* (2008) Analysis and evaluation on energy utilization of main crop straw resources in China. *Transactions from the Chinese Society of Agricultural Engineering*, **12**, 291–295.

Edmonds J, Clarke L, Lurz J, Wise M (2008) Stabilizing CO_2 concentrations with incomplete international cooperation. *Climate Policy*, **8**, 355–376.

Elmore AJ, Shi X, Gorence NJ, Li X, Jin H, Wang F, Zhang X (2008) Spatial distribution of agricultural residue from rice for potential biofuel production in China. *Biomass and Bioenergy*, **32**, 22–27.

Eom J, Calvin K, Clarke L *et al.* (2012) Exploring the future role of Asia utilizing a scenario matrix architecture and shared socio-economic pathways. *Energy Economics*, **34**(Supplement 3), S325–S338.

Gregg JS, Izaurralde RC (2009) Effect of crop residue harvest on long-term crop yield, soil erosion and nutrient balance: trade-offs for a sustainable bioenergy feedstock. *Biofuels*, **1**, 69–83.

Gregg J, Smith S (2010) Global and regional potential for bioenergy from agricultural and forestry residue biomass. *Mitigation and Adaptation Strategies for Global Change*, **15**, 241–262.

Guest G, Bright RM, Cherubini F, Strømman AH (2013) Consistent quantification of climate impacts due to biogenic carbon storage across a range of bio-product systems. *Environmental Impact Assessment Review*, **43**, 21–30.

Gustavsson L, Holmberg J, Dornburg V, Sathre R, Eggers T, Mahapatra K, Marland G (2007) Using biomass for climate change mitigation and oil use reduction. *Energy Policy*, **35**, 5671–5691.

He G, Bluemling B, Mol APJ, Zhang L, Lu Y (2013) Comparing centralized and decentralized bio-energy systems in rural China. *Energy Policy*, **63**, 34–43.

IEA (2007) *Bioenergy Project Development & Biomass Supply*. International Energy Agency, Paris.

IEA (2010) *World Energy Outlook 2010*. International Energy Agency, Paris.

IEA (2011) *Combining Bioenergy with CCS Reporting and Accounting for Negative Emissions under UNFCCC and the Kyoto Protocol*. International Energy Agency, Paris.

Jiang KJ, Zhuang X, Liu Q (2009) China's low-carbon scenarios and roadmap for 2050. *Sino-Global Energy*, **24**, 28–30.

Jing WR (2006) *Power Generation Utilizing Biological Energy Sources*. North China Power Engineering Co., Ltd, Beijing.

Kim SH, Edmonds J, Lurz J, Smith S, Wise M (2006) The object-oriented energy climate technology systems (ObjECTS) framework and hybrid modeling of transportation in the MiniCAM Long-term, global integrated assessment model. *The Energy Journal Special Issue: Hybrid Modeling of Energy-Environment Policies: Reconciling Bottom-up and Top-down*, **2**, 63–91.

Klein D, Bauer N, Bodirsky B, Dietrich JP, Popp A (2011) Bio-IGCC with CCS as a long-term mitigation option in a coupled energy-system and land-use model. *Energy Procedia*, **4**, 2933–2940.

Kyle GP, Luckow P, Calvin K, Emanuel W, Nathan M, Zhou Y (2011) *GCAM 3.0 Agriculture and Land Use: Data Sources and Methods. PNNL-21025*. Pacific Northwest National Laboratory, Richland, WA.

Li Q, Hu S, Chen D, Zhu B (2012) System analysis of grain straw for centralised industrial usages in China. *Biomass and Bioenergy*, **47**, 277–288.

Liao CP, Yan YJ, Wang CZ, Huang H-T (2004) Study on the distribution and quantity of biomass residues resource in China. *Biomass and Bioenergy*, **27**, 111–117.

Liu TF (1986) *Technology Economic Manual*. LiaoNing Peopes's Publishing House, ShengYang.

Lu N (1997) *Introduction of New Energy*, pp. 152–153. China Agriculture Press, Beijing.

Luckow P, Wise MA, Dooley JJ, Kim SH (2010) Large-scale utilization of biomass energy and carbon dioxide capture and storage in the transport and electricity sectors under stringent CO_2 concentration limit scenarios. *International Journal of Greenhouse Gas Control*, **4**, 865–877.

Lv ZW, Wang JL (1998) Evaluation of the nutritional value on Highland barley straw in Tibet. *Shaanxi Journal of Agricultural Sciences*, **2**, 5–8.

Ministry of Agriculture (MOA) of the P.R.C Project Expert Team (1998) *Assessment of Biomass Resource Availability in China*. China Environmental Science Press, Beijing.

Ministry of Agriculture (MOA) of the P.R.C Project Expert Team (2006) Preliminary research of wood energy development potential in China. *China Forestry Industry*, **1**, 12–21.

National Bureau of Statistics (NBS) of China of the P.R.C (2003–2007) *China Statistical Year Book*. China Statistics Press, Beijing.

Nan ZD, Yang M, Han W *et al.* (2008) Fast pyrolysis oil crops straw and characteristics of bio-oil. *Chinese Journal of Oil Crop Sciences*, **30**, 501–505.

National Bureau of Statistics of China of the P.R.C (2004–2008) *China Statistical Yearbook*. China Statistics Press, Beijing.

Niu RF, Liu TF (1984) *Agricultural Technology Economic Manual*, Revised edn. China Agriculture Press, Beijing.

Okello C, Pindozzi S, Faugno S, Boccia L (2013) Bioenergy potential of agricultural and forest residues in Uganda. *Biomass and Bioenergy*, **56**, 515–525.

Pimentel D (1994) Renewable energy: economic and environmental issues. *Bioscience*, **44**, 536–547.

Shi D (2013) Changes in global energy supply landscape and implications to China's energy security. *Sino-Global Energy*, **18**, 1–7.

Smith JO, Smith P, Wattenbach M *et al.* (2007) Projected changes in the organic carbon stocks of cropland mineral soils of European Russia and the Ukraine, 1990–2070. *Global Change Biology*, **13**, 342–356.

State Forestry Administration (SFA) of the P.R.C (2009) *The Seventh National Forest Resources Inventory and the Status of Forest Resources (2004-2008)*. State Forestry Administration (SFA) of the P.R.C, Beijing.

Sun J, Chen J, Xi Y, Hou J (2011) Mapping the cost risk of agricultural residue supply for energy application in rural China. *Journal of Cleaner Production*, **19**, 121–128.

Tan T, Shang F, Zhang X (2010) Current development of biorefinery in China. *Biotechnology Advances*, **28**, 543–555.

Thomson A, Calvin K, Smith S *et al.* (2011) RCP4.5: a pathway for stabilization of radiative forcing by 2100. *Climatic Change*, **109**, 77–94.

Wang XZ, Liu YM (1984) Study of beet residue utilization. *Heilongjiang Animal Science and Veterinary Medicine*, **5**, 13–15.

Wang YJ, Bi YY, Gao CY (2010) The assessment and utilization of straw resources in China. *Agricultural Sciences in China*, **9**, 1807–1815.

Wise M, Calvin KV (2011) GCAM 3.0 Agriculture and Land Use: Technical Description of Modeling Approach. Pacific Northwest National Laboratory. *PNNL-20971*.

Wise MK, Calvin A, Thomson L *et al.* (2009) Implications of limiting CO_2 concentrations for land use and energy. *Science*, **324**, 1183–1186.

Wise MA, Calvin KV, Kyle GP, Luckow P, Edmonds JE (2014) Economic and physical modeling of land use in GCAM 3.0 and an application to agricultural productivity, land, and terrestrial carbon. *Climate Change Economics*, **5**, 1450003.

Xie GH, Han DQ, Wang XY, Lu RH (2011) Harvest index and residue factor of cereal crops in China. *Journal of China Agricultural University*, **16**, 1–8.

Yan H, Cao M, Liu J, Tao B (2007) Potential and sustainability for carbon sequestration with improved soil management in agricultural soils of China. *Agriculture, Ecosystems & Environment*, **121**, 325–335.

Yu H, Wang Q, Ileleji KE, Yu C, Luo Z, Cen K, Gore J (2012) Design and analysis of geographic distribution of biomass power plant and satellite storages in China. Part 1: straight-line delivery. *Biomass and Bioenergy*, **46**, 773–784.

Yu Y, Huang Y, Zhang W (2013) Projected changes in soil organic carbon stocks of China's croplands under different agricultural managements, 2011–2050. *Agriculture, Ecosystems & Environment*, **178**, 109–120.

Yuan ZH (2002) Research and development on biomass energy in China. *International Journal of Energy Technology and Policy*, **2**, 108–144.

Zhang XL, Lv W, Zhang CH *et al.* (2008) *China Forest Energy*. China Agriculture Press, Beijing.

Zhang SH, Yang ZX, Wang XH, Chen HP (2012) Experiment on agglomeration characteristics during fluidized bed combustion of tobacco stem. *Transactions of the Chinese Society for Agricultural Machinery*, **43**, 97–101.

Zhou XP, Wang F, Hu H, Yang L, Guo P, Xiao B (2011) Assessment of sustainable biomass resource for energy use in China. *Biomass and Bioenergy*, **35**, 1–11.

Zhou S, Kyle GP, Yu S *et al.* (2013) Energy use and CO_2 emissions of China's industrial sector from a global perspective. *Energy Policy*, **58**, 284–294.

3

Land-use change to bioenergy: grassland to short rotation coppice willow has an improved carbon balance

ZOE M. HARRIS[1], GIORGIO ALBERTI[1,2], MAUD VIGER[1], JOE R. JENKINS[1], REBECCA ROWE[3], NIALL P. MCNAMARA[3] and GAIL TAYLOR[1]

[1]*University of Southampton, Southampton, SO17 1BJ, UK,* [2]*University of Udine, Via delle Scienze 206, 33100 Udine, Italy,* [3]*Centre for Ecology & Hydrology, Lancaster Environment Centre, Library Avenue, Bailrigg, Lancaster, UK*

Abstract

The effect of a transition from grassland to second-generation (2G) bioenergy on soil carbon and greenhouse gas (GHG) balance is uncertain, with limited empirical data on which to validate landscape-scale models, sustainability criteria and energy policies. Here, we quantified soil carbon, soil GHG emissions and whole ecosystem carbon balance for short rotation coppice (SRC) bioenergy willow and a paired grassland site, both planted at commercial scale. We quantified the carbon balance for a 2-year period and captured the effects of a commercial harvest in the SRC willow at the end of the first cycle. Soil fluxes of nitrous oxide (N_2O) and methane (CH_4) did not contribute significantly to the GHG balance of these land uses. Soil respiration was lower in SRC willow (912 ± 42 g C m^{-2} yr^{-1}) than in grassland (1522 ± 39 g C m^{-2} yr^{-1}). Net ecosystem exchange (NEE) reflected this with the grassland a net source of carbon with mean NEE of 119 ± 10 g C m^{-2} yr^{-1} and SRC willow a net sink, -620 ± 18 g C m^{-2} yr^{-1}. When carbon removed from the ecosystem in harvested products was considered (Net Biome Productivity), SRC willow remained a net sink (221 ± 66 g C m^{-2} yr^{-1}). Despite the SRC willow site being a net sink for carbon, soil carbon stocks (0–30 cm) were higher under the grassland. There was a larger NEE and increase in ecosystem respiration in the SRC willow after harvest; however, the site still remained a carbon sink. Our results indicate that once established, significant carbon savings are likely in SRC willow compared with the minimally managed grassland at this site. Although these observed impacts may be site and management dependent, they provide evidence that land-use transition to 2G bioenergy has potential to provide a significant improvement on the ecosystem service of climate regulation relative to grassland systems.

Keywords: carbon balance, bioenergy, climate regulation, ecosystem services, grassland, greenhouse gas, soil carbon

Introduction

Dedicated second-generation (2G) nonfood feedstocks offer an opportunity to provide biomass for bioenergy-derived heat, electricity and biofuels without competing with land for food (Dornburg *et al.*, 2010; Stoof *et al.*, 2015). However, evidence is still required to support this assertion, particularly with respect to soil properties (Kort *et al.*, 1998), greenhouse gas (GHG) emissions (see refs within Rowe *et al.*, 2009) and a whole basket of associated ecosystem services (Holland *et al.*, 2015). Although recent reports suggest that energy and food may be produced in a multifunctional landscape in a sustainable way (Manning *et al.*, 2015; Souza *et al.*, 2015), many of these positive effects are dependent on land management, vegetation type, and in particular, the land-use change (LUC) implemented when the bioenergy crop is planted (Milner *et al.*, 2015). It is therefore important to consider how these crops will be placed within the landscape (Dauber *et al.*, 2010) and the impacts of particular land-use transitions on ecosystem services, of which climate regulation is of outstanding importance (Anderson-Teixeira *et al.*, 2012). In 2013, 51×10^3 ha (0.8% total arable land) were used to grow bioenergy in the UK (DEFRA, 2014) and, at the same time, it is estimated that there are still 3.5×10^6 ha of land currently available to grow bioenergy crops without impacting food production (Lovett *et al.*, 2014), with estimated yields ranging from 6 to 12 t ha^{-1} yr^{-1} for SRC willow (Hastings *et al.*, 2014). Adoption of bioenergy will inevitably result in large scale LUC; therefore, it is important to consider which land classes are most suited to the conversion to minimize environmental damage and competition with food crops.

Correspondence: Gail Taylor
e-mail: G.Taylor@soton.ac.uk

Land-use change, irrespective of crop type, may have many direct consequences on climate regulation, such as altered GHG emissions (IPCC, 2007a), changes in soil carbon (Guo & Gifford, 2002) as well as impacts on other ecosystem services and biodiversity (Sala *et al.*, 2000). Additionally for bioenergy crops, the impacts of indirect land-use change (iLUC; Searchinger *et al.*, 2008; Melillo *et al.*, 2009; Finkbeiner, 2014) and those of quantifying the counterfactual land use (DECC, 2014; Mathews *et al.*, 2014) are increasingly recognized and considered in land-use conversions. St. Clair *et al.* (2008) found that former land use is the most important consideration determining whether a transition to 2G bioenergy will result in a net source or net sink of carbon. A number of studies and meta-analyses have suggested that, although dependant on site, LUC from arable cropping to 2G bioenergy is most likely to result in neutral or net increases in soil carbon (Dimitriou *et al.*, 2012; Don *et al.*, 2012; Harris *et al.*, 2015; Qin *et al.*, 2015). Similarly, reductions in other GHG emissions have also been reported for LUC from arable to 2G bioenergy (Drewer *et al.*, 2012; Gauder *et al.*, 2012; Zona *et al.*, 2013a; Palmer *et al.*, 2014), a proportion of which is attributable to change in management and land-use intensity. However, there is much more uncertainty surrounding the effects of LUC from grassland to 2G bioenergy crops (Harris *et al.*, 2015; Qin *et al.*, 2015), partly reflecting the considerable variability that is found amongst grassland types with significant differences in management which can dictate GHG balance (Soussana *et al.*, 2010). Although grasslands may be managed to encourage a carbon sink (defined here as an ecosystem in which the net gain of carbon is greater than the net loss; Smith, 2014), other management practices such as fertilizer addition and grazing may lead to large emissions of nitrous oxide (N_2O) and methane (CH_4). Ciais *et al.* (2010) suggested that emissions of N_2O and CH_4 following management practices may offset approximately 70–80% of the net carbon sink in European grasslands. This indicates that conversion to 2G bioenergy cropping may result in additional GHG savings. Moreover, Styles & Jones (2007) demonstrated that initial cultivation emissions associated with LUC from grassland to SRC willow could be offset by GHG emissions savings from replacing fossil fuel usage. The timescale for this 'payback', as calculated from current research is uncertain, varying between 0 and 423 years depending on former land use, management and bioenergy crop cultivated (Fargione *et al.*, 2008; Don *et al.*, 2012; Ter-Mikaelian *et al.*, 2015).

Two limitations are apparent when considering much of the literature in current LUC and bioenergy research. The first is that many studies rely entirely on modelled data with extremely limited or no validation (Cherubini *et al.*, 2009) and this is worrying, given that outputs from such models, often parameterized for non-bioenergy 'exemplar' arable, grass and tree ideotypes, may be used to develop sustainability criteria and policy instruments (Creutzig *et al.*, 2012; Buchholz *et al.*, 2014). Secondly, when empirical data have been captured for model validation, they have often been small research-scale plots of limited commercial relevance (e.g. Nikiema *et al.*, 2012; Zatta *et al.*, 2014). Additionally, there are methodological considerations which may affect the conclusions drawn about LUC, such as soil sampling depth (Dolan *et al.*, 2006; Blanco-Canqui & Lal, 2008) and calculation of soil carbon stocks using a fixed depth method (Walter *et al.*, 2015).

Given the need for empirical data, which is critical for LUC evaluation and model validation, here we present the results from a paired-site evaluation of LUC to bioenergy. The aim of this study was to quantify the impacts of a LUC at commercial scale from a grassland with limited management intervention, to that of SRC willow and to quantify the ecosystem GHG balance of this change 7 years after conversion. During 3 years of measurement, the SRC willow was harvested at commercial scale, and the impact of this activity on GHG balance and whole ecosystem carbon balance was also quantified. These findings will add to our understanding of the effects of LUC to bioenergy in temperate climates and contribute to the parameterization and testing of models to predict effect out to future climates.

Materials and methods

The aim of this side-by-side comparison was to develop an intensive data set for all components of the ecosystem GHG balance from a commercial plantation over a period of 3 years, including bioenergy SRC harvest. Figure 1 outlines the different components which were measured to assess the ecosystem GHG balance. The experimental set-up was established in November 2011 and measurements continued through until December 2014 (see Fig. S1 for experimental timeline).

Site description and management

This study was conducted in the south of England (50°58′N, 0°27′W) in an established SRC willow plantation (8.1 ha) and permanent grassland with low inputs (7.4 ha).

Mixed commercial genotypes of SRC willow were planted in June 2008 on a grassland field, previously defined as set-aside (2000–2007) at a density of 15 000 stems ha^{-1} in double rows with distances of 0.75 m in the row and 1.4 m between the rows (Forestry Commission, 2002). Prior to planting, the site was ploughed to 0.25 m in September 2007 and treated with herbicide (1.6 kg ha^{-1} glyphosate) and insecticide (0.75 kg ha^{-1} chlorpyrifos). In April 2008, the site was power harrowed to 0.10 m depth and there was a further application of herbicide (1.6 kg ha^{-1} glyphosate in June 2008). At pre-emergence the site was treated with herbicides (0.25 kg ha^{-1} isoxaben, 1.5 kg ha^{-1} pendimethalin) and insecticide (0.75 kg ha^{-1} chlorpyrifos). The

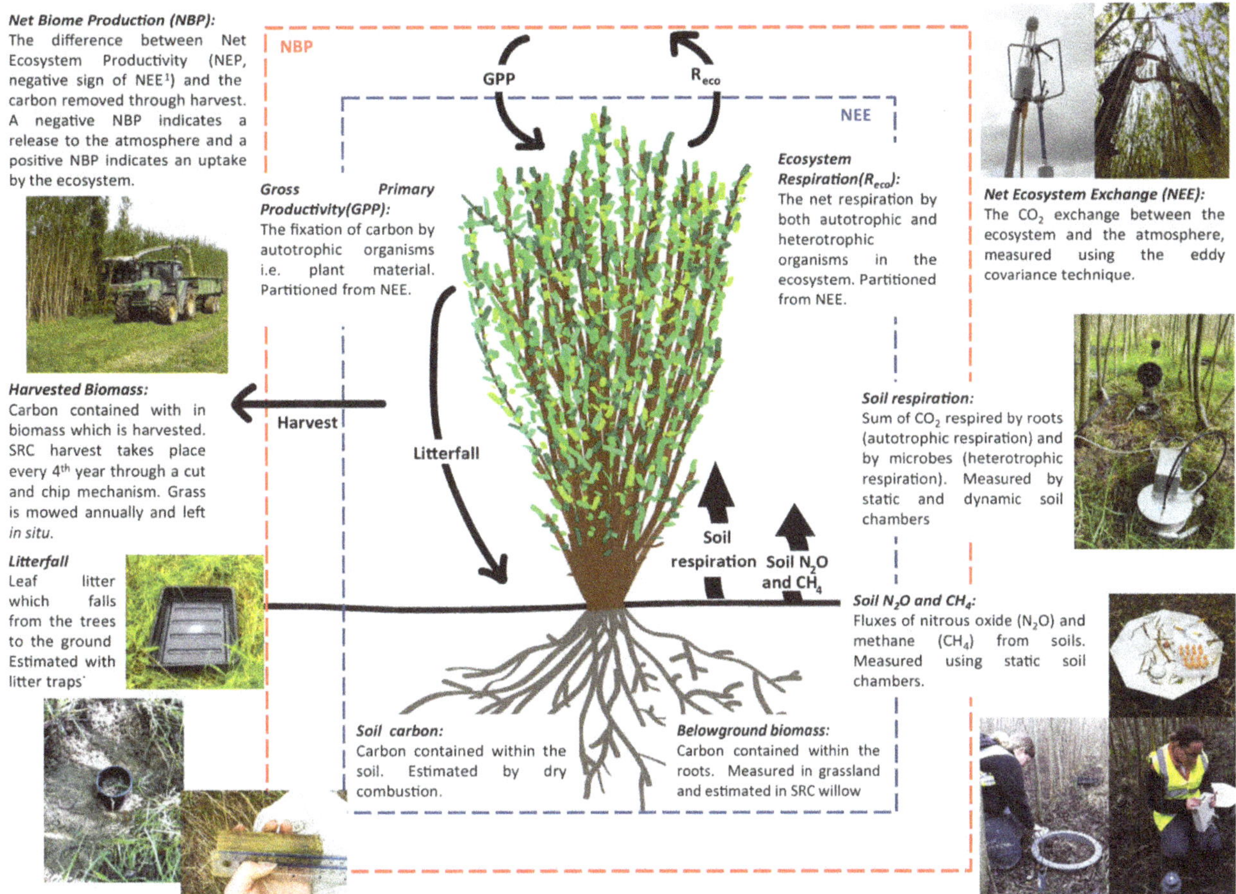

Fig. 1 Measurements taken to establish whole ecosystem greenhouse gas balance showing the main flows of carbon through the ecosystem. Crop shown represents short rotation coppice willow but cycle is applicable to any vegetation type.[1]Reichstein *et al.* (2005).

SRC willow was cut back in March 2009, further treated with herbicide (2.3 kg ha^{-1} aminotriazole) and then underwent a rotation of 5 years prior to harvest in April 2014.

The grassland site was enlisted in the set-aside scheme until 2004 and was maintained as low input grassland thereafter. The site was a mixed grassland including *Lolium* spp., *Schedonorus* spp., *Dactylis* spp. and other cultivated species. There were no inputs to the site other than an addition of a total of 10 t of manganese lime across the site in April 2011. Management was variable year to year, with grazing by sheep once per year (2–4 weeks), or if this did not occur, the grass was mown to control grass height. During the experiment, the site was grazed for 2 weeks in 2012 and the grass was mowed in August/September in 2013 and 2014. Mowed grass was left on the site.

Mean annual rainfall at the sites is 794 mm, and mean annual temperature is 11.0 °C (1960–2010; Met Office, 2015). The soil is the same at both sites, silt loam (Table 1) with a pH of 5.5. Root exploration depth was 0.30 m in grassland, with the majority of root biomass found in the top 0.15 m and SRC willow roots were found to 1 m, with the majority of biomass in the 0.50–1.00 m horizon (Table S1). The dominant wind direction is from the southwest; therefore, eddy covariance towers were established in the north-easterly corner of the grassland and SRC willow in order to ensure enough fetch (Fig. 2).

Table 1 Soil texture for grassland, short rotation coppice willow and initial grassland site

Site	Depth (cm)	Clay (%)	Silt (%)	Sand (%)	Soil type
SRC willow	0–15	7.38	59.44	33.19	Silt loam
	15–30	6.93	60.06	33.02	Silt loam
Grass	0–15	5.54	65.27	29.19	Silt loam
	15–30	14.06	62.79	23.15	Silt loam
Initial grassland	0–15	6.43	69.62	23.94	Silt loam
	15–30	15.26	66.69	18.04	Silt loam

Micrometeorological measurements

A meteorological station was installed in SRC willow in August 2011 and in grassland in November 2011 (Fig. S1). Each station measured soil temperature and heat flux at three depths (5, 10 and 15 cm; TCAV; Campbell Scientific, Logan, UT, USA; HFP01SC heat flux plates; Campbell Scientific), soil water

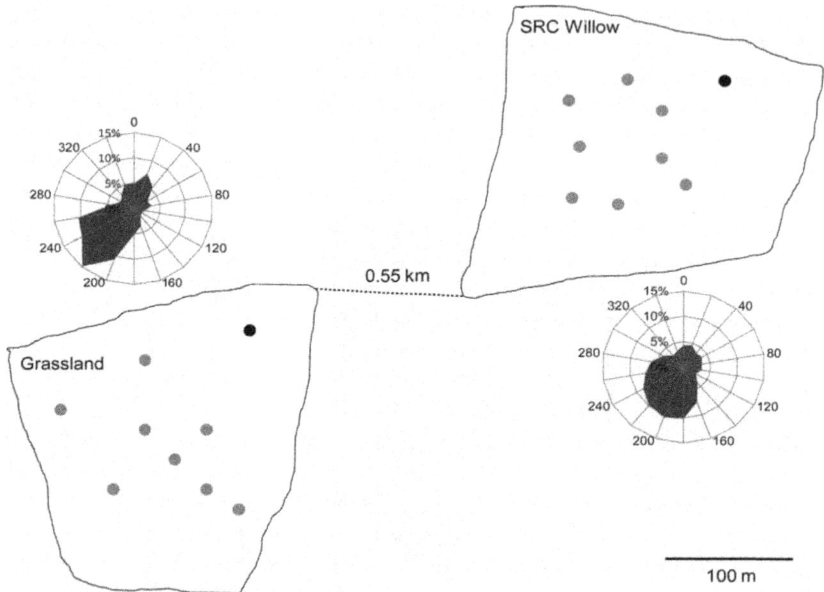

Fig. 2 Site maps of grassland and short rotation coppice willow, including wind rose for each site showing a predominant south-westerly wind. Black circle indicates location of eddy covariance tower and meteorological station. Grey circles indicate experimental plots where soil greenhouse gas, litter fall, litter decomposition measurement were taken. 100 m rule indicated for scale.

content using time-domain reflectometers (CS616; Campbell Scientific), incoming photo flux density (SKP215 quantum sensor; Skye Instruments, Powys, UK), net radiation (NR-LITE; Kipp and Zonen, Delft, the Netherlands), air temperature and humidity (HMP155A; Vaisala, Vantaa, Finland). Additionally, precipitation (52203; Young, Traverse City, MI, USA) and wind speed and direction (05103-5; Young) were measured at the SRC willow site only. At both stations, variables were measured at 0.1 Hz and then collected and averaged half-hourly using a CR1000 datalogger (Campbell Scientific). The 50-year (1960–2010) average monthly temperature and rainfall for the region were obtained from the UK Met Office (Met Office, 2015).

Soil GHGs fluxes

Eight plots were established in random locations in the SRC willow and grassland in November 2011 to measure soil GHGs, soil chemistry, aboveground and belowground biomass, litter fall and litter decomposition (Fig. 2). Randomization took place within a few metres of field edges to avoid any edge effects. Within these plots, soil CO_2 efflux was measured at monthly intervals using a portable chamber (SRC-1; PP Systems, Amesbury, MA, USA) coupled with an IRGA (EMG-4; PP Systems). Every effort was made at each sampling date to avoid the inclusion of significant amounts of vegetation in the sampling chamber, since this would reflect both plant shoot as well as soil and root efflux. However, small amounts of shoot vegetation remained inside the chamber, and therefore, soil CO_2 flux may be overestimated. Air temperature, soil temperature (stab probe; Testo, Alton, Hampshire, UK; 0–10 cm) and soil moisture (Theta probe; Delta-T, Burwell, Cambridge, UK; 0–6 cm) were also measured around the chamber at the time of sampling. As soil temperature is generally a good predictor of soil respiration, annual soil respiration was computed using an exponential function between monthly soil respiration data (SR) and continuous soil temperature data (T_{soil}) measured at each weather station (Raich & Schlesinger, 1992; Raich *et al.*, 2002):

$$SR = a \times \exp^{b \times T_{soil}}.$$

At each of the eight sampling locations, N_2O and CH_4 soil fluxes were measured using closed vented static chambers (Smith & Mullins, 2000) made of PVC base rings (8 cm high with a diameter of 40 cm), inserted in the soil to 5 cm depth, and chamber lids (20 cm high with a diameter of 40 cm). To determine GHG fluxes, headspace gas (10 ml) was sampled from a self-sealing septa in the chamber lid using gas-tight syringes, at 0, 15, 30 and 50 min after closure; it was immediately stored in pre-evacuated gas-tight vials (3 ml; Labco Ltd, Lampeter, Ceredigion, UK). Gas samples were analysed on a PerkinElmer Autosystem XL Gas Chromatograph (GC) fitted with a flame ionization detector for CH_4 and an electron capture detector for N_2O. All results were calibrated against certified gas standards (BOC, Guildford, UK; Case *et al.*, 2014). N_2O and CH_4 flux rates were determined by linear regression of the four sampling time points for each chamber and by applying a temperature and pressure correction (Holland *et al.*, 1999). The analytical precision of the GC for standards at ambient concentration was approximately 2%, using two standard deviations as a measure of mean error. Sampling for soil GHG fluxes took place every month, from November 2011 until December 2014 (Fig. S1). Sampling of the grassland initially took place in a smaller grassland site from November 2011 until August 2012, when sampling was moved to another larger site (to accommodate eddy covariance equipment). Grassland sites were both sampled for GHG fluxes for the next 3 months to compare fluxes and there was no significant difference between the sites ($t_{(4)} = -0.06$, $P = 0.95$). Non-CO_2 GHG fluxes were first converted into CO_2 equivalents using the global warming potentials over a 100-year horizon of 298 for N_2O and 25 for CH_4 and then to carbon equivalents using a conversion factors of 0.2727 (IPCC, 2007b). Linear interpolation between measurements dates (i.e. trapezoidal integration) was used to compute annual cumulative GHG fluxes.

Six (two per plot: one root excluded, one total respiration) automated soil chambers were also established in the SRC willow in February 2012 (Ventura *et al.*, 2015). These chambers measured soil CO_2 flux every 4 h, and three of the chambers were placed in root exclusion chambers to allow the partitioning of autotrophic and heterotrophic respiration. Data from automated chambers were used to validate periodic measurements.

Soil analysis

Soil carbon was measured at 0–30 cm (15 cores) and to 1 m depth (three cores) in both grassland and SRC willow (and initial grassland). Samples were only taken once during the experiment in October 2012. Five plots were randomly selected in each field; from each of these plots, three within-plot soil cores were taken using a split-tube soil sampler (Eijkelkamp Agrisearch Equipment BV, Giesbeek, the Netherlands) with an inner diameter of 4.8 cm to a depth of 30 cm. This gave a total of 15 spatially nested samples per field, accounting for both field-scale (between sampling plots) and plot-scale (cores within plots) variability. One of the five sampling plots was randomly selected and three additional 1 m cores were taken. In the case of both the 1 m and 0–30 cm core, one core was taken from the centre of the plot, with two further cores taken at distances of 1 and 1.5 m in random compass directions from the centre. The 1 m cores were taken using a window sampler system with a 4.4 cm cutting diameter (Eijkelkamp Agrisearch Equipment BV), allowing a full 1 m core to be extracted and subsequently transported in one section. If coring to the full depth was not possible, for example when large stones or bedrock were encountered, the precise depth of the cored hole was recorded (see Rowe *et al.*, 2016, for full methods). Fresh soil was sieved to 2 mm before being frozen at −80 °C and subsequently freeze-dried for minimum of 24 h. A subsample of the freeze-dried soil (20–30 ml) was milled to a fine powder in a ball mill (Planetary Mill; *FRITSCH*, Idar-Oberstein, Germany). A 200 mg subsample of the milled soil was used for the assessment of carbon concentration using an elemental analyser (Leco Truspec CN, Milan, Italy). Total soil carbon stock for the 0–30 and 0–100 cm fractions was calculated on an equivalent soil mass basis (Keith *et al.*, 2015).

Aboveground and belowground biomass and net primary production (NPP)

Aboveground biomass. In SRC willow, aboveground biomass was estimated from the stem : volume index (Pontallier *et al.*, 1997) which was calculated for all shoots of 160 stumps distributed in eight plots using stem diameter (22 cm from ground height; Rae *et al.*, 2004) and dominant stem height. Nondestructive sampling took place every year in winter during the experiment (Fig. S1). Destructive sampling of SRC willow was also conducted prior to commercial harvest in November 2013, to allow an estimation of actual biomass from stem : volume index values. A linear regression of stem:volume index against fresh weight allowed estimation of total dry weight (kg tree^{-1}) from trees which were nondestructively sampled. Total carbon contained in aboveground biomass was calculated by assuming that the amount of carbon contained in woody biomass was approximately 49.3 ± 1.2% (mean ± SD), calculated from an assessment of measured values in the literature for SRC (Fahmi *et al.*, 2007; Bridgeman *et al.*, 2008; Sannigrahi *et al.*, 2010; Gudka, 2012). Willow leaf litter was collected in trays during the months of litter fall, July–December, to quantify leaf biomass. Leaf litter was oven-dried at 80 °C for 48 h, weighed and extrapolated from tray to tonnes per hectare. Litter decomposition was measured over 2 years in SRC willow. Mesh bags (20 × 10 cm; 1 mm aperture) each containing 5 g leaf litter (picked green leaves) were placed by each of the GHG chambers in November 2011. Bags were collected at several points postinsertion – 2 weeks then 1, 2, 4, 6, 9, 12, 18 and 24 months. Leaf litter was gently washed with distilled water, then dried at 85 °C for 24 h before dry weight was recorded.

Aboveground biomass was estimated in the grassland from four randomized plots by cutting all biomass within a 50 × 50 cm quadrat with hand shears flush to ground. Samples were taken twice during the experiment, in August 2013 and August 2014 prior to the mowing of the field. Samples were oven-dried at 80 °C for 48 h, weighed and extrapolated from quadrat to tonnes per hectare.

Belowground biomass. Belowground biomass in SRC willow was estimated using equations for aboveground stool and stem and belowground biomass found in Pacaldo *et al.* (2013a). Then, the ratio of belowground : aboveground (i.e. aboveground stool + stem) was calculated as 0.99 for our site. In the grassland, belowground biomass was measured using 5 cm diameter auger and taken at three depths (0–10, 10–20 and 20–30 cm) across four randomized plots. Roots were sieved consecutively through sieves of decreasing mesh size (3350, 2000 and 500 μm), oven-dried at 85 °C for 24 h, weighed and extrapolated from core to tonnes per hectare. Total biomass was calculated by summing total above ground biomass and belowground biomass; for SRC willow the aboveground components included stem, stool, branches and leaf biomass. Net primary production was calculated on an annual basis using two consecutive biomass measurement data sets (nondestructive for SRC willow and destructive biomass harvest for grassland). Standard error was calculated for all components of biomass, as well as for NPP.

Eddy covariance measurements

Eddy covariance towers were installed in SRC willow in April 2012 and in grassland in August 2012 to measure ecosystem CO_2 fluxes. Each system consisted of an open path infrared gas analyser (Li-7500*A*; Licor, Lincoln, NE, USA) and a sonic anemometer (Windmaster Pro; Gill, Hampshire, UK). Data were logged at 20 Hz to an industrial grade USB stick in the LiCor interface box. Instrument height was 2.5 m from the ground for the grassland site. For SRC willow, instrument height was 8 m at the start of the experiment and extended as the crop grew to a maximum instrument height of 9.3 m in March 2014. After harvest, the instrument height was reduced to 3.6 m aboveground level.

Eddy covariance data were processed using EddyPro (Licor) and averaged over 30-min intervals. The applied methodology was based on the EuroFlux protocol (Aubinet *et al.*, 2000). Data were then elaborated and quality-checked using Stata IC 10 (StataCorp LP, College Station, TX, USA). Most of the data were discarded during nigh-time as the assumptions using for eddy covariance measurements (i.e. turbulence) were not fulfilled. Data were rejected when fluxes came from outside the flux footprint which was between 135° and 262° for SRC willow and 140–290° for grassland. Data were also discarded during rain and fog. Energy balance closure at each site was estimated only using measured data. Gapfilling to estimate Net Ecosystem Exchange (NEE) and flux partitioning into Ecosystem Respiration (R_{eco}) and Gross Primary Production (GPP) were done according to the standard methodology used in Fluxnet (http://www.bgc-jena.mpg.de/~MDIwork/eddyproc/ ; Reichstein *et al.*, 2005). NEE uncertainty (i.e. standard deviation) was computed according to the FLUXNET methodology using the online software, whilst error terms were unable to be calculated for R_{eco} and GPP as these are modelled terms.

Ecosystem GHG balance

A conceptual model was constructed to represent the whole system GHG balance for both grassland and SRC willow for two whole years during the measurement period, January 2013–December 2014 (Fig. S1, blue box). All gas flux data were expressed as g C m^{-2} yr^{-1} and soil storage terms presented as standing stock (g C m^{-2}). The terminology used is as defined by Chapin *et al.* (2006); however, we assigned a positive sign to emissions of carbon to the atmosphere and a negative sign to an uptake of C by the ecosystem, as generally used in micrometeorology. Briefly, NEE was defined as the CO_2 exchange between the ecosystem and the atmosphere, measured using the eddy covariance technique. Gross Primary Productivity was defined as the fixation of carbon by autotrophic organisms and Ecosystem Respiration (R_{eco}) is the net respiration by both autotrophic and heterotrophic organisms in the ecosystem. Soil respiration is the sum of CO_2 respired by roots (autotrophic respiration) and by microbes (heterotrophic respiration). Net Primary Production is the accumulation of biomass within the study system, measured used in litter fall and biomass estimates. Net Biome Production (NBP) describes the difference between Net Ecosystem Productivity (NEP, negative sign of NEE; Reichstein *et al.*, 2012) and the carbon removed through harvest. The sign of NBP is opposite to measures described above, where negative indicates a release to the atmosphere and a positive value indicates an uptake by the ecosystem. NBP was only calculated for SRC willow where harvested biomass was removed from the system.

Statistical analysis

A basic *t*-test was performed to detect any significant difference in soil carbon stocks at 0–30 cm (n = 15) and 0–100 cm (n = 3) between land uses, using SigmaPlot 12.5. All statistical analyses for GHG and eddy data were conducted in the R programming environment (R version 3.1.3; R Core Team, 2015). GHG data were analysed using linear mixed models (Bates *et al.*, 2014) where fixed effects were treatment, year, soil temperature and soil moisture. Air temperature and soil temperature exhibited collinearity so could not both be included in the model. Chamber number was used as a random factor to account for repeat sampling over time. Main effects were tested in addition to all second-order interactions. Analysis of N_2O and CH_4 reveals normality of residuals and homoscedasticity; however, there was heteroscedasticity detected in the CO_2 data; therefore, log-transformation was performed. Model selection was performed according to Crawley (2007) using AIC to construct the minimum adequate model (see Table S2).

For eddy covariance data, a global model was constructed to assess the effects of land use and climate variables [fixed effects: treatment, photosynthetically active radiation (PAR), wind speed, rain, soil temperature, relative humidity (RH) and soil water content; random effect: date] on NEE using daily averaged data (see Table S3 for full model). There was collinearity between air temperature and soil temperature so only one was used in the model, likewise for soil water content at both depths. Data were then partitioned by site and two separate models were constructed for each data set to see whether the drivers of NEE differed between fields. The aim of this analysis was to try to identify the drivers of NEE to environmental variables which were measured on site.

Results

Weather patterns

Air temperature in 2012 were close to average values for the region. Spring of 2013 was cooler than average, whereas winter 2013 and spring of 2014 experienced higher than average temperatures (Fig. 3). The spring/ summer of 2012 and winter 2013 are notably wet years with above average rainfall for the region, whilst in contrast the spring and summer of 2013 were drier than average (Fig. 3). Air temperature in 2013 was cooler and much drier than both 2012 and 2014 with an average air temperature of 9.9 °C and rainfall of 673.3 mm. 2012 was slightly cooler but wetter (10.6 °C and 1318 mm) than 2014 which experienced an average temperature of 11.1 °C and 1023 mm rainfall.

Net primary production

Total aboveground biomass in SRC willow increased from the first measurement, March 2012, to the final

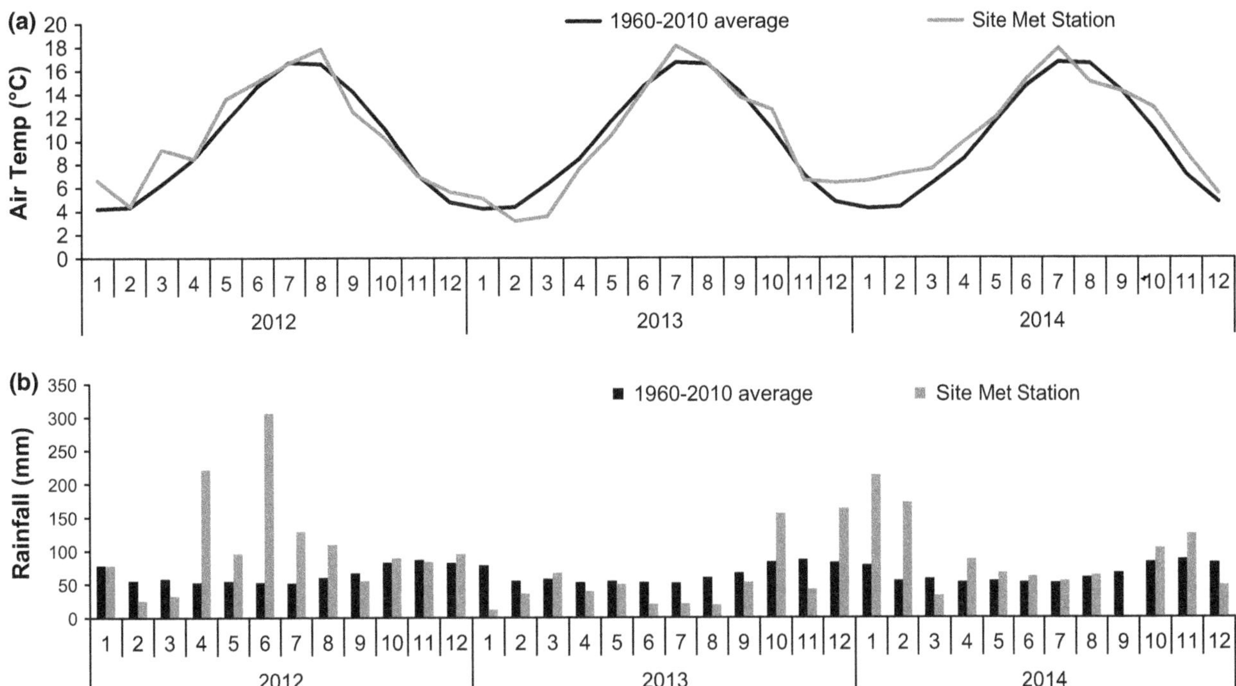

Fig. 3 (a) Monthly mean values of air temperature (°C) showing 50-year average (1960–2010; black line) and values measured by site meteorological station (grey line). (b) Sum of rainfall (mm) for 50-year average (1960–2010; black line) and measured on site (grey line).

measurement before the harvest, November 2013 (Fig. 4a). Biomass was rapidly accumulated after the harvest in April 2014 with total aboveground woody biomass reaching 11.4 ± 1.1 t ha^{-1} (mean ± SE; n = 8) by the end of 2014. Leaf litter was similar for 2012 and 2013 with 5.6 ± 0.2 and 5.8 ± 0.2 t ha^{-1} yr^{-1}, respectively. There was a decrease in leaf litter fall after the harvest in 2014 with only 2.1 ± 0.2 t ha^{-1} yr^{-1}. The majority of SRC willow leaf litter decomposed within the first year, with only 17% leaf litter remaining after 12 months and only 8% remaining after 2 years (Fig. S2). Total grassland biomass was over double that in 2014 compared to 2013, for both aboveground and belowground biomass (Fig. 4b). Total biomass in 2013 was higher in SRC willow (96.2 ± 3.6 t ha^{-1}; n = 4) than grassland (8.7 ± 1.5 t ha^{-1}), and owing to the remaining belowground biomass, total biomass remained higher in SRC willow in 2014 after harvest (69.8 ± 2.8 and 20.8 ± 1.6 t ha^{-1} for SRC willow and grassland, respectively). There was a decrease in NPP in SRC willow from 2012 to 2013, which corresponds to year 4 and year 5 of the rotation (Fig. 4c; 14.6 ± 2.1 and 10.8 ± 2.4 t C ha^{-1} yr^{-1}, respectively). There was an increase in NPP postharvest to 12.4 ± 0.8 t C ha^{-1} yr^{-1} (Fig. 4c). In 2014, the NPP in grassland (4.9 ± 1.0 t C ha^{-1} yr^{-1}) was less than that of SRC willow, 12.4 ± 0.8 t C ha^{-1} yr^{-1} (Fig. 4c).

Soil respiration

CO_2 accounted for the majority of soil GHG flux, c.96% and c.99% for grassland and SRC willow, respectively. Mean soil respiration (2012–2014) was significantly higher in grassland (1522 ± 39 g C m^{-2} yr^{-1}; mean ± SE; n = 8) than in SRC willow (912 ± 42 g C m^{-2} yr^{-1}; Fig. 5, Table S4, P = 0.03). Year, soil temperature and soil moisture were all significant factors affecting soil respiration (P < 0.001), as well as second-order interactions for treatment and year (P < 0.001), treatment and soil temperature (P < 0.001), and year and soil moisture (P = 0.007). According to the continuous soil respiration measurements (also reported in Ventura *et al.*, 2015), heterotrophic respiration accounted for 84% of total soil respiration in the SRC willow.

Eddy flux measurements

For the eddy covariance data, after quality control checks and footprint analysis the data remaining were 40% for grassland and 37% for SRC willow in 2013. In 2014, the remaining data for each site was 46% and 20% for grassland and SRC willow, respectively. The energy balance closure for the sites, based on measured data only, was a 73% for grassland (Fig. 6a) and 77% for SRC willow (Fig. 6b).

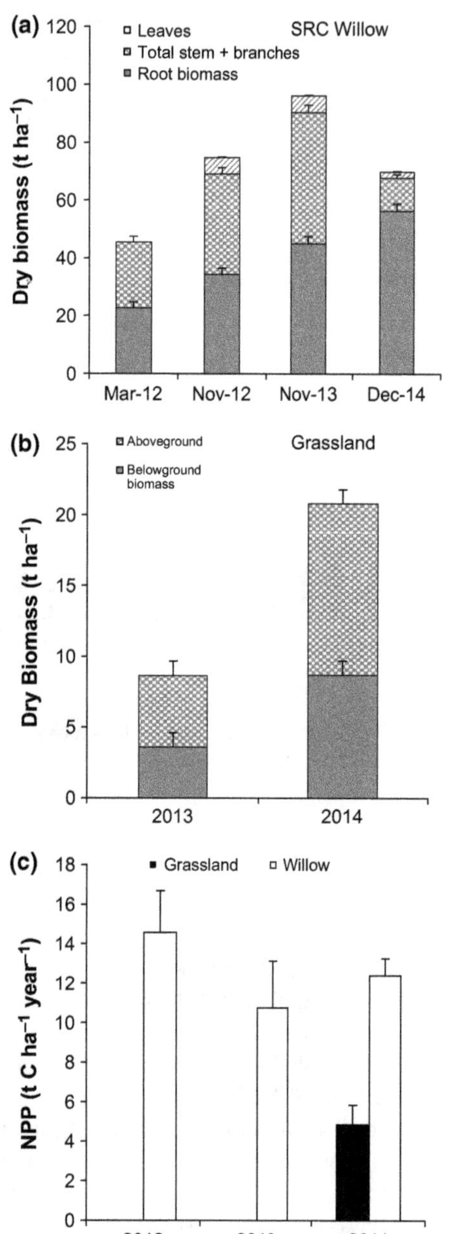

Fig. 4 (a) Total biomass for short rotation coppice willow (t ha^{-1}; mean ± SE; n = 8) including measured stem (checkerboard hatching) and leaf biomass (diagonal hatching), and estimated root biomass (grey fill). (b) Grassland biomass from measured aboveground (checkerboard hatching) and belowground sampling (grey fill) (t ha^{-1}; mean ± SE; n = 4). (c) Net primary productivity for short rotation coppice willow (white bars) and grassland (black bars).

For grassland, the mean NEE over 2 years (2013–2014) was 119 ± 10 g C m^{-2} yr^{-1} (mean ± SD). In year one (2013), the grassland was a net source of carbon, 246 ± 11 g C m^{-2} yr^{-1}, whereas in year two (2014) it was a net sink, −9 ± 16 g C m^{-2} yr^{-1}. In year one, there was a small uptake of carbon during the growing season from June 2013 to the end of July 2013 (Fig. 7a); however, in year two there is a more defined uptake period starting from March 2014. This early onset of carbon fixation could be attributed to the higher mean monthly temperature in January–March 2014 compared to 2013 (Fig. 3). SRC willow was a C sink for the 2-year duration of the experiment with a mean annual NEE of −620 ± 18 g C m^{-2} yr^{-1} (Fig. 7b). In the first year, which corresponded to the 4th year of growth, the site was a large sink of carbon (−901 ± 23 g C m^{-2} yr^{-1}). The NEE for the second year was smaller due to the harvest in April 2014 (−339 ± 27 g C m^{-2} yr^{-1}). NEE was lower in the SRC willow than in grassland during the second year ($P < 0.001$). Analyses of eddy covariance data also revealed that NEE in grassland and SRC willow were driven by different components (Table S3). In the grassland, PAR, year, soil (and air) temperature, wind speed and rain were factors affecting NEE, whilst in the SRC willow only PAR, year and soil water content were affecting NEE. Relative humidity was not found to be a factor affecting NEE at either site.

There were also differences in R_{eco} and GPP between grassland and SRC willow. R_{eco} was 33% higher in 2014 than in 2013 in grassland (1261 and 1675 g C m^{-2} yr^{-1} for year one and year two, respectively). R_{eco} in SRC willow in year one was lower than both years in grassland at 971 g C m^{-2} yr^{-1}. In 2014, R_{eco} was larger than year one in SRC willow and both years in the grassland site at 1971 g C m^{-2} yr^{-1}. Mean R_{eco} over 2 years was similar for grassland and SRC willow, 1468 and 1471 g C m^{-1} yr^{-1}, respectively. GPP in grassland was 1015 and 1683 g C m^{-2} yr^{-1} for year one and two, respectively. In SRC willow, GPP was higher than the grassland for both years at 1873 and 2309 g C m^{-2} yr^{-1} for year one and year two, respectively. Over 2 years, mean GPP was higher in SRC willow than in grassland, 2091 and 1349 g C m^{-2} yr^{-1}, respectively.

Belowground carbon pools

Soil carbon stocks (Table 2) were higher in the grassland than in the SRC willow for both the 0–30 and 0–100 cm profiles, but for the latter this effect was only significant to $P = 0.062$, despite a clear trend. For 0–30 cm, we found 63.4 ± 3.5 t C ha^{-1} in grassland and 42.6 ± 1.8 t C ha^{-1} in SRC willow (mean ± SE; $t_{(28)} = -5.30$, $P < 0.001$). And for the 0–100 cm profile, there was 107.6 ± 1.8 and 77.3 ± 7.7 t C ha^{-1} for grassland and SRC willow, respectively ($t_{(4)} = -3.84$, $P = 0.062$). The grassland which was used initially for chamber measurements had a similar carbon stocks to grassland in the upper 30 cm (61.2 ± 2.8 t C ha^{-1}),

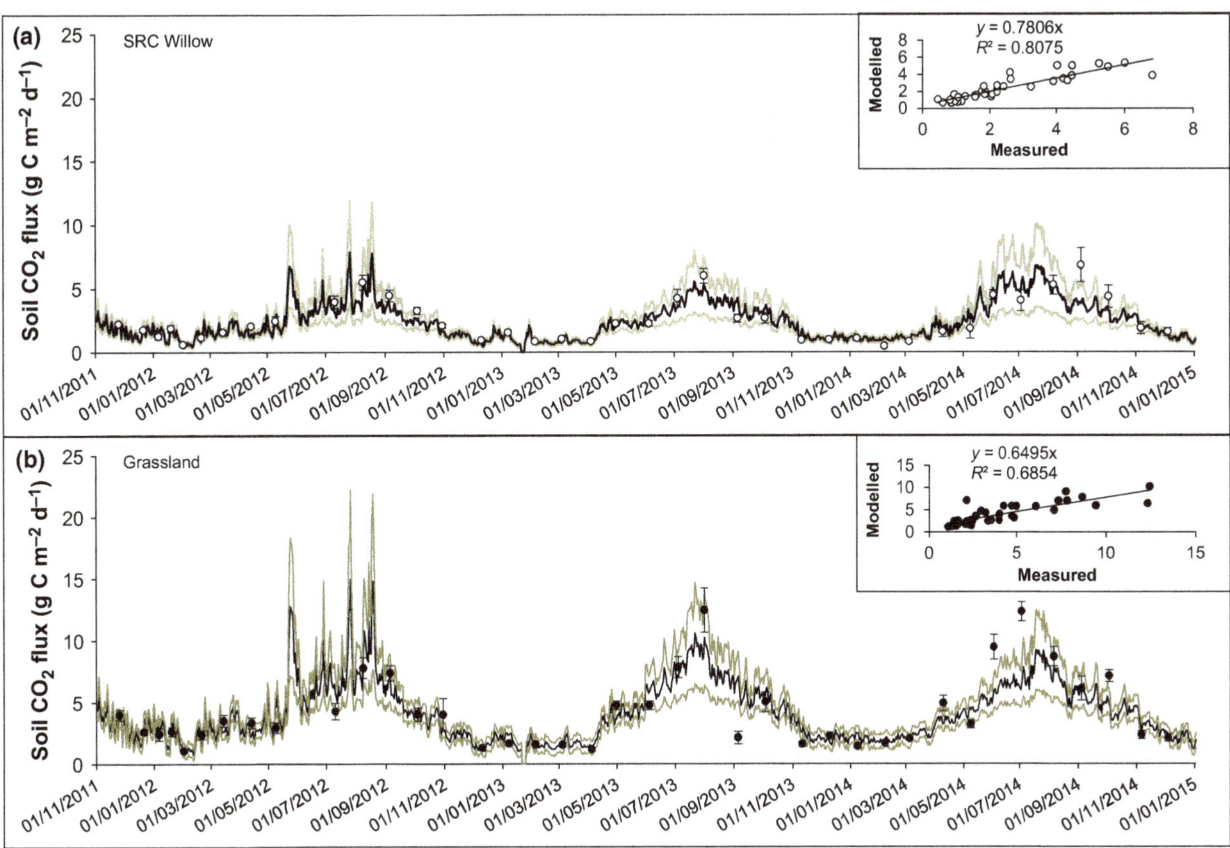

Fig. 5 Soil CO_2 flux (g C m^{-2} day^{-1}) for short rotation coppice willow (a; white circles) and grassland (b; black circles) (mean ± SE; n = 8). Periodic sampling events (circles) and modelled CO_2 flux (black line; using soil temperature) are shown. Green line indicates 5th and 95th percentiles around the modelled values. Additionally, modelled CO_2 data are regressed against measured CO_2 data for both sites and the relationship shown on the graph as R^2.

with slightly less carbon in the 100 cm profile than the SRC willow (63.8 ± 4.1 t C ha^{-1}; Table 2).

Soil GHG fluxes

N_2O and CH_4 were not important contributors to the whole GHG balance of these two particular sites, accounting for less than 4% (3.4% N_2O and 0.4% CH_4) for grassland and less than 1% (0.77% N_2O and 0.07% CH_4) for SRC willow. Mean N_2O fluxes at both sites (2012–2014) were very low (within detection limit of equipment) with emissions of 1.2 ± 0.3 and 1.9 ± 0.6 g C m^{-2} yr^{-1} for grassland and SRC willow, respectively (Fig. S3; 4.4 ± 1.1 and 7.0 ± 2.2 g CO_2-eq m^{-2} yr^{-1} for grass and SRC, respectively). There was no difference between N_2O fluxes between the sites (P = 0.81; Table S5). N_2O flux was significantly affected by year across both sites (P = 0.003), as well as an interaction between year and soil moisture (P = 0.007). CH_4 was also very low at both sites; however, there was a difference between the sites with an emission of 0.2 ± 0.2 g C m^{-2} yr^{-1} from grassland and uptake of −0.2 ± 0.1 g C m^{-2} yr^{-1} in SRC willow (P = 0.003; Table S6, Fig. S4; 0.7 ± 0.7 and −0.7 ± 0.4 g CO_2-eq m^{-2} yr^{-1} for grass and SRC, respectively). For both sites, soil temperature significantly affected CH_4 flux (P < 0.001), as well an interaction between soil moisture and soil temperature (P = 0.02).

Conceptual model

Data from January 2013 to December 2014 were summarized in a conceptual model to allow comparison of the grassland and SRC willow (Fig. 8). This figure shows the movement of carbon through the ecosystem, highlighting major fluxes and stocks. The harvested carbon was expressed on annual basis (i.e. total harvested biomass was divided by the rotation length in the willow) and is shown, 445 ± 68 and 399 ± 23 g C m^{-2} yr^{-1} (mean ± SE) for grassland and SRC willow, respectively. However, as the mowed grass was not removed

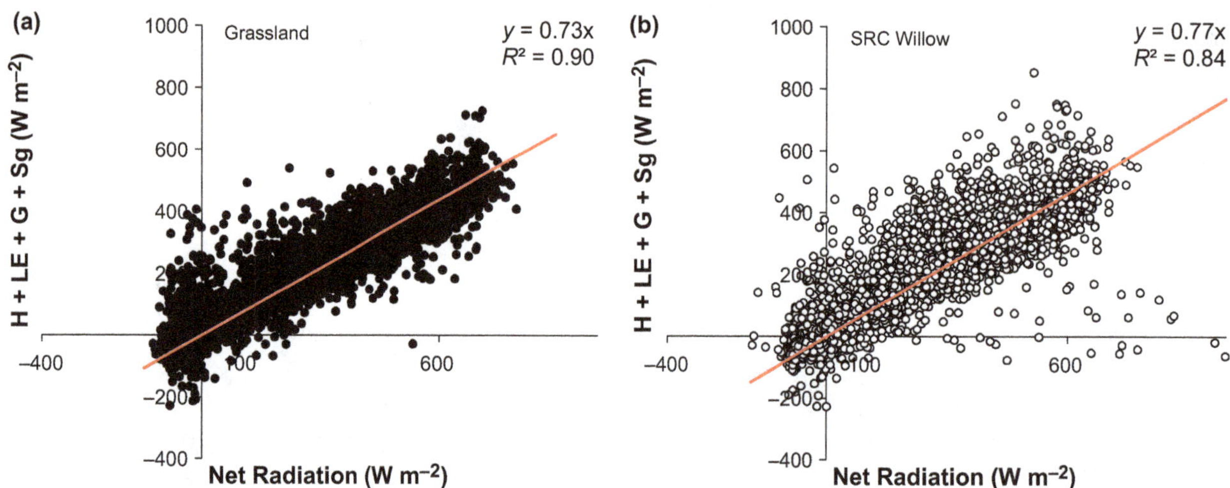

Fig. 6 Energy balance closure for grassland (a; black circles) and short rotation coppice willow (b; white circles) for 2013–2014, where H is sensible heat flux, LE is latent heat flux, G is soil heat flux and Sg is soil heat storage. Strength of regression indicted on graph by R^2 value.

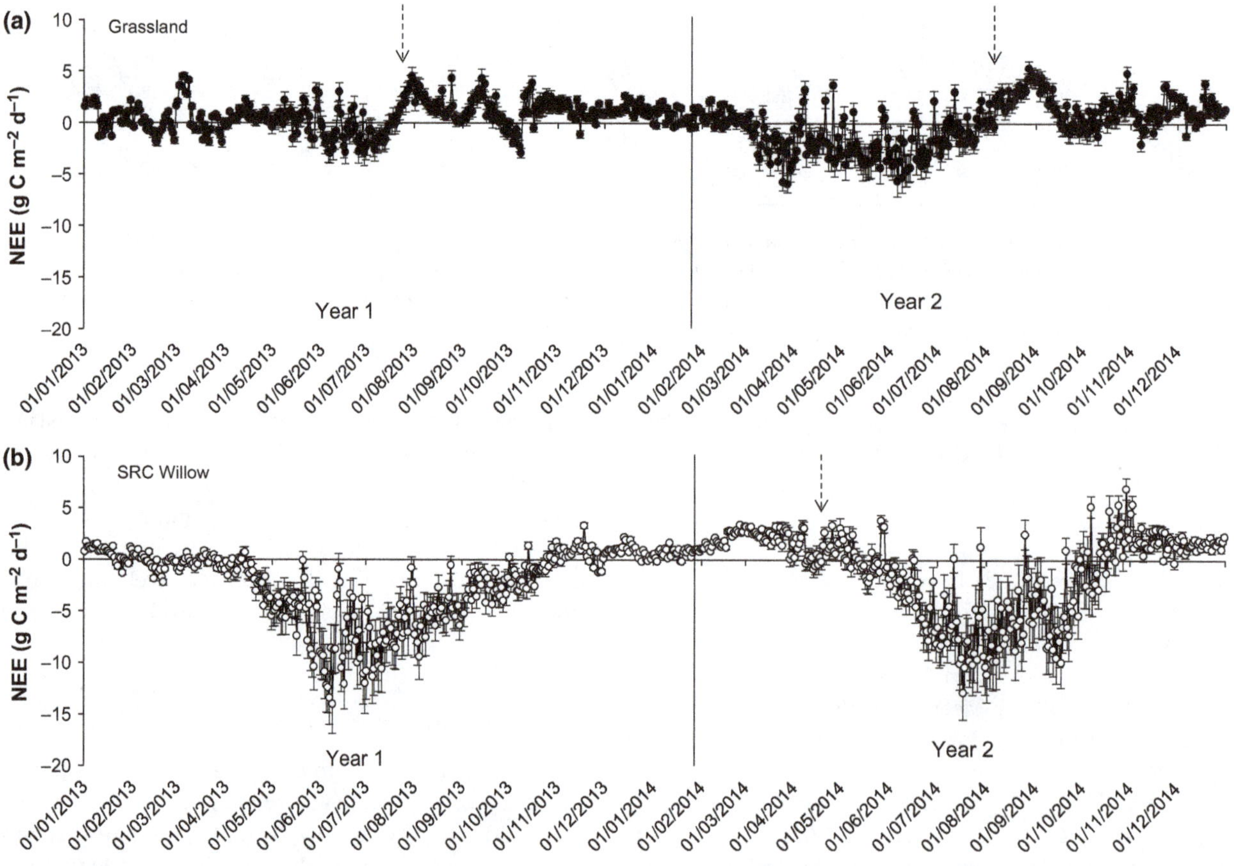

Fig. 7 Net Ecosystem Exchange (NEE; g C m^{-2} day^{-1}; mean ± SD) for grassland (a; black circles) and short rotation coppice willow (b; white circle) for 2013–2014. Harvest events at both sites are indicated by dashed arrows.

from the site, NBP is equal to NEE. Thus, mean NBP (2013–2014) was −118 ± 10 g C m^{-2} yr^{-1} for grassland and 221 ± 66 g C m^{-2} yr^{-1} for SRC willow (mean ± SD), which, despite the removal of 399 g C m^{-2} yr^{-1} biomass from the SRC field, remained a net sink for carbon.

Impact of harvest in SRC willow

The SRC willow was harvested in April 2014 which corresponded to year 5 of the first rotation. There was no detectable effect of the harvest on soil moisture or soil temperature in the SRC willow, compared to preharvest measurements. The effect of the harvest on the NEE can be seen in Fig. 7b (dashed arrow indicated harvest date), where NEE decreased prior to harvest and then quickly increased after harvest. The smaller NEE and increased R_{eco} observed in SRC willow in 2014 compared to 2013 is likely attributable to the disturbance caused by the harvest. The site quickly became a net C sink again as there was a rapid re-sprout of willow stumps and understory vegetation. There was no noticeable effect on soil CO_2 and CH_4 emissions as a result of the harvest. There was a large one-off emission of N_2O in June 2014, 2 months postharvest, which may have arisen as a result of the harvesting process (Fig. S3b).

Table 2 Soil carbon stocks (t C ha^{-1}) under grassland and short rotation coppice willow, calculated on an equivalent soil mass basis, for 0–30 cm and 0–100 cm. Initial grassland refers to site where static chamber measurements were taken prior to installation of eddy covariance monitoring equipment. Samples collected in October 2012. n = 15 for 0–30 samples and n = 3 for 0–100 cm samples

Soil depth (cm)	Grassland	SRC willow	Initial grassland
	Mean ± SE (t C ha^{-1})		Mean ± SE (t C ha^{-1})
0–30	63.4 ± 3.5	42.6 ± 1.8*	61.2 ± 2.8
0–100	107.6 ± 1.8	77.3 ± 7.7 (10%)	63.8 ± 4.1

*Significance to 0.05 (5%) and significance to 0.1 (10%).

Discussion

Understanding the consequences of LUC for ecosystem GHG balance is important if we are to tackle the impact of agricultural practices on global GHG emissions. This research addressed a critical – the provision of empirical GHG balance data from commercial-scale operations, where bioenergy has been deployed for a period of years. It has demonstrated that over a 2-year period (including the harvest operation in SRC willow), during a side-by-side commercial-scale comparison, an SRC willow field was a net sink for carbon, whilst the minimally managed grassland field was a net source for carbon. N_2O and CH_4 emissions were generally low for both sites, contributing little to the total GHG balance for these contrasting land-use types in southern England. Thus, we can conclude that 7 years postland-use transition, this SRC bioenergy crop had an improved GHG balance relative to the adjacent grassland. This

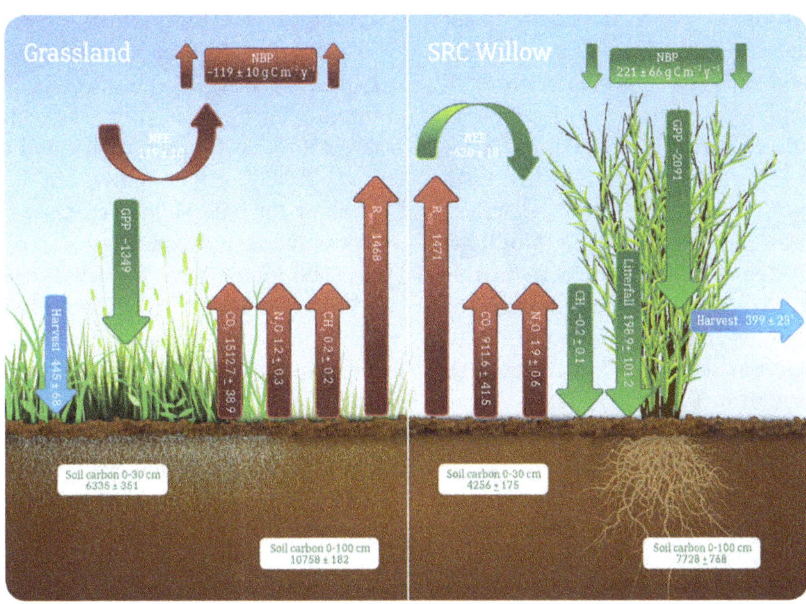

Fig. 8 Annual greenhouse gas budget for grassland and short rotation coppice willow for measurement period January 2013 to December 2014. All fluxes are in g C m^{-2} yr^{-1} in square boxes and soil storage terms presented as standing stock (g m^{-2}) in oval boxes. Measured values are presented as mean ± SE, except for net ecosystem exchange (NEE) and Net Biome Production (NBP) where measured values are presented as mean ± SD (see Materials and methods for details on uncertainty calculation for NEE). Note for all fluxes, apart from NBP, a negative flux indicated a gain to the ecosystem and a positive flux indicates a loss to the atmosphere. [1]Harvest data have been annualized from the total biomass taken off the field during coppicing at year 5.

suggests that not only did this LUC provide bioenergy as a net provisioning ecosystem service, but was also able to contribute to improved climate regulation through the generation of a net carbon sink relative to the original land use. In the area of bioenergy science, this is an important empirical finding and suggests that in temperate climates, where reasonable land-use transitions are considered, bioenergy may add positively to the multifunctional landscape, as suggested recently by those such as Manning *et al.* (2015) and Souza *et al.* (2015). These results coupled with the potential carbon and GHG savings made by replacing fossil fuels demonstrate the potential of bioenergy for climate change mitigation and improved energy security (Cannell, 2003; Styles & Jones, 2007).

Improved grasslands are important sources of terrestrial carbon storage, holding the second largest store after bogs, with approximately 274×10^6 t C (Ostle *et al.*, 2009) and here we hypothesized that LUC from grassland to SRC willow would lead to a significant reduction in GHG emissions as proposed in previous modelling studies in this temperate bioenergy system (Hillier *et al.*, 2009; Hastings *et al.*, 2014; Milner *et al.*, 2015). In a UK context, conversion of semipermanent, permanent or managed grassland to bioenergy cropping systems represents one of the most significant potential land-use transitions, since grassland is a considerable part of the UK landscape (4–5 $\times 10^6$ ha; DEFRA *et al.*, 2007) and because management of grasslands can vary widely in the UK, particularly with respect to fertilizer input and grazing. This can have a dramatic effect on consequential GHG and carbon balance as a result of LUC. For the grassland studied here, we found that over a 2-year measurement period, grassland was a net source for carbon and SRC willow was a net sink. Even when considering the carbon removed from the system scaled on an annual basis in harvest biomass (NBP), the SRC willow site remained a sink for carbon. In this experiment, we observed high biomass yields for SRC willow, comparable to those found in some other studies (Laurent *et al.*, 2015) but within the range reported by Allwright & Taylor (2016). To our knowledge, there has been only one previous limited study of eddy covariance measurements over SRC willow for bioenergy (Drewer *et al.*, 2012), though much research attention has been focussed on SRC poplar. These studies have generally found that SRC poplar is a sink at the ecosystem level (Arevalo *et al.*, 2011; Jassal *et al.*, 2013; Sabbatini *et al.*, 2015), even as soon as 2 years postestablishment of the crop (Verlinden *et al.*, 2013).

One question from our study is the relevance of the grassland considered here, since retention of cut grass on the surface, which resulted in no C exports from the system, could be considered uncommon with grazing and mowing for hay or silage much more likely as a management option (Smit *et al.*, 2008). As a result, grassland NBP was equal to NEE at -119 ± 10 g C m^{-1} yr^{-1} at our site. Qun & Huizhi (2013) investigated similarly managed grassland with no exports of carbon and found that the site was similarly a net source of carbon, with a NBP of -138 g C m^{-2} yr^{-1}. Thus, we can conclude that management of the grazing and mowing regime might be central to the carbon balance of such a system and determine net source or sink status. We identified PAR and soil moisture to be the main climatic drivers of NEE in grassland and SRC willow, which has been found in other studies (Ruimy *et al.*, 1995; Qun & Huizhi, 2013; Shao *et al.*, 2015). In contrast, some studies have identified leaf area index (LAI) to be the main biophysical driver of NEE in SRC poplar (Broeckx *et al.*, 2014; Zenone *et al.*, 2015), but our data for willow do not support this. Data syntheses from a network of sites such as FLUXNET have already begun identifying driving factors of NEE, GPP and R_{eco} over a number of biomes (Law *et al.*, 2002), and as the amount of flux data from bioenergy crops increases, there is potential for syntheses in these biomes in future.

In this experiment, we found that CO_2 was the main contributor to soil GHG emissions in both sites, supporting the observations of Drewer *et al.* (2012) who also found CO_2 to be the dominant soil GHG for SRC willow at a second UK site. In the SRC willow, we were able to observe the partitioning of soil CO_2 flux which revealed that 84% of total soil respiration was heterotrophic in origin (Ventura *et al.*, 2015). Since heterotrophic respiration can vary from 10% to 90% depending on vegetation type and time of year (Hanson *et al.*, 2000), our data fall within this wide range. Future work at this site should measure autotrophic and heterotrophic respiration in grassland for a direct comparison and inference on the effects of LUC to bioenergy.

Grasslands can vary in both space and time for GHG emissions and carbon balance (Soussana *et al.*, 2007; Imer *et al.*, 2013), as found here where the grassland in this study was a net source of carbon in 2013 and a net sink in 2014, possibly attributable to the higher temperatures observed in January–March 2014 compared to 2013. Grass begins growing when air temperature exceeds 5 °C (Robson *et al.*, 1988), which was achieved earlier in 2014, providing an extended season for carbon fixation. This combination of increased temperature with an increase in winter rainfall (which resulted in increased soil moisture) could explain the higher aboveground biomass in grassland and consequently why the site was a net sink in 2014 (Pitt & Heady, 1978).

As well as large variability, there are also large uncertainties surrounding the overall GHG balance of temperate grasslands (Janssens *et al.*, 2003). Within the literature, there are reports that grasslands are acting as both carbon sources and carbon sinks (Scurlock & Hall, 1998; Bellamy *et al.*, 2005; Soussana *et al.*, 2007; Ciais *et al.*, 2010; Merbold *et al.*, 2014; Schipper *et al.*, 2014; Rutledge *et al.*, 2015), with the balance tightly linked to management regime, including fertilizer application, rotation and grazing regime (Smith, 2014), with changes in management causing grasslands to switch from a source to a sink (Merbold *et al.*, 2014). Grassland management practices such as fertilization, grazing and mowing lead to large N_2O and CH_4 emissions which counterbalance this CO_2 sink (Ciais *et al.*, 2010; Imer *et al.*, 2013). For our particular site, N_2O and CH_4 contributed little to GHG balance of either land use and both were present in small quantities. Interestingly, grassland was a net source of CH_4, whereas SRC willow was a net sink, but the fluxes were small. SRC willow has been found to be a net sink for CH_4 in other studies to a similar extent to that found here (Drewer *et al.*, 2012; Kern *et al.*, 2012). For both sites, there was an effect of soil moisture, and a significant interaction for soil moisture and soil temperature on CH_4 fluxes, confirming a number of other studies in bioenergy crops (Drewer *et al.*, 2012; Kern *et al.*, 2012) and grasslands (Kammann *et al.*, 2001; Imer *et al.*, 2013). Future climate changes may result in the need for fertilizer to maintain yields in SRC willow, which may lead to an altered GHG balance due to subsequent N_2O emissions.

Here, we found that grassland had significantly higher soil carbon stocks than the SRC willow up to 30 cm with a similar trend at 1 m depth. Sampling depth is a recurrent problem in studies which attempt to quantify soil carbon (Dolan *et al.*, 2006; Blanco-Canqui & Lal, 2008) and it is essential that the whole profile is sampled to draw robust conclusions (Harrison *et al.*, 2011). At this particular site, the higher soil carbon observed in grassland may be attributable to the amount of organic material left on the soil surface after mowing (Post & Kwon, 2000) and may not be widely representative of much managed rotational grassland. In grassland, on average, 445 ± 48 g C m^{-2} yr^{-1} of organic material was left on the soil surface after mowing however in SRC willow, annual litter fall reached a maximum of 292 ± 12.5 g C m^{-2} yr^{-1} in 2013. There have been reports in the literature of both increased soil carbon under SRC compared to grassland (Zan *et al.*, 2001; Arevalo *et al.*, 2009), as well as others which have found no significant difference (Grigal & Berguson, 1998; Walter *et al.*, 2015). Walter *et al.* (2015), from a chronosequence of SRC sites, suggested that this transition results in a redistribution of carbon through the profile, despite total SOC stock not being significantly different. After 7 years postconversion, we may be beginning to see redistribution of C in the soil profile. We found that at the two grasslands sites 59% and 96% carbon was stored in the top 30 cm, whereas in SRC willow 54% carbon was stored in the top 30 cm of the whole 100 cm profile. Whilst these differences are not large, the transition may still be at the early stages of C redistribution through the soil profile, though further data would be required to confirm this postulation. Chronosequence data also suggest that after initial conversion from grassland to SRC willow, there can be a loss of soil carbon for up to 5 years, which is followed by recovery up to 19 years (Pacaldo *et al.*, 2013b). Our site is only 7 years postconversion and therefore is likely still in the recovery phase with respect to soil carbon.

One limitation of this study is the lack of measured root biomass in the SRC willow system, which may have resulted in an underestimation of the SRC willow sink postharvest. However, the calculated values in this study are in line with empirical findings recently published by Cunniff *et al.* (2015); therefore, we are able to use these estimated with some confidence. This demonstrates one of the challenges of working in a commercial system where restrictions to experimental measurements are imposed by the commercial regime.

Capturing the effects of a commercial harvest on the soil and ecosystem GHG balance was important since harvesting is recognized as one of the most energy intensive stages of the SRC willow life cycle due to the large consumption of diesel fuel (Murphy *et al.*, 2014) and relatively little is known about the effects on the GHG balance in SRC willow (Vanbeveren *et al.*, 2015). From our study, we have shown that whilst there is an increase in R_{eco}, and subsequently NEE after the harvest, within 3 weeks of harvest, the site was returned to being a sink for carbon. The observed increase in NEE is comparable to that observed by Zenone *et al.* (2015) for the 2nd year postestablishment of an SRC poplar plantation; indicating the effect of disturbance on NEE. LCA findings have shown that whilst the harvest can increase emissions due to the harvest machinery, the carbon sink created by SRC willow is able to offset these emissions and result in a negative GHG balance (reported in the range of −138 to −53 kg CO_2-eq. per odt biomass; Caputo *et al.*, 2014). We also observed a one-off peak in N_2O emissions, 2 months postharvest, which was the largest emission, observed across both sites for the duration of the experiment. In contrast, other studies have observed little effect of harvest on N_2O emissions from SRC cultures (Zona *et al.*, 2013b). It is possible that this emission arose as a result of increased soil exposure after harvest and increased rainfall in May and June 2014, relative to 2013. It is also possible that there was some compaction due to the harvest machinery which can cause a reduction in soil porosity, in turn resulting in increased N_2O emissions. Soil N_2O fluxes are known to vary spatially and temporally and to arise quickly after changes in rainfall, temperature and management (Skiba & Smith, 2000). N_2O emissions, therefore, require more intense monitoring to be able to capture these emissions, since a single large emission can account for a large proportion of total N_2O fluxes over a measurement period (Zona *et al.*, 2013b).

In conclusion, we have shown that LUC to SRC willow from grassland can result in reduced GHG emissions. In the minimally managed site studied here, where harvested grass remained on the field, we found that grassland was a net carbon source and SRC willow a net carbon sink, 7 years after land conversion. However, soil carbon stocks were likely still in recovery as soil C at the SRC site remained significantly lower than grassland, even after this amount of time postestablishment. Whilst grasslands have been shown to be highly variable, there is evidence that this LUC may result in climate mitigation advantages and may be considered a viable bioenergy option for the future.

Acknowledgements

We would like to thank Andrew Ramsden and Richard Ramsden for allowing us to establish our experiment on their farm. We would also like to thank Mathew Tallis, Maud Viger, Caitriona Murray, Suzanne Milner, Billy Valdes, Alan Foy, Emily Clark (CEH) and members of Taylorlab for their technical assistance. This work was funded by Energy Technologies Institute (ETI), Carbo-BioCrop (www.carbobiocrop.ac.uk; a NERC funded project; NE/H010742/1), UKERC (funded as part of the flexible research fund of UKERC, NERC; NE/H013237/1), MAGLUE (www.maglue.ac.uk; an EPSRC funded project; EP/M013200/1) and as part of the Seventh Framework For Research Programme of the EU, within the EUROCHAR project (N 265179).

References

Allwright MR, Taylor G (2016) Molecular breeding for improved second generation bioenergy crops. *Trends in Plant Science*, **21**, 43–54.

Anderson-Teixeira KJ, Snyder PK, Twine TE, Cuadra SV, Costa MH, Delucia EH (2012) Climate-regulation services of natural and agricultural ecoregions of the Americas. *Nature Climate Change*, **2**, 177–181.

Arevalo CBM, Bhatti JS, Chang SX, Sidders D (2009) Ecosystem carbon stocks and distribution under different land-uses in north central Alberta, Canada. *Forest Ecology and Management*, **257**, 1776–1785.

Arevalo CBM, Bhatti JS, Chang SX, Sidders D (2011) Land use change effects on ecosystem carbon balance: from agricultural to hybrid poplar plantation. *Agriculture, Ecosystems & Environment*, **141**, 342–349.

Aubinet M, Grelle A, Ibrom A (2000) Estimates of the annual net carbon and water exchange of European forests: the EUROFLUX methodology. *Advances in Ecological Research*, **30**, 114–175.

Bates D, Maechler M, Bolker B, Walker S (2014) lme4: Linear mixed-effects models using Eigen and S4. R package version 1.1-7. Available at: http://CRAN.R-project.org/package=lme4 (accessed May 2015).

Bellamy PH, Loveland PJ, Bradley RI, Lark RM, Kirk GJD (2005) Carbon losses from all soils across England and Wales 1978–2003. *Nature*, **437**, 245–248.

Blanco-Canqui H, Lal R (2008) No-tillage and soil-profile carbon sequestration: an on-farm assessment. *Soil Science Society of America Journal*, **27**, 693–701.

Bridgeman TG, Jones JM, Shield I, Williams PT (2008) Torrefaction of reed canary grass, wheat straw and willow to enhance solid fuel qualities and combustion properties. *Fuel*, **87**, 844–856.

Broeckx LS, Verlinden MS, Berhongaray G, Zona D, Fichot R, Ceulemans R (2014) The effect of a dry spring on seasonal carbon allocation and vegetation dynamics in a poplar bioenergy plantation. *Global Change Biology Bioenergy*, **6**, 473–487.

Buchholz T, Prisley S, Marland G, Canham C, Sampson N (2014) Uncertainty in projecting GHG emissions from bioenergy. *Nature Climate Change*, **4**, 1045–1047.

Cannell MGR (2003) Carbon sequestration and biomass energy offset: theoretical, potential and achievable capacities globally, in Europe and the UK. *Biomass and Bioenergy*, **24**, 97–116.

Caputo J, Balogh S, Volk T, Johnson L, Puettmann M, Lippke B, Oneil E (2014) Incorporating uncertainty into a life cycle assessment (LCA) model of short-rotation willow biomass (*Salix* spp.) crops. *BioEnergy Research*, **7**, 48–59.

Case SDC, McNamara NP, Reay DS, Whitaker J (2014) Can biochar reduce soil greenhouse gas emissions from a Miscanthus bioenergy crop? *Global Change Biology Bioenergy*, **6**, 76–89.

Chapin FS III, Woodwell GM, Randerson JT *et al.* (2006) Reconciling carbon-cycle concepts, terminology, and methods. *Ecosystems*, **9**, 1041–1050.

Cherubini F, Bird ND, Cowie A, Jungmeier G, Schlamadinger B, Woess-Gallasch S (2009) Energy- and greenhouse gas-based LCA of biofuel and bioenergy systems: key issues, ranges and recommendations. *Resources, Conservation and Recycling*, **53**, 434–447.

Ciais P, Soussana JF, Vuichard N *et al.* (2010) The greenhouse gas balance of European grasslands. *Biogeosciences Discussions*, **7**, 5997–6050.

Crawley MJ (2007) *The R Book*. Wiley Publishing, England. ISBN 13: 978-0-470-51024-7.

Creutzig F, Popp A, Plevin R, Luderer G, Minx J, Edenhofer O (2012) Reconciling top-down and bottom-up modelling on future bioenergy deployment. *Nature Climate Change*, **2**, 320–327.

Cunniff J, Purdy SJ, Barraclough TJP *et al.* (2015) High yielding biomass genotypes of willow (*Salix* spp.) show differences in below ground biomass allocation. *Biomass and Bioenergy*, **80**, 114–127.

Dauber J, Jones MB, Stout JC (2010) The impact of biomass crop cultivation on temperate biodiversity. *Global Change Biology Bioenergy*, **2**, 289–309.

DECC (2014) *Life Cycle Impacts of Biomass Electricity in 2020*. Department of Energy & Climate Change, London; 24 July 2014. 154p URN 14D/243.

DEFRA (2014) *Area of Crops Grown for Bioenergy in England and the UK: 2008 – 2013*. Department for Environment, Food and Rural Affairs, York, UK.

DEFRA, SEERAD, DARD, DEPC (2007) *Agriculture in the United Kingdom 2006*. The Stationery Office, London.

Dimitriou I, Mola-Yudego B, Aronsson P, Eriksson J (2012) Changes in organic carbon and trace elements in the soil of willow short-rotation coppice plantations. *BioEnergy Research*, **5**, 563–572.

Dolan MS, Clapp CE, Allmaras RR, Baker JM, Molina JAE (2006) Soil organic carbon and nitrogen in a Minnesota soil as related to tillage, residue and nitrogen management. *Soil and Tillage Research*, **89**, 221–231.

Don A, Osborne B, Hastings A *et al.* (2012) Land-use change to bioenergy production in Europe: implications for the greenhouse gas balance and soil carbon. *Global Change Biology Bioenergy*, **4**, 372–391.

Dornburg V, Van Vuuren D, Van De Ven G *et al.* (2010) Bioenergy revisited: key factors in global potentials of bioenergy. *Energy & Environmental Science*, **3**, 258–267.

Drewer J, Finch JW, Lloyd CR, Baggs EM, Skiba U (2012) How do soil emissions of N_2O, CH_4 and CO_2 from perennial bioenergy crops differ from arable annual crops? *Global Change Biology Bioenergy*, **4**, 408–419.

Fahmi R, Bridgwater AV, Darvell LI, Jones JM, Yates N, Thain S, Donnison IS (2007) The effect of alkali metals on combustion and pyrolysis of Lolium and Festuca grasses, switchgrass and willow. *Fuel*, **86**, 1560–1569.

Fargione J, Hill J, Tilman D, Polasky S, Hawthorne P (2008) Land clearing and the biofuel carbon debt. *Science*, **319**, 1235–1238.

Finkbeiner M (2014) Indirect land use change – help beyond the hype? *Biomass and Bioenergy*, **62**, 218–221.

Forestry Commission (2002) *Practice Note: Establishment and Management of Short Rotation Coppice. FCPN7 (REVISED)*. Forestry Commission, Edinburgh.

Gauder M, Butterbach-Bahl K, Graeff-Honninger S, Claupein W, Wiegel R (2012) Soil-derived trace gas fluxes from different energy crops – results from a field experiment in Southwest Germany. *Global Change Biology Bioenergy*, **4**, 289–301.

Grigal DF, Berguson WE (1998) Soil carbon changes associated with short-rotation systems. *Biomass and Bioenergy*, **14**, 371–377.

Gudka BA (2012) Combustion Characteristics of some Imported Feedstocks and Short Rotation Coppice (SRC) Willow for UK Power Stations. Unpublished, Doctor of Philosophy, The University of Leeds.

Guo LB, Gifford RM (2002) Soil carbon stocks and land use change: a meta analysis. *Global Change Biology*, **8**, 345–360.

Hanson PJ, Edwards NT, Garten CT, Andrews JA (2000) Separating root and soil microbial contributions to soil respiration: a review of methods and observations. *Biogeochemistry*, **48**, 115–146.

Harris ZM, Spake R, Taylor G (2015) Land use change to bioenergy: a meta-analysis of soil carbon and GHG emissions. *Biomass and Bioenergy*, **82**, 27–39.

Harrison RB, Footen PW, Strahm BD (2011) Deep soil horizons: contribution and importance to soil carbon pools and in assessing whole-ecosystem response to management and global change. *Forest Science*, **57**, 67–76.

Hastings A, Tallis MJ, Casella E *et al.* (2014) The technical potential of Great Britain to produce ligno-cellulosic biomass for bioenergy in current and future climates. *Global Change Biology Bioenergy*, **6**, 108–122.

Hillier J, Whittaker C, Dailey G *et al.* (2009) Greenhouse gas emissions from four bioenergy crops in England and Wales: integrating spatial estimates of yield and soil carbon balance in life cycle analyses. *GCB Bioenergy*, **1**, 267–281.

Holland EA, Robertson GP, Greenberg J, Groffman PM, Boone RD, Gosz JR (1999) Soil CO_2, N_2O and CH_4 exchange. In: *Standard Soil Methods for Long-Term Ecological Research* (eds Robertson GP, Coleman DC, Bledsoe CS, Sollins P), pp. 185–201. Oxford University press, Oxford, UK.

Holland RA, Eigenbrod F, Muggeridge A, Brown G, Clarke D, Taylor G (2015) A synthesis of the ecosystem services impact of second generation bioenergy crop production. *Renewable and Sustainable Energy Reviews*, **46**, 30–40.

Imer D, Merbold L, Eugster W, Buchmann N (2013) Temporal and spatial variations of soil CO_2, CH_4 and N_2O fluxes at three differently managed grasslands. *Biogeosciences*, **10**, 5931–5945.

IPCC (2007a) Fourth Assessment Report, Working Group III Report "Mitigation of Climate Change". Metz B, Davidson OR, Bosch PR, Dave R, Meyer LA (eds), Cambridge University Press, Cambridge, United Kingdom and New York, NY, USA.

IPCC (2007b) Climate Change 2007: Working Group I: The Physical Science Basis. Solomon S, Qin D, Manning M *et al.* (eds), Cambridge University Press, Cambridge, United Kingdom and New York, NY, USA.

Janssens IA, Freibauer A, Ciais P *et al.* (2003) Europe's terrestrial biosphere absorbs 7 to 12% of European anthropogenic CO_2 emissions. *Science*, **300**, 1538–1542.

Jassal RS, Black TA, Arevalo C, Jones H, Bhatti JS, Sidders D (2013) Carbon sequestration and water use of a young hybrid poplar plantation in north-central Alberta. *Biomass and Bioenergy*, **56**, 323–333.

Kammann C, Grünhage L, Jäger HJ, Wachinger G (2001) Methane fluxes from differentially managed grassland study plots: the important role of CH_4 oxidation in grassland with a high potential for CH_4 production. *Environmental Pollution*, **115**, 261–273.

Keith AM, Rowe RL, Parmar K, Perks MP, Mackie E, Dondini M, Mcnamara NP (2015) Implications of land-use change to Short Rotation Forestry in Great Britain for soil and biomass carbon. *Global Change Biology Bioenergy*, **7**, 541–552.

Kern J, Hellebrand H, Gömmel M, Ammon C, Berg W (2012) Effects of climatic factors and soil management on the methane flux in soils from annual and perennial energy crops. *Biology and Fertility of Soils*, **48**, 1–8.

Kort J, Collins M, Ditsch D (1998) A review of soil erosion potential associated with biomass crops. *Biomass and Bioenergy*, **14**, 351–359.

Laurent A, Pelzer E, Loyce C, Makowski D (2015) Ranking yields of energy crops: a meta-analysis using direct and indirect comparisons. *Renewable and Sustainable Energy Reviews*, **46**, 41–50.

Law BE, Falge E, Gu L *et al.* (2002) Environmental controls over carbon dioxide and water vapor exchange of terrestrial vegetation. *Agricultural and Forest Meteorology*, **113**, 97–120.

Lovett A, Sünnenberg G, Dockerty T (2014) The availability of land for perennial energy crops in Great Britain. *Global Change Biology Bioenergy*, **6**, 99–107.

Manning P, Taylor G, Hanley M (2015) Bioenergy, food production and biodiversity – an unlikely alliance? *Global Change Biology Bioenergy*, **7**, 570–576.

Mathews R, Sokka L, Soimakallio S *et al.* (2014) Review of literature on biogenic carbon and life cycle assessment of forest bioenergy. Final Task 1 report, DG ENER project, 'Carbon impacts of biomass consumed in the EU'. 300 pp.

Melillo JM, Reilly JM, Kicklighter DW *et al.* (2009) Indirect emissions from biofuels: how important? *Science*, **326**, 1397–1399.

Merbold L, Eugster W, Stieger J, Zahniser M, Nelson D, Buchmann N (2014) Greenhouse gas budget (CO_2, CH_4 and N_2O) of intensively managed grassland following restoration. *Global Change Biology*, **20**, 1913–1928.

Met Office (2015) Climate Summaries – Regional Values. Available at: http://www.metoffice.gov.uk/climate/uk/summaries/datasets (accessed 5 June 2015).

Milner S, Holland RA, Lovett A *et al.* (2015) Potential impacts on ecosystem services of land use transitions to second generation bioenergy crops in GB. *Global Change Biology Bioenergy*, **8**, 317–333.

Murphy F, Devlin G, Mcdonnell K (2014) Energy requirements and environmental impacts associated with the production of short rotation willow (*Salix* sp.) chip in Ireland. *Global Change Biology Bioenergy*, **6**, 727–739.

Nikiema P, Rothstein DE, Miller RO (2012) Initial greenhouse gas emissions and nitrogen leaching losses associated with converting pastureland to short-rotation woody bioenergy crops in northern Michigan, USA. *Biomass and Bioenergy*, **39**, 413–426.

Ostle NJ, Levy PE, Evans CD *et al.* (2009) UK land use and soil carbon sequestration. *Land Use Policy*, **26S**, S274–S283.

Pacaldo R, Volk T, Briggs R (2013a) Greenhouse gas potentials of shrub willow biomass crops based on below-and aboveground biomass inventory along a 19-year chronosequence. *BioEnergy Research*, **6**, 252–262.

Pacaldo RS, Volk TA, Briggs RD (2013b) No significant differences in soil organic carbon contents along a chronosequence of shrub willow biomass crop fields. *Biomass and Bioenergy*, **58**, 136–142.

Palmer MM, Forrester JA, Rothstein DE, Mladenoff DJ (2014) Establishment phase greenhouse gas emissions in short rotation woody biomass plantations in the Northern Lake States, USA. *Biomass and Bioenergy*, **62**, 26–36.

Pitt MD, Heady HF (1978) Responses of annual vegetation to temperature and rainfall patterns in northern California. *Ecology*, **59**, 336–350.

Pontallier JY, Ceulemans R, Guittet J, Mau F (1997) Linear and non-linear functions of volume index to estimate woody biomass in high density young poplar stands. *Annals of Forest Science*, **54**, 335–345.

Post WM, Kwon KC (2000) Soil carbon sequestration and land-use change: processes and potential. *Global Change Biology*, **6**, 317–327.

Qin Z, Dunn JB, Kwon H, Mueller S, Wander MM (2015) Soil carbon sequestration and land use change associated with biofuel production: empirical evidence. *Global Change Biology Bioenergy*, **8**, 66–80.

Qun D, Huizhi L (2013) Seven years of carbon dioxide exchange over a degraded grassland and a cropland with maize ecosystems in a semiarid area of China. *Agriculture, Ecosystems & Environment*, **173**, 1–12.

R Core Team (2015) *R: A Language and Environment for Statistical Computing*. R Foundation for Statistical Computing, Vienna, Austria. Available at: http://www.R-project.org/.

Rae AM, Robinson KM, Street NR, Taylor G (2004) Morphological and physiological traits influencing biomass productivity in short-rotation coppice poplar. *Canadian Journal of Forest Research*, **34**, 1488–1498.

Raich JW, Schlesinger WH (1992) The global carbon dioxide flux in soil respiration and its relationship to vegetation and climate. *Tellus*, **44B**, 81–99.

Raich JW, Potter CS, Bhagawati D (2002) Interannual variability in global soil respiration, 1980–94. *Global Change Biology*, **8**, 800–812.

Reichstein M, Falge E, Baldocchi D *et al.* (2005) On the separation of net ecosystem exchange into assimilation and ecosystem respiration: review and improved algorithm. *Global Change Biology*, **11**, 1424–1439.

Reichstein M, Stoy PC, Desai AR, Richardson AD (2012) Partitioning of net fluxes. In: *Eddy Covaiance: A Practical Guide to Measurement and Data Analysis* (eds Aubinet M, Vesala T, Papale D). Springer Atmospheric Sciences, Dordrecht, Netherlands.

Robson MJ, Ryle GJA, Woledge J (1988) The grass plant – its form and function. In: *The Grass Crop* (eds Jones M, Lazenby A), pp. 25–84. Springer, Dordrecht, Netherlands.

Rowe RL, Street NR, Taylor G (2009) Identifying potential environmental impacts of large-scale deployment of dedicated bioenergy crops in the UK. *Renewable and Sustainable Energy Reviews*, **13**, 271–290.

Rowe RL, Keith AM, Elias D, Dondini M, Smith P, Oxley J, McNamara NP (2016) Initial soil C and land use history determine soil C sequestration under perennial bioenergy crops. *Global Change Biology Bioenergy*, doi: 10.1111/gcbb.12311.

Ruimy A, Jarvis PG, Baldocchi DD, Saugier B (1995) CO_2 fluxes over plant canopies and solar radiation. A review. *Advances in Ecological Research*, **26**, 1–68.

Rutledge S, Mudge PL, Campbell DI *et al.* (2015) Carbon balance of an intensively grazed temperate dairy pasture over four years. *Agriculture, Ecosystems & Environment*, **206**, 10–20.

Sabbatini S, Arriga N, Bertolini T *et al.* (2015) Greenhouse gas balance of cropland conversion to bioenergy poplar short rotation coppice. *Biogeosciences Discussions*, **12**, 8035–8084.

Sala OE, Chapin FS III, Armesto JJ *et al.* (2000) Global biodiversity scenarios for the year 2100. *Science*, **287**, 1770–1774.

Sannigrahi P, Ragauskas AJ, Tuskan GA (2010) Poplar as a feedstock for biofuels: a review of compositional characteristics. *Biofuels, Bioproducts and Biorefining*, **4**, 209–226.

Schipper LA, Parfitt RL, Fraser S, Littler RA, Baisden WT, Ross C (2014) Soil order and grazing management effects on changes in soil C and N in New Zealand pastures. *Agriculture, Ecosystems & Environment*, **184**, 67–75.

Scurlock JMO, Hall DO (1998) The global carbon sink: a grassland perspective. *Global Change Biology*, **4**, 229–233.

Searchinger T, Heimlich R, Houghton RA *et al.* (2008) Use of U.S. croplands for biofuels increases greenhouse gases through emissions from land-use change. *Science*, **319**, 1238–1240.

Shao J, Zhou X, Luo Y *et al.* (2015) Biotic and climatic controls on interannual variability in carbon fluxes across terrestrial ecosystems. *Agricultural and Forest Meteorology*, **205**, 11–22.

Skiba U, Smith KA (2000) The control of nitrous oxide emissions from agricultural and natural soils. *Chemosphere – Global Change Science*, **2**, 379–386.

Smit HJ, Metzger MJ, Ewert F (2008) Spatial distribution of grassland productivity and land use in Europe. *Agricultural Systems*, **98**, 208–219.

Smith P (2014) Do grasslands act as a perpetual sink for carbon? *Global Change Biology*, **20**, 2708–2711.

Smith KA, Mullins CE (eds) (2000) *Soil and Environmental Analysis: Physical Methods* (2nd edn). CRC Press, New York.

Soussana JF, Allard V, Pilegaard K *et al.* (2007) Full accounting of the greenhouse gas (CO_2, N_2O, CH_4) budget of nine European grassland sites. *Agriculture, Ecosystems & Environment*, **121**, 121–134.

Soussana JF, Tallec T, Blanfort V (2010) Mitigating the greenhouse gas balance of ruminant production systems through carbon sequestration in grasslands. *Animal*, **4**, 334–350.

Souza GM, Victoria R, Joly C, Verdade L (eds) (2015). *Bioenergy & Sustainability: Bridging the Gaps*, Vol **72**. SCOPE, Paris. ISBN 978-2-9545557-0-6.

St. Clair S, Hillier J, Smith P (2008) Estimating the pre-harvest greenhouse gas costs of energy crop production. *Biomass and Bioenergy*, **32**, 442–452.

Stoof C, Richards B, Woodbury P *et al.* (2015) Untapped potential: opportunities and challenges for sustainable bioenergy production from marginal lands in the Northeast USA. *BioEnergy Research*, **8**, 482–501.

Styles D, Jones MB (2007) Energy crops in Ireland: quantifying the potential life-cycle greenhouse gas reductions of energy-crop electricity. *Biomass and Bioenergy*, **31**, 759–772.

Ter-Mikaelian MT, Colombo SJ, Lovekin D *et al.* (2015) Carbon debt repayment or carbon sequestration parity? Lessons from a forest bioenergy case study in Ontario, Canada. *GCB Bioenergy*, **7**, 704–716.

Vanbeveren SPP, Schweier J, Berhongaray G, Ceulemans R (2015) Operational short rotation woody crop plantations: manual or mechanised harvesting? *Biomass and Bioenergy*, **72**, 8–18.

Ventura M, Alberti G, Viger M *et al.* (2015) Biochar mineralization and priming effect on SOM decomposition in two European short rotation coppices. *Global Change Biology Bioenergy*, doi: 10.1111/gcbb.12219.

Verlinden MS, Broeckx LS, Zona D *et al.* (2013) Net ecosystem production and carbon balance of an SRC poplar plantation during its first rotation. *Biomass and Bioenergy*, **56**, 412–422.

Walter K, Don A, Flessa H (2015) No general soil carbon sequestration under Central European short rotation coppices. *Global Change Biology Bioenergy*, doi: 10.1111/gcbb.12177.

Zan CS, Fyles JW, Girouard P, Samson RA (2001) Carbon sequestration in perennial bioenergy, annual corn and uncultivated systems in southern Quebec. *Agriculture, Ecosystems & Environment*, **86**, 135–144.

Zatta A, Clifton-Brown J, Robson P, Hastings A, Monti A (2014) Land use change from C3 grassland to C4 *Miscanthus*: effects on soil carbon content and estimated mitigation benefit after six years. *Global Change Biology Bioenergy*, **6**, 360–370.

Zenone T, Fischer M, Arriga N *et al.* (2015) Biophysical drivers of the carbon dioxide, water vapor, and energy exchanges of a short-rotation poplar coppice. *Agricultural and Forest Meteorology*, **209–210**, 22–35.

Zona DJ, Janssens IA, Aubinet M, Gioli B, Vicca S, Fichot R, Ceulemans R (2013a) Fluxes of the greenhouse gases (CO_2, CH_4 and N_2O) above a short-rotation poplar plantation after conversion from agricultural land. *Agricultural and Forest Meteorology*, **169**, 100–110.

Zona D, Janssens IA, Gioli B, Jungkunst HF, Serrano MC, Ceulemans R (2013b) N_2O fluxes of a bio-energy poplar plantation during a two years rotation period. *Global Change Biology Bioenergy*, **5**, 536–547.

4

Full carbon and greenhouse gas balances of fertilized and nonfertilized reed canary grass cultivations on an abandoned peat extraction area in a dry year

JÄRVI JÄRVEOJA[1], MATTHIAS PEICHL[2], MARTIN MADDISON[1], ALAR TEEMUSK[1] and ÜLO MANDER[1,3]

[1]*Department of Geography, Institute of Ecology and Earth Sciences, University of Tartu, 46 Vanemuise St, Tartu 51014, Estonia,* [2]*Department of Forest Ecology and Management, Swedish University of Agricultural Sciences Skogsmarksgränd 1, 90183 Umeå, Sweden,* [3]*Hydrosystems and Bioprocesses Research Unit, National Research Institute of Science and Technology for Environment and Agriculture (Irstea), 1 rue Pierre-Gilles de Gennes CS 10030, F92761 Antony Cedex, France*

Abstract

Bioenergy crop cultivation on former peat extraction areas is a potential after-use option that provides a source of renewable energy while mitigating climate change through enhanced carbon (C) sequestration. This study investigated the full C and greenhouse gas (GHG) balances of fertilized (RCG-F) and nonfertilized (RCG-C) reed canary grass (RCG; *Phalaris arundinacea*) cultivation compared to bare peat (BP) soil within an abandoned peat extraction area in western Estonia during a dry year. Vegetation sampling, static chamber and lysimeter measurements were carried out to estimate above- and belowground biomass production and allocation, fluxes of carbon dioxide (CO_2), methane (CH_4) and nitrous oxide (N_2O) in cultivated strips and drainage ditches as well as the dissolved organic carbon (DOC) export, respectively. Heterotrophic respiration was determined from vegetation-free trenched plots. Fertilization increased the above- to belowground biomass production ratio and the autotrophic to heterotrophic respiration ratio. The full C balance (incl. CO_2, CH_4 and DOC fluxes from strips and ditches) was 96, 215 and 180 g C m^{-2} yr^{-1} in RCG-F, RCG-C and BP, respectively, suggesting that all treatments acted as C sources during the dry year. The C balance was driven by variations in the net CO_2 exchange, whereas the combined contribution of CH_4 and DOC fluxes was <5%. The GHG balances were 3.6, 7.9 and 6.6 t CO_2 eq ha^{-1} yr^{-1} in RCG-F, RCG-C and BP, respectively. The CO_2 exchange was also the dominant component of the GHG balance, while the contributions of CH_4 and N_2O were <1% and 1–6%, respectively. Overall, this study suggests that maximizing plant growth and the associated CO_2 uptake through adequate water and nutrient supply is a key prerequisite for ensuring sustainable high yields and climate benefits in RCG cultivations established on organic soils following drainage and peat extraction.

Keywords: bioenergy, biomass production, carbon dioxide, carbon sequestration, dissolved organic carbon, land management, methane, nitrous oxide, organic soils, *Phalaris arundinacea*

Introduction

Commercial peat extraction for energy production and horticultural use is an important industry in many countries within northern Europe (e.g., Finland, Sweden, Ireland, Estonia, Belarus) and other parts of the world (i.e., USA, Canada, Russia, Indonesia) (Waddington *et al.*, 2002; Lemus & Lal, 2005; Couwenberg *et al.*, 2011; Don *et al.*, 2012). Peat extraction has taken place on more than 50 000 km^2 of northern peatland areas which accounts for 10% of the total loss of natural peatlands due to human use (Joosten & Clarke, 2002). Peat extraction can commonly be sustained over several decades depending on the depth of the peat deposit but eventually ceases, resulting in large abandoned areas of drained and degraded peat soils. Furthermore, due to the expansion of peat extraction activities into pristine peatlands and the vast areas of current peat extraction sites being annually abandoned, the extent of these degraded areas will likely increase in the future, resulting in a growing demand for developing appropriate after-use strategies (Tuittila *et al.*, 2000; Maljanen *et al.*, 2010). In addition, the large carbon dioxide (CO_2) emission from these drained organic soils is a major concern from the climate change perspective (Gorham, 1991; Waddington *et al.*, 2002; Mäkiranta *et al.*, 2007; Salm *et al.*, 2012), which further amplifies the need for adequate management strategies.

Correspondence: Järvi Järveoja
e-mail: Jarvi.Jarveoja@ut.ee

Among different after-use options, cultivating dedicated high-yielding energy crops (e.g., RCG, *Phalaris arundinacea*) on abandoned peat soils has been suggested as a promising strategy to increase the proportion of renewable energy supply while creating a sink for atmospheric CO_2 without competing for productive agricultural land required for food crop production (Lewandowski *et al.*, 2003; Lemus & Lal, 2005; Shurpali *et al.*, 2010; Don *et al.*, 2012). Furthermore, cultivation of perennial crops has the advantage of reducing the need for regular tillage commonly associated with annual crop cultivation, which could reduce the associated C losses and therefore increase the soil C sequestration potential (Adler *et al.*, 2007; Jones *et al.*, 2015). Although RCG grows well on most kinds of soils, highest biomass is commonly reached on wet and humus-rich soils (e.g., in abandoned peat extraction areas) (Don *et al.*, 2012). In addition, RCG is adapted to short growing seasons and low temperatures. Thus, the potential of RCG cultivation is especially high in Northern Europe given the climatic conditions and vast occurrence of organic soils.

In most bioenergy cropping systems, fertilizer is applied to maximize biomass production (Maljanen *et al.*, 2010; Don *et al.*, 2012). Increased plant growth and nutrient supply may, however, also alter the above- and belowground allocation of plant biomass and trigger structural changes in the belowground biomass due to its contrasting effects on root and rhizome production (Kätterer & Andrén, 1999; Xiong & Kätterer, 2010; Kinmonth-Schultz & Kim, 2011; Jones *et al.*, 2015). Furthermore, enhanced plant growth and changes in biomass allocation following fertilization may also modify the partitioning of the ecosystem CO_2 exchange into its component fluxes of gross and net primary production as well as soil heterotrophic and plant autotrophic respiration (Shurpali *et al.*, 2008; Kandel *et al.*, 2013). To date, detailed information of fertilization effects on biomass allocation and CO_2 balance partitioning is limited for perennial cropping systems; however, this knowledge is crucial to confidently predict variations in annual yields and the C cycle of bioenergy cropping systems on abandoned organic soils under future climate change and management scenarios (Strand *et al.*, 2008; Gong *et al.*, 2014; Jones *et al.*, 2015).

Within the climate change context, it is imperative to understand fertilization effects on not only biomass production but also on the full ecosystem C balance which is determined by the net CO_2 exchange of vegetation and soil as well as by non-CO_2 carbon fluxes such as the exchange of methane (CH_4) and the aquatic export of dissolved organic carbon (DOC). Furthermore, due to the relatively small area coverage by drainage ditches, most studies focus on the C balance of the cultivated strips ignoring potentially high emissions from ditches (Sundh *et al.*, 2000; Schrier-Uijl *et al.*, 2010; Hyvönen *et al.*, 2013). Although individual studies have previously investigated the patterns and magnitudes of the different C balance components (Sundh *et al.*, 2000; Shurpali *et al.*, 2009; Strack *et al.*, 2011; Hyvönen *et al.*, 2013), a comprehensive assessment of all major C fluxes from both strips and ditches and their contrasting contributions to the full annual C balance in fertilized and nonfertilized RCG cultivations on organic soil is currently lacking.

A major drawback of nitrogen (N) fertilizer application is that it may cause high emissions of the greenhouse gas (GHG) nitrous oxide (N_2O) (Crutzen *et al.*, 2008). Considering that N_2O has a much larger (by 298 times) global warming potential compared to CO_2 (IPCC, 2013), the positive fertilizer effects on plant growth and C sequestration could be partly or entirely offset when considering the GHG balance of bioenergy cultivations (Don *et al.*, 2012). A recently proposed strategy for reducing N_2O emissions is to substitute annual crops with perennial crops, such as RCG, which have a lower N-demand and higher N-use efficiency and hence emit 40–99% less N_2O compared to conventional annual crops (Don *et al.*, 2012). To date, the number of studies investigating the trade-off between the increased CO_2 uptake due to stimulated plant growth and the enhanced N_2O emission following fertilization is, however, limited and its implication for the GHG balance of bioenergy cultivations therefore highly uncertain.

To address these knowledge gaps, this study investigated the full C and GHG balance in fertilized and nonfertilized RCG cultivations compared to bare peat soil in an abandoned peat extraction area in western Estonia. The study addressed the following specific questions:

1. How does fertilization affect the above- and belowground biomass production and allocation
2. How does the net ecosystem CO_2 exchange (NEE) and its partitioning into production and respiration component fluxes differ among the fertilized RCG, nonfertilized RCG and bare peat treatments
3. What are the full C and GHG balances and the contributions of their respective component fluxes from cultivated strips and drainage ditches in the fertilized RCG, nonfertilized RCG and bare peat treatments

Material and methods

Study site

The study site (58°34′20″N, 24°23′15″E; Halinga Parish, Pärnu County) is located within the largest Estonian peat extraction area, Lavassaare, which is situated in the northern part of the

Pärnu Lowland. The region has a temperate climate with a 30-year (1981–2010) mean annual temperature of 6.3 °C and annual precipitation of 746 mm (Estonian Weather Service). The peat extraction area is divided into 20 m wide strips separated by 1 m wide drainage ditches. Commercial peat extraction at the site started in the 1960s and lasted until 2006. In 2007, the abandoned strips were tilled and sown with seeds of the Estonian-bred RCG variety 'Pedja'. Since the pH_{KCL} of the remaining peat was above 5.0 (Table 1), no liming was carried out. The remaining peat deposit is approximately 0.45–0.60 m deep and consists of well-mineralized *Phragmites–Carex* peat with a degree of humification of H7 according to the von Post scale. The main soil properties are summarized in Table 1. No fertilization or biomass harvest was carried out between the seeding in 2007 and spring 2012.

Experimental design

The study was designed as a replicated field experiment with six experimental plots (2.5 × 10 m) of which four were located within the cultivated strips and two within the abandoned bare peat strips. The cultivated plots consisted of two fertilized plots and two adjacent nonfertilized control plots. The fertilized plots received 72 kg N, 18 kg P and 36 kg K of mineral fertilizer per hectare once per year during the early growing season since 2012. Thus, the experiment consisted of two replicate plots for each of the three treatments: reed canary grass cultivation with fertilization (RCG-F), reed canary grass cultivation control (RCG-C) and bare peat (BP).

Above- and belowground biomass and net primary production

Above- and belowground biomass stocks were measured in the cultivated RCG-F and RCG-C plots. Aboveground biomass was harvested on five subplots (50 × 50 cm) within each plot in September 2014. Belowground biomass (roots and rhizomes) was determined in April and September from five soil cores per plot taken to a depth of 30 cm using an 8.5 cm diameter corer. Each soil core was divided into three 10 cm sections (0–10, 10–20 and 20–30 cm). After manually washing each core over a 0.5 mm mesh sieve to remove the bulk soil, roots and rhizomes were manually picked with tweezers from the residual soil. Both above- and belowground biomass was oven-dried at 70 °C to a constant weight and analyzed for C concentrations at the Tartu Laboratory of the Estonian Environmental Research Centre.

Table 1 Mean values of the soil chemical and physical properties in 0–20 cm depth for reed canary grass with fertilization (RCG-F), reed canary grass control (RCG-C) and bare peat (BP) treatments; numbers in parentheses indicate standard error

Soil property*	RCG-F	RCG-C	BP
pH	5.15 (0.02)	5.11 (0.04)	5.47 (0.17)
Bulk density (g cm^{-3})	0.17 (0.01)	0.18 (0.01)	0.19 (0.01)
C (%)	46.5 (0.7)	45.5 (0.6)	44.3 (0.9)
N (%)	2.8 (0.1)	2.9 (0.1)	2.4 (0.1)
C : N	16.6	16.0	18.6
Soil C stock (kg C m^{-2})	15.6	16.6	16.4
Total P (mg g^{-1})	0.32 (0.01)	0.32 (0.01)	0.25 (0.01)
K (mg g^{-1})	0.26 (0.01)	0.12 (0.02)	0.09 (0.02)
Ca (mg g^{-1})	0.69 (0.03)	1.03 (0.03)	0.29 (0.02)
NH_4-N (mg L^{-1})	<1.5	<1.5	<1.5
NO_3-N (mg L^{-1})	85.0 (24.4)	33.0 (3.7)	49.5 (4.8)

*All variables were measured in 2014, except for C % and N % which were measured in 2012.

Annual aboveground net primary production (ANPP) was calculated by multiplying the harvested biomass with its C concentration. The annual belowground net primary production (BNPP) was estimated with the maximum–minimum method (McClaugherty *et al.*, 1982) as the difference between the maximum (September sampling) and minimum (April sampling) belowground biomass stocks. The sum of ANPP and BNPP resulted in the annual total net primary production (NPP_B) estimate based on biomass sampling (Eqn 1).

$$NPP_B = ANPP + BNPP \quad (1)$$

Vegetation greenness index

To estimate plant development, we derived a vegetation greenness index from digital images using repeat photography (Sonnentag *et al.*, 2012; Peichl *et al.*, 2015). A Wingscapes TimelapseCam 8.0 camera (model WSCT01; Wingscapes, Calera, AL, USA) was installed on a vertical pole at 3 m height above the ground surface viewing the experimental plots with a downward looking viewing angle of 15° from a southerly direction. The white balance was set to 'sunlight'. The camera was programmed to take images at half-hourly intervals from mid-April to late October 2014. The vegetation greenness index was derived from the green chromatic coordinate (g_{cc}) (Eqn 2):

$$g_{cc} = \frac{G}{R + G + B} \quad (2)$$

where R, G and B are the digital numbers (0–255) of the red, green and blue image channels. The RGB digital numbers were calculated for each pixel and averaged over selected regions of interest (ROI) representing fertilized and nonfertilized RCG cultivations as well as bare peat areas, respectively. Three-day mean g_{cc} time series were then created by assigning the 90th percentile to all values within a 3-days window to the center day of a discrete (nonoverlapping) moving window following Sonnentag *et al.* (2012). To estimate the collar-specific g_{cc}, images were also taken from above each collar on July 19 and analyzed for their g_{cc}. The time series of the mean g_{cc} in RCG-F and RCG-C were then multiplied with the relative difference in g_{cc} among collars to obtain collar-specific g_{cc} estimates for the growing season as input for the regression models described below.

Meteorological and soil environmental measurements

During every sampling event, soil volumetric water content (VWC) was measured on each plot in two depths (0–5 and

15–20 cm) using a handheld Decagon GS3 soil moisture sensor (Decagon Devices Inc., Pullman, WA, USA). In addition, soil temperature (T_s) was recorded on each plot in four different depths (10, 20, 30 and 40 cm) by a handheld Comet S0141 temperature logger with Pt1000TG8 sensors (Comet Systems Ltd., Rožnov pod Radhoštěm, Czech Republic). At each plot, continuous 30 min records of water table level (WTL) position relative to the soil surface were obtained with submerged HOBO Water Level Loggers (model U20-001-01; Onset Computer Corporation, Bourne, MA, USA). These automatic WTL records were calibrated with manual WTL measurements taken at the same locations.

A meteorological station was installed on-site in June to continuously measure air temperature (T_a) using a shielded temperature sensor (model CS 107; Campbell Scientific Inc., Logan, UT, USA) as well as photosynthetically active radiation (PAR) and precipitation (PPT) using a LI-190SL Quantum Sensor (LI-COR Inc., Lincoln, NE, USA) and a Young 52202 tipping bucket (R. M. Young Company, Traverse City, MI, USA), respectively. All meteorological instruments were mounted on a pole at 1.2 m height above the ground. Soil temperature, T_s (model CS 107; Campbell Scientific Inc.), was recorded at 5 and 30 cm depths. The soil volumetric water content, VWC (model CS615; Campbell Scientific Inc.), was measured at 10 cm depth. One sensor for each of the T_s and VWC measurements was placed in one of the RCG-F, RCG-C and BP plots. All automated meteorological and soil environmental data were collected in 1 min intervals and stored as 10 min averages on a CR1000 datalogger (Campbell Scientific Inc.). On-site meteorological measurements were complemented by hourly data of T_a, global radiation and PPT from the nearby (~20 km away) Pärnu meteorological station (Estonian Weather Service) for 2014. Global radiation was converted to PAR based on its linear regression relationship to on-site PAR during periods when both measurements were available.

Net ecosystem CO_2 exchange, respiration and production measurements

Net ecosystem CO_2 exchange (NEE) and ecosystem respiration (RE) on the RCG-F and RCG-C plots were measured using the closed dynamic chamber method at a weekly to biweekly interval from May to December 2014. Three PVC collars (Ø 50 cm) with a water-filled ring for airtight sealing were permanently installed into the soil to a depth of 10 cm within each plot. For NEE measurements, a transparent (95% transparency) Plexiglas chamber (h 50 cm, V 65 L) was used. The chamber was equipped with a TRP-2 probe (PP Systems, Hitchin, UK) which measured PAR and T_a inside the chamber. An additional sensor was placed on the chamber outside to record the ambient T_a. The chamber was climate-controlled via internal and external metal thermoelectric cooling ribs powered by a rechargeable 12-V battery. In addition, frozen cooling packs were fixed inside the chamber to further limit the T_a increase inside the chamber during measurements. The headspace air was continuously mixed by a low-speed fan. Immediately after each NEE measurement, RE measurements were conducted on the same collar with the chamber covered by an opaque, light reflecting shroud that blocked 100% of the incoming PAR.

During each chamber deployment period (3 and 4 min for RE and NEE measurements, respectively), CO_2 concentration, PAR, T_a, pressure and relative humidity inside the chamber were monitored with a portable infrared gas analyzer (IRGA, EGM-4; PP Systems) connected to the chamber in a closed loop through 1.5 m inlet and outlet tubing (inner Ø 0.3 cm, flow rate 350 ml min^{-1}). Measurements were conducted in a random plot order between 10:00 and 14:00 to avoid diurnal effects on the fluxes. Gross primary production (GPP) was derived from the difference between NEE and RE (Eqn 3):

$$\text{GPP} = \text{NEE} - \text{RE} \quad (3)$$

In addition, hourly net primary production rates based on the CO_2 flux measurements (NPP_F) were derived for each sampling date from the difference between NEE and heterotrophic respiration (Rh; see below) (Eqn 4):

$$\text{NPP}_F = \text{NEE} - \text{Rh} \quad (4)$$

Due to the absence of vegetation, NEE and RE at the BP plots were represented by the static chamber measurements described below.

Heterotrophic and autotrophic respiration measurements

Heterotrophic respiration (Rh) was measured concurrently with NEE and RE fluxes on trenched plots (A 0.07 m^2) which were established in the RCG-F and RCG-C plots in late March 2014. Within each plot, three subplots were trenched to 0.5 m depth and a water-permeable cloth was inserted vertically to prevent lateral root in-growth. All living plants were clipped and removed from inside these trenched plots and the plots were kept vegetation-free for the remaining year. One PVC collar (Ø 17.5 cm) was permanently installed to 10 cm depth in each of the trenched plots. During the measurement, an opaque chamber (h 30 cm, V 0.065 L) equipped with a low-speed fan was placed onto the collar and the CO_2 concentration as well as headspace T_a was measured every 4.8 s with the EGM-4 IRGA during 3 min. Autotrophic respiration (Ra) was derived as the difference between the measured RE and Rh (Eqn 5):

$$\text{Ra} = \text{RE} - \text{Rh} \quad (5)$$

Relative humidity was also measured in the BP plots using the same collars and chamber. However, trenching and vegetation removal were not necessary, and as Rh measurements represent RE in vegetation-free ecosystems, Ra was not determined in the BP plots.

CH_4, N_2O and nongrowing season CO_2 flux measurements

Methane (CH_4) and nitrous oxide (N_2O) fluxes were measured weekly to biweekly during the 2014 growing season (May 1 to October 31) and once per month outside the growing season with the closed static chamber method at the same measurement locations (i.e., same collars) of the NEE and RE measurements. During each 1-h chamber deployment period, four evenly timed (0, 0.33, 0.66 and 1 h) air samples were drawn from the chamber headspace (h 50 cm, V 65 L; white opaque PVC chambers) with polypropylene syringes through a plastic tube into pre-evacuated (0.3 mbar) 100 mL glass bottles. The air samples were analyzed within a week for CH_4 and N_2O concentrations using a Shimadzu GC-2014 gas chromatograph (GC) combined with a Loftfield automatic sample injection system (Loftfield *et al.*, 1997), a flame ionization detector (FID) and an electron capture detector (ECD). To obtain RE estimates during the nongrowing season months of January to April, the same air samples were also analyzed for their CO_2 concentrations on the same GC using the ECD detector. These RE measurements were also assumed to represent NEE from January to April in all treatments.

Ditch CO_2, CH_4 and N_2O flux measurements

To account for the spatial variation between fluxes from strips and drainage ditches, CO_2, CH_4 and N_2O fluxes were measured within the same peat extraction area from drainage ditches bordering RCG-F, RCG-C and BP strips at a monthly interval from June to December 2011 (with an annual PPT of 826 mm, 2011 was wetter than 2014). Three collars were permanently installed at the bottom of each ditch. The measurements were conducted using same chamber and measurement protocol as described above for the static chamber measurements in the strips.

Flux calculation and quality control

Fluxes of CO_2, CH_4 and N_2O were calculated from the change in gas concentrations in the chamber headspace volume corrected for air density using the ideal gas law (Eqn 6):

$$F_{\mathrm{dyn,stat}} = S \times \frac{p \times V \times M \times t}{R \times T_a \times A} \tag{6}$$

where F_{dyn} and F_{stat} are fluxes measured by the dynamic chamber (i.e., CO_2 in mg CO_2-C m^{-2} h^{-1}) and the static chamber method (i.e., CH_4 in μg CH_4-C m^{-2} h^{-1}, N_2O in μg N_2O-N m^{-2} h^{-1} and CO_2 in mg CO_2-C m^{-2} h^{-1}), respectively, S is the linear slope fitted to the concentration change over time (ppm s^{-1} for the dynamic and ppm h^{-1} or ppb h^{-1} for the static chamber method), p is the air pressure (measured by the EGM-4 instrument or approximated by a constant value of 1013 kPa in the dynamic and static chamber methods, respectively), V is chamber headspace volume, M is the molar mass of the gas, R is the universal gas constant of 8.3143 (J mol^{-1} K^{-1}), T_a is the mean headspace air temperature during the measurement (°K), A is the collar area, and t converts the time unit from seconds to hour (i.e., t = 3600 for the dynamic chamber method and t = 1 for the static chamber method). In the dynamic chamber method, S was the slope with the best R^2 from the individual slopes determined for windows of 25 measurement points (i.e., 2 min) moving stepwise (with one-point increments) over the measurement period after discarding the first two measurement points (i.e., applying a 9.6 s 'dead band'). In the static chamber method, S was calculated over all four data points. The headspace volume was corrected for changes in effective chamber height due to frost heave (resulting in uplifting of the collars) or snow/ice buildup.

All dynamic chamber CO_2 fluxes with a coefficient of determination (R^2) ≥ 0.90 (P < 0.001) were accepted as good fluxes. However, as low fluxes generally result in a lower R^2 (which is especially critical for NEE measurements), fluxes with $S \leq \pm$ 0.15 ppm s^{-1} were always accepted. The S threshold was determined based on a regression between S and R^2 values. For static chamber measurements, the R^2 threshold for accepting CO_2, CH_4 and N_2O fluxes was 0.90 (P < 0.05), 0.80 (P < 0.1) and 0.80 (P < 0.1), respectively, except no filtering criterion was used when the maximum difference in the concentration values was less than the gas-specific GC detection limit (i.e., <20 ppm for CO_2, <20 ppb for CH_4 and <20 ppb for N_2O). This study used the atmospheric sign convention in which positive (e.g., RE) and negative (e.g., GPP and NPP) fluxes represent emission and uptake, respectively.

Model development for estimating annual CO_2, CH_4 and N_2O fluxes

Nonlinear regression models following Kandel *et al.* (2013) were used to estimate annual RE and GPP fluxes based on T_a, PAR and vegetation development. Specifically, GPP fluxes from each collar were fitted to PAR inside the chamber using a hyperbolic function with an additional parameter describing the seasonal changes in vegetation biomass (expressed by the collar-specific g_{cc} estimates) (Eqn 7):

$$\mathrm{GPP} = \frac{a \times A_{\max} \times \mathrm{PAR} \times g_{cc_{norm}}}{a \times \mathrm{PAR} \times A_{\max} \times g_{cc_{norm}}} \tag{7}$$

where GPP is gross primary production (mg CO_2-C m^{-2} h^{-1}), PAR is the photosynthetically active radiation (μmol m^{-2} s^{-1}) inside the chamber, α is the light-use efficiency of photosynthesis (i.e., the initial slope of the light response curve, mg CO_2-C μmol $photons^{-1}$), A_{max} is maximum photosynthesis at light saturation (mg CO_2-C m^{-2} h^{-1}), and $g_{cc_{norm}}$ is the collar-specific chromatic greenness index normalized to scale between 0 and 1.

RE fluxes were fitted to headspace T_a accounting for effects from vegetation biomass using an exponential function (Eqn 8):

$$\mathrm{RE} = R_0 \times \exp^{(b \times T_a)} + (\beta \times g_{cc_{norm}}) \times \exp^{(b \times T_a)} \tag{8}$$

where RE is ecosystem respiration (mg CO_2-C m^{-2} h^{-1}), T_a is air temperature (°C), R_0 is the soil respiration (mg CO_2-C m^{-2} h^{-1}) at 0 °C, b is the sensitivity of respiration to T_a, and β is a scaling parameter representing the contribution of plant respiration to ecosystem respiration. Using the respective model coefficients, hourly GPP and RE were modeled for the entire year using hourly T_a, PAR and g_{cc} as input variables. Annual GPP and RE were then estimated from the cumulative sums of these modeled estimates. The balance between annual GPP and RE estimates resulted in the annual NEE in RCG-F and RCG-C.

In the BP plots, RE was modeled based on an exponential relationship to T_a only (Eqn 9):

$$\mathrm{RE} = R_0 \times \exp^{(b \times T_a)} \quad (9)$$

The cumulative RE model estimates also represented annual NEE at the BP treatment. The GPP and RE model parameters for the different treatments are summarized in Table 2.

Due to weak relationships with environmental variables, the annual CH_4 exchange was estimated by linear interpolation. The annual N_2O exchange, however, was calculated by scaling the median of the measured fluxes to an annual sum as the occurrence of episodic high peak fluxes would have caused annual sums to be overestimated in the linear interpolation method. Annual ditch emissions were estimated by scaling the mean CO_2 and CH_4 fluxes as well as the median N_2O flux to the entire year. The annual GHG balances were estimated by converting the cumulative strip and ditch fluxes to CO_2 equivalents (CO_2 eq) using the global warming potentials (GWP, over a 100-year time frame including carbon–climate feedbacks) of 34 and 298 for CH_4 and N_2O, respectively (IPCC, 2013).

Concentrations and fluxes of dissolved organic carbon

Starting in February 2014, dissolved organic carbon (DOC) concentrations were determined for water samples taken at each flux sampling location from groundwater wells (perforated PVC pipes; Ø 7.5 cm) which collected soil solution from the soil surface down to 50 cm depth. From late June onward, DOC concentrations were estimated at 30 cm soil depth (below the main rooting zone) on every flux sampling date using stainless steel plate lysimeters with a collecting area of 625 cm^2 (Uri *et al.*, 2011). All water samples were analyzed for their DOC concentrations within 1 day after collection. The DOC export was calculated by multiplying the DOC concentration with the water leaching rate which was assumed to be 50% of the monthly precipitation based on previous literature estimates (Kløve *et al.*, 2010; Hyvönen *et al.*, 2013). To assess the sensitivity of the annual DOC export to the choice of this assumed leaching rate, DOC export was also estimated using rates of 25% and 75% of the monthly precipitation to provide a minimum–maximum range which very likely encompasses the true leaching rate.

Table 2 Parameters for the gross primary production (GPP) (Eqn 7) and ecosystem respiration (RE) (Eqns 8 and 9) models for reed canary grass with fertilization (RCG-F), reed canary grass control (RCG-C) and bare peat (BP) treatments; α is the quantum-use efficiency of photosynthesis (mg CO_2-C μmol photons); A_{max} is the maximum rate of photosynthesis at light saturation (mg CO_2-C m^{-2} h^{-1}); R_0 is the soil respiration (mg CO_2-C m^{-2} h^{-1}) at 0 °C; b is the sensitivity of respiration to air temperature; and β is a scaling parameter representing the contribution of plant respiration to ecosystem respiration; numbers in parentheses indicate standard error; Adj. R^2 = adjusted R^2

Model properties	RCG-F	RCG-C	BP
GPP model			
α	−1.21 (0.73)	−0.33 (0.22)	n.a.
A_{max}	−386 (56)	−120 (25)	n.a.
Adj. R^2	0.71	0.67	n.a.
RE model			
R_0	11.9 (8.8)	12.8 (5.5)	13.0 (3.6)
b	0.045 (0.008)	0.035 (0.007)	0.045 (0.010)
β	61 (22)	37 (12.2)	n.a.
Adj. R^2	0.90	0.89	0.61

n.a., not applicable.

Statistical analysis

Collar flux data were averaged for each plot before conducting further statistical analysis to avoid pseudoreplication. The nonparametric Friedman one-way analysis of variance (ANOVA) by ranks test for dependent samples was used to account for repeated measurements in time when testing for treatment effects (i.e., fertilized RCG, nonfertilized RCG and bare peat) on the growing season or annual means of the various component fluxes. This analysis was followed by a Bonferroni *post hoc* comparison to determine significant differences among treatment means. The Mann–Whitney *U*-test was used when comparing only the fertilized and nonfertilized treatments for significant effects (i.e., on GPP, NPP and Ra fluxes and biomass pools). The significance level was $P < 0.05$ unless stated otherwise. All calculations and statistics were computed using the MATLAB software (MATLAB Student version, 2013a; Mathworks, Natick, MA, USA).

Results

Environmental conditions

The annual mean T_a and total PPT for the study year 2014 were 6.9 °C and 525 mm, respectively, which indicates warmer and drier conditions relative to the long-term climate normal (6.3 °C and 745 mm). PAR and T_a peaked in the first week of July and in the first week of August, respectively (Fig. 1a, b). The growing season included two warm and dry periods: one from mid-May to mid-June and the other from early July to early August. Total rainfall during these two periods was only 22.8 and 37.7 mm, respectively (Fig. 1c). The dry summer was interrupted by intermittent rainfall events in mid- to late June and eventually terminated by 2 weeks of heavy rainfall (179.5 mm) from early to late August.

Soil temperatures at 5 and 10 cm depths were similar among the three treatments throughout the year (Fig. 1d). The WTL, however, was higher in RCG-F and RCG-C compared to BP for most of the measurement period and dropped below the depth of the peat layer (i.e., <−45 cm) in all treatments during the two dry and warm summer periods (mid-May to mid-June and early July to early August) (Fig. 1e). The intermittent rainfall in late June and the onset of heavy rainfall in early August were reflected by rapid increases in the WTL.

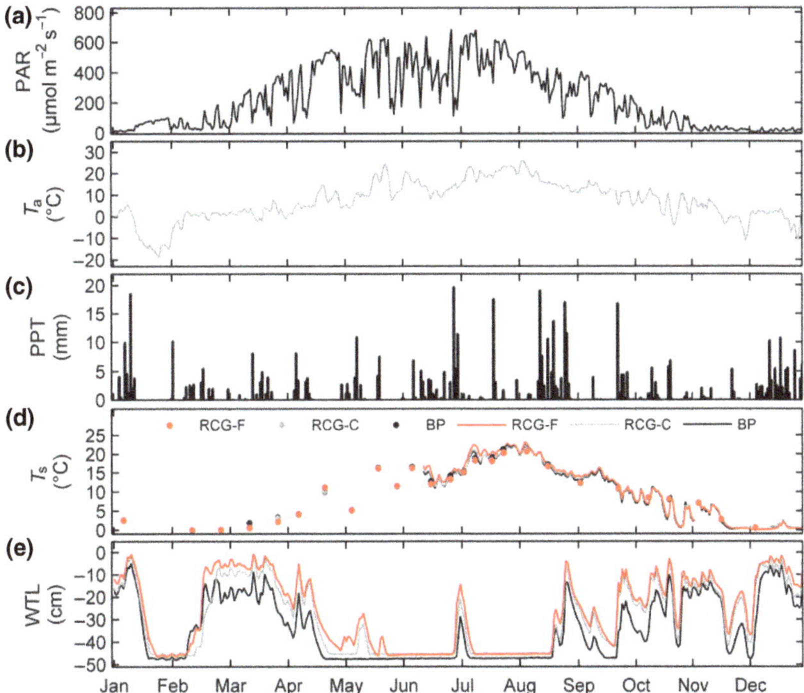

Fig. 1 Seasonal patterns of (a) photosynthetically active radiation (PAR), (b) air temperature (T_a), (c) precipitation (PPT), (d) soil temperature (T_s) at 5 and 10 cm depths (dots and lines represent manual and automated measurements, respectively) and (e) water table level (WTL) for reed canary grass with fertilization (RCG-F), reed canary grass control (RCG-C) and bare peat (BP) treatments in 2014. Data for PAR, T_a and PPT were taken from the Pärnu meteorological station for January 1 to June 11 and measured at the study site from June 12 to December 31.

Vegetation greenness index

The g_{cc} was higher in RCG-F than in RCG-C throughout the growing season (Fig. 2). Its temporal patterns suggest that plant growth started in mid-May and that full canopy development was reached by the second week of June in both RCG-F and RCG-C. Furthermore, the dry period in late June coincided with a temporary reduction in g_{cc}. The start of the senescence period in early August, as indicated by the decline in g_{cc}, coincided with decreasing T_a and T_s. The g_{cc} eventually decreased to its pregrowing season values after the harvest cut on September 3. The g_{cc} did not show any seasonal patterns in BP, except for small fluctuations related to illumination noise and color changes due to soil moisture variations.

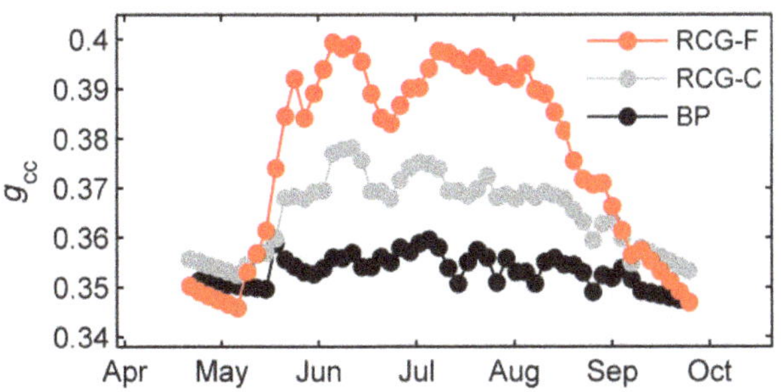

Fig. 2 Temporal pattern of 3-day means of the vegetation greenness index (g_{cc}) for reed canary grass with fertilization (RCG-F), reed canary grass control (RCG-C) and bare peat (BP) treatments in 2014.

Above- and belowground biomass production and allocation

At the time of harvest, mean aboveground biomass (± standard error) in RCG-F and RCG-C was 234 ± 19 and 42 ± 6 g m^{-2}, respectively (Fig. 3). The belowground biomass increased from 536 ± 17 and 364 ± 22 g m^{-2} in April to 646 ± 23 and 416 ± 29 g m^{-2} in September in RCG-F and RCG-C, respectively, and was significantly greater in RCG-F than in RCG-C on both sampling dates (Fig. 3). Both root and rhizome biomass decreased significantly with soil depth. The upper 0–10 cm layer contained 57% and 58% of total belowground biomass, 92% and 93% of rhizome biomass and 46% and 43% of the root biomass in RCG-F and RCG-C, respectively. While root biomass in RCG-F was significantly greater than in RCG-C for all soil depth classes and sampling dates, rhizome biomass was significantly

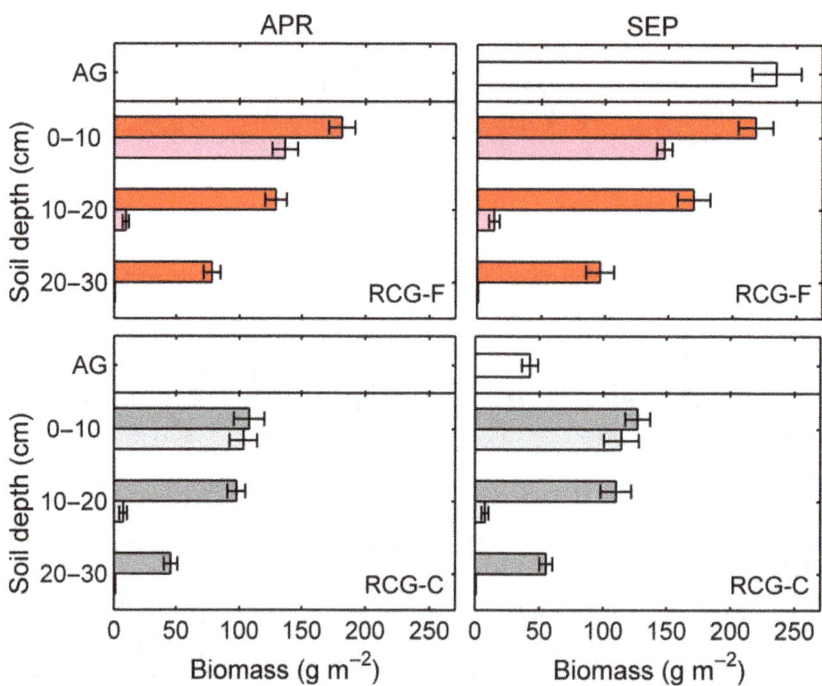

Fig. 3 Aboveground (AG) biomass (white bars) in September and the vertical distribution (0–30 cm depth) of belowground root (dark red and gray bars) and rhizome (light red and gray bars) biomass (dry weight) as a mean (±standard error) of April and September sampling dates in reed canary grass with fertilization (RCG-F) and reed canary grass control (RCG-C) treatments.

greater (on both sampling dates) in the upper 10 cm layer only in RCG-F compared to RCG-C. The root to rhizome ratio was 2.84 and 2.31 in RCG-F and RCG-C, respectively.

The C concentration of harvested aboveground biomass was 48% and 47%, while it was 50% for belowground biomass in both RCG-F and RCG-C, respectively, without any significant difference between the April and September sampling dates.

Both above- and belowground NPP_B derived from biomass sampling were greater in RCG-F (−115 and −55 g C m^{-2} yr^{-1}) than in RCG-C (−20 and −26 g C m^{-2} yr^{-1}) (Table 3). The ratio of above- to belowground NPP_B decreased from 2.1 in RCG-F to 0.8 in RCG-C, respectively.

Seasonal net ecosystem CO_2 exchange, respiration and production

A negative midday NEE, that is, CO_2 uptake, was observed in RCG-F from late May to the end of September, with a maximum uptake rate of 162 mg C m^{-2} h^{-1} noted in early July (Fig. 4a). In comparison, midday NEE remained close to zero during the early growing season (May and June) and switched to positive values suggesting CO_2 emission of up to 77 mg C m^{-2} h^{-1} during the late growing season (July and August) in RCG-C. Both RCG-F and RCG-C were small CO_2 sources for most of the nongrowing season. Continuous midday CO_2 emission occurred throughout the year in BP, reaching a maximum rate of 71 mg C m^{-2} h^{-1} in early July. The annual mean midday NEE was significantly lower in RCG-F than in RCG-C and BP (Fig. 5a).

During the growing season, midday RE was highest in RCG-F and lowest in BP, reaching peak values of 268, 149 and 71 mg C m^{-2} h^{-1} during late July in RCG-F, RCG-C and BP, respectively (Fig. 4b). The annual mean midday RE was significantly higher in RCG-F than in BP (Fig. 5b).

Midday GPP was consistently greater (i.e., more negative) in RCG-F than in RCG-C during the growing season (Fig. 4c). In RCG-F, GPP peaked with −359 mg C m^{-2} h^{-1} simultaneously with T_a in late July whereas in RCG-C, values remained in the range of −50 to −110 mg C m^{-2} h^{-1} throughout most of the growing season following the pattern of g_{cc}. The seasonal patterns in midday NPP were similar to those of GPP, reaching maximum rates of −211 and −85 mg C m^{-2} h^{-1} in RCG-F and RCG-C, respectively (Fig. 4d). The mean midday GPP and NPP were significantly lower (i.e., suggesting greater production) in RCG-F than in RCG-C (Fig. 5c, d).

Heterotrophic and autotrophic respiration

Similar seasonal patterns of Rh were observed in RCG-F and RCG-C, with maximum rates of 119 and 92 mg C m^{-2} h^{-1}, respectively, occurring in late July (Fig. 6a).

Table 3 The full carbon (C) balance (g C m^{-2} yr^{-1}) and the annual sums of its components: ecosystem respiration (RE), gross primary production (GPP), above- and belowground net primary production (ANPP, BNPP), total net primary production (NPP_B; based on biomass sampling), net ecosystem CO_2 exchange (NEE) and methane (CH_4), export of dissolved organic carbon (DOC) and the ditch CO_2 and CH_4 fluxes for reed canary grass with fertilization (RCG-F), reed canary grass control (RCG-C) and bare peat (BP) treatments. Negative and positive fluxes represent C uptake and emission, respectively

C flux component	RCG-F	RCG-C	BP
Strips			
RE	512	326	170
GPP	−433	−125	n.a.
ANPP	−115	−20	n.a.
BNPP	−55	−26	n.a.
NPP_B	−170	−46	n.a.
NEE	79	201	170*
CH_4	0.014	0.018	0.020
DOC export	4.2	4.2	4.1
Ditches			
CO_2	360	415	298
CH_4	0.134	0.120	0.063
Total C balance†	96	215	180

n.a., not applicable.

*GPP for BP was assumed to be zero.

†The total C balance is the sum of area-weighted (strip width = 20 m; ditch width = 1 m) fluxes of NEE, CH_4 and DOC as well as the ditch CO_2 and CH_4 fluxes.

In RCG-F and RCG-C, Rh was higher ($P < 0.01$) from early July to mid-August than in BP where maximum Rh rates of 77 mg C m^{-2} h^{-1} were observed (Fig. 6a). In comparison, Ra was consistently higher in RCG-F than in RCG-C throughout the growing season (Fig. 6b). Maximum Ra during July was approximately 2.5 times higher in RCG-F (~150 mg C m^{-2} h^{-1}) than in RCG-C (~60 mg C m^{-2} h^{-1}). Except for one sampling date (May 30), the Ra to Rh ratio was always ≥1 in RCG-F while it was ≤1 in RCG-C for most of the growing season (Fig. 6c). The mean growing season Rh was not significantly different between RCG-F and RCG-C (Fig. 5e). In contrast, mean Ra was significantly higher in RCG-F than in RCG-C (Fig. 5f). Averaged over all sampling dates, Rh accounted for 42% and 62% of RE in RCG-F and RCG-C, respectively.

CH_4 and N_2O exchanges

Throughout the growing season, CH_4 emission occurred in the range of 0.01 to 9.3 μg C m^2 h^{-1} in all three treatments (Fig. 7a). Between mid-June and early September, the mean CH_4 emission was approximately 1.5 times higher in BP than in RCG-F and RCG-C ($P = 0.052$). During the nongrowing season, the CH_4 exchange was close to zero with small rates (−2.3 to 3.7 μg C m^2 h^{-1}) of both uptake and emission occurring at individual collars and sampling dates, with no differences among the three treatments. The annual mean CH_4 exchange was

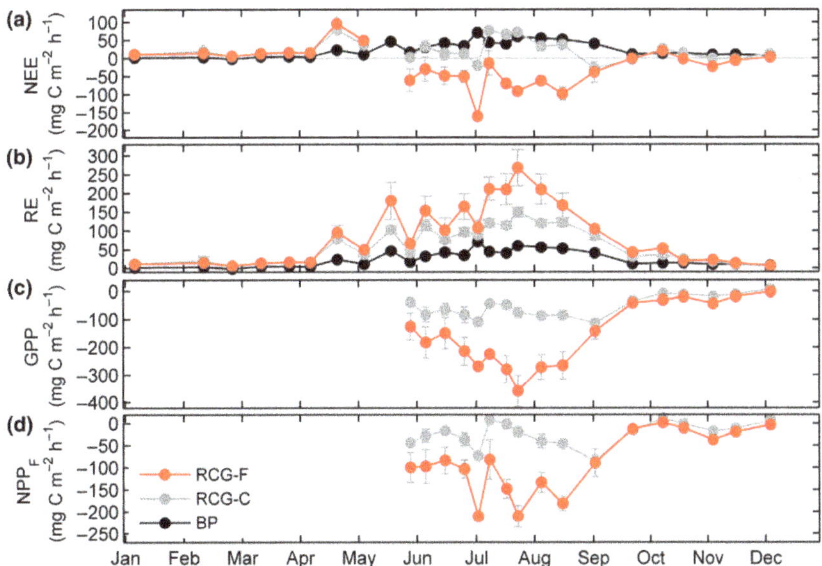

Fig. 4 (a) Net ecosystem CO_2 exchange (NEE), (b) ecosystem respiration (RE), (c) gross primary production (GPP) and (d) net primary production (NPP_F) for reed canary grass with fertilization (RCG-F), reed canary grass control (RCG-C) and bare peat (BP) treatments; error bars indicate standard error; the horizontal dotted line in (a) visualizes the zero line above and below which emission and uptake occur, respectively.

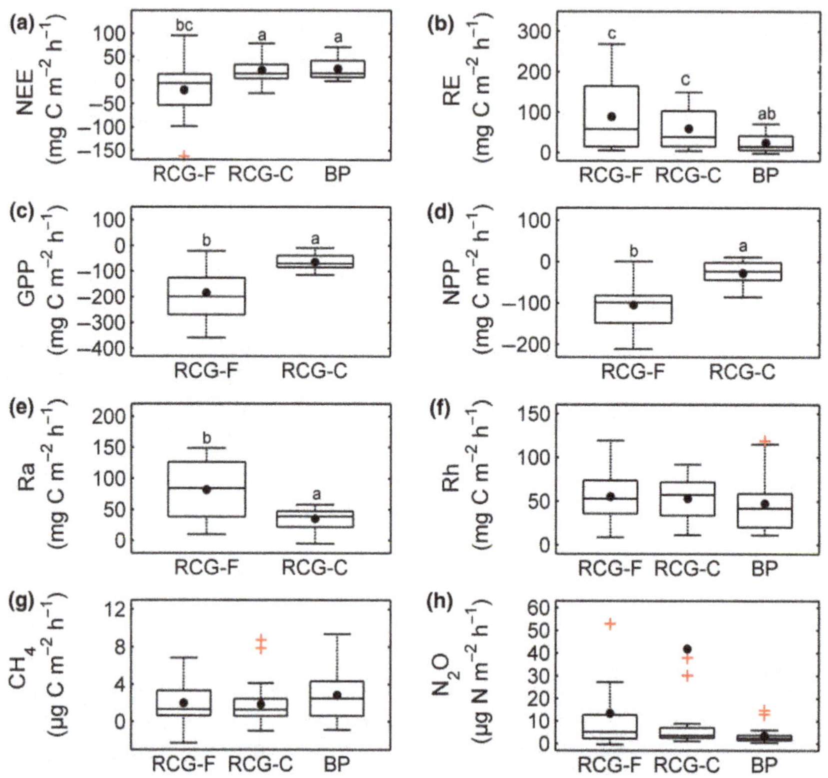

Fig. 5 Box plots for (a) net ecosystem CO_2 exchange (NEE), (b) ecosystem respiration (RE), (c) gross primary production (GPP), (d) net primary production (NPP_F; derived from flux measurements), (e) autotrophic respiration (Ra), (f) heterotrophic respiration (Rh), (g) methane (CH_4) and (h) nitrous oxide (N_2O) fluxes for reed canary grass with fertilization (RCG-F), reed canary grass control (RCG-C) and bare peat (BP) treatments; boxplots for NEE, RE, CH_4 and N_2O are based on annual data; boxplots for GPP, NPP, Ra and Rh are based on growing season data. Note that outliers of 391 and 421 µg N m-2 h-1 for RCG-F and 171 µg N m-2 h-1 for RCG-C are not shown in panel (h). The central line and dot are the median and mean, respectively, the edges of the box are the 25th and 75th percentiles, the whiskers extend to the most extreme data points which are not considered outliers, red cross symbols indicate outliers defined as data points exceeding a standard deviation of 2.7, and different letters indicate significant ($P < 0.05$) differences among treatments.

not significantly different among the three treatments (Fig. 5g).

In all three treatments, N_2O fluxes were within the range of −0.4 to 25 μg N_2O-N m^{-2} h^{-1} for most of the year, with the exception of large emission peaks of up to 420 μg N_2O-N m^{-2} h^{-1} in RCG-F on July 4 and September 3 (Fig. 7b). On September 3, an N_2O emission peak of 172 μg N_2O-N m^{-2} h^{-1} was also observed in RCG-C. These peak emission events on July 4 and September 3 coincided with large rainfall events occurring just prior to both sampling dates (compare with Fig. 1c). The annual means and medians of the N_2O exchange were not significantly different among the three treatments (Fig. 5h). The N_2O emission factor (i.e., the % of N fertilizer lost as N_2O) was 0.63% in RCG-F.

Ditch emissions of CO_2, CH_4 and N_2O

Mean CO_2, CH_4 and N_2O emissions from the drainage ditches were not significantly different among treatments (Fig. 8a–c). Ditch CO_2 emissions accounted for >99% of the total ditch C flux (Table 3) and >98% of the total GHG flux from ditches (Table 4) in all three treatments.

DOC concentrations and export

Concentrations of DOC ranged within 10–19 mg L^{-1} in the nongrowing season months and increased during the early growing season to a maximum of 25 mg L^{-1} (Fig. 9). No soil solution samples could be retrieved from the dried out soil during the two dry summer periods. Averaged over all sampling dates and depths, the mean DOC concentrations were 17, 16 and 16 mg L^{-1} in RCG-F, RCG-C and BP, respectively, with no significant differences among treatments (see inset figure in Fig. 9). Assuming the leaching rate to be 50% of the annual precipitation, the annual DOC export was estimated at 4.2, 4.2 and 4.1 g C m^{-2} yr^{-1} in RCG-F, RCG-C and BP, respectively (Table 3). In comparison,

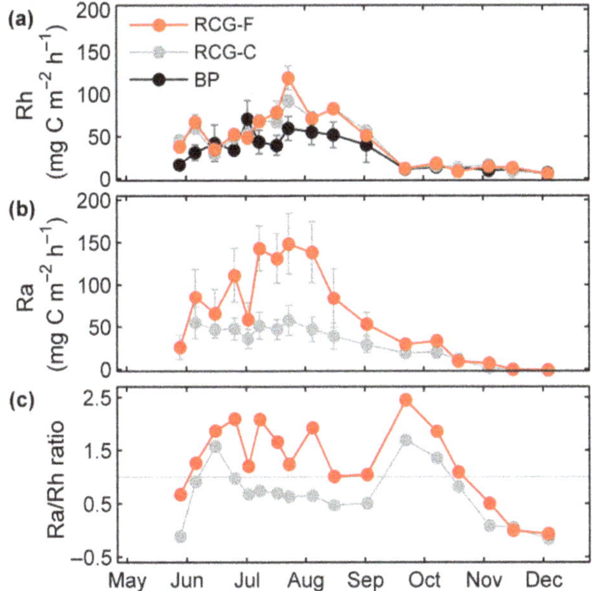

Fig. 6 (a) Heterotrophic respiration (Rh), (b) autotrophic respiration (Ra) and (c) the Ra/Rh ratio for reed canary grass with fertilization (RCG-F), reed canary grass control (RCG-C) and bare peat (BP) treatments; error bars indicate standard error; the horizontal dotted line in (c) indicates Ra/Rh = 1 (i.e., Ra = Rh).

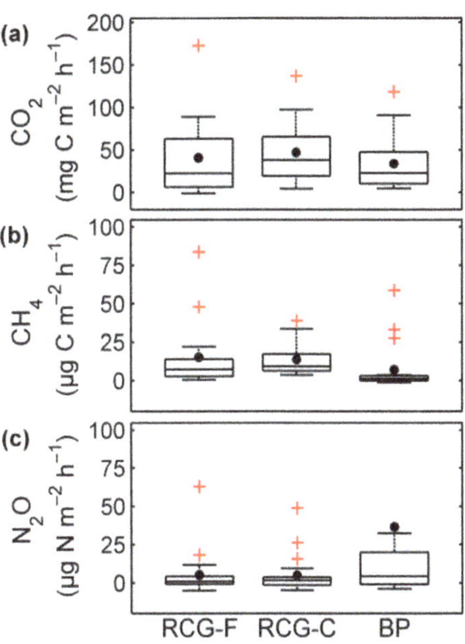

Fig. 8 Box plots for ditch emissions of (a) carbon dioxide (CO_2), (b) methane (CH_4) and (c) nitrous oxide (N_2O) for reed canary grass with fertilization (RCG-F), reed canary grass control (RCG-C) and bare peat (BP) treatments (see Fig. 5 for a description of the box plot features).

the annual DOC export was 2.1, 2.1 and 2.1 g C m^{-2} yr^{-1} assuming a leaching rate of 25% and 6.3, 6.2 and 6.2 g C m^{-2} yr^{-1} assuming a leaching rate of 75% in RCG-F, RCG-C and BP, respectively.

The full carbon balance

The model estimates of cumulative annual RE were 512, 326 and 170 g C m^{-2} yr^{-1} in RCG-F, RCG-C and BP, respectively, and −433 and −125 g C m^{-2} yr^{-1} for GPP in RCG-F and RCG-C, respectively (Table 3). The carbon-use efficiencies (i.e., the ratio of NPP_B to GPP) were 0.39 and 0.37 in RCG-F and RCG-C, respectively. The cumulative NEE based on model estimates suggested annual CO_2 emission of 79, 201 and 170 g C m^{-2} yr^{-1} in RCG-F, RCG-C and BP, respectively (Table 3).

The cumulative CH_4 exchange resulted in annual CH_4 emission of 0.01, 0.02 and 0.02 g C m^{-2} yr^{-1} in RCG-F, RCG-C and BP, respectively (Table 3). The

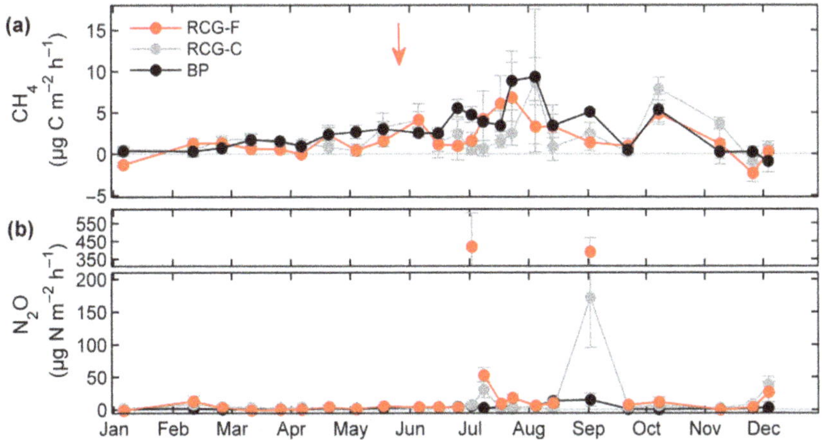

Fig. 7 Exchanges of (a) methane (CH_4) and (b) nitrous oxide (N_2O) for reed canary grass with fertilization (RCG-F), reed canary grass control (RCG-C) and bare peat (BP) treatments; error bars indicate standard error; arrow indicates timing of fertilizer application; the horizontal dotted line visualizes the zero line above and below which emission and uptake occur, respectively.

Table 4 The greenhouse gas (GHG) balance (t CO_2 eq ha^{-1} yr^{-1}) and its component fluxes net ecosystem CO_2 exchange (NEE), methane (CH_4) and nitrous oxide (N_2O) within the strips and carbon dioxide (CO_2), CH_4 and N_2O emissions from the ditches adjusted for their global warming potentials (34 and 298 for CH_4 and N_2O, respectively) for reed canary grass with fertilization (RCG-F), reed canary grass control (RCG-C) and bare peat (BP) treatments

GHG flux component	RCG-F	RCG-C	BP
Strips			
NEE	2.9	7.4	6.2
CH_4	0.006	0.008	0.009
N_2O	0.21	0.15	0.10
Ditches			
CO_2	13.2	15.2	10.9
CH_4	0.061	0.055	0.029
N_2O	0.04	0.07	0.17
Total GHG balance*	3.6	7.9	6.6

*The total GHG balance is the sum of area-weighted (strip width = 20 m; ditch width = 1 m) fluxes of CO_2, CH_4 and N_2O in strips and ditches.

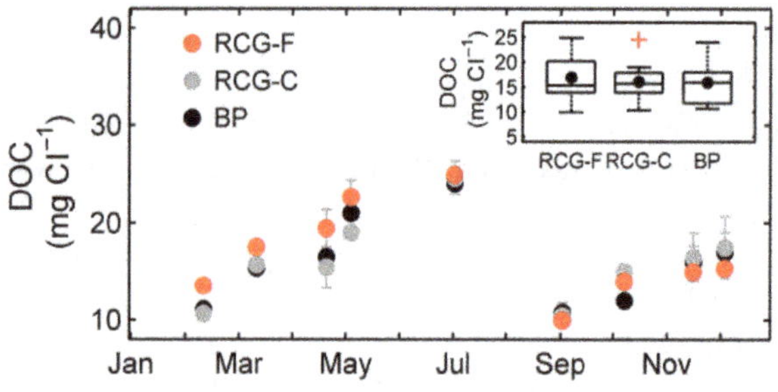

Fig. 9 Concentrations of dissolved organic carbon (DOC) in soil solution (0–50 cm depth) for reed canary grass with fertilization (RCG-F), reed canary grass control (RCG-C) and bare peat (BP) treatments; error bars indicate standard error. The inset figure shows the treatment means from all sampling dates (see Fig. 5 for a description of the box plot features).

cumulative CO_2 and CH_4 fluxes from the ditches were 360, 415 and 298 g C m^{-2} yr^{-1} and 0.13, 0.12 and 0.06 g C m^{-2} yr^{-1} in RCG-F, RCG-C and BP, respectively (Table 3). Combining the area-weighted annual CO_2 and CH_4 exchanges from the 20 m wide strips and the 1 m wide drainage ditches with the DOC export resulted in full C balances of 96, 215 and 180 g C m^{-2} yr^{-1} in RCG-F, RCG-C and BP, respectively (Table 3). The relative contribution of DOC export to the full C balance was 4.1%, 1.9% and 2.2% in RCG-F, RCG-C and BP, respectively, while the relative contribution of the CH_4 exchange was <1% for all treatments.

The greenhouse gas balance

Combining the area-weighted annual CO_2, CH_4 and N_2O exchanges from the strips and drainage ditches resulted in total GHG balances of 3.6, 7.9 and 6.6 t CO_2 eq ha^{-1} yr^{-1} in RCG-F, RCG-C and BP, respectively (Table 4). The contribution of the combined ditch CO_2, CH_4 and N_2O emissions to the total GHG balance in RCG-F (18%) was two times greater than in RCG-C (9%) and BP (8%). The sum of CH_4 and N_2O fluxes from both strips and ditches accounted for 5.9, 1.8 and 1.6% of the total GHG balance in RCG-F, RCG-C and BP, respectively. The N_2O to biomass yield ratio (i.e., the ratio of the N_2O flux to yield in CO_2 eq) increased from 0.05 in RCG-F to 0.21 in RCG-C.

Discussion

Above- and belowground biomass production and allocation

The yields in both RCG-F (2.3 t ha^{-1}) and RCG-C (0.4 t ha^{-1}) were at the bottom end of the range of 2.0 to 13.9 t ha^{-1} and 1.0 to 11.0 t ha^{-1} previously reported for fertilized and nonfertilized RCG cultivations, respectively (Shurpali *et al.*, 2010; Heinsoo *et al.*, 2011; Mander *et al.*, 2012; Kandel *et al.*, 2013; Karki *et al.*, 2014). The low yields in this study were likely due to water stress constraining plant growth during an exceptionally dry summer. Our findings therefore suggest that RCG cultivation on abandoned peat extraction areas has limited potential for economically sustainable biomass production during dry years without proper WTL management.

While aboveground biomass is harvested and exported from the system, a significant fraction of belowground biomass C is permanently incorporated into the soil C pool and BNPP is therefore important with regard to long-term C sequestration (Xiong & Kätterer, 2010). Specifically, RCG cultivations have been previously highlighted as systems with a higher potential for C input into the soil compared to annual crops and nonrhizomatous perennial leys due to their expansive rootstocks and high root turnover rates (Hansson & Andren, 1986; Xiong & Kätterer, 2010; Don *et al.*, 2012). In our study, BNPP (20–56 g C m^{-2} yr^{-1}) was at the lower end of the 80–235 g C m^{-2} yr^{-1} reported for irrigated RCG cultivations in Sweden (Kätterer & Andrén, 1999) and a RCG cultivation in the same abandoned peat extraction area during wet years (Mander *et al.*, 2012). Thus, BNPP and its contribution to soil C sequestration in RCG cultivation systems might be considerably reduced during dry years.

Fertilizer effects on plant growth and soil nutrient status might affect not only the total biomass production but also its allocation into above- and belowground components (Xiong & Kätterer, 2010). For instance, the greater above- to belowground biomass ratio in RCG-F than in RCG-C suggests that fertilization resulted in greater biomass yields available for bioenergy production, however, at the cost of C allocation and long-term storage belowground. Nevertheless, given the greater absolute magnitudes of BNPP, increased C input to the soil may still occur in fertilized compared to nonfertilized RCG cultivations.

Belowground biomass dynamics have been proposed as one of the least understood aspects of plant functioning (Strand *et al.*, 2008). Specifically, a good understanding of the partitioning of belowground biomass into roots and rhizomes is imperative as it has important implications for nutrient uptake and storage and provides information on plant survival strategies (Don *et al.*, 2012; Jones *et al.*, 2015). The increase in the root to rhizome biomass ratio observed in our study suggests that fertilized RCG plants invest less into storage organs, that is, rhizomes, under nutrient-rich conditions. As roots and rhizomes have different turnover times and longevity (Xiong & Kätterer, 2010), structural changes in belowground biomass due to fertilization may affect the potential for long-term soil C storage. Our study therefore highlights the need to consider fertilization effects on belowground biomass structure and allocation patterns, for example, in process-based models, to improve predictions of C sequestration under future climate and management scenarios (Strand *et al.*, 2008; Shurpali *et al.*, 2010; Gong *et al.*, 2014).

Seasonal net ecosystem CO_2 exchange, respiration and production

Midday net CO_2 uptake occurred only in RCG-F whereas RCG-C and BP were consistent net CO_2 sources even during the midday. However, midday net CO_2 uptake also occurred at RCG-C after intermittent rainfall at the end of June which indicates that nonfertilized RCG cultivations might also sequester CO_2 given sufficient water supply. In comparison, both fertilized and nonfertilized RCG cultivations with VWC > 55% provided midday net CO_2 uptake for the entire growing season in a Danish study (Kandel *et al.*, 2013). Daily net CO_2 uptake rates also decreased by about half from wet to dry years in a RCG cultivation in eastern Finland (Shurpali *et al.*, 2009). The combined findings from this and other studies indicate that soil water availability is a major control of the CO_2 sink–source strength of RCG cultivations on drained peat soils.

Understanding the controls and seasonal patterns of the NEE component fluxes GPP, NPP and RE is essential to explain the variations in NEE. Overall, the independent estimates of annual NPP and modeled GPP based on biomass and flux sampling, respectively, agreed reasonably well in both RCG-F and RCG-C given that NPP commonly represents between 39% and 58% of GPP (Waring *et al.*, 1998; Vicca *et al.*, 2012). The enhanced potential for GPP and NPP due to fertilization resulted in a more pronounced response to abiotic conditions (e.g., T_a and PPT) in RCG-F than in RCG-C which may explain why the timing of peak production was decoupled from that of plant development and instead more closely related to that of T_a in RCG-F. In contrast, GPP and NPP in RCG-C were considerably lower and their seasonal patterns followed closely that of the plant development. Thus, GPP and NPP might be controlled primarily by abiotic controls in fertilized RCG cultivations whereas reduced photosynthetic capacity might be the main constraint on plant production in nonfertilized (i.e., low productive) cultivations.

The greater midday net CO_2 uptake in RCG-F relative to RCG-C was due to variations in GPP as the increase in GPP (by 69%) was larger than the increase in RE (by 37%) in RCG-F relative to RCG-C. Similarly, GPP was also reported as the main driver for interannual variations in NEE during wet and dry years in a Finnish RCG cultivation (Shurpali *et al.*, 2009). This suggests that ensuring optimum growing conditions is essential not only for achieving economically sustainable yields but also to maximize the CO_2 sequestration potential in RCG cultivations.

Heterotrophic and autotrophic respiration

The difference in the RE partitioning into its components Rh and Ra between RCG-C (Rh > Ra) and RCG-F (Rh < Ra) was the result of enhanced plant growth due to fertilization and the subsequent increase of Ra in RCG-F. Meanwhile, fertilization had no significant effect on Rh. Previous studies reported a decrease or no effect on mineralization rates following fertilization, with the contrasting findings primarily related to the indirect effects of fertilization on soil pH (e.g., Fog, 1988; Aerts & Toet, 1997). Overall, the contribution of Rh to RE in RCG-F (42%) was similar to the 45% reported for a fertilized RCG cultivation in Finland (Shurpali *et al.*, 2008) but lower than the 55–75% observed in other drained and natural peatlands (Silvola *et al.*, 1996; Riutta *et al.*, 2007; Biasi *et al.*, 2012). Thus, fertilization of RCG systems may reduce the relative contribution of Rh to RE which has important implications for the response of RE to management and climate impacts as the two respiratory component fluxes Ra and Rh respond to different controls, that is, to biotic vs. abiotic variables, respectively.

During the warmest summer period (July to August), Rh was consistently higher in RCG-F and RCG-C than in the bare peat soil. Increased mineralization of organic matter in drained peat soils following cultivation and its negative implications for the C and GHG balances have been previously highlighted in several studies (Kasimir-Klemedtsson *et al.*, 1997; Drösler *et al.*, 2008; Maljanen *et al.*, 2010; Schrier-Uijl *et al.*, 2014). Although Rh rates were not significantly different among treatments over the entire measurement period, we estimated that the cumulative CO_2 loss due to mineralization during the warmest summer period (July and August) was 50 and 43 g C m^{-2} yr^{-1} greater in RCG-F and RCG-C, respectively, relative to BP. Thus, a substantial additional C input from plant CO_2 uptake is required to outbalance these CO_2 losses due to enhanced mineralization in cultivated organic soils.

The full carbon balance

Our finding that both RCG-F and RCG-C as well as the bare peat soil were considerable annual CO_2 sources is in contrast to previous work conducted in the same abandoned peat extraction area in which both fertilized and nonfertilized RCG cultivations were CO_2 sinks during a year with above-normal precipitation (911 826 mm) (Mander *et al.*, 2012). This indicates a switch from a CO_2 sink in a previous wetter year to a CO_2 source during the dry year in the current study. Similarly, the CO_2 sink strength of a fertilized RCG cultivation established on organic soil in Finland substantially decreased from −127 and −211 g C m^{-2} yr^{-1} during two wet years to −9 and −52 g C m^{-2} yr^{-1} in two dry years (Shurpali *et al.*, 2009). Thus, these results highlight the risk that future increases in drought frequency (IPCC, 2013) might considerably reduce the potential of RCG cultivations for C sequestration.

Furthermore, while fertilized and nonfertilized RCG systems in our study and in another study in Denmark (Kandel *et al.*, 2013) were large CO_2 sources, the Finnish RCG cultivation remained a CO_2 sink even during dry years (Shurpali *et al.*, 2009). These contrasting CO_2 source–sink strength potentials could be related to differences in the annual mean T_a (6.9 and 7.3 for this study and the Danish site vs. 3.7 °C for the Finnish site, respectively) and its effect on Rh. For instance, annual Rh from BP in the current study (170 g C m^{-2} yr^{-1}) was about twice that of bare peat soil at the Finnish site (72 g C m^{-2} yr^{-1}) during a dry year (Shurpali *et al.*, 2008). Thus, a latitudinal effect from the positive correlation between T_a and Rh might determine the CO_2 sink–source strength of RCG cultivation during dry years with reduced plant CO_2 uptake.

The annual CH_4 emissions of <0.02 g C m^{-2} yr^{-1} from RCG-F and RCG-C were much smaller compared to the ranges of 3 to 14 g C m^{-2} yr^{-1} reported for pristine peatlands (Roulet *et al.*, 2007; Nilsson *et al.*, 2008) and of 0.5 to 1.5 g C m^{-2} yr^{-1} observed in restored or cultivated cutaway peatlands (Tuittila *et al.*, 2000; Hyvönen *et al.*, 2009). These low CH_4 emissions were likely the result of an exceptionally low WTL which reduced the potential for anaerobic CH_4 production. In comparison, CH_4 emissions of 18 to 31 g C m^{-2} yr^{-1} were observed in an Irish RCG cultivation on cutaway peatland in which the WTL remained mostly close to the surface (i.e., within 10 cm) (Wilson *et al.*, 2009). Thus, while the contribution of the CH_4 exchange to the full C balance was negligible in these dry RCG systems, cultivation techniques which raise the WTL to sustain high biomass yields during dry years, for example, paludiculture (Wichtmann & Schäfer, 2007; Karki *et al.*, 2014), might considerably increase the potential for CH_4 emissions in cultivated organic soils.

The annual DOC export (~4.2 g C m^{-2} yr^{-1}) at this site was slightly lower than the 5.7 and 6.2 g C m^{-2} yr^{-1} reported from a Finnish RCG cultivation in an abandoned peat extraction area (Hyvönen *et al.*, 2013) and from a Canadian cutover peatland (Strack *et al.*, 2011), respectively. Together, these studies suggest that the DOC export from cultivated peatlands is considerably lower in comparison with the 12–15 g C m^{-2} yr^{-1} reported for natural peatlands (Roulet *et al.*, 2007; Nilsson *et al.*, 2008; Koehler *et al.*, 2009). Nevertheless, despite the relatively small contribution to the full C balance (~2–4%) during the dry year in this study, DOC export might increase under management and climate scenarios that alter soil hydrology and runoff (Freeman *et al.*, 2004).

After incorporating all major C fluxes of CO_2, CH_4 and DOC from strips and ditches, all three treatments represented net C sources. Although the net C emission was lowest in RCG-F, its source strength would further increase when including the additional C loss via the export of harvested biomass (115 g C m^{-2} yr^{-1}). Meanwhile, RCG-C was a considerably greater C source than RCG-F and BP, regardless of whether or not the additional export of harvested biomass was accounted for. This indicates that fertilized RCG cultivations may to some extent mitigate the negative effects on the full C balance commonly observed following cultivation of drained peat soils (Maljanen *et al.*, 2010).

Ditch CO_2 and CH_4 emissions are rarely included in C balance estimates due to the small area coverage by ditches (Maljanen *et al.*, 2010). Previous studies, however, reported considerable CH_4 emissions specifically from water-logged drainage ditches in abandoned peat extraction areas (Maljanen *et al.*, 2010; Hyvönen *et al.*, 2013). In contrast, the CO_2 flux constituted the dominant ditch C emission component in this study likely as the result of the exceptionally dry conditions at this study site at which ditches dry out completely during the summer months. Overall, ditch emissions accounted for 8–18% of the full C balance although ditches represented <5% of the area in this study. Thus, the combined CO_2 and CH_4 emissions from ditches might represent a significant component of the full C balance not only in wet but also in dry cropping systems.

The greenhouse gas balance

N_2O emissions from the strips were low (0.03–0.07 g N_2O m^{-2} yr^{-1}) in all treatments when compared to the range of 0.2–5.5 g N_2O m^{-2} yr^{-1} reported for agricultural systems (Klemedtsson *et al.*, 2005; Maljanen *et al.*, 2010; Don *et al.*, 2012). However, high peak fluxes were observed after rainfall which indicates that annual N_2O emission might be greater during wetter years with more frequent rainfall events. Furthermore, fertilizer application did not result in an immediate increase in N_2O emissions although it likely enhanced the potential for the high peak N_2O emissions occurring after the heavy rainfall in late June. Overall, the combined contribution of N_2O and CH_4 emission from strips and ditches to the GHG balance was small (<6%) during the dry year in this study. To some extent, this low contribution might be due to using the median when extrapolating N_2O fluxes to the annual scale. Using the mean would have increased the contribution of N_2O (3%, 6% and 30%, respectively, for BP, RCG-C and RCG-F) as the mean is highly sensitive to the few extremely large peak emission rates measured in RCG-F and RCG-C. Given the highly non-normal distribution of N_2O fluxes, the choice of the median over the mean appeared therefore more reasonable. Nevertheless, the large uncertainty related to upscaling of N_2O fluxes in agricultural systems remains problematic and can only be reduced through measurements with high temporal (i.e., hourly to daily) resolution. In support of our findings, other studies also found a relatively small contribution of N_2O and CH_4 to the GHG balance of cultivated organic soils (Hyvönen *et al.*, 2009; Shurpali *et al.*, 2010; Mander *et al.*, 2012; Karki *et al.*, 2015a). We therefore conclude that the net CO_2 exchange determines both the C and GHG balances in RCG cultivations on organic soils. Management practices need to be therefore carefully evaluated with respect to their direct and indirect impacts on the ecosystem CO_2 exchange.

The recently proposed management strategy of cultivating perennial bioenergy crops with low N-demand, such as RCG, to reduce N_2O emissions compared to conventional crop cultivation (Don *et al.*, 2012) is supported by the low N-emission factor (0.63) observed in this current study. However, in contrast to other perennial cropping systems (e.g., *Miscanthus*) which may produce high yields and greater climate benefits without fertilization (Strullu *et al.*, 2011), the lower N_2O to yield ratio in the fertilized compared to the nonfertilized RCG system in our study suggests that the increase in biomass production and net CO_2 uptake largely exceed the increase in N_2O emissions (in CO_2 eq) following moderate fertilization, even after accounting for the additional emissions of about 0.5 t CO_2 eq ha^{-1} yr^{-1} occurring during fertilizer production, transport and application (Järveoja *et al.*, 2013). Thus, moderate fertilization could still be a beneficial management practice to maximize yield and climate benefits of RCG cultivation given the limited land resources available for reaching national bioenergy production targets. Nevertheless, other aspects such as economic constraints, effects on combustion quality and ecological concerns (e.g., groundwater eutrophication) must be considered when evaluating optimum fertilizer rates (Smith & Slater, 2010; Verhoeven & Setter, 2010; Don *et al.*, 2012).

In contrast to previous studies suggesting that RCG cultivations provide negative GHG balances and thus mitigate the GHG emissions from drained organic soils (Shurpali *et al.*, 2010; Mander *et al.*, 2012), both the fertilized and nonfertilized RCG systems had positive GHG balances in the current study due to the exceptionally dry conditions during the studied year. However, previous studies indicate that a negative GHG balance could be achieved by cultivating RCG in agricultural systems with elevated WTL and sufficient soil water availability (Kätterer & Andrén, 1999; Freibauer *et al.*, 2004; Schrier-Uijl *et al.*, 2014; Karki *et al.*, 2015b). Although raising the WTL in drained organic soils might result in increased CH_4 emissions, these increases have been estimated to be modest (Komulainen *et al.*, 1998; Tuittila *et al.*, 2000; Wilson *et al.*, 2009; Karki *et al.*, 2014), and are therefore unlikely to compromise the benefits gained from increased plant growth and CO_2 uptake due to sufficient water supply. Thus, we conclude that, when converting abandoned peat extraction areas into RCG cultivations, management strategies need to ensure optimum plant growth through adequate water and nutrient supply to maximize the net ecosystem CO_2 uptake as its benefits are likely to considerably exceed the associated potentially negative effects from increased CH_4 and N_2O emissions.

Acknowledgements

This study was supported by the European Regional Development Fund through ENVIRON (Centre of Excellence in Environmental Adaption), by the Ministry of Education and Research of the Republic of Estonia (grant IUT2-16) and by the Estonian Environmental Observatory Biosphere-Atmosphere Science and Development Programme: BioAtmos (KESTA,

SLOOM12022T). We would like to thank Dr. Ivika Ostonen for comments and discussions, Dr. Marika Truu for participation in setting up the experiment and Dr. Veiko Uri for providing the plate lysimeters. We are grateful to Mae Uri, Kristi Jänesmägi, Kaarel Kukk and Kevin Kesküla for their help in the laboratory. We also thank the three anonymous reviewers for their constructive comments on the original version of the manuscript.

References

Adler PR, Grosso SJD, Parton WJ (2007) Life-cycle assessment of net greenhouse-gas flux for bioenergy cropping systems. *Ecological Applications*, **17**, 675–691.

Aerts R, Toet S (1997) Nutritional controls on carbon dioxide and methane emission from *Carex*-dominated peat soils. *Soil Biology and Biochemistry*, **29**, 1683–1690.

Biasi C, Pitkamaki AS, Tavi NM, Koponen HT, Martikainen PJ (2012) An isotope approach based on C-13 pulse-chase labelling vs. the root trenching method to separate heterotrophic and autotrophic respiration in cultivated peatlands. *Boreal Environment Research*, **17**, 184–192.

Couwenberg J, Thiele A, Tanneberger F *et al.* (2011) Assessing greenhouse gas emissions from peatlands using vegetation as a proxy. *Hydrobiologia*, **674**, 67–89.

Crutzen PJ, Mosier AR, Smith KA, Winiwarter W (2008) N_2O release from agro-biofuel production negates global warming reduction by replacing fossil fuels. *Atmospheric Chemistry and Physics*, **8**, 389–395.

Don A, Osborne B, Hastings A *et al.* (2012) Land-use change to bioenergy production in Europe: implications for the greenhouse gas balance and soil carbon. *GCB Bioenergy*, **4**, 372–391.

Drösler M, Freibauer A, Christensen TR, Friborg T (2008) Observations and status of peatland greenhouse gas emissions in Europe. In: *The Continental-Scale Greenhouse Gas Balance of Europe* (eds Dolman AJ, Valentini R, Freibauer A), pp. 243–261. Springer, New York.

Fog K (1988) The effect of added nitrogen on the rate of decomposition of organic matter. *Biological Reviews*, **63**, 433–462.

Freeman C, Fenner N, Ostle NJ *et al.* (2004) Export of dissolved organic carbon from peatlands under elevated carbon dioxide levels. *Nature*, **430**, 195–198.

Freibauer A, Rounsevell MDA, Smith P, Verhagen J (2004) Carbon sequestration in the agricultural soils of Europe. *Geoderma*, **122**, 1–23.

Gong J, Kellomäki S, Shurpali NJ *et al.* (2014) Climatic sensitivity of the CO_2 flux in a cutaway boreal peatland cultivated with a perennial bioenergy crop (*Phalaris arundinaceae*, L.): beyond diplotelmic modeling. *Agricultural and Forest Meteorology*, **198–199**, 232–249.

Gorham E (1991) Northern peatlands: role in the carbon cycle and probable responses to climatic warming. *Ecological Applications*, **1**, 182–195.

Hansson A-C, Andren O (1986) Below-ground plant production in a perennial grass ley (*Festuca pratensis* Huds.) assessed with different methods. *Journal of Applied Ecology*, **23**, 657–666.

Heinsoo K, Hein K, Melts I, Holm B, Ivask M (2011) Reed canary grass yield and fuel quality in Estonian farmers' fields. *Biomass and Bioenergy*, **35**, 617–625.

Hyvönen NP, Huttunen JT, Shurpali NJ, Tavi NM, Repo ME, Martikainen PJ (2009) Fluxes of nitrous oxide and methane on an abandoned peat extraction site: effect of reed canary grass cultivation. *Bioresource Technology*, **100**, 4723–4730.

Hyvönen NP, Huttunen JT, Shurpali NJ, Lind SE, Marushchak ME, Heitto L, Martikainen PJ (2013) The role of drainage ditches in greenhouse gas emissions and surface leaching losses from a cutaway peatland cultivated with a perennial bioenergy crop. *Boreal Environment Research*, **18**, 109–126.

IPCC (2013) The physical science basis. In: *Contribution of Working Group I to the Fifth Assessment Report of the Intergovernmental Panel on Climate Change* (eds Stocker TF, Qin D, Plattner G-K, Tignor M, Allen SK, Boschung J, Nauels A, Xia Y, Bex V, Midgley PM). Cambridge University Press, Cambridge.

Järveoja J, Laht J, Maddison M, Soosaar K, Ostonen I, Mander Ü (2013) Mitigation of greenhouse gas emissions from an abandoned Baltic peat extraction area by growing reed canary grass: life-cycle assessment. *Regional Environmental Change*, **13**, 781–795.

Jones MB, Finnan J, Hodkinson TR (2015) Morphological and physiological traits for higher biomass production in perennial rhizomatous grasses grown on marginal land. *GCB Bioenergy*, **7**, 375–385.

Joosten H, Clarke D (2002) *Wise Use of Mires and Peatlands: Background and Principles Including a Framework for Decision-Making*. International Mire Conservation Group and Internatonal Peat Society, Saarijärvi, Finland.

Kandel TP, Elsgaard L, Karki S, Lærke PE (2013) Biomass yield and greenhouse gas emissions from a drained fen peatland cultivated with reed canary grass under different harvest and fertilizer regimes. *BioEnergy Research*, **6**, 883–895.

Karki S, Elsgaard L, Audet J, Lærke PE (2014) Mitigation of greenhouse gas emissions from reed canary grass in paludiculture: effect of groundwater level. *Plant and Soil*, **383**, 217–230.

Karki S, Elsgaard L, Kandel TP, Laerke PE (2015a) Full GHG balance of a drained fen peatland cropped to spring barley and reed canary grass using comparative assessment of CO_2 fluxes. *Environmental Monitoring and Assessment*, **187**, 62.

Karki S, Elsgaard L, Lærke PE (2015b) Effect of reed canary grass cultivation on greenhouse gas emission from peat soil at controlled rewetting. *Biogeosciences*, **12**, 595–606.

Kasimir-Klemedtsson Å, Klemedtsson L, Berglund K, Martikainen P, Silvola J, Oenema O (1997) Greenhouse gas emissions from farmed organic soils: a review. *Soil Use and Management*, **13**, 245–250.

Kätterer T, Andrén O (1999) Growth dynamics of reed canarygrass (*Phalaris arundinacea* L.) and its allocation of biomass and nitrogen below ground in a field receiving daily irrigation and fertilisation. *Nutrient Cycling in Agroecosystems*, **54**, 21–29.

Kinmonth-Schultz H, Kim S (2011) Carbon gain, allocation, and storage in rhizomes in response to elevated atmospheric carbon dioxide and nutrient supply in a perennial C 3 grass, *Phalaris arundinacea*. *Functional Plant Biology*, **38**, 797–807.

Klemedtsson L, Von Arnold K, Weslien P, Gundersen P (2005) Soil CN ratio as a scalar parameter to predict nitrous oxide emissions. *Global Change Biology*, **11**, 1142–1147.

Kløve B, Sveistrup TE, Hauge A (2010) Leaching of nutrients and emission of greenhouse gases from peatland cultivation at Bodin, Northern Norway. *Geoderma*, **154**, 219–232.

Koehler A-K, Murphy K, Kiely G, Sottocornola M (2009) Seasonal variation of DOC concentration and annual loss of DOC from an Atlantic blanket bog in South Western Ireland. *Biogeochemistry*, **95**, 231–242.

Komulainen V-M, Nykänen H, Martikainen PJ, Laine J (1998) Short-term effect of restoration on vegetation change and methane emissions from peatlands drained for forestry in southern Finland. *Canadian Journal of Forest Research*, **28**, 402–411.

Lemus R, Lal R (2005) Bioenergy crops and carbon sequestration. *Critical Reviews in Plant Sciences*, **24**, 1–21.

Lewandowski I, Scurlock JMO, Lindvall E, Christou M (2003) The development and current status of perennial rhizomatous grasses as energy crops in the US and Europe. *Biomass and Bioenergy*, **25**, 335–361.

Loftfield N, Flessa H, Augustin J, Beese F (1997) Automated gas chromatographic system for rapid analysis of the atmospheric trace gases methane, carbon dioxide, and nitrous oxide. *Journal of Environment Quality*, **26**, 560.

Mäkiranta P, Hytönen J, Lasse A *et al.* (2007) Soil greenhouse gas emissions from afforested organic soil croplands and cutaway peatlands. *Boreal Environment*, **12**, 159–175.

Maljanen M, Sigurdsson BD, Guðmundsson J, Óskarsson H, Huttunen JT, Martikainen PJ (2010) Greenhouse gas balances of managed peatlands in the Nordic countries – present knowledge and gaps. *Biogeosciences*, **7**, 2711–2738.

Mander Ü, Järveoja J, Maddison M, Soosaar K, Aavola R, Ostonen I, Salm J-O (2012) Reed canary grass cultivation mitigates greenhouse gas emissions from abandoned peat extraction areas. *GCB Bioenergy*, **4**, 462–474.

McClaugherty CA, Aber JD, Melillo JM (1982) The role of fine roots in the organic matter and nitrogen budgets of two forested ecosystems. *Ecology*, **63**, 1481–1490.

Nilsson M, Sagerfors J, Buffam I *et al.* (2008) Contemporary carbon accumulation in a boreal oligotrophic minerogenic mire – a significant sink after accounting for all C-fluxes. *Global Change Biology*, **14**, 2317–2332.

Peichl M, Sonnentag O, Nilsson MB (2015) Bringing color into the picture: using digital repeat photography to investigate phenology controls of the carbon dioxide exchange in a boreal mire. *Ecosystems*, **18**, 115–131.

Riutta T, Laine J, Tuittila E-S (2007) Sensitivity of CO2 exchange of fen ecosystem components to water level variation. *Ecosystems*, **10**, 718–733.

Roulet NT, Lafleur PM, Richard PJH, Moore TR, Humphreys ER, Bubier J (2007) Contemporary carbon balance and late Holocene carbon accumulation in a northern peatland. *Global Change Biology*, **13**, 397–411.

Salm J-O, Maddison M, Tammik S, Soosaar K, Truu J, Mander Ü (2012) Emissions of CO_2, CH_4 and N_2O from undisturbed, drained and mined peatlands in Estonia. *Hydrobiologia*, **692**, 41–55.

Schrier-Uijl AP, Kroon PS, Leffelaar PA, van Huissteden JC, Berendse F, Veenendaal EM (2010) Methane emissions in two drained peat agro-ecosystems with high and low agricultural intensity. *Plant and Soil*, **329**, 509–520.

Schrier-Uijl AP, Kroon PS, Hendriks DMD, Hensen A, Van Huissteden J, Berendse F, Veenendaal EM (2014) Agricultural peatlands: towards a greenhouse gas sink – a synthesis of a Dutch landscape study. *Biogeosciences*, **11**, 4559–4576.

Shurpali NJ, Hyvönen NP, Huttunen JT, Biasi C, Nykänen H, Pekkarinen N, Martikainen PJ (2008) Bare soil and reed canary grass ecosystem respiration in peat extraction sites in Eastern Finland. *Tellus Series B*, **60**, 200–209.

Shurpali NJ, Hyvönen NP, Huttunen JT *et al.* (2009) Cultivation of a perennial grass for bioenergy on a boreal organic soil – carbon sink or source? *GCB Bioenergy*, **1**, 35–50.

Shurpali NJ, Strandman H, Kilpeläinen A *et al.* (2010) Atmospheric impact of bioenergy based on perennial crop (reed canary grass, *Phalaris arundinaceae*, L.) cultivation on a drained boreal organic soil. *GCB Bioenergy*, **2**, 130–138.

Silvola J, Alm J, Ahlholm U, Nykanen H, Martikainen PJ (1996) CO_2 fluxes from peat in boreal mires under varying temperature and moisture conditions. *Journal of Ecology*, **84**, 219–228.

Smith R, Slater FM (2010) The effects of organic and inorganic fertilizer applications to *Miscanthus×giganteus*, *Arundo donax* and *Phalaris arundinacea*, when grown as energy crops in Wales, UK. *GCB Bioenergy*, **2**, 169–179.

Sonnentag O, Hufkens K, Teshera-Sterne C *et al.* (2012) Digital repeat photography for phenological research in forest ecosystems. *Agricultural and Forest Meteorology*, **152**, 159–177.

Strack M, Tóth K, Bourbonniere R, Waddington JM (2011) Dissolved organic carbon production and runoff quality following peatland extraction and restoration. *Ecological Engineering*, **37**, 1998–2008.

Strand AE, Pritchard SG, McCormack ML, Davis MA, Oren R (2008) Irreconcilable differences: fine-root life spans and soil carbon persistence. *Science*, **319**, 456–458.

Strullu L, Cadoux S, Preudhomme M, Jeuffroy M-H, Beaudoin N (2011) Biomass production and nitrogen accumulation and remobilisation by *Miscanthus×giganteus* as influenced by nitrogen stocks in belowground organs. *Field Crops Research*, **121**, 381–391.

Sundh I, Nilsson M, Mikkela C, Granberg G, Svensson BH (2000) Fluxes of methane and carbon dioxide on peat-mining areas in Sweden. *Ambio*, **29**, 499–503.

Tuittila E-S, Komulainen V-M, Vasander H, Nykänen H, Martikainen PJ, Laine J (2000) Methane dynamics of a restored cut-away peatland. *Global Change Biology*, **6**, 569–581.

Uri V, Lõhmus K, Mander Ü *et al.* (2011) Long-term effects on the nitrogen budget of a short-rotation grey alder (*Alnus incana* (L.) Moench) forest on abandoned agricultural land. *Ecological Engineering*, **37**, 920–930.

Verhoeven JTA, Setter TL (2010) Agricultural use of wetlands: opportunities and limitations. *Annals of Botany*, **105**, 155–163.

Vicca S, Luyssaert S, Peñuelas J *et al.* (2012) Fertile forests produce biomass more efficiently. *Ecology Letters*, **15**, 520–526.

Waddington JM, Warner KD, Kennedy GW (2002) Cutover peatlands: a persistent source of atmospheric CO_2. *Global Biogeochemical Cycles*, **16**, 1–7.

Waring RH, Landsberg JJ, Williams M (1998) Net primary production of forests: a constant fraction of gross primary production? *Tree Physiology*, **18**, 129–134.

Wichtmann W, Schäfer A (2007) Alternative management options for degraded fens – utilisation of biomass from rewetted peatlands.

Wilson D, Alm J, Laine J, Byrne KA, Farrell EP, Tuittila E-S (2009) Rewetting of cutaway peatlands: are we re-creating hot spots of methane emissions? *Restoration Ecology*, **17**, 796–806.

Xiong S, Kätterer T (2010) Carbon-allocation dynamics in reed canary grass as affected by soil type and fertilization rates in northern Sweden. *Acta Agriculturae Scandinavica, Section B — Soil & Plant Science*, **60**, 24–32.

5

Plant roots and GHG mitigation in native perennial bioenergy cropping systems

JACOB M. JUNGERS[1], JAMES O. ECKBERG[1], KEVIN BETTS[1], MARGARET E. MANGAN[2], DONALD L. WYSE[1] and CRAIG C. SHEAFFER[1]

[1]*Department of Agronomy and Plant Genetics, University of Minnesota, 411 Borlaug Hall, 1991 Upper Buford Circle, Saint Paul, MN 55108, USA*, [2]*Minnesota Department of Agriculture, 625 Robert Street N, Saint Paul, MN 55155, USA*

Abstract

Native perennial bioenergy crops can mitigate greenhouse gases (GHG) by displacing fossil fuels with renewable energy and sequestering atmospheric carbon (C) in soil and roots. The relative contribution of root C to net GHG mitigation potential has not been compared in perennial bioenergy crops ranging in species diversity and N fertility. We measured root biomass, C, nitrogen (N), and soil organic carbon (SOC) in the upper 90 cm of soil for five native perennial bioenergy crops managed with and without N fertilizer. Bioenergy crops ranged in species composition and were annually harvested for 6 (one location) and 7 years (three locations) following the seeding year. Total root biomass was 84% greater in switchgrass (*Panicum virgatum* L.) and a four-species grass polyculture compared to high-diversity polycultures; the difference was driven by more biomass at shallow soil depth (0–30 cm). Total root C (0–90 cm) ranged from 3.7 Mg C ha^{-1} for a 12-species mixture to 7.6 Mg C ha^{-1} for switchgrass. On average, standing root C accounted for 41% of net GHG mitigation potential. After accounting for farm and ethanol production emissions, net GHG mitigation potential from fossil fuel offsets and root C was greatest for switchgrass (−8.4 Mg CO2e ha^{-1} yr^{-1}) and lowest for high-diversity mixtures (−4.5 Mg CO2e ha^{-1} yr^{-1}). Nitrogen fertilizer did not affect net GHG mitigation potential or the contribution of roots to GHG mitigation for any bioenergy crop. SOC did not change and therefore did not contribute to GHG mitigation potential. However, associations among SOC, root biomass, and root C : N ratio suggest greater long-term C storage in diverse polycultures vs. switchgrass. Carbon pools in roots have a greater effect on net GHG mitigation than SOC in the short-term, yet variation in root characteristics may alter patterns in long-term C storage among bioenergy crops.

Keywords: biofuel, carbon sequestration, diversity, greenhouse gases, prairie, species richness, switchgrass

Introduction

The concentration of carbon dioxide (CO_2) and other greenhouse gases (GHG) in the atmosphere are increasing and altering global climate (IPCC, 2014). Two pathways to mitigate atmospheric GHG emissions include (1) reducing fossil fuel consumption and CO_2 related emissions and (2) increasing terrestrial carbon (C) sinks (Pacala & Socolow, 2004). Perennial grassland bioenergy crops can mitigate GHG emissions via both pathways by displacing fossil fuel use with renewable biofuels and sequestering C in roots and soils (Gelfand *et al.*, 2013; Tilman *et al.*, 2006). However, the relative contribution of fossil fuel displacement and belowground C sequestration to net GHG mitigation potential is not well known, especially in response to management options such as plant diversity and N fertilization.

Perennial grassland bioenergy systems – ranging from grass monocultures to diverse polycultures – can produce bioenergy with greater net energy yields compared to conventional biofuel crops such as corn (*Zea mays* L.; Tilman *et al.*, 2006; Schmer *et al.*, 2008). Switchgrass (*Panicum virgatum* L.) was identified as a widely adapted, high-yielding native perennial bioenergy crop during the late 20th century and is currently undergoing selection and breeding for greater biomass yields throughout the United States (Casler & Vogel, 2014). Mixing switchgrass with other grasses, legumes, and forbs in polycultures can increase diversity and other ecosystem services including soil C sequestration (Fornara & Tilman, 2008) compared with conventional biofuel systems (Tilman *et al.*, 2006). While ecological theory and experimentation also show that communities with more species should produce more biomass (Tilman *et al.*, 1996; Hector *et al.*, 1999; Loreau *et al.*, 2001), switchgrass monocultures often yield more biomass than polycultures in agricultural settings (Picasso *et al.*,

Correspondence: Jacob M. Jungers, e-mail: junge037@umn.edu

2008; Zamora *et al.*, 2013; Jungers *et al.*, 2015b; Zilverberg *et al.*, 2014). The high performance of switchgrass monocultures is partly explained by the selection and breeding of germplasm along with advancements in agronomics (Casler, 2005). There is a persisting need to improve species selection and agronomic management of diversity and fertility of such polycultures to efficiently produce both bioenergy and ecosystem services including C sequestration.

Nitrogen is a limiting nutrient in most native grassland ecosystems, and thus, N fertilization has been frequently used to increase biomass yields in perennial grassland bioenergy systems (Lemus *et al.*, 2008; Kering *et al.*, 2011; Jarchow & Liebman, 2013; Jungers *et al.*, 2015a,b). Fertilizing grasslands also enhances net energy gain because greater biomass yields from fertilization provide more energy than is used in producing and applying N fertilizer (Schmer *et al.*, 2008). Nitrogen fertilization had a similar affect on the GHG balance of perennial grassland bioenergy systems; fertilized grasslands offset more C with increased biofuel yields despite additional emissions related to production and application of N fertilizer (Gelfand *et al.*, 2013). Negative environmental consequences related to N fertilization are less for perennial than annual bioenergy crops. Once established, perennial bioenergy crops leached ten times less nitrate (McIsaac *et al.*, 2010) and emitted significantly less nitrous oxide (Smith *et al.*, 2013) compared with a corn–soybean rotation in the Upper Midwest, USA. However, long-term N addition can reduce grassland diversity and have potential consequences for carbon storage (Fornara & Tilman, 2012) and productivity (Isbell *et al.*, 2013), which emphasizes the need to better understand these interactions.

Understanding the mechanisms and consequences of belowground C and nutrient dynamics in grassland bioenergy systems can improve agronomic management to enhance GHG mitigation. Soil organic carbon (SOC) has been identified as an important C sink for GHG mitigation (Lemus & Lal, 2005) and is dependent on the quantity and quality of root biomass inputs (Agostini *et al.*, 2015). During the early years of perennial bioenergy crop establishment, a significant amount of C can be transferred from the atmosphere to the relatively labile belowground pool of roots (Frank *et al.*, 2004). Root decomposition results in a complex set of feedbacks and interactions that can have both positive and negative consequences for GHG mitigation. Nutrient mineralization from root decomposition stimulates plant productivity (De Graaff *et al.*, 2010), which is especially important for sustaining yields in annually harvested, low-input systems (Garten & Wullschleger, 2000). However, GHGs are emitted in the form of CO_2 during decomposition, which can increase in the presence of mineralized nutrients and result in a net loss of ecosystem C (Mack *et al.*, 2004; Dijkstra *et al.*, 2013). During decomposition, some fraction of root C is also transferred to recalcitrant SOC for long-term storage (Horwath, 2007). Therefore, the relative importance of these processes in relation to net GHG mitigation partly depends on the timescale of assessment (Knops & Tilman, 2000). A range of SOC change values have been reported in both short-and long-term studies (Ma *et al.*, 2000; McLauchlan *et al.*, 2006; Adler *et al.*, 2009; Gelfand *et al.*, 2013) which emphasizes the importance of our need to understand the factors driving these processes (e.g., root C to N ratio (C : N), root and shoot C allocation, management activities) and how they interact through time.

The relative effects of plant species composition and N fertilization – alone and in combination – on net GHG mitigation potential in bioenergy feedstock systems are not well known. Long-term research shows that diverse polycultures of native perennial grassland plant species accumulated more soil C than monocultures of the same species (Fornara & Tilman, 2008; Steinbeiss *et al.*, 2008; Cong *et al.*, 2014). Complementarity among species of varying functional traits can enhance SOC storage by increasing organic matter inputs from above- and belowground plant communities (Fornara & Tilman, 2008). Nitrogen fertilization increases aboveground, but not belowground, biomass in grassland bioenergy crops (Jarchow *et al.*, 2012), thus improving GHG mitigation by increasing fossil fuel offsets rather than belowground C storage. Besides altering biomass quantity, both species diversity and N fertilizer can alter SOC dynamics by influencing biomass quality. The C : N of biomass inputs like roots is important for determining decomposition rate and SOC storage and has been negatively correlated with species diversity and N fertilizer rates (Mueller *et al.*, 2013). As species composition and N fertilizer are important management factors in grassland bioenergy systems, it is essential to understand how they interact to influence above- and belowground productivity and the relative impact of these variables on net GHG mitigation.

Our objectives were to (1) determine the effects of plant species composition and N fertilization on root biomass and soil C; (2) determine the relative and absolute contribution of C stored in soil, belowground biomass, and fossil fuel offsets (i.e., aboveground biomass) to net GHG mitigation potential for all systems; and (3) compare the net GHG mitigation potential of five bioenergy cropping systems ranging in species diversity and managed with vs. without N fertilizer. Rather than comparing productivity and C dynamics from random assemblages of species across a diversity gradient like most dedicated diversity experiments, this study presents findings from a dedicated bioenergy experiment that was designed and managed (no hand weeding) to resemble practical bioenergy cropping systems, replicated across a range of growing conditions, and span 7–8 years of production.

Methods

Study sites

This study was established in the spring of 2006 at Becker, Lamberton, and Waseca, Minnesota; and in 2007 at Crookston, Minnesota, USA. Data used for this study were collected from a subset of locations and treatments of a larger study (See Mangan *et al.*, 2011). We selected these locations because they represent a range of growing conditions including soil type, precipitation, and temperature gradients (Table 1). Data were analyzed for all available years at each location (2006–2013 or 2007–2013). Soils at Becker are sandy, mixed, frigid Entic Haploboroll, while soils at Crookston are coarse-silty over clayey, mixed over smectitic, superactive, frigid Aeric Calciaquolls. Soils at Lamberton and Waseca are Aauic Hapludolls and Typic Endoaquolls, respectively, and both described as fine-loamy, mixed, superactive, and mesic. The previous crop was corn (*Zea mays* L.) at Waseca and Lamberton, rye (*Secale cereale* L.) at Becker, and spring wheat (*Triticum aestivum* L.) at Crookston.

The experiment was a randomized complete block in a split-plot arrangement with three replications per location. Nitrogen fertilizer was applied to the main plots at either 0 or 67 kg N ha^{-1} as ammonium nitrate annually in spring from 2008 to 2013. Fertilization was delayed until 1 or 2 years after the seeding year to ensure establishment of sown species and prevent invasion of noxious weeds. Species mixture treatments were assigned to subplots (9 m^2; 3 m × 3 m) and included: switchgrass monoculture; four-species grass mix; eight species grass/legume mix; 12 species grass, forb, and legume mix; and high-diversity 24 species mix of species from all three functional groups (i.e., grasses, legumes and nonleguminous forbs; See Table S1 for complete list). All species represent unimproved ecotype seed collected in the upper Midwest. For information on establishment success and plant densities for each treatment and location, see Mangan *et al.* (2011). Annual biomass yields and species composition are available in Jungers *et al.* (2015b).

The research sites were tilled in May prior to seeding. Sites at Becker, Lamberton, and Waseca received Vapam (Sodium N-methyldithiocarbamate) herbicide at a rate of 84 kg active ingredient ha^{-1} to suppress weeds prior to seeding. Plots were broadcast seeded at a rate of 12 g pure live seed m^{-2}. Weeds were manually removed in 2006 and 2007, controlled using a pre-emergent herbicide (acetochlor [2-Chloro-*N*-(ethoxymethyl)-*N*-(2-ethyl-6-methylphenyl) acetamide]) in 2008 and 2009, and then by hand weeding and infrequent spot-spraying of glyphosate [N-(phosphomonomethyl) glycine] to control sporadic outbreaks of Canada thistle (*Cirsium arvense* (L.) Scop.) for the remainder of the experiment.

Soil measurements

Prior to seeding (2006 or 2007), three replicates of ten soils cores (1.5 cm diameter) were collected to a depth of 60 cm at each site. Soil cores were divided into four 15-cm increments and combined by depth for each replicate. In the fall of 2013, three soil cores were collected per subplot and divided into the same depth increments. The three cores were combined by depth for each subplot. Samples were dried at 33 °C and passed through a 2-mm sieve. Carbonates were measured with an acid test consisting of 1 M HCL. Soil organic C and N were determined via dry combustion (Nelson & Sommers, 1996) using an Elementar Vario Max CN analyzer (Hanau, Germany).

Root measurement

In October 2013, three soil cores (4.1 cm diameter) were collected (equally spaced) along a 1-m transect in the middle of each subplot to a depth of 90 cm. Soil cores were divided into three 30-cm segments and frozen until processed for root mass determination. Plant biomass was separated from soil using a hydropneumatic elutriation system (Smucker *et al.*, 1982). This system uses pressurized air to force water through soil and a series of mesh screens to remove soil from biomass. Samples were further cleaned by hand sieving to remove sand and debris. Belowground biomass was dried at 60 ° until constant weight and then manually

Table 1 Soil and climate conditions at each experimental site in Minnesota, USA

		Precipitation (cm)*		Temperature (°C)*		Soil texture (%)			OM†	TOC‡
Location	Location	May–July	August–October	May–July	August–October	> 2.0 mm	0.25–2.0 mm	< 0.25 mm	g kg^{-1}	
Becker	45.3870, −93.8818	30.0	25.3	18.6	14.8	12.1	80.7	7.2	14	9
Crookston	47.8103, −96.6145	23.0	17.8	17.6	13.5	NA	NA	NA	27	NA
Lamberton	44.2373, −95.3020	27.2	20.0	19.2	15.1	23.7	42.5	33.9	36	22
Waseca	44.0678, −93.5258	32.1	26.2	18.8	15.0	28.6	43.5	27.8	43	23

*30-year average (1971–2000).
†Organic matter.
‡Total organic carbon.

screened to remove all aboveground structures (i.e., leaves and stems). Some samples included crowns and rhizomes, which were included in the belowground biomass sample. Dry mass was recorded for all belowground biomass and hereafter referred to as roots. As we do not have repeated measures of root biomass across time, root biomass is presented as the net difference in root growth and root mortality after 7 (Crookston) or 8 (Becker, Lamberton, and Waseca) years since seeding.

After dried root samples were weighed separately, the three replicates were combined by depth for each subplot and ground to pass through a 1-mm screen. The samples were then analyzed for C and N using a combustion analyzer.

GHG mitigation potential by roots was calculated as the product of root biomass, root C content, and the CO_2 e constant, and then annualized by dividing the value by the number of years since seeding (7 years for Crookston and 8 years for Becker, Lamberton, and Waseca; Eqn S1). The GHG mitigation potential from root biomass was annualized so that it can be compared to other components of the net GHG mitigation potential equation (Eqn S7).

Fossil fuel offsets

Biomass yield was measured annually after the seeding year (from 2007 or 2008 to 2013) in all plots from each location following a killing frost as described by Jungers *et al.* (2015yb). In 2008, all dried biomass samples were ground and scanned using near infrared spectroscopy (NIRS) to estimate the concentration of cell wall carbohydrates using calibrated equations based on wet chemistry results. A subset of biomass samples was processed using wet chemistry to validate NIRS results. A detailed explanation of this method is described by Vogel *et al.* (2010). Cell wall carbohydrate concentrations were used to predict theoretical ethanol yield (Eqn S2) and adjusted to assume 90% conversion efficiency. GHG mitigation from fossil fuel offsets of ethanol production was determined using published estimates from the GREET life cycle analysis model (Gelfand *et al.*, 2013; Eqn S3).

GHG emissions from bioenergy crop management

GHG emissions from managing bioenergy crops come from farm inputs (N fertilizer production and application), biomass handling (transporting and processing biomass for cellulosic energy production), and N_2O emissions from farmland. For GHG emission estimates related to N fertilizer production and application, biomass harvest and handling, and cellulosic ethanol conversion, we used GREET model estimates published in Gelfand *et al.* (2013; Eqn S4). GHG emissions from N_2O emissions were estimated using the empirical model MGRASS, which is based on N fertilizer rate and the mean temperature (°C) in January, February, and March (Roelandt *et al.*, 2005; Eqn S5). We estimated N_2O emissions using daily temperature readings at each location for each year. Therefore, estimates are reported by location and do not include subplot level variation from species mixture treatments.

Net GHG mitigation

Net GHG mitigation (Eqn S7) is the average annual net total of the GHG sinks minus GHG emissions measured as Mg CO_2e ha^{-1} yr^{-1}. GHG sinks include (1) root C: the annualized difference in root growth and root mortality, and (2) fossil fuel offsets: average annual GHG mitigation potential from offsetting fossil fuels. Greenhouse gas sources include average annual GHG emissions related to N fertilizer production and application, biomass harvest and handling, cellulosic ethanol conversion, and N_2O emissions from management for each bioenergy crop/N fertilizer treatment combination. As not all of these GHG components were measured at the same spatial scale, we chose not to conduct statistical tests on the sum of these components (i.e., net GHG mitigation), but rather statistically test only the variables measured for each experimental unit (subplot).

Statistics

Data were analyzed using the open-source statistical software R (version 3.1.3; R Development Core Team, 2014). We used mixed-effects analysis of variance models to explain variation in root biomass, C, N, and C : N ratio by fitting the fixed effects of species mixture and N fertilizer treatments as categorical variables, and depth as a continuous variable, along with all possible interactions. We used this same fixed effects structure to explain variation in final SOC, and changes in SOC from the first year to 2013. When there were interactions between depth and either of the other main effects (species mixture or N fertilizer treatment), we fit separate models with depth as a categorical variable to evaluate interactions.

For all response variables, we fit a nested random effects structure that included species mixture subplots nested within fertilizer treatment main plots, within block, within site to account for spatial autocorrelation and pseudo-replication. Mixed-effects models were conducted using the R package LME4 (Bates *et al.*, 2014). We fit this model 9999 times to bootstrapped datasets for each response variable. Each bootstrapped dataset included *n* values that were randomly sampled with replacement from the original dataset of size *n*. From the 9999 models, we used the distribution of the coefficients to estimate the mean effect size and 95% confidence intervals (CIs) for all main effect treatment levels for each response variable. *Post hoc* comparisons of treatment level means were made by comparing overlap of CIs across treatment combinations of interest and verified with Tukey adjusted P values ($\alpha = 0.05$) to control the experimentwise type 1 error rate (Ruxton & Beauchamp, 2008). Mean comparisons by CI overlap and Tukey were mostly consistent, and when they differed, we reference the Tukey P value in the text. Pairwise comparisons were made with tools in the R package multcomp (Torsten *et al.*, 2008).

We tested for relationships among response variables to further explore observed patterns in the data. We used mixed-effects linear regression models to test for associations among root biomass, root C : N, and changes in SOC. The same random effects structure was used as in the ANOVA models (i.e., subplot within plot within block within site).

Table 2 Analysis of variance table for root biomass (Mg ha^{-1}), root carbon (C) content (g kg^{-1}), root nitrogen (N) content (g kg^{-1}), and root carbon to nitrogen ratio (C : N)

	Root biomass			Root C			Root N		Root C : N	
Source	DF*	*F* ratio	*P* value	DF	*F* ratio	*P* value	*F* ratio	*P* value	*F* ratio	*P* value
Trt	4, 88	22.52	**<0.001**	4, 88	6.17	**<0.001**	34.45	**<0.001**	39.72	**<0.001**
Nfert	1, 11	0.77	0.4	1, 11	1.12	0.313	10.65	**0.008**	19.36	**0.001**
Depth	2, 709	1301.45	**<0.001**	2, 229	117.13	**<0.001**	250.28	**<0.001**	335.20	**<0.001**
TrtXNfert	4, 88	0.83	0.507	4, 88	0.71	0.586	0.44	0.777	2.62	**0.040**
TrtXDepth	4, 709	21.06	**<0.001**	4, 229	2.15	0.076	1.03	0.394	7.99	**<0.001**
NfertXDepth	1, 709	4.47	**0.035**	1, 29	0.04	0.848	4.92	0.168	0.68	0.411
TrtXNfertXDepth	4, 709	2.84	**0.024**	4, 229	0.74	0.569	2.82	**0.026**	0.36	0.837

*Number of degrees of freedom in the numerator and denominator of *F* test.
Bold values indicate significance at $P = 0.05$.

Results

Belowground biomass production

The effect of species mixtures and N fertilization on root biomass varied by depth as indicated by the two-way and three-way interactions among these variables (Table 2). Root biomass was greater in the switchgrass, grass mix, and grass/legume mix compared to the 12-species and high-diversity mixtures in the 0–30 cm soil profile, but root biomass was similar for switchgrass and the mixtures in the 30–60 and 60–90 cm profiles (Fig. 1). Switchgrass root biomass from 0 to 30 cm was greater in unfertilized compared with fertilized plots ($t_{11} = 4.51$; $P = 0.019$), but root biomass for the other species mixtures were similar across fertilizer treatments and depths.

Root C and N concentration

There was a significant effect of species mixture treatments and depth on root C concentration, but root C was not affected by N fertilizer. There were no two-way or three-way interactions among variables (Table 2). Root C concentration increased with depth for all species mixture treatments (Fig. S1). At the shallow depth (0–30) where roots were most abundant, mean root C concentration was largest in switchgrass (mean = 40.5%; SE = 0.6) compared with the other species mixture treatments (mean = 37.8%; SE = 0.3).

Species mixture, N fertilizer, and depth all affected root N concentration alone and in combination (Table 2). Root N concentration was greater in grass/legume, 12-species, and high-diversity mixture treatments compared to the grass mix and switchgrass

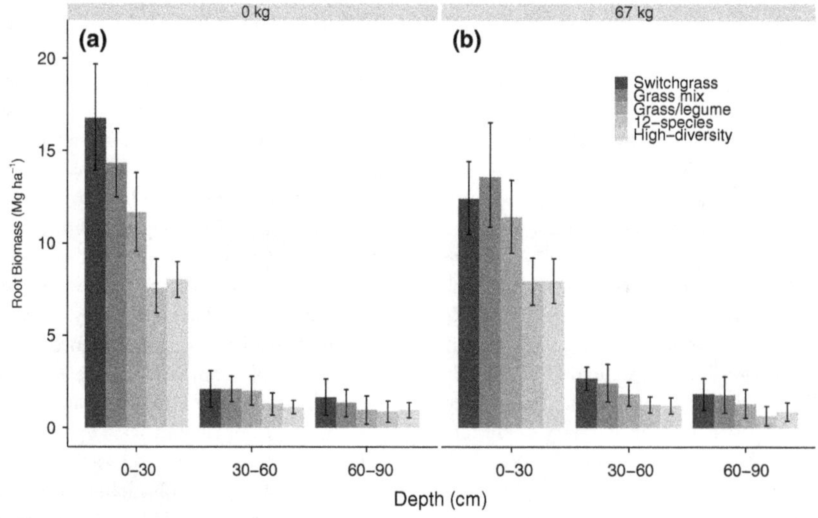

Fig. 1 Root biomass (95% CIs) at three depth intervals for five native perennial bioenergy crops managed with 0 (a) and 67 (b) kg N ha^{-1}. Measurements reflect seven (one site) or eight (three sites) years of root growth after seeding.

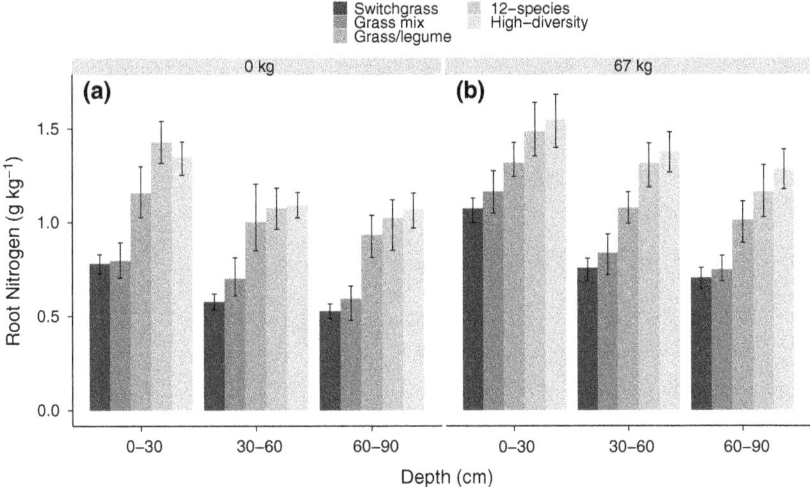

Fig. 2 Root nitrogen content (95% CIs) at three depth intervals for five native perennial bioenergy crops managed with 0 (a) and 67 (b) kg N ha^{-1}.

treatments at most depth/N fertilizer rate combinations (Fig. 2). Root N concentration increased with N fertilizer and decreased with depth. Root N varied more within and across treatments than root C and therefore contributed more to variation in the root C : N ratio (Fig. 3).

Species mixture and N fertilization affected root C : N, both alone and interactively; species mixture effects also varied by depth (Table 2). Root C : N was greatest in the switchgrass and grass mix and lowest in the 12-species and high-diversity mixtures at most depths. Root C : N was lower in the fertilized vs. unfertilized switchgrass treatment at all depths, and lower in the grass mix at 0–30 and 60–90 cm depths. Root C : N was similar across fertilizer treatments for the grass/legume, 12-species, and high-diversity mixtures. Root C : N increased with depth, but more so for switchgrass and the grass mix (Fig. 3). Root C : N was positively associated with root biomass at the 0–30 ($P < 0.001$; Fig. 4a), 30–60 ($P < 0.001$), and the 60–90 ($P < 0.001$) cm depth increments.

Soil organic carbon

Final SOC was similar across all treatments but decreased with depth ($F_{1,347} = 597$; $P < 0.001$). Averaged across treatments final SOC was 22.5, 17.7, 14.5, and 11.3 g kg^{-1} at the 0–15, 15–30, 30–45, and 45–60 cm

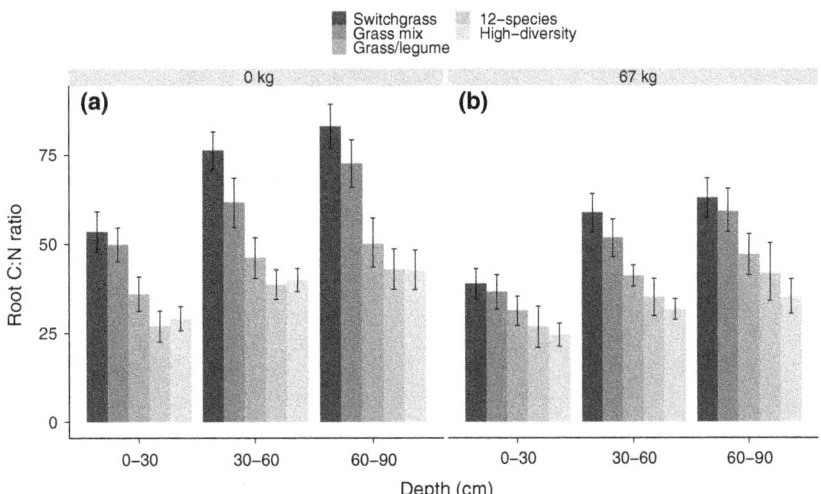

Fig. 3 Root carbon to nitrogen ratio (95% CIs) at three depth intervals for five native perennial bioenergy crops managed with 0 (a) and 67 (b) kg N ha^{-1}.

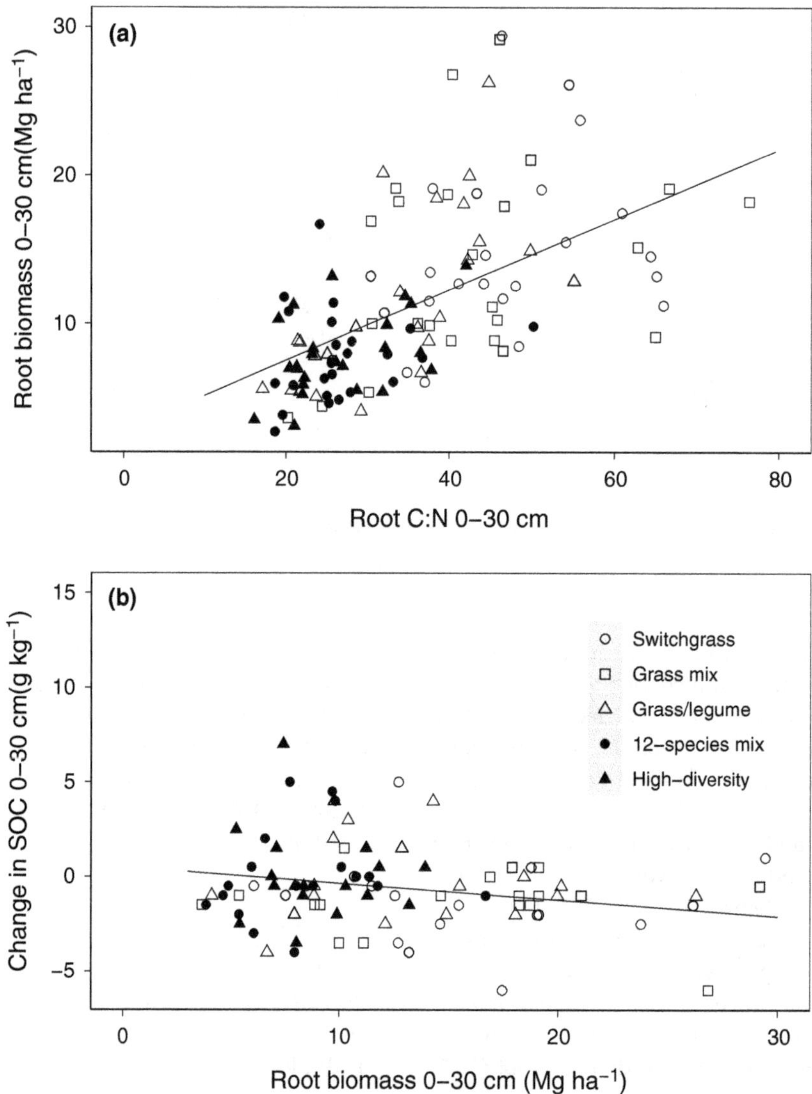

Fig. 4 Relationship between root biomass and C : N from 0 to 30 cm (a) and relationship between the change in shallow soil organic carbon (SOC; average of 0–15 and 15–30 cm intervals for each subplot) and total root biomass from 0 to 30 cm (b) in five native perennial bioenergy crops.

depth increments, respectively. Comparing final to initial SOC values showed no change across treatments (Fig. S3). When analyzing the data by discrete depth increments, SOC decreased in the 15–30 cm interval for all treatments in the unfertilized plots, and for all treatments except the high-diversity in the fertilized plots (mean = 2.66 g kg^{-1}; SE = 0.22). Change in SOC averaged across 0–15 and 15–30 cm depth intervals was negatively associated with root biomass from 0 to 30 cm (Fig. 4b; P = 0.026; Y-intercept = 2.781; β = 0.237).

Greenhouse gas mitigation potential: root biomass

The GHG mitigation potential by roots varied by species mixture treatment (F = 27.824, P < 0.001, Fig. 5) but not N fertilizer (F = 0.452, P = 0.549). Averaged across fertilizer treatments, the GHG mitigation potential of carbon storage in root biomass was largest for switchgrass (mean = −3.6 Mg CO_2 e ha^{-1} yr^{-1}, SE = 0.2) and the grass mix (mean = −3.2 Mg CO_2 e ha^{-1} yr^{-1}, SE = 0.3). The grass mix was similar to the grass/legume mix (mean = −2.6 Mg CO_2 e ha^{-1} yr^{-1}, SE = 0.2), but different than the 12-species (mean = −1.8 Mg CO_2 e ha^{-1} yr^{-1}, SE = 0.1) and high-diversity (mean = −1.8 Mg CO_2 e ha^{-1} yr^{-1}, SE = 0.1) mixtures. Variation in root biomass with soil depth (Fig. 1) contributed significantly to GHG mitigation potential with roots from shallow soil depth (0–30 cm) contributing considerably more to GHG mitigation than other soil depths.

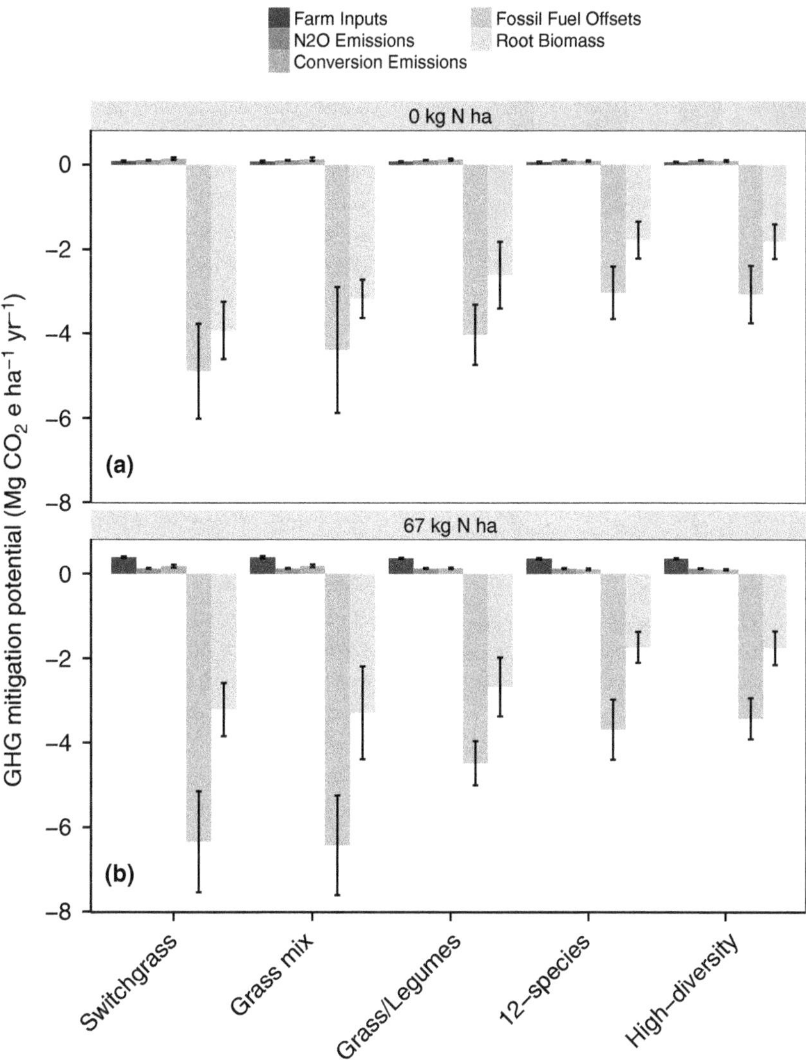

Fig. 5 Average annual greenhouse gas (GHG) mitigation potential of five native perennial bioenergy crop mixtures managed with 0 (a) and 67 (b) kg N ha^{-1}. Farm inputs include fossil fuel requirements for fertilizer production and application and biomass harvest. N_2O emissions are modeled using winter temperatures and N inputs. Conversion emissions include GHGs related to converting biomass to cellulosic ethanol. Fossil fuel offsets are based on harvested biomass yield and fermentable carbohydrate content. Root biomass includes root carbon from 0 to 90 cm depth.

Greenhouse gas mitigation potential: fossil fuel offsets

The GHG mitigation potential by fossil fuel offsets varied by species mixture treatment ($F = 29.204$; $P < 0.001$) and N fertilizer ($F = 11.453$; $P = 0.043$; Fig. 5). There was a significant interaction between species mixture and N fertilizer ($F = 3.141$; $P = 0.019$). In unfertilized plots, GHG mitigation by fossil fuel offsets was similar for switchgrass, grass mix, and the grass/legume mixture (Fig. 5). In fertilized plots, GHG mitigation by fossil fuel offsets was greatest in switchgrass and the grass mix compared to grass/legumes, 12-species, and high-diversity mixtures. This is a result of lower biomass yields and ethanol potential of the 12-species and high-diversity mixtures compared to switchgrass (Table S2 and S3). Within 12-species and high-diversity mixtures, fossil fuel offsets were similar for fertilized and unfertilized plots.

GHG sources and net mitigation

On average, GHG emissions estimated at the site level or from literature sources were 7% of the GHG mitigated from fossil fuel offsets and root biomass. In fertilized plots, farm inputs (which include N fertilizer production and application) were 570% greater compared with unfertilized plots. Nitrous oxide emissions were estimated at the site level annually

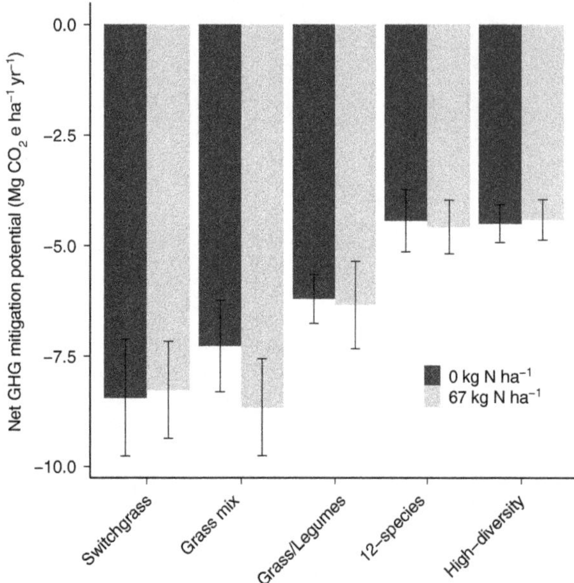

Fig. 6 Average annual net GHG mitigation potential based on the sum of all components (Fig. 5) for five native perennial bioenergy crops managed with 0 and 67 kg N ha^{-1}. Error bars are 95% CI of the mean.

based on N fertilizer rate and late winter temperatures, and averages ranged from 81 to 139 kg CO_2e ha^{-1} yr^{-1}.

Net GHG mitigation was largely driven by fossil fuel offsets and total root C. Net GHG mitigation potential was similar for switchgrass, grass, and grass/legume mixtures, and greater than the 12-species and high-diversity mixtures (Fig. 6). Within species mixture treatments, GHG mitigation potential did not vary by N fertilizer rate.

Discussion

This study is among the first to show the relative contribution of standing root biomass to GHG mitigation in perennial grassland bioenergy systems. Our results show that after seven or eight years of production, root carbon in unfertilized grassland monocultures and mixtures had nearly an equal effect on GHG mitigation as fossil fuel offsets by converting aboveground biomass to cellulosic ethanol (Fig. 5). Accounting for GHG emissions and fossil fuel offsets, root C contributions to net GHG mitigation ranged from 13 to 69%. On average, there were 13.7 and 22.5 Mg of CO_2 e ha^{-1} in the form of root C after seven or eight years of perennial bioenergy crop production.

Our estimates of root biomass are larger than estimates from other studies on native perennial bioenergy crops, which may be related to the longer timespan of our study. For example, our estimates of switchgrass root biomass were three times larger than estimates by Jung & Lal (2011) for the 0–30 cm layer, a study that sampled roots 4 years after seeding. Similarly, our estimates for grass mixture root biomass in the 0–100 cm layer were two times greater than those by Jarchow & Liebman (2012) who sampled 2 years after seeding. Root biomass and C storage in roots increased linearly through time for up to 12 years after restoring formally cultivated fields to native perennial grasses (Baer *et al.*, 2002); therefore, it is likely that the extended time period of our study contributed to the relatively high root biomass yields compared with more abbreviated studies.

An alternative explanation for why we observed more root biomass compared with other studies may be related to soil texture. Although we focus on the average effect of species mixture and N fertilizer treatments on root biomass across locations, there were site-level differences in root biomass responses to treatments (Fig. S2). The sites with the coarser soils – Becker and Crookston – tended to have more root biomass than the Lamberton and Waseca sites with finer soils, but this also depended on species mixture treatments. Root biomass values in the fine-loamy soils at Lamberton and Waseca were closer to those measured by Jarchow & Liebman (2012), who also conducted their study on fine-loamy soils. Root density tends to increase in coarse- compared to fine-textured soils because water-holding capacity is usually less in coarse-textured soil (Schenk & Jackson, 2002).

Variation in root biomass across studies can also be related to planting configuration and root sampling. Seeds were broadcast across plots in this study, while many studies measuring switchgrass biomass planted seeds in rows and sampled for roots between rows where crowns and rhizomes are minimal (Jung & Lal, 2011; Garten *et al.*, 2011). As crowns and rhizomes can account for 10–20% of switchgrass total plant biomass, comparing root biomass samples with and without these organs could result in very different results (Garten *et al.*, 2011).

Variation in root biomass across species mixture treatments may have been related to differences in root morphology among species of various functional groups. Root biomass samples included fine, coarse, live and dead roots, and rhizomes. Rhizomes were infrequently observed, but may have contributed to the greater root biomass values in switchgrass and grass mixtures in the shallow depth compared to the 12-species and high-diversity treatments because the forbs in the polycultures did not form rhizomes. In 2013 when the roots were sampled, 49% of the 12-species and 57% of the high-diversity mixtures were composed of forbs. Forbs in our mixtures included Maximilian sunflower

(*Helianthus maximiliani* Shrad.), yellow coneflower (*Ratibida pinnata* (Vent.) Barnhart), and stiff goldenrod (*Solidago rigida* L.), which produce long taproots instead of shallow fibrous systems that are known to colonize soils well beyond 1 m deep (Weaver, 1958). Our study was consistent with others in that it restricted sampling to a 90 cm depth, but this may not have captured total root biomass beneath treatments with deep-rooting forbs (Garten *et al.*, 2011).

Net GHG mitigation represented the total annual contribution from all observed and literature-based sources of GHG emissions and sinks in native bioenergy cropping systems. The GHG sinks were fossil fuel offsets from converting annual aboveground biomass to cellulosic ethanol and root C accumulation. Our measure of GHG mitigation potential of fossil fuel offsets was similar to those reported by Gelfand *et al.* (2013) for fertilized (6.2 Mg of CO_2 e ha^{-1} yr^{-1}) and unfertilized prairie (4.6 Mg of CO_2 e ha^{-1} yr^{-1}), but more than those measured by Adler *et al.* (2009) for switchgrass (about 1.5 Mg of CO_2 e ha^{-1} yr^{-1}). In our study and those by Gelfand *et al.* (2013) and Adler *et al.* (2009), fossil fuel offset values were the largest contributors to net GHG mitigation. However, our results differ from those of Gelfand *et al.* (2013) and Adler *et al.* (2009) in that we did not observe changes in SOC, which they report as a substantial sink for GHG mitigation. Instead, we found that C in root biomass was the next most significant GHG mitigation pathway, which was nearly similar in its effect as SOC in the Gelfand *et al.* (2013) study. Potential reasons for this are discussed below.

Nitrogen fertilization did not affect net GHG mitigation potential for any of the bioenergy cropping systems (Fig. 6). Although N fertilization significantly increased aboveground biomass yields (Jungers *et al.*, 2015b), the associated GHG benefit was somewhat offset as a result of increased fossil fuel emissions related to producing and applying N fertilizer, as well as decreases in belowground biomass in some species composition treatment × depth combinations. Even though N fertilization did not have a positive or negative effect on net GHG mitigation, a shift in biomass from belowground to aboveground components from N fertilization should be economically beneficial for producers as aboveground biomass has marketable value (Mooney *et al.*, 2009). However, bioenergy crops with less root biomass may be more susceptible to drought stress, which also has economic implications if aboveground biomass is reduced (Mann *et al.*, 2013).

Net GHG mitigation potential varied by species mixture treatments, which was mostly due to treatment variations in above- and belowground biomass. Farm inputs, N_2O, and conversion emissions did not vary across species mixtures. Without fertilizer, there was little variation in fossil fuel offsets among mixture treatments despite a difference in biomass yield (Table S2). For example, average biomass yields were greater in switchgrass and the grass/legume mixture compared to other treatments without fertilizer (Jungers *et al.*, 2015b) yet these treatments did not have greater fossil fuel offsets. Differences in the concentration of fermentable sugars across locations and species mixture treatments may have introduced variability to the fossil fuel offset potential (Table S3). Although short-term studies have shown that annual fossil fuel offsets are mostly dependent on biomass yield and not theoretical ethanol yield (Jungers *et al.*, 2013; Jarchow & Liebman, 2012), the cumulative effect of relatively small alterations in theoretical ethanol yield compared with biomass yield over 7 years of production suggests that theoretical ethanol yield is important when considering long-term bioenergy production.

Although the potential for perennial bioenergy crop roots to offset GHGs is similar to converting aboveground biomass to bioenergy in the absence of fertilizer, the actual impact of these GHG mitigation pathways is uncertain. First, it is not known if an equal volume of fossil fuels will actually be displaced with the production of cellulosic ethanol. The availability of cellulosic ethanol could increase overall energy use, which could limit fossil fuel displacement and subsequent GHG mitigation. Second, the C sink of roots is highly susceptible to decomposition in response to land conversion, thus the actual GHG mitigation potential of this sink is dependent on long-term land use decisions and management practices. Tillage to replace the perennial bioenergy crop with an annual crop could emit a large volume of GHGs (Gelfand *et al.*, 2011), and it is not clear how long perennial bioenergy crops will remain in a field before being converted. Nonetheless, results from this study could inform future policies designed to maximize GHG mitigation in bioenergy cropping systems.

Soil organic C accumulation is regarded as a significant contributor to GHG mitigation (Pacala & Socolow, 2004). Similar studies of perennial bioenergy crops show that SOC can mitigate from 4.0 to 7.3 Mg CO_2 e ha^{-1} yr^{-1} for 13-yr-old prairie and 9-yr-old switchgrass stands, respectively (Gelfand *et al.*, 2013; Follett *et al.*, 2012). However, SOC did not change much in this study, especially across treatments, during 7 or 8 growing seasons (Fig. 4b; Fig. S3). Previous studies agree that changes in SOC as a result of converting row crop fields to perennial cover are highly variable and dependent on, among other variables, initial soil conditions (Bransby *et al.*, 1998; Zan *et al.*, 2001; Sartori *et al.*, 2006; Adler *et al.*, 2009; Schmer *et al.*, 2011; Bandaru *et al.*, 2013; Agostini *et al.*, 2015). Although all our research sites were tilled following row crop production prior to establishment of native perennials, initial soil C levels were relatively high at most sites (Table 1). High initial C levels may have limited C accumulation by perennial crops (Tiemann & Grandy, 2015). In other studies where initial SOC levels were low and aboveground biomass was not removed, SOC increases occurred as early as 5 years after converting land from row crops to native perennial cover (Reeder *et al.*, 1998; Chimento *et al.*, 2014). In even more studies from other environments, researchers reported no change in SOC beneath switchgrass during 5 years of growth (Garten & Wullschleger, 2000; Ma *et al.*, 2000), and only after 10 years of switchgrass production did Follett *et al.* (2012) observe an increase in SOC on marginal land. Even in soils with low initial SOC values, short-term changes (<5 yr) in SOC are unlikely to be observed as the mean residence time can range from 1.2 to 2.4 years for roots (Agostini *et al.*, 2015).

Because we did not find differences in SOC change across treatments, we omitted this component from the GHG mitigation potential equation. Even if we had included SOC change to the GHG mitigation equation, the best-case N fertilizer/species mixture treatment response would have resulted in −0.14 kg CO2e ha^{-1} of storage, which is <0.5% of the net GHG mitigation potential for any treatment combination. Summing SOC changes across soil depths for each treatment combination usually resulted in net SOC loss, which in the worst-case scenario would have been equal to 15% of net GHG mitigation potential. Steinbeiss *et al.* (2008) found that converting agricultural fields to perennial grasslands resulted in a net C loss 2 years after conversion, but C stocks recovered and exceeded initial values by year 4. Spring tillage and decomposition of the previous years' crop residue combined with low biomass production during the seeding year of perennial bioenergy crops may have led to SOC losses (Zhang *et al.*, 2010). As C inputs from aboveground biomass are less in an annually harvested bioenergy system, SOC recovery may take longer compared with an unharvested system (like those studied in Steinbeiss *et al.*, 2008).

Mechanisms responsible for the negative or neutral SOC change beneath the perennial bioenergy cropping systems could be related to the complex interactions between plant litter inputs and SOM decomposition known as rhizosphere priming effects (Dijkstra *et al.*, 2013). Positive priming effects are when SOM decomposition is accelerated by the addition of plant organic matter, such as roots (Kuzyakov, 2002). Perennial bioenergy cropping systems with more root biomass exude more organic substances that fuel microbial growth, which then begin to consume old SOM. This could explain the negative association in SOC change and root biomass we observed (Fig. 4b). However, net SOC losses by root-mediated SOM decomposition is often short term, and sequestration of root-derived C to humified SOM can offset those losses within one growing season (Kuzyakov, 2002). Research quantifying respiration and SOM turnover through time in long-term experiments is needed to understand these complex plant–soil–microbe interactions and their consequences for GHG mitigation in bioenergy systems.

Soil organic C could be increasing in the grass/legume, 12-species, and high-diversity treatments more than the switchgrass and grass mix in the 0–15 cm layer (Fig. 4), but large variation in measurements led to statistically insignificant results when analyzing treatments as factored variables with $\alpha = 0.05$. Initial SOC measurements occurred at the site level, not subplot, which may have reduced our ability to detect changes in SOC among species mixtures. We did, however, find statistically significant relationships between root quality and SOC change that support this trend. Individual subplots with SOC increases in the top 30 cm of soil typically had fewer total roots (Fig. 4b), and the roots that were in the soil had a lower C : N ratio (Fig. 4a). The 12-species and high-diversity polycultures had less root biomass and lower C : N ratios than the other treatments (Table 2, Fig. 3 and 4). If SOC increases are mainly driven by root C inputs (because aboveground litter inputs are less in these annually harvested systems; Rasse *et al.*, 2005), it is expected that roots with a lower C : N ratio would be more susceptible to decomposition compared with high C : N roots (Johnson *et al.*, 2007). In turn, plots with lower C : N root tissue (e.g., 12-species and high-diversity mixtures) should have less root biomass and potentially more SOC buildup if there is also greater decomposition.

Our data suggest that the low C : N ratio roots in diverse mixture bioenergy grasslands may be decomposing faster than high C : N ratio roots of grass-dominated systems. This could reduce the short-term C stock in standing root biomass under diverse polyculture systems, but may lead to greater SOC in the long-term compared with grass-dominated systems (Johnson *et al.*, 2007). Nutrient removal during annual biomass harvest is substantial (Jungers *et al.*, 2015a; Gillitzer *et al.*, 2012) and could affect soil nutrient stocks and subsequent productivity (Schmer *et al.*, 2011). Low-input systems with lower root mean retention times may have greater mineralization rates that help maintain yields longer than those with more recalcitrant roots in the absence of fertilizers. Therefore, the benefit of species diversity for SOC accumulation and yield may not be realized until a decade or more after establishment.

This study is one of the first to compare above- and belowground factors contributing to GHG mitigation by grasslands of varying diversity levels managed for bioenergy. Without N fertilizer, root C contributed as much GHG mitigation as fossil fuel offsets by aboveground biomass converted to cellulosic ethanol for some species mixtures. N fertilizer increased aboveground biomass and fossil fuel offsets, but generally had no effect on root biomass except for switchgrass, and the overall effect of N fertilizer on net GHG mitigation was small. Grass-dominated plantings (e.g., switchgrass monocultures, grass mixtures, and grass/legume mixtures) showed greater net GHG mitigation potential than more diverse mixtures. Changes in SOC did not vary across species mixture or N fertilizer treatments and did not influence net GHG mitigation. However, the 12-species and high-diversity mixtures had lower root C : N ratios, which were associated with less root biomass and increases in SOC. This study underscores the importance of short-term (roots) and long-term (SOC) belowground C pools in relation to fossil fuel offsets in native perennial bioenergy systems.

Acknowledgements

This project was funded by the Minnesota Environment and Natural Resources Trust Fund, the University of Minnesota Agricultural Experiment Station, and the Minnesota Department of Agriculture. We thank Joshua Larson, Antonio Airton Lima Serra Jr., and Glen Bengtson for their help with this study.

References

Adler PR, Sanderson MA, Weimer PJ, Vogel KP (2009) Plant species composition and biofuel yields of conservation grasslands. *Ecological Applications*, **19**, 2202–2209.

Agostini F, Gregory AS, Richter GM (2015) Carbon sequestration by perennial energy crops: is the jury still out? *Bioenergy Research*, **8**, 1057–1080.

Baer SG, Kitchen DJ, Blair JM, Rice CW (2002) Changes in ecosystem structure and function along a chronosequence of restored grasslands. *Ecological Applications*, **12**, 1688–1701.

Bandaru V, Izaurralde RC, Manowitz D, Link R, Zhang X, Post WM (2013) Soil carbon change and net energy associated with biofuel production on marginal lands: a regional modeling perspective. *Journal of Environmental Quality*, **42**, 1802–1814.

Bates D, Maechler M, Bolker B, Walker S (2014) Linear mixed-effects models using Eigen and S4. R package version 1.1-7.

Bransby DI, McLaughlin SB, Parrish DJ (1998) A review of carbon and nitrogen balances in switchgrass grown for energy. *Biomass and Bioenergy*, **14**, 379–384.

Casler MD (2005) Ecotypic variation among switchgrass populations from the Northern USA. *Crop Science*, **45**, 388–398.

Casler MD, Vogel KP (2014) Selection for Biomass Yield in Upland, Lowland, and Hybrid Switchgrass. *Crop Science*, **54**, 626–636.

Chimento C, Almagro M, Amaducci S (2014) Carbon sequestration potential in perennial bioenergy crops: the importance of organic matter inputs and its physical protection. *GCB Bioenergy*. doi:10.1111/gcbb.12232.

Cong WF, van Ruijven J, Mommer L, De Deyn GB, Berendse F, Hoffland E (2014) Plant species richness promotes soil carbon and nitrogen stocks in grasslands without legumes. *Journal of Ecology*, **102**, 1163–1170.

De Graaff MA, Classen AT, Castro HF, Schadt CW (2010) Labile soil carbon inputs mediate the soil microbial community composition and plant residue decomposition rates. *New Phytologist*, **188**, 1055–1064.

Development Core Team R (2014) *R: A Language and Environment for Statistical Computing*. Austria, Vienna.

Dijkstra FA, Carrillo Y, Pendall E, Morgan JA (2013) Rhizosphere priming: a nutrient perspective. *Frontiers in Microbiology*, **4**, 1–8.

Follett RF, Vogel KP, Varvel GE, Mitchell RB, Kimble J (2012) Soil carbon sequestration by switchgrass and no-till maize grown for bioenergy. *BioEnergy Research*, **5**, 866–875.

Fornara DA, Tilman D (2008) Plant functional composition influences rates of soil carbon and nitrogen accumulation. *Journal of Ecology*, **96**, 314–322.

Fornara DA, Tilman D (2012) Soil carbon sequestration in prairie grasslands increased by chronic nitrogen addition. *Ecology*, **93**, 2030–2036.

Frank AB, Berdahl JD, Hanson JD, Liebig MA, Johnson HA (2004) Biomass and carbon partitioning in switchgrass. *Crop Science*, **44**, 1391–1396.

Garten CT, Wullschleger SD (2000) Soil carbon dynamics beneath switchgrass as indicated by stable isotope analysis. *Journal of Environmental Quality*, **29**, 645–653.

Garten CT, Brice DJ, Castro HF, *et al.* (2011) Response of "Alamo" switchgrass tissue chemistry and biomass to nitrogen fertilization in West Tennessee, USA. *Agriculture, Ecosystems & Environment*, **140**, 289–297.

Gelfand I, Zenone T, Jasrotia P, Chen J, Hamilton SK, Robertson GP (2011) Carbon debt of Conservation Reserve Program (CRP) grasslands converted to bioenergy production. *Proceedings of the National Academy of Sciences of the United States of America*, **108**, 10–15.

Gelfand I, Sahajpal R, Zhang X, Izaurralde RC, Gross KL, Robertson GP (2013) Sustainable bioenergy production from marginal lands in the US Midwest. *Nature*, **493**, 514–517.

Gillitzer PA, Wyse DL, Sheaffer CC, Taff SJ, Lehman CC (2012) Biomass production potential of grasslands in the oak savanna region of Minnesota, USA. *BioEnergy Research*, **6**, 131–141.

Hector A, Schmid B, Beierkuhnlein C, *et al.* (1999) Plant diversity and productivity experiments in European grasslands. *Science*, **286**, 1123–1126.

Horwath W (2007) Carbon cycling and formation of soil organic matter. In: *Soil Microbiology, Ecology, and Biochemistry* (ed Paul EA), pp. 303–339. Elsevier, Oxford, UK.

IPCC (2014) *Climate Change 2014: Impacts, Adaptation, and Vulnerability. Part B: Regional Aspects. Contribution of Working Group II to the Fifth Assessment Report of the Intergovernmental Panel on Climate Change* (eds Barros VR, Field CB, Dokken DJ, Mastrandrea MD, Mach KJ, Bilir TE, Chatterjee M, Ebi KL, Estrada YO, Genova RC, Girma B, Kissel ES, Levy AN, MacCracken S, Mastrandrea PR, White LL), pp. 688. Cambridge University Press, Cambridge, United Kingdom and New York, NY, USA.

Isbell F, Reich PB, Tilman D, Hobbie SE, Polasky S, Binder S (2013) Nutrient enrichment, biodiversity loss, and consequent declines in ecosystem productivity. *Proceedings of the National Academy of Sciences of the United States of America*, **110**, 11911–11916.

Jarchow ME, Liebman M (2012) Tradeoffs in biomass and nutrient allocation in prairies and corn managed for bioenergy production. *Crop Science*, **52**, 1330–1342.

Jarchow ME, Liebman M (2013) Nitrogen fertilization increases diversity and productivity of prairie communities used for bioenergy. *Global Change Biology: Bioenergy*, **5**, 281–289.

Jarchow ME, Liebman M, Rawat V, Anex RP (2012) Functional group and fertilization affect the composition and bioenergy yields of prairie plants. *Global Change Biology: Bioenergy*, **4**, 671–679.

Johnson JM-F, Barbour NW, Weyers SL (2007) Chemical composition of crop biomass impacts its decomposition. *Soil Science Society of America Journal*, **71**, 155–162.

Jung JY, Lal R (2011) Impacts of nitrogen fertilization on biomass production of switchgrass (Panicum Virgatum L.) and changes in soil organic carbon in Ohio. *Geoderma*, **166**, 145–152.

Jungers JM, Fargione JE, Sheaffer CC, Wyse DL, Lehman C (2013) Energy potential of biomass from conservation grasslands in Minnesota, USA. *PLoS One*, **8**, e61209.

Jungers JM, Sheaffer CC, Lamb JA (2015a) The Effect of nitrogen, phosphorus, and potassium fertilizers on prairie biomass yield, ethanol yield, and nutrient harvest. *BioEnergy Research*, **8**, 279–291.

Jungers JM, Clark AT, Betts K, Mangan ME, Sheaffer CC, Wyse DL (2015b) Long-term biomass yield and species composition in native perennial bioenergy cropping systems. *Agronomy Journal*, **107**, 1627–1640.

Kering MK, Butler TJ, Biermacher JT, Guretzky JA (2011) Biomass yield and nutrient removal rates of perennial grasses under nitrogen fertilization. *BioEnergy Research*, **5**, 61–70.

Knops JMH, Tilman D (2000) Dynamics of soil nitrogen and carbon accumulation for 61 years after agricultural abandonment. *Ecology*, **81**, 88–98.

Kuzyakov Y (2002) Review: factors affecting rhizosphere priming effects. *Journal of Plant Nutrition and Soil Science*, **165**, 382–396.

Lemus R, Lal R (2005) Bioenergy crops and carbon sequestration. *Critical Reviews in Plant Sciences*, **24**, 1–21.

Lemus R, Parrish DJ, Abaye O (2008) Nitrogen-use dynamics in switchgrass grown for biomass. *BioEnergy Research*, **1**, 153–162.

Loreau M, Naeem S, Inchausti P, *et al.* (2001) Biodiversity and ecosystem functioning: current knowledge and future challenges. *Science*, **294**, 804–808.

Ma Z, Wood C, Bransby D (2000) Soil management impacts on soil carbon sequestration by switchgrass. *Biomass and Bioenergy*, **18**, 469–477.

Mack MC, Schuur EAG, Bret-Harte MS, Shaver GR, Chapin FS (2004) Ecosystem carbon storage in arctic tundra reduced by long-term nutrient fertilization. *Nature*, **431**, 440–443.

Mangan ME, Sheaffer C, Wyse DL, Ehlke NJ, Reich PB (2011) Native perennial grassland species for bioenergy: establishment and biomass productivity. *Agronomy Journal*, **103**, 509–519.

Mann JJ, Barney JN, Kyser GB, DiTomaso JM (2013) Root system dynamics of Miscanthus x giganteus and Panicum virgatum in response to rainfed and irrigated conditions in California. *Bioenergy Research*, **6**, 678–687.

McIsaac GF, David MB, Mitchell CA (2010) Miscanthus and Switchgrass production in Central Illinois: impacts on hydrology and inorganic nitrogen leaching. *Journal of Environment Quality*, **39**, 1790–1799.

McLauchlan KK, Hobbie SE, Post WM (2006) Conversion from agriculture to grassland builds soil organic matter on decadal timescales. *Ecological Applications*, **16**, 143–153.

Mooney DF, Roberts RK, English BC, Tyler DD, Larson JA (2009) Yield and breakeven price of "Alamo" switchgrass for biofuels in tennessee. *Agronomy Journal*, **101**, 1234–1242.

Mueller KE, Hobbie SE, Tilman D, Reich PB (2013) Effects of plant diversity, N fertilization, and elevated carbon dioxide on grassland soil N cycling in a long-term experiment. *Global Change Biology*, **19**, 1249–1261.

Nelson DW, Sommers LE (1996) Total carbon, organic carbon, and organic matter. In: *Methods of Soil Analysis. Part 3. Chemical Methods* (ed Sparks DL), pp. 961–1010. SSSA and ASA, Madison, WI.

Pacala S, Socolow R (2004) Stabilization wedges: solving the climate problem for the next 50 years with current technologies. *Science*, **305**, 968–972.

Picasso VD, Brummer EC, Liebman M, Dixon PM, Wilsey BJ (2008) Crop species diversity affects productivity and weed suppression in perennial polycultures under two management strategies. *Crop Science*, **48**, 331–342.

Rasse DP, Rumpel C, Dignac MF (2005) Is soil carbon mostly root carbon? Mechanisms for a specific stabilisation. *Plant and Soil*, **269**, 341–356.

Reeder J, Schuman G, Bowman R (1998) Soil C and N changes on conservation reserve program lands in the Central Great Plains. *Soil and Tillage Research*, **47**, 339–349.

Roelandt C, Van Wesemael B, Rounsevell M (2005) Estimating annual N2O emissions from agricultural soils in temperate climates. *Global Change Biology*, **11**, 1701–1711.

Ruxton GD, Beauchamp G (2008) Time for some a priori thinking about post hoc testing. *Behavioral Ecology*, **19**, 690–693.

Sartori F, Lal R, Ebinger M, Parrish D (2006) Potential soil carbon sequestration and CO2 offset by dedicated energy crops in the USA. *Critical Reviews in Plant Sciences*, **25**, 441–472.

Schenk HJ, Jackson RB (2002) The global biogeography of roots. *Ecological Monographs*, **72**, 311–328.

Schmer MR, Vogel KP, Mitchell RB, Perrin RK (2008) Net energy of cellulosic ethanol from switchgrass. *Proceedings of the National Academy of Sciences of the United States of America*, **105**, 464–469.

Schmer MR, Liebig MA, Vogel KP, Mitchell RB (2011) Field-scale soil property changes under switchgrass managed for bioenergy. *Global Change Biology: Bioenergy*, **3**, 439–448.

Smith CM, David MB, Mitchell CA, Masters MD, Anderson-Teixeira KJ, Bernacchi CJ, Delucia EH (2013) Reduced nitrogen losses after conversion of row crop agriculture to perennial biofuel crops. *Journal of Environmental Quality*, **42**, 219–228.

Smucker AJM, Mcburney SL, Srivastava AK (1982) Quantitative separation of roots from compacted soil profiles by the hydropneumatic elutriation system. *Agronomy Journal*, **74**, 500–503.

Steinbeiss S, Beßler H, Engels C, *et al.* (2008) Plant diversity positively affects short-term soil carbon storage in experimental grasslands. *Global Change Biology*, **14**, 2937–2949.

Tiemann LK, Grandy AS (2015) Mechanisms of soil carbon accrual and storage in bioenergy cropping systems. *GCB Bioenergy*, **7**, 161–174.

Tilman D, Wedin D, Knops JMH (1996) Productivity and sustainability influenced by biodiversity in grassland ecosystems. *Nature*, **379**, 718–720.

Tilman D, Hill J, Lehman C (2006) Carbon-negative biofuels from low-input high-diversity grassland biomass. *Science*, **314**, 1598–1600.

Torsten H, Bretz F, Westfall P (2008) Simultaneous inference in general parametric models. *Biometrical Journal*, **50**, 346–363.

Vogel KP, Dien BS, Jung HG, Casler MD, Masterson SD, Mitchell RB (2010) Quantifying actual and theoretical ethanol yields for switchgrass strains using NIRS analyses. *BioEnergy Research*, **4**, 96–110.

Weaver JE (1958) Classification of root systems of forbs of grassland and a consideration of their significance. *Ecology*, **39**, 393–401.

Zamora DS, Wyatt GJ, Apostol KG, Tschirner U (2013) Biomass yield, energy values, and chemical composition of hybrid poplars in short rotation woody crop production and native perennial grasses in Minnesota, USA. *Biomass and Bioenergy*, **49**, 222–230.

Zan CS, Fyles JW, Girouard P, Samson RA (2001) Carbon sequestration in perennial bioenergy, annual corn and uncultivated systems in southern Quebec. *Agriculture, Ecosystems & Environment*, **86**, 135–144.

Zhang K, Dang H, Tan S, Cheng X, Zhang Q (2010) Change in soil organic carbon following the "grain-for-green" programme in China. *Land Degradation and Development*, **21**, 13–23.

Zilverberg CJ, Johnson WC, Owens V, *et al.* (2014) Biomass yield from planted mixtures and monocultures of native prairie vegetation across a heterogeneous farm landscape. *Agriculture, Ecosystems and Environment*, **186**, 148–159.

6

Range and uncertainties in estimating delays in greenhouse gas mitigation potential of forest bioenergy sourced from Canadian forests

JÉRÔME LAGANIÈRE[1], DAVID PARÉ[1], EVELYNE THIFFAULT[2] and PIERRE Y. BERNIER[1]

[1]*Natural Resources Canada, Canadian Forest Service, Laurentian Forestry Centre, Québec, QC G1V 4C7, Canada,* [2]*Département des sciences du bois et de la forêt, Université Laval, Québec, QC G1K 7P4, Canada*

Abstract

Accurately assessing the delay before the substitution of fossil fuel by forest bioenergy starts having a net beneficial impact on atmospheric CO_2 is becoming important as the cost of delaying GHG emission reductions is increasingly being recognized. We documented the time to carbon (C) parity of forest bioenergy sourced from different feedstocks (harvest residues, salvaged trees, and green trees), typical of forest biomass production in Canada, used to replace three fossil fuel types (coal, oil, and natural gas) in heating or power generation. The time to C parity is defined as the time needed for the newly established bioenergy system to reach the cumulative C emissions of a fossil fuel, counterfactual system. Furthermore, we estimated an uncertainty period derived from the difference in C parity time between predefined best- and worst-case scenarios, in which parameter values related to the supply chain and forest dynamics varied. The results indicate short-to-long ranking of C parity times for residues < salvaged trees < green trees and for substituting the less energy-dense fossil fuels (coal < oil < natural gas). A sensitivity analysis indicated that silviculture and enhanced conversion efficiency, when occurring only in the bioenergy system, help reduce time to C parity. The uncertainty around the estimate of C parity time is generally small and inconsequential in the case of harvest residues but is generally large for the other feedstocks, indicating that meeting specific C parity time using feedstock other than residues is possible, but would require very specific conditions. Overall, the use of single parity time values to evaluate the performance of a particular feedstock in mitigating GHG emissions should be questioned given the importance of uncertainty as an inherent component of any bioenergy project.

Keywords: carbon debt, carbon dioxide emissions, carbon parity time, climate change, forest ecosystems, life cycle assessment, logging residues, renewable energy, salvage logging, wood pellets

Introduction

The use of forest-based bioenergy to replace fossil fuels in heat and electricity generation has the potential to reduce greenhouse gas (GHG) emissions. Under sustainable forest management practices, forests can provide renewable feedstock for bioenergy as the CO_2 released during wood combustion is later recaptured by photosynthesis as the forest regrows. However, the presumed 'C neutrality' of forest bioenergy has been the subject of much debate recently (Searchinger *et al.*, 2009; Manomet, 2010) because of the three following points: (i) Wood emits more CO_2 than fossil fuel per unit of energy released (Gómez *et al.*, 2006); (ii) the release of CO_2 is much faster when wood is combusted than when wood undergoes natural decomposition; and (iii) CO_2 recapture by vegetation is not immediate and is usually achieved on decade- to century-long timescales. Therefore, there is a period of variable length during which cumulative CO_2 emissions to the atmosphere from an energy plant are greater for bioenergy than for fossil fuel. The delay before atmospheric GHG benefits are achieved has been referred to as C payback time (or C debt repayment time) when preharvest C levels are reached (absolute C balance), or as time to C parity when C levels of a reference case are reached (relative C balance) (see Lamers & Junginger, 2013, for a thorough discussion on terminology).

Canada is among the largest producers and exporters of solid bioenergy (Lamers *et al.*, 2012; Goh *et al.*, 2013). To date, case studies assessing the C debt and potential CO_2 emission savings of different forest bioenergy projects in Canada have yielded varying results, from instant atmospheric benefits to C payback/parity times of over 100 years. For example, cofiring pellets with coal

Correspondence: Jérôme Laganière
e-mail: jerome.laganiere@canada.ca

in Ontario for electricity generation resulted in C debt repayment times of 16 and 38 years when pellets were made from harvest residues and green trees, respectively (McKechnie *et al.*, 2011). Using eddy covariance flux towers in Saskatchewan and Quebec to estimate net ecosystem exchanges, Bernier & Paré (2013) obtained a multidecadal time to C parity (>90 years) for a scenario that used wood chips from green trees to replace diesel oil in heat generation. A study in British Columbia forests impacted by the mountain pine beetle (MPB) (*Dendroctus ponderosae* Hopkins) showed that some scenarios had immediate atmospheric benefits (no C debt) and that using harvest residues and nonmerchantable trees for pellet production was more C beneficial than a stand protection alternative with no harvest (Lamers *et al.*, 2014).

Factors regulating the GHG mitigation potential of bioenergy projects and the underlying large variation in C parity times include biomass feedstock source and processing, the type of fossil fuel replaced, energy conversion efficiency, tree growth rate, and the definition of the counterfactual 'reference' scenario, that is, what would have happened to the forest land if biomass had not been sourced and used for bioenergy? (Lamers *et al.*, 2013; Buchholz *et al.*, 2014, 2015). Because many of those factors usually differ among studies, it is often difficult to compare C parity times among a variety of forest bioenergy uses. This situation stresses the need for a common accounting system to support decision-making (Buchholz *et al.*, 2015).

Furthermore, Buchholz *et al.* (2015) recommend that future studies assessing the C balance of bioenergy pathways consider quantifying and reporting uncertainties, which have rarely been addressed in past life cycle assessment (LCA) studies (e.g., Johnson *et al.*, 2011; Caputo *et al.*, 2014; Cherubini *et al.*, 2014; Röder *et al.*, 2015). Indeed, sources of uncertainty are encountered all along the supply chain as well as within the forest ecosystem, where various ecological factors may impact tree regeneration and decay rates. Understanding how variability in key parameters affects the mitigation potential of a bioenergy system is necessary to appreciate the full range of possible outcomes and make informed decisions and establish the right policies.

The aim of this study was therefore to compare, using a common framework, the mitigation potential and timing of atmospheric benefits for different bioenergy deployment scenarios sourcing their biomass from Canadian forests. Specific objectives were to quantify the uncertainties associated with such scenarios and identify how such uncertainties could be reduced to increase confidence in the timing and scale of GHG benefits for major forest bioenergy pathways. To this end, we developed a landscape-scale GHG emission calculator based on a LCA approach in which sources of variation and uncertainty are explicitly identified. Carbon parity times and their associated uncertainty were calculated for scenarios sourcing biomass from different feedstock types (harvest residues, salvaged trees (i.e., trees killed by natural disturbances), or green trees) typical of biomass production in Canada used to replace three fossil fuel types (coal, oil or natural gas) in heating or power generation. Results from this study may provide guidance for defining policies aimed at promoting the best forest bioenergy pathways for GHG mitigation. A free Web-based version of the calculator will be made available at https://apps-scf-cfs.rncan.gc.ca/calc/en (section GHG Bioenergy).

Materials and methods

Study area description

Our study focuses on the Canadian managed forest, which is estimated at 153 million ha (NRCan, 2014b). The area encompasses five terrestrial ecozones (i.e., Atlantic maritime, Boreal shield, Mixedwood plain, Montane cordillera, and Pacific maritime), where mean annual temperature (MAT) and mean annual precipitation range from −1 to 5 °C and from 400 to 3000 mm, respectively (Environment Canada, 2015). On average (1990–2013), forest harvesting occurs on 1.0 million ha annually, whereas fire and insects disturb 3.1 and 19.1 million ha, respectively (NRCan, 2014a). Frequency and severity of natural disturbances are expected to increase in the future (Soja *et al.*, 2007; Boulanger *et al.*, 2014), potentially making salvage wood an increasing feedstock source for harvested wood products, which include bioenergy.

Model framework for GHG accounting

The components of our LCA for GHG accounting include emissions from feedstock production and use, forest C dynamics, and energy conversion efficiency (Fig. S1). The GHG mitigation potential over time for a given bioenergy scenario needs to be assessed relative to a baseline, or counterfactual, scenario, which implies the use of fossil fuel. The GHG mitigation potential is calculated as follows:

$$\Delta GHG_t = \frac{GHG_{t\,BIO} + FC_{t\,BIO}}{CE_{BIO}} - \frac{GHG_{t\,FOSSIL} + FC_{t\,FOSSIL}}{CE_{FOSSIL}}, \quad (1)$$

where ΔGHG_t is the cumulative difference in CO_2eq emissions between the bioenergy and fossil fuel scenarios at time t (in kg CO_2 emitted per GJ of bioenergy produced), $GHG_{t\ BIO}$ and $GHG_{t\ FOSSIL}$ are cumulative emissions from the bioenergy and fossil fuel systems (production and use) at time t, respectively, $FC_{t\ BIO}$ and $FC_{t\ FOSSIL}$ are the forest C status (reported in CO_2) of the bioenergy and fossil fuel systems at time t, respectively, and CE_{BIO} and CE_{FOSSIL} are the energy conversion efficiency of bioenergy and fossil fuel, respectively. When ΔGHG_t reaches zero, the C parity time has been reached and GHG mitigation benefits begin to occur (Fig. 1). All emissions were derived

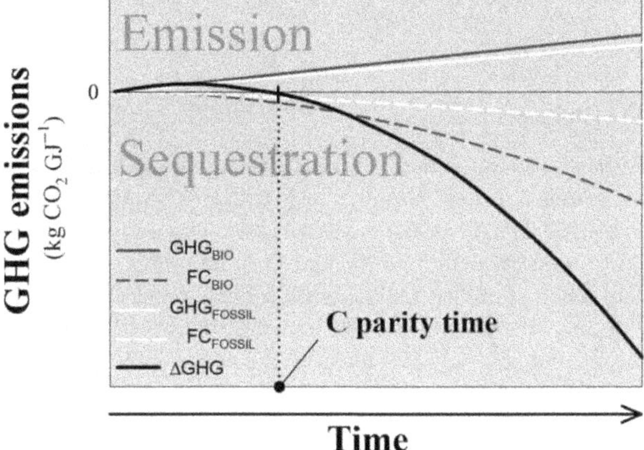

Fig. 1 The C parity time concept illustrated using the current model framework for C accounting. GHG_{BIO} and GHG_{FOSSIL} are emissions from the bioenergy and fossil fuel systems (production and use), respectively; FC_{BIO} and FC_{FOSSIL} are the forest C status of the bioenergy and fossil fuel systems, respectively; ΔGHG is the difference in CO_2 emissions between the bioenergy and fossil fuel scenarios. When ΔGHG reaches zero, C parity time has been reached and GHG mitigation benefits begin to occur.

from the production of 1 GJ of energy per year for a 100-year period (landscape-scale analysis). All modeling was performed using SAS 9.3 (SAS Institute Inc., Cary, NC, USA).

Our forest C analysis assumes constant soil C stocks although in theory a certain fraction of the deadwood decaying on the ground should eventually contribute to the soil C stock. Hence, some modeling results suggest that continual residue removal may permanently reduce forest floor C storage and delay time to C parity (Repo *et al.*, 2011, 2012). However, there is little empirical support for systematic and significant long-term mineral soil C changes following harvesting across the boreal and temperate forest biomes (Johnson & Curtis, 2001; Nave *et al.*, 2010; Thiffault *et al.*, 2011). In addition, forest floor C is usually quickly replenished as the forest regenerates (Nave *et al.*, 2010). In our opinion, additional research assessing long-term impact of residue removal on soil C is still warranted to consider with confidence soil C dynamics in forest bioenergy C accounting studies.

Forest carbon dynamics in bioenergy and counterfactual scenarios

Harvest residues. Harvest residues are defined as all woody debris generated in harvest operations for traditional wood products (e.g., branches, tree tops, bark), excluding stumps and downed nonmerchantable trees. Harvest residues can be left on site to decompose or, as it is still the practice in parts of Canada, they can be piled by the roadside and burned under controlled conditions to reduce the fire hazard. In the case in which unused residues are burned by the roadside, the CO_2 release from these residues happens nearly at the same time whether the energy is generated from biomass or from fossil fuel, with no consequences for time to C parity. Combustion of the residues burned by the roadside is assumed to be complete although some fraction may contribute to the soil C pool in the form of charcoal. In the case in which harvest residues are not harvested for bioenergy and left on site to decompose, the multiyear delay in the release of CO_2 must be accounted for in the GHG comparison between bioenergy and fossil fuel scenarios. In our calculator, we used the following exponential decay function to express CO_2 release over time:

$$C_{t\,\mathrm{WD}} = C_{0\,\mathrm{WD}} \times e^{-k\times t}, \tag{2}$$

where $C_{t\ \mathrm{WD}}$ is the quantity of C (kg CO_2) stored in woody debris at time t (years), $C_{0\ \mathrm{WD}}$ is the initial quantity of C stored in woody debris (kg CO_2), and k is the decomposition rate of woody debris (year^{-1}). Because temperature is the main driver of decomposition rates in these forests (Litton & Giardina, 2008; Laganière *et al.*, 2012), we used the temperature-dependent decay function of the Canadian forest C budget model CBM-CFS3 (Kurz *et al.*, 2009) to compute the decay rate (year^{-1}) across the range of temperatures found in the Canadian managed forest:

$$k = \mathrm{BDR}_k \times \mathrm{TempMod}, \tag{3}$$

where BDR_k is the base decomposition rate of woody debris (aboveground fast pool = 0.1435 year^{-1}) at a reference MAT of 10 °C, and TempMod is the temperature modifier that reduces the decay rate for MAT below the reference MAT and is calculated as:

$$\mathrm{TempMod} = e^{((\mathrm{MAT}_f - \mathrm{RefMAT})\times \ln(Q_{10})\times 0.1)}, \tag{4}$$

where MAT_f is the MAT of the forest area (−1 to 5 °C in Canada's managed forest), RefMAT is the reference MAT of 10 °C, and Q_{10} is the temperature sensitivity of decomposition set at 2. Because BDR_k varies markedly among tree species (Tarasov & Birdsey, 2001; Brais *et al.*, 2006; Shorohova & Kapitsa, 2014), we performed a sensitivity analysis on this parameter.

Salvaged trees. In scenarios sourcing their biomass from salvaged trees (i.e., standing trees killed by natural disturbances), the stemwood is harvested for bioenergy while the residues are left on site (i.e., the fate of the residues is not considered in the accounting). In the counterfactual scenario, the standing dead trees (i.e., snags) are assumed to start decaying immediately after tree death at a BDR_k of 0.0187 year^{-1} (Kurz *et al.*, 2009) following Eqn. 2, until they fall to the ground following Eqn 5, where they start to decay at a BDR_k of 0.0374 year^{-1} (Kurz *et al.*, 2009) following Eqn 2. The equation for snag C transfer to the ground is as follows:

$$C_{t\,\mathrm{snag}} = C_{0\,\mathrm{snag}} \times e^{-\mathrm{CTR}\times t}, \tag{5}$$

where $C_{t\ \mathrm{snag}}$ is the quantity of C (kg CO_2) stored in snags (standing woody debris) at time t (years), $C_{0\ \mathrm{snag}}$ is the initial quantity of C stored in snags (kg CO_2), and CTR is the C transfer rate of snags (year^{-1}) that varies between 0.04 and 0.10 (Hilger *et al.*, 2012).

Green trees. In scenarios sourcing their biomass from green trees (living biomass), we assume that only the stemwood is harvested for bioenergy (tree tops and branches are left on site) and that no harvesting is carried out and there is only a negligible risk of disturbance in the reference forest in the counterfactual scenario. Because we consider harvesting of green trees for bioenergy to complement, not to compete with, that for traditional forest products, harvesting of green trees for bioenergy is viewed as 'additional harvesting' meaning that this feedstock would not be used in the counterfactual scenario due to various reasons (e.g., species unused by the traditional forest industry, fiber quality unsuitable for traditional products but suitable for bioenergy). Scenarios where the feedstock competes for its use (bioenergy vs. traditional products) were not explored in the current study.

The time required for the forest C of the bioenergy system to balance itself with that of the fossil fuel system depends on the regeneration rate of the harvested forest and also on the rate at which the forest continues to grow in the counterfactual scenario. We define three generic forest growth curves: fast, medium, and slow, reaching an age of maximum mean annual increment (MAI) at 45, 75, and 120 years, respectively (Fig. S2). We assume that a forest is harvested at age of maximum MAI. The time required to reach maximum MAI following harvesting is the time required for the harvested forest to recapture all of the biogenic CO_2 emitted in a year from the combustion of 1 GJ of biomass (112 kg CO_2). Using this approach, we can convert absolute stand volume (m^3 ha^{-1}) into relative measures of time required to reach the original stand volume in units of % of initial harvestable volume. To account for the growth of the reference forest that is not harvested and thus continues to sequester C, we use the portion of the curves that follows maximum MAI, that is, after reaching 100% harvested stand biomass regeneration (Fig. S2).

Upstream emissions

Biomass production in bioenergy scenarios. The GHG emissions associated with biomass production include those related to biomass collection (harvesting, forest stand renewal, and road construction/maintenance), processing (chipping and pelletization), and transportation (transport to processing plant and to local or international market). We used an emission factor of 2.63 kg CO_2eq GJ^{-1} for roundwood collection (salvaged and green trees), averaged from values found in studies on Canadian forests (i.e., Magelli *et al.*, 2009; Meil *et al.*, 2009; McKechnie *et al.*, 2011; Pa *et al.*, 2012; Lamers *et al.*, 2014). For harvest residue collection, we used 0.84 kg CO_2eq GJ^{-1}, as in McKechnie *et al.* (2011). For roundwood and harvest residue chipping, we used 0.76 kg CO_2eq GJ^{-1} and 0.05 kg CO_2eq GJ^{-1}, respectively, as in Lamers *et al.* (2014). For the pelletization process, which includes drying, milling, and pelletizing, we used 2.14 kg CO_2eq GJ^{-1} for pellets made from harvest residues, and 10.45 kg CO_2eq GJ^{-1} for pellets made from roundwood (i.e., salvaged and green trees), as in Lamers *et al.* (2014).

Fossil fuel production in counterfactual scenarios. Upstream emissions for fossil fuels include extraction, distribution and storage, production, transmission, land-use changes, gas leaks, and flares. Emission factors used for coal, oil, and natural gas were 6.4, 14.9, and 9.0 kg CO_2eq GJ^{-1}, respectively ($(S\&T)^2$, 2015).

Energy use

For coal, oil, and natural gas combustion, we used the following emission factors: 90.6, 71.1, and 50.3 kg CO_2eq GJ^{-1}, respectively ($(S\&T)^2$, 2015). For wood biomass, we used the default IPCC emission factor of 112.0 kg CO_2eq GJ^{-1} IPCC (2006). The conversion efficiency factors used for heat and electricity were 75% and 26% for biomass, 80% and 33% for coal, 82% and 35% for oil, and 85% and 45% for natural gas, respectively ($(S\&T)^2$, 2015).

Scenario development (parameters and definition of uncertainty)

We calculated C parity time (in years) and potential emission reductions (in kg CO_2 GJ^{-1}) of forest bioenergy sourced from different feedstocks (harvest residues, salvaged trees, or green trees) to replace three fossil fuel types (coal, oil, or natural gas) for two uses (heating or power generation). An uncertainty period was defined as the range in C parity times between predefined best-case (shortest C parity time) and worst-case (longest C parity time) scenarios for each scenario, with several potential cases lying in between (Fig. 2). To define the two end cases, we varied model parameters, including transportation distance to final users (local use or exportation), biomass processing (chips or pellets), and environmental characteristics (i.e., MAT, C transfer rate from snags to the ground). For example, for scenarios using harvest residues as feedstock, the best case implied: (i) collection of residues in the warmer part of our study area (MAT = 5 °C; the decomposition rate of residues left on site in the counterfactual scenario is high); (ii) processing into wood chips; and (iii) local use of wood chips (100 km of truck transport to final user). The worst case implied: (i) collection of residues in the colder part of the study area (MAT = −1 °C, which translates into a slow decomposition rate for biomass left on site in the counterfactual scenario); (ii) processing into pellets, which produces additional emissions

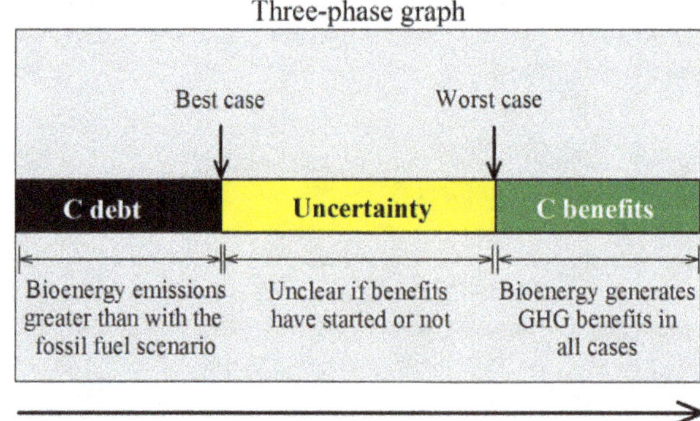

Fig. 2 Three-phase graph used in the current study to represent estimates of C parity time and the associated uncertainty phase.

relative to wood chips; and (iii) transoceanic shipping from British Columbia to the United Kingdom (100 km by truck, 1000 km by train, and 16 000 km by vessel). Therefore, there are two types of parameters contributing to uncertainty: those based on choices related to the supply chain (i.e., transportation distance and biomass processing), and those based on variable ecological processes or environmental characteristics. Feedstock-specific details on the parameters defining the different cases are found above.

Sensitivity analysis

A sensitivity analysis was performed on a set of bioenergy scenarios substituting coal in power production. We investigated how silviculture, energy conversion efficiency, and deadwood decay rate affected the performance of these scenarios (timing and uncertainty).

Silviculture. Because silvicultural operations (e.g., site preparation, tree planting, weed control) that increase tree growth following harvesting are widespread in Canada, we added scenarios where tree growth rate in the bioenergy system was 1.5 (Growth ×1.5), 2 (Growth ×2), and 2.5 (Growth ×2.5) times higher than that in the counterfactual fossil fuel system. In other words, age of maximum MAI of the forest is reduced by 1.5, 2, or 2.5 times in the bioenergy system relative to that in the counterfactual one. These estimates of potential growth increases via silviculture are conservative considering that the average timber yield in Canada forest is around 1 m^3 ha^{-1} yr^{-1} while that of extensive plantations in Canada usually reaches 2–6 m^3 ha^{-1} yr^{-1} (Messier *et al.*, 2003; Paquette & Messier, 2010). Although regeneration failure (i.e., when predisturbance biomass levels are never recovered without proper forest management) may happen following clear-cut or natural disturbance (Lecomte *et al.*, 2006; Thiffault *et al.*, 2013), this possibility was not explored in the present study.

Conversion efficiency. We investigated how electricity conversion efficiency may affect timing and the uncertainty period by increasing the parameter from 26% to 35%, by 3% increments.

Decay rate of woody debris. Because the default base decomposition rate of CBM-CFS3 represents an average value that does not necessarily capture all the variability in decay rates among tree species across Canada, we performed a sensitivity analysis on selected scenarios (i.e., harvest residues and salvaged trees replacing coal in electricity generation) with elevated BDR_k (i.e., decomposition rate doubled or tripled) to reflect the faster decay rates of intolerant hardwood species such as aspen and birch (Tarasov & Birdsey, 2001; Brais *et al.*, 2006; Shorohova & Kapitsa, 2014). These tree species usually have a low economic value and are often viewed as nonmerchantable by the industrial forest sector of timber and pulp.

Results

The uncertainty phase

The estimate of C parity time follows three temporal phases (Fig. 2): (i) a phase of C debt representing the period of time during which all cases for a given scenario, even the best case, do not provide any C benefits; (ii) a phase of C parity uncertainty, representing the range of C parity values between the best and the worst cases; and (iii) a phase of C benefits for all cases, during which even the worst cases provide C benefits. The length of the second phase, C parity uncertainty, during which it is unclear whether the benefits have started or not, varies from a few years to several decades and depends on the bioenergy feedstock, the type of fossil fuel replaced, silvicultural practices, energy conversion efficiency, and environmental characteristics. As shown in Fig. 3, the

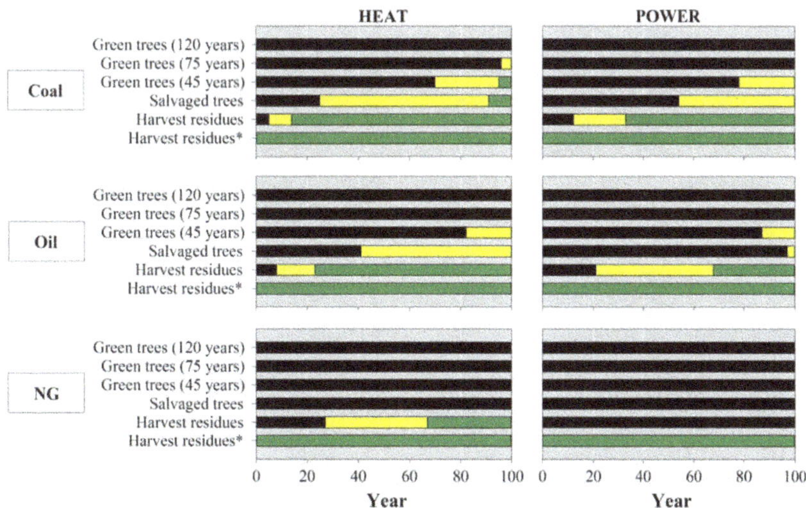

Fig. 3 Length of the C debt (black), uncertainty (yellow), and C benefit (green) phases for scenarios using different bioenergy feedstock to replace different fossil fuels for heat and power production. The asterisk indicates that harvest residues are burned by the roadside instead of left to decompose on the harvest site in the counterfactual scenario. NG: natural gas.

uncertainty is usually small for harvest residues, intermediate for green trees, and large for salvaged trees, and it increases with the efficiency of the fossil fuel in the following order: coal < oil < natural gas.

The effect of biomass feedstock, type of fossil fuel replaced, and energy use

Substitution of coal by forest bioenergy generates GHG emission savings over the shortest time frame, followed by oil and natural gas (Table 1; Fig. 3). Except for some residue-based cases (i.e., heat generation), substitution of natural gas by forest bioenergy does not provide any atmospheric benefits within a 100-year period.

Immediate C benefits occur when bioenergy is sourced from residues normally burned by the roadside, irrespective of the choices in model parameters (Table 1; Fig. 3). Cumulative CO_2 emissions saved after 100 years vary from 4.6 to 11.8 Mg GJ^{-1} for heat generation and from 6.5 to 28.6 Mg GJ^{-1} for power production, depending on

Table 1 Range of C parity time, uncertainty phase, and C balance for each best- and worst-case bioenergy scenario

Scenario			Carbon parity time	Uncertainty	Carbon balance (Kg GJ^{-1})		
Feedstock	Fossil fuel	Use	(year)	phase (year)	25 years	50 years	100 years
Harvest residues*	Coal	Heat	0	0	**−2449** to **−2 962**	**−4897** to **−5 923**	**−9795** to **−11 846**
Harvest residues	Coal	Heat	5–14	9	**−571** to **−1520**	**−2478** to **−4 152**	**−6839** to **−9 600**
Salvaged trees	Coal	Heat	25–91	66	1130 to **−11**	1249 to **−1327**	**−426** to **−5592**
Green trees (45 years)	Coal	Heat	70–95	25	1914–1124	2487–907	**−778** to **−3938**
Green trees (75 years)	Coal	Heat	96–>100	>4	1980–1190	3379–1799	2894 to **−265**
Green trees (120 years)	Coal	Heat	>100	>0	1886–1096	3768–2189	4917–1757
Harvest residues*	Coal	Power	0–0	0	**−5668** to **−7148**	**−11 336** to **−14 295**	**−22 672** to **−28 590**
Harvest residues	Coal	Power	12–33	21	604 to **−1932**	**−2030** to **−6561**	**−8626** to **−16 264**
Salvaged trees	Coal	Power	54–>100	>46	4893–1764	7337–364	7585 to **−6394**
Green trees (45 years)	Coal	Power	78–>100	>22	6540–4261	8764–4207	**−67** to **−9181**
Green trees (75 years)	Coal	Power	>100	>0	6871–4593	11 767–7209	11 651–2536
Green trees (120 years)	Coal	Power	>100	>0	6713–4434	13 243–8686	18 484–9370
Harvest residues*	Oil	Heat	0	0	**−2039** to **−2552**	**−4078** to **−5104**	**−8156** to **−10 208**
Harvest residues	Oil	Heat	8–23	15	**−116** to **−1054**	**−1535** to **−3194**	**−4908** to **−7652**
Salvaged trees	Oil	Heat	41–>100	>59	1552–420	2118 to **−434**	1384 to **−3734**
Green trees (45 years)	Oil	Heat	82–>100	>18	2304–1514	3243–1663	680 to **−2480**
Green trees (75 years)	Oil	Heat	>100	>0	2377–1587	4157–2578	4412–1252
Green trees (120 yrs)	Oil	Heat	>100	>0	2289–1499	4565–2986	6487–3328
Harvest residues*	Oil	Power	0	0	**−4462** to **−5941**	**−8923** to **−11 882**	**−17 847** to **−23 765**
Harvest residues	Oil	Power	21–68	47	2068 to **−408**	1083 to **−3359**	**−2140** to **−9679**
Salvaged trees	Oil	Power	97–>100	>3	6171–3091	10 034–3198	13 382 to **−318**
Green trees (45 years)	Oil	Power	87–>100	>13	7633–5354	10 815–6258	3734 to **−5380**
Green trees (75 years)	Oil	Power	>100	>0	8007–5728	13 947–9390	15 790–6676
Green trees (120 years)	Oil	Power	>100	>0	7882–5604	15 529–10 972	22 923–13 809
Harvest residues*	Gas	Heat	0	0	**−1162** to **−1675**	**−2324** to **−3350**	**−4649** to **−6700**
Harvest residues	Gas	Heat	27–67	40	825 to **−98**	393 to **−1244**	**−988** to **−3707**
Salvaged trees	Gas	Heat	>100	>0	2447–1327	3943–1425	5133–85
Green trees (45 years)	Gas	Heat	>100	>0	3152–2363	4907–3327	3933–773
Green trees (75 years)	Gas	Heat	>100	>0	3236–2446	5854–4274	7749–4589
Green trees (120 years)	Gas	Heat	>100	>0	3157–2367	6288–4708	9899–6740
Harvest residues*	Gas	Power	0	0	**−1615** to **−3095**	**−3230** to **−6189**	**−6460** to **−12 378**
Harvest residues	Gas	Power	>100	>0	5859–3604	9343–5231	15 337–8158
Salvaged trees	Gas	Power	>100	>0	9281–6379	16 767–10 438	28 334–15 653
Green trees (45 years)	Gas	Power	>100	>0	10 063–7785	15 183–10 626	11 365–2 251
Green trees (75 years)	Gas	Power	>100	>0	10 594–8 315	18 788–14 231	24 661–15 547
Green trees (120 years)	Gas	Power	>100	>0	10 593–8315	20 760–16 203	32 896–23 782

The '>' sign is used when C parity time or uncertainty phase has reached the 100-year time boundary of this study and therefore cannot be estimated precisely. C balance with a negative sign (in bold) indicates that the bioenergy scenario generates net atmospheric benefit (sequestration) relative to the counterfactual scenario.

*Harvest residues are normally burned by the roadside in the counterfactual scenario.

the type of fossil fuel replaced (Table 1). When bioenergy is sourced from harvest residues normally left to decompose *in situ*, C parity times range from 5 to 67 years for heat generation and from 12 to over 100 years for power production, depending on the type of fossil fuel replaced (Table 1; Fig. 3). Cumulative CO_2 emissions saved after 100 years are slightly lower than in the burned residues scenarios, that is, from 0.9 to 9.6 Mg GJ^{-1} for heat generation and from no savings to 16.3 Mg GJ^{-1} for power generation (Table 1).

When bioenergy is sourced from salvaged trees, C parity times range from 25 to over 100 years for heat production and from 54 to over 100 years for power production (Table 1; Fig. 3). Cumulative CO_2 emissions saved after 100 years for salvaged trees range from no savings to 5.6 Mg GJ^{-1} for heat production and from no savings to 6.4 Mg GJ^{-1} for power production (Table 1).

When bioenergy is sourced from fast-growing trees (age of maximum MAI = 45 years), C parity times range from 70 to 95 years for heat production and from 78 to 100 years for power production (Table 1; Fig. 3). Cumulative CO_2 emissions saved after 100 years vary from 0.8 to 3.9 Mg GJ^{-1} for heat production and from 0.1 to 9.2 Mg GJ^{-1} for power production (Table 1). When medium- or slow-growing trees are used (maximum MAI of 75 and 120 years, respectively), no emission savings generally occur on a 100-year time frame, except for medium-growing trees in the coal-heating scenario.

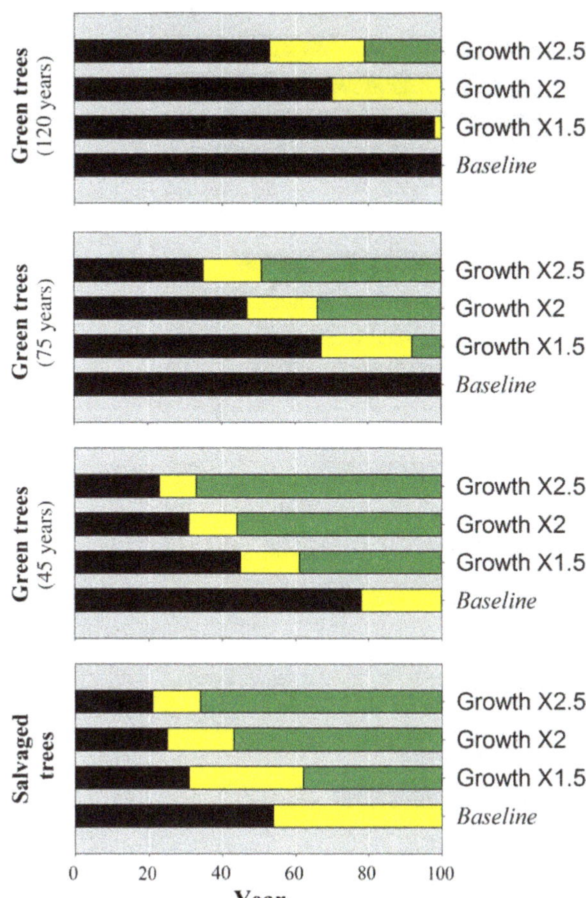

Fig. 4 Timing of GHG benefits and length of the uncertainty phase of scenarios using different bioenergy feedstock to replace coal for power production. For each feedstock, scenarios show the effect of silvicultural operations that increase the growth rate in regenerating forest stands by 1.5- (Growth ×1.5), 2- (Growth ×2), and 2.5-fold (Growth ×2.5) relative to the reference growth rate of forests in the counterfactual scenario. The 'no silviculture' scenario (baseline), in which growth rates are equal to the reference growth rate, is also shown.

Sensitivity analysis

When silvicultural operations resulting in 1.5-, 2-, and 2.5-fold increases in tree growth rate are carried out, time to C parity and the length of the uncertainty phase are reduced (Fig. 4). Parity times of bioenergy sourced from salvaged trees to replace coal in power generation are under 62 years for 'Growth ×1.5', under 43 years for 'Growth ×2', and under 34 years for 'Growth ×2.5' (Fig. 4), with cumulative CO_2 emissions saved reaching 26.1, 40.2, and 54.6 Mg GJ^{-1}, respectively (data not shown). When silvicultural operations are carried out, fast- and medium-growing trees may also become suitable feedstock options to achieve short- to medium-term mitigation benefits. Parity times for bioenergy sourced from fast-growing green trees are under 61 years for 'Growth ×1.5', under 44 years for 'Growth ×2', and under 33 years for 'Growth ×2.5', while parity times for bioenergy sourced from medium-growing green trees are under 92 years for 'Growth ×1.5', under 66 years for 'Growth ×2', and under 51 years for 'Growth ×2.5' (Fig. 4). Cumulative CO_2 emissions saved for 'Growth ×1.5', 'Growth ×2', and 'Growth ×2.5' reach 32.7, 54.0, and 77.6 Mg GJ^{-1}, respectively, for fast-growing trees, while they reach 12.7, 26.9, and 41.2 Mg GJ^{-1}, respectively, for medium-growing trees (data not shown).

Increasing energy conversion efficiency decreases time to parity of all bioenergy scenarios, but more so for salvaged trees (Fig. 5). Parity times of best-case scenarios using salvaged trees decrease from 54 years (without efficiency improvement) to 34 years with 3% improvement, to 21 years with 6% improvement, and to 12 years with 9% improvement. Moreover, improving efficiency by 9% allows the best case of the harvest residue scenario to achieve immediate benefits compared with 12 years without efficiency improvement (baseline scenario).

Increasing the basal decay rate (BDR) of the model by two and three times reduces parity time and the length

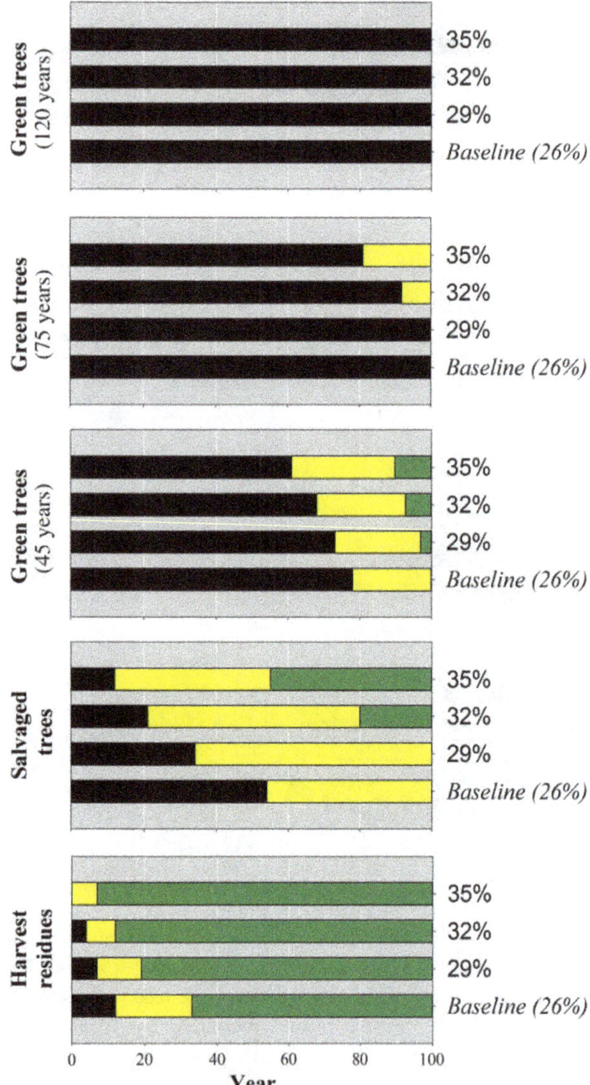

Fig. 5 Timing of GHG benefits and length of the uncertainty phase of scenarios using different bioenergy feedstock to replace coal for power production. For each feedstock, scenarios show the effect of enhanced energy conversion efficiency of biomass relative to a baseline value.

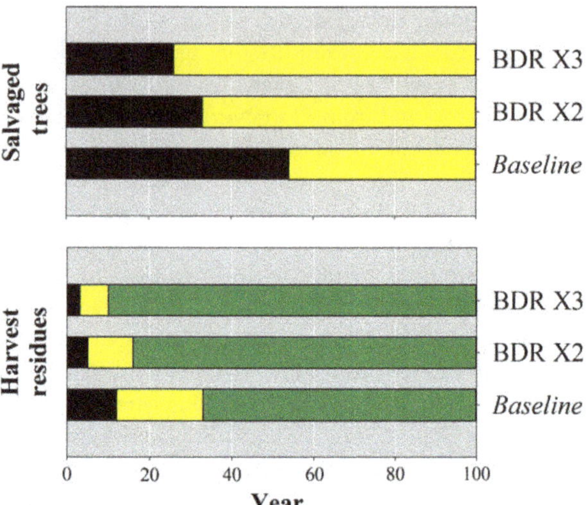

Fig. 6 Timing of GHG benefits and length of the uncertainty phase of scenarios using salvaged trees and harvest residues as bioenergy feedstock to replace coal for power production. For each feedstock, scenarios show the effect of doubling and tripling the basal decay rate (BDR) of the model relative to the baseline rate to account for tree species decaying faster than average. BDR of the baseline scenario at a reference temperature of 10 °C is 0.144 and 0.037 year^{-1} for residues and salvaged trees, respectively.

of the uncertainty phase (Fig. 6), but not as much as does the improvement of conversion efficiency (Fig. 5). Increasing conversion efficiency by 9% has a more beneficial effect on the reduction of C parity time of harvest residues than tripling the BDR.

Decomposing the uncertainty

In a scenario using salvaged trees to replace coal in power generation, the key parameters to reducing the length of the uncertainty phase and C parity time in the worst case are, in decreasing order of importance, transportation distance (local use vs. export), feedstock processing (chips vs. pellets), and mean annual temperature (MAT = 5 vs. −1 °C), whereas the rate of C transfer from snag to the ground (CTR = 0.10 vs. 0.04 year^{-1}) only has a minor effect (Fig. 7). This ranking is also true for scenarios involving different feedstock sources, fossil fuel types, and uses (results not shown).

Discussion

Mitigation potential and timing of bioenergy sourced from Canadian forests

Biomass feedstock and the type of fossil fuel replaced greatly affect the GHG mitigation potential and timing of forest bioenergy scenarios. The results indicate short-to-long ranking of parity times for residues < salvaged < green trees and for replacing the less efficient fossil fuels (coal < oil < natural gas). Not surprisingly, bioenergy sourced from harvest residues yielded the fastest atmospheric benefits. The uncertainty around the estimate of C parity time was also the smallest. Most studies documented parity times <20 years for bioenergy sourced from harvest residues excluding stumps (Repo *et al.*, 2011, 2012; Lamers & Junginger, 2013; Lamers *et al.*, 2014). Branches and tree tops are small woody debris that quickly decompose on the forest floor (Tarasov & Birdsey, 2001; Palviainen *et al.*, 2004; Preston *et al.*, 2012), and the parity time between the

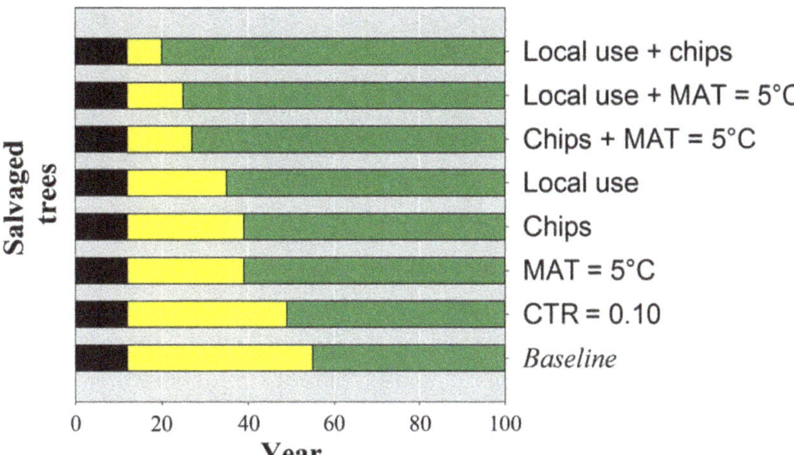

Fig. 7 Length of the uncertainty phase of scenarios using salvaged trees as bioenergy feedstock (with elevated conversion efficiency of 35%) to replace coal for power production. Uncertainty is generated through choices in parameter values for snag C transfer rates (CTR = 0.10 or 0.04 year^{-1}), mean annual temperatures (MAT = 5 °C or −1 °C), biomass processing (chips or pellets), and transportation distances (local use or export). Scenarios are identified by which parameters are set as fixed while all others are varied to generate the uncertainty. For the baseline, all parameters are varied.

bioenergy system, in which biomass emits C to the atmosphere to produce energy, and the reference fossil fuel system, in which biomass is left to decompose in the forest, is therefore quickly reached. Furthermore, in the case of harvest residues that are normally burned by the roadside to reduce the fire hazard, the use of bioenergy to replace fossil fuel generates immediate atmospheric benefits (C parity time = 0 year). Likewise, increasing biomass conversion efficiency to 35% can generate immediate benefits in some cases of harvest residues normally left to decompose *in situ*. Given that the environmental cost of delaying GHG emission reductions is increasingly being recognized (IPCC, 2014), residue-based bioenergy therefore is a suitable feedstock for mitigating GHG emissions in a short time frame.

By contrast, using medium- and slow-growing green trees showed little to no atmospheric benefits over the 100-year period. In northern forests, trees grow slowly and harvested lands usually take many decades to regenerate and regain C levels that are similar to pre-harvest levels (Seely *et al.*, 2002; Kurz *et al.*, 2013). Furthermore, when the reference forest is assumed to be unharvested in the counterfactual scenario, CO_2 may still be taken up from the atmosphere while the land harvested for bioenergy slowly starts to regenerate. Accordingly, C parity time for procuring biomass from living trees takes many decades to be reached. Bernier & Paré (2013) obtained a time to C parity of over 90 years for a scenario that used wood chips from boreal tree species to replace oil in heat generation. Other studies also documented multidecadal parity times (or payback times) for bioenergy made from green trees in northern forests (McKechnie *et al.*, 2011; Holtsmark, 2012; Mitchell *et al.*, 2012; Ter-Mikaelian *et al.*, 2015). However, using silviculture to increase tree growth rate in the regenerating stand can improve the performance of this feedstock source and generate atmospheric benefits within a shorter time frame. Silvicultural practices as seen in Canada may increase timber yield from two to six times relative to natural forests (Paquette & Messier, 2010). Higher tree productivity and faster C capture through silviculture allow to reach parity time faster. Similar conclusions were obtained by the Ter-Mikaelian *et al.* (2015) study, in which coal was replaced by wood pellets sourced from Ontario forests. Moreover, Lamers *et al.* (2014) assumed faster tree growth (2×) for replanted sites relative to natural forests and obtained a parity time of 84 years for slow-growing spruce-fir stands, which falls well within our range of parity times for a comparable scenario (the one in which bioenergy is obtained from slow-growing green trees, i.e., 120 years). In summary, while using trees is most often associated with long-term parity time, some specific cases may show parity time <50 years. These cases would involve growth enhancement by silviculture, which would happen only with bioenergy scenarios and also good growing conditions (productive stand types with relatively short rotation periods).

Salvaged trees had intermediate parity times between that of harvest residues and that of green trees. This feedstock also had a very wide phase of uncertainty, indicating that some cases present reasonable parity times that meet short- and medium-term GHG emission reduction targets, while others do not. For example, using bioenergy sourced entirely from slow-decomposing dead stemwood (e.g., pine species in cold regions) without regeneration improvement through silviculture is perhaps not an option to prioritize, given that the parity time would likely always be over 75 years. However, if silviculture is performed in the regenerating stand following harvesting and biomass procurement, this feedstock source may become more interesting in terms of C savings, with several cases falling below 40 years before achieving atmospheric benefits. Results from Lamers *et al.* (2014) also highlighted the potential of using salvage wood from MPB-impacted stands to mitigate GHG emissions. Relative to a reference 'no harvest' scenario, they obtained immediate benefits and a parity time of 22 years when pine-only (85% dead trees) and pine-dominated (62% dead trees) stands were first harvested for pellets, replanted (assuming a twofold growth yield in plantations relative to natural forest), and then harvested for sawlog timber with the residues used for pellets. Jonker *et al.* (2014) varied the forest management intensity levels and obtained >50% reduction in time to C parity in high-intensity management scenarios relative to low-intensity ones. In summary, as is the case for green trees, specific conditions need to be present to reduce the time to parity in salvaged wood scenarios. These conditions often involve silviculture. An interesting example is given in Barrette *et al.* (2013), where black spruce (*Picea mariana*) stands showed little regeneration 8 years after fire while jack pine (*Pinus banksiana*) stands showed a good regeneration. Harvesting biomass for bioenergy in the black spruce site would facilitate the silvicultural treatment carried out to restore forest productivity, while it would probably not enhance forest productivity in the jack pine site.

Increasing the base decomposition rate (BDR) of the model to account for tree species decaying faster than average indicates that sourcing bioenergy from fast-decomposing species such as intolerant hardwoods (e.g., aspen, birch) would be another potentially suitable GHG mitigation option, especially if the feedstock is collected in warmer regions. Although our knowledge of logs' decomposition rate is limited (Weedon *et al.*, 2009), empirical observations in northern forests showed that the logs of such species may achieve almost complete decomposition (85–95%) within 57 years, while pine and spruce species may take over 80 years (Tarasov & Birdsey, 2001; Brais *et al.*, 2006; Shorohova & Kapitsa, 2014).

Salvaged trees have the potential to generate relatively fast atmospheric benefits, but would require a good tracking system to reduce uncertainty and meet precise time frames. As shown in our analysis, favoring wood chips over pellets and local use over transoceanic export are good options to prioritize in order to reduce the uncertainty period. Moreover, the speed at which parity time is reached is also impacted by the regional climate and tree species, which regulate the decomposition rate of deadwood (in the counterfactual scenario). Performing silviculture and improving energy conversion efficiency can also greatly reduce the time to GHG mitigation of bioenergy sourced from dead trees.

Overall, our results are coherent with the perspective of Haberl *et al.* (2012) on C emission reduction by bioenergy. Short- to medium-term atmospheric benefits (<50 years C parity time) must involve the use of 'additional biomass', defined as biomass from additional vegetation growth or biomass that would decay rapidly if not used for bioenergy. Such parity times are possible in some cases under salvaged tree scenarios, but more likely under specific conditions involving important gains in forest productivity (silviculture) under either green tree or salvage tree scenarios.

Taking uncertainty into account

To our knowledge, few studies have addressed the uncertainty around the estimation of C debt in a forest bioenergy context (Johnson *et al.*, 2011; Caputo *et al.*, 2014; Röder *et al.*, 2015). To date, studies have mostly focused on estimating a unique and precise C debt repayment time or C parity time for particular case studies without addressing any sources of variation. For correct accounting, however, estimates need to take uncertainty into account, from variations in the biomass supply chain to the realism of the counterfactual scenario (Johnson *et al.*, 2011; Bowyer *et al.*, 2012; Buchholz *et al.*, 2014, 2015). We found that the length of the uncertainty period can be short and inconsequential for some scenarios (e.g., harvest residues). However, for other scenarios, it can be large enough to cast doubts as to whether a particular feedstock should be considered in GHG mitigation efforts in the short term. In the current study, the length of the uncertainty phase depends on how we define the best and worst cases, that is, which parameters will vary and to what extent. In our scenarios involving green trees as feedstock, only upstream emissions (processing, transport) could affect uncertainty. By contrast, for salvaged trees, upstream emissions, MAT (which impacts the decomposition rate), and snag C transfer rate all are elements whose range of possible values contributed to uncertainty. These additional sources of variation explained the longer phase of C debt uncertainty in the salvaged tree scenarios relative to the green tree scenarios, while the slower decay rate of stemwood (salvaged trees) than of branches (residues) explained the longer uncertainty phase relative to harvest residues. Varying the tree growth rate in a scenario involving green trees (instead of making separate scenarios) or adding natural disturbances (Buchholz *et al.*, 2015) would push the length of the uncertainty period for green tree-based scenarios beyond that of scenarios involving salvaged trees.

Not all sources of uncertainty were tested in our analysis. Some parameters including the emission factor for combustion of biomass and fossil fuel, fossil energy conversion efficiency, and temperature sensitivity of decomposition (Q_{10}) were set as constants, based on averaged values found in the literature. The IPCC default emission factor for biomass combustion is 112 kg CO_2 GJ^{-1}, but its 95% confidence limits range from 95 to 132 kg CO_2 GJ^{-1} (IPCC, 2006). The heating value of wood also varies among tree species (Singh & Kostecky, 1986; Telmo & Lousada, 2011; Barrette *et al.*, 2013). Similarly, energy conversion efficiency for a given fossil fuel may vary substantially depending on factors such as generator capacity, age, and technology (Koop *et al.*, 2010). Jonker *et al.* (2014) varied the conversion efficiency of a coal power plant from 35% to 46% and observed decadal differences in payback and parity times of bioenergy under low- and high-efficiency scenarios. Röder *et al.* (2015) also pointed out the impact of wood chip storage duration on methane emissions, which greatly affect the C balance of forests and sawmill residues. As we gain confidence in understanding belowground processes and long-term impact of forest harvesting intensification, soil C (which was set as constant here) may become an important parameter to consider in the C balance of forest bioenergy, given the large share of ecosystem C that resides in soils (Laganière *et al.*, 2015). These additional sources of uncertainty could make the uncertainty phase even longer than what is presented here. Evidently, proper knowledge of both the bioenergy and the reference fossil fuel systems is required in order to accurately evaluate the potential of a bioenergy project to mitigate GHG emissions.

Key to reducing uncertainty around estimates of C parity time is a better assessment of ecological processes (e.g., forest regeneration and growth rate, decomposition dynamics), as also pointed out by Caputo *et al.* (2014). Favoring local use of wood chips over the export of wood pellets can also reduce the length of the uncertainty period. Potential economic feedback between biomass procurement practices and other forest management activities should also be considered: Adding bioenergy to the basket of products that can be sourced from a given stand or landscape may increase the profitability of overall forest operations and foresters' belief in future markets, creating new incentives for forest management (Bellassen & Luyssaert, 2014). Overall, the current study brings into question the use of single parity time values to evaluate the performance of a particular feedstock to mitigate GHG emissions given the importance of uncertainty as an inherent component of every bioenergy project. More specifically, it suggests that some feedstock, such as green or salvaged trees that are usually associated with long and uncertain time to parity, can, under some very specific circumstances, show shorter and less uncertain parity times.

Acknowledgements

We thank Xiao Jing Guo for her assistance with SAS programming and Isabelle Lamarre for text editing. We also thank Nicolas Mansuy and three anonymous reviewers for providing helpful comments on the article. This study was funded through the Program of Energy Research and Development (PERD) and the ecoENERGY Innovation Initiative (ecoEII) of the Government of Canada. We also acknowledge the support from the Natural Sciences and Engineering Research Council of Canada (NSERC) through the Visiting Fellowships in Canadian Government Laboratories Program.

References

Barrette J, Thiffault E, Paré D (2013) Salvage harvesting of fire-killed stands in northern Quebec: analysis of bioenergy and ecological potentials and constraints. *J-FOR*, **3**, 16–25.

Bellassen V, Luyssaert S (2014) Carbon sequestration: managing forests in uncertain times. *Nature*, **506**, 153–155.

Bernier P, Paré D (2013) Using ecosystem CO_2 measurements to estimate the timing and magnitude of greenhouse gas mitigation potential of forest bioenergy. *GCB Bioenergy*, **5**, 67–72.

Boulanger Y, Gauthier S, Burton PJ (2014) A refinement of models projecting future Canadian fire regimes using homogeneous fire regime zones. *Canadian Journal of Forest Research*, **44**, 365–376.

Bowyer C, Baldock D, Kretschmer B, Polakova J (2012) *The GHG Emissions Intensity of Bioenergy: Does Bioenergy have a Role to Play in Reducing GHG Emissions of Europe's Economy?*Institute for European Environmental Policy (IEEP), London, UK.

Brais S, Paré D, Lierman C (2006) Tree bole mineralization rates of four species of the Canadian eastern boreal forest: implications for nutrient dynamics following stand-replacing disturbances. *Canadian Journal of Forest Research*, **36**, 2331–2340.

Buchholz T, Prisley S, Marland G, Canham C, Sampson N (2014) Uncertainty in projecting GHG emissions from bioenergy. *Nature Climate Change*, **4**, 1045–1047.

Buchholz T, Hurteau MD, Gunn J, Saah D (2015) A global meta-analysis of forest bioenergy greenhouse gas emission accounting studies. *GCB Bioenergy* (in press).

Caputo J, Balogh S, Volk T, Johnson L, Puettmann M, Lippke B, Oneil E (2014) Incorporating uncertainty into a life cycle assessment (LCA) model of short-rotation willow biomass (*Salix* spp.) crops. *BioEnergy Research*, **7**, 48–59.

Cherubini F, Gasser T, Bright RM, Ciais P, Stromman AH (2014) Linearity between temperature peak and bioenergy CO2 emission rates. *Nature Climate Change*, **4**, 983–987.

Environment Canada (2015) *Canadian Climate Normals or Averages, 1971–2000*. Environment Canada, Fredericton, NB, Canada.

Goh CS, Junginger M, Cocchi M *et al.* (2013) Wood pellet market and trade: a global perspective. *Biofuels, Bioproducts and Biorefining*, **7**, 24–42.

Haberl H, Sprinz D, Bonazountas M *et al.* (2012) Correcting a fundamental error in greenhouse gas accounting related to bioenergy. *Energy Policy*, **45**, 18–23.

Hilger AB, Shaw CH, Metsaranta JM, Kurz WA (2012) Estimation of snag carbon transfer rates by ecozone and lead species for forests in Canada. *Ecological Applications*, **22**, 2078–2090.

Holtsmark B (2012) Harvesting in boreal forests and the biofuel carbon debt. *Climatic Change*, **112**, 415–428.

IPCC (2006) Chapter 2: Stationary combustion. In: *IPCC Guidelines for National Greenhouse Gas Inventories, Prepared by the National Greenhouse Gas Inventories Programme*, Vol 2 (eds Eggleston HS, Buendia L, Miwa K, Ngara T, Tanabe K), pp. 2.1–2.53. IGES, Japan.

IPCC (2014) Summary for policymakers. In: *Climate Change 2014: Mitigation of Climate Change. Contribution of Working Group III to the Fifth Assessment Report of the Inter-*

governmental Panel on Climate Change (eds Edenhofer O, Pichs-Madruga R, Sokona Y, Farahani E, Kadner S, Seyboth K, Adler A, Baum I, Brunner S, Eickemeier P, Kriemann B, Savolainen J, Schlömer S, Von Stechow C, Zwickel T, Minx JC), pp. 1–30. Cambridge University Press, Cambridge, UK and New York, NY, USA.

Johnson DW, Curtis PS (2001) Effects of forest management on soil C and N storage: meta analysis. *Forest Ecology and Management*, **140**, 227–238.

Johnson DR, Willis HH, Curtright AE, Samaras C, Skone T (2011) Incorporating uncertainty analysis into life cycle estimates of greenhouse gas emissions from biomass production. *Biomass and Bioenergy*, **35**, 2619–2626.

Jonker JGG, Junginger M, Faaij A (2014) Carbon payback period and carbon offset parity point of wood pellet production in the South-eastern United States. *GCB Bioenergy*, **6**, 371–389.

Koop K, Koper M, Bijsma R, Wonink S, Ouwens JD (2010) *Evaluation of Improvements in End-Conversion Efficiency for Bioenergy Production*. Ecofys, Utrecht, The Netherlands.

Kurz WA, Dymond CC, White TM *et al.* (2009) CBM-CFS3: A model of carbon-dynamics in forestry and land-use change implementing IPCC standards. *Ecological Modelling*, **220**, 480–504.

Kurz WA, Shaw CH, Boisvenue C *et al.* (2013) Carbon in Canada's boreal forest – a synthesis. *Environmental Reviews*, **21**, 260–292.

Laganière J, Paré D, Bergeron Y, Chen HYH (2012) The effect of boreal forest composition on soil respiration is mediated through variations in soil temperature and C quality. *Soil Biology and Biochemistry*, **53**, 18–27.

Laganière J, Cavard X, Brassard BW, Paré D, Bergeron Y, Chen HYH (2015) The influence of boreal tree species mixtures on ecosystem carbon storage and fluxes. *Forest Ecology and Management*, **354**, 119–129.

Lamers P, Junginger M (2013) The 'debt' is in the detail: a synthesis of recent temporal forest carbon analyses on woody biomass for energy. *Biofuels, Bioproducts and Biorefining*, **7**, 373–385.

Lamers P, Junginger M, Hamelinck C, Faaij A (2012) Developments in international solid biofuel trade – an analysis of volumes, policies, and market factors. *Renewable and Sustainable Energy Reviews*, **16**, 3176–3199.

Lamers P, Thiffault E, Paré D, Junginger M (2013) Feedstock specific environmental risk levels related to biomass extraction for energy from boreal and temperate forests. *Biomass and Bioenergy*, **55**, 212–226.

Lamers P, Junginger M, Dymond CC, Faaij A (2014) Damaged forests provide an opportunity to mitigate climate change. *GCB Bioenergy*, **6**, 44–60.

Lecomte N, Simard M, Fenton N, Bergeron Y (2006) Fire severity and long-term ecosystem biomass dynamics in coniferous boreal forests of eastern Canada. *Ecosystems*, **9**, 1215–1230.

Litton CM, Giardina CP (2008) Below-ground carbon flux and partitioning: global patterns and response to temperature. *Functional Ecology*, **22**, 941–954.

Magelli F, Boucher K, Bi HT, Melin S, Bonoli A (2009) An environmental impact assessment of exported wood pellets from Canada to Europe. *Biomass and Bioenergy*, **33**, 434–441.

Manomet Center for Conservation Sciences (2010) *Massachusetts Biomass Sustainability and Carbon Policy Study: Report to the Commonwealth of Massachusetts Department of Energy Resources*. Natural Capital Initiative Report NCI-2010-03, Brunswick, Maine.

McKechnie J, Colombo S, Chen J, Mabee W, MacLean HL (2011) Forest bioenergy or forest carbon? Assessing trade-offs in greenhouse gas mitigation with wood-based fuels. *Environmental Science and Technology*, **45**, 789–795.

Meil J, Bushi L, Garrahan P, Aston R, Gingras A, Elustondo D (2009) *Status of Energy Use in the Canadian Wood Products Sector*. Canadian Industry Program for Energy Conservation, Natural Resources Canada, Ottawa, ON.

Messier C, Bigué B, Bernier L (2003) Using fast-growing plantations to promote forest ecosystem protection in Canada. *Unasylva*, **54**, 59–63.

Mitchell SR, Harmon ME, O'Connell KEB (2012) Carbon debt and carbon sequestration parity in forest bioenergy production. *GCB Bioenergy*, **4**, 818–827.

Nave LE, Vance ED, Swanston CW, Curtis PS (2010) Harvest impacts on soil carbon storage in temperate forests. *Forest Ecology and Management*, **259**, 857–866.

NRCan (2014a) *National Forestry Database*. Natural Resources Canada, Canadian Forest Service, Ottawa, ON. Available at http://nfdp.ccfm.org/index_e.php. (accessed 1 December 2015).

NRCan (2014b) *The State of Canada's Forests: Annual Report 2014*, Natural Resources Canada. Canadian Forest Service, Ottawa, ON.

Pa A, Craven J, Bi X, Melin S, Sokhansanj S (2012) Environmental footprints of British Columbia wood pellets from a simplified life cycle analysis. *The International Journal of Life Cycle Assessment*, **17**, 220–231.

Palviainen M, Finér L, Kurka AM, Mannerkoski H, Piirainen S, Starr M (2004) Decomposition and nutrient release from logging residues after clear-cutting of mixed boreal forest. *Plant and Soil*, **263**, 53–67.

Paquette A, Messier C (2010) The role of plantations in managing the world's forests in the Anthropocene. *Frontiers in Ecology and the Environment*, **8**, 27–34.

Preston CM, Trofymow JA, Nault JR (2012) Decomposition and change in N and organic composition of small-diameter Douglas-fir woody debris over 23 years. *Canadian Journal of Forest Research*, **42**, 1153–1167.

Repo A, Tuomi M, Liski J (2011) Indirect carbon dioxide emissions from producing bioenergy from forest harvest residues. *GCB Bioenergy*, **3**, 107–115.

Repo A, Känkänen R, Tuovinen J-P, Antikainen R, Tuomi M, Vanhala P, Liski J (2012) Forest bioenergy climate impact can be improved by allocating forest residue removal. *GCB Bioenergy*, **4**, 202–212.

Röder M, Whittaker C, Thornley P (2015) How certain are greenhouse gas reductions from bioenergy? Life cycle assessment and uncertainty analysis of wood pellet-to-electricity supply chains from forest residues. *Biomass and Bioenergy*, **79**, 50–63.

$(S\&T)^2$ (2015) *GHGenius, Model Version 4.03*. $(S\&T)^2$ Consultants Inc. for Natural Resources Canada, Ottawa, ON, Canada.

Searchinger TD, Hamburg SP, Melillo J *et al.* (2009) Fixing a critical climate accounting error. *Science*, **326**, 527–528.

Seely B, Welham C, Kimmins H (2002) Carbon sequestration in a boreal forest ecosystem: results from the ecosystem simulation model, FORECAST. *Forest Ecology and Management*, **169**, 123–135.

Shorohova E, Kapitsa E (2014) Influence of the substrate and ecosystem attributes on the decomposition rates of coarse woody debris in European boreal forests. *Forest Ecology and Management*, **315**, 173–184.

Singh T, Kostecky MM (1986) Calorific value variations in components of 10 Canadian tree species. *Canadian Journal of Forest Research*, **16**, 1378–1381.

Soja AJ, Tchebakova NM, French NHF *et al.* (2007) Climate-induced boreal forest change: predictions versus current observations. *Global and Planetary Change*, **56**, 274–296.

Tarasov ME, Birdsey RA (2001) Decay rate and potential storage of coarse woody debris in the Leningrad region. *Ecological Bulletins*, **49**, 137–147.

Telmo C, Lousada J (2011) Heating values of wood pellets from different species. *Biomass and Bioenergy*, **35**, 2634–2639.

Ter-Mikaelian MT, Colombo SJ, Lovekin D *et al.* (2015) Carbon debt repayment or carbon sequestration parity? Lessons from a forest bioenergy case study in Ontario, Canada. *GCB Bioenergy*, **7**, 704–716.

Thiffault E, Hannam KD, Paré D, Titus BD, Hazlett PW, Maynard DG, Brais S (2011) Effects of forest biomass harvesting on soil productivity in boreal and temperate forests – a review. *Environmental Reviews*, **19**, 278–309.

Thiffault N, Fenton N, Munson A *et al.* (2013) Managing understory vegetation for maintaining productivity in black spruce forests: a synthesis within a multi-scale research model. *Forests*, **4**, 613–631.

Weedon JT, Cornwell WK, Cornelissen JHC, Zanne AE, Wirth C, Coomes DA (2009) Global meta-analysis of wood decomposition rates: a role for trait variation among tree species? *Ecology Letters*, **12**, 45–56.

7

Tissue chemistry and morphology affect root decomposition of perennial bioenergy grasses on sandy soil in a sub-tropical environment

XI LIANG[1,2], JOHN E. ERICKSON[1], MARIA L. SILVEIRA[3], LYNN E. SOLLENBERGER[1] and DIANE L. ROWLAND[1]

[1]*Agronomy Department, University of Florida, 3105 McCarty Hall B, Gainesville, FL 32611, USA,* [2]*Department of Plant, Soil and Entomological Sciences, University of Idaho, 1693 S 2700 W, Aberdeen, ID 83210, USA,* [3]*Department of Soil and Water Sciences, University of Florida, 3401 Experimental Station, Ona, FL 33865, USA*

Abstract

Second-generation biofuels and bio-based products derived from lignocellulosic biomass are likely to replace current fuels derived from simple sugars and starch because of greater yield potential and less competition with food production. Besides the high aboveground biomass production, these bioenergy grasses also exhibit extensive root systems. The decomposition of root biomass greatly influences nutrient cycling and microbial activity and subsequent accumulation of carbon (C) in the soil. The objective of this research was thus to characterize root morphological and chemical differences in six perennial grass species in order to better understand root decomposition and belowground C cycling of these bioenergy cropping systems. Giant reed (*Arundo donax*), elephantgrass (*Pennisetum purpureum*), energycane (*Saccharum* spp.), sugarcane (*Saccharum* spp.), sweetcane (*Saccharum arundinaceum*), and giant miscanthus (*Miscanthus* × *giganteus*) were established in Fall 2008 in research plots near Gainesville, Florida. Root decomposition rates were measured *in situ* from root decomposition bags over 12 months along with initial and final root tissue composition. Root potential decomposition rate constant (K) was higher in elephantgrass (3.64 g kg^{-1} day^{-1}) and sweetcane (2.77 g kg^{-1} day^{-1}) than in sugarcane (1.62 g kg^{-1} day^{-1}) and energycane (1.48 g kg^{-1} day^{-1}). Notably, K was positively related to initial root tissue total C (Total C), total fiber glucose (TFG), total fiber xylose (TFX), and total fiber carbohydrate (TFC) concentrations, but negatively related to total fiber arabinose (TFA) and lignin (TL) concentrations and specific root volume (SRV). Among the six species, elephantgrass exhibited root traits most favorable for fast decomposition: high TFG, high TFX, high TFC, high specific root length (SRL), and a low SRV, whereas giant reed, sugarcane, and energycane exhibited slow decomposition rates and the corresponding root traits. Thus, despite similar aboveground biomass yields in many cases, these species are likely to differentially affect soil C accumulation.

Keywords: Arundo donax, bioenergy grasses, chemical composition, *Miscanthus* × *giganteus*, *Pennisetum purpureum*, root decomposition, root morphology, *Saccharum arundinaceum*, *Saccharum* spp.

Introduction

Recent attention focused on biomass crops to increase and diversify energy production and help mitigate greenhouse gas emissions has identified perennial grasses as potential dedicated energy feedstocks (Lemus & Lal, 2005; Carroll & Somerville, 2009; Davis *et al.*, 2010; Somerville *et al.*, 2010; Don *et al.*, 2012; Drewer *et al.*, 2012). Although large plantings of perennial energy grasses could help to supplement biofuel production, the implications for other ecosystems services such as carbon (C) sequestration and nutrient cycling are not well understood. The decomposition of plant tissues, particularly belowground pools, is an important process that affects soil microbial activity and C accumulation (Six *et al.*, 2000; Jones & Donnelly, 2004; Rasse *et al.*, 2005). As the main source of C inputs to the soil, the quantity and quality of root and rhizome biomass influences their decomposition rate and their residence time with subsequent impacts on C stored in the soil (Fontaine *et al.*, 2007; Hättenschwiler & Jørgensen, 2010; Amougou *et al.*, 2011; Knoll *et al.*, 2012; Sun *et al.*, 2013). Therefore, understanding the decomposition of belowground biomass is critical in investigating C fluxes in terrestrial ecosystems.

Root decomposition is greatly affected by climatic and edaphic conditions and is thus expected to be unique to each specific environment. In general, litter decomposition rates are positively correlated with mean

Correspondence: John E. Erickson
e-mail: jerickson@ufl.edu

annual temperature and precipitation (Silver & Miya, 2001; Zhang *et al.*, 2008; Prescott, 2010). Similarly, saturated soil moisture conditions or extreme low moisture levels tend to limit decomposition of plant litter (von Haden & Dornbush, 2014; Lee *et al.*, 2014). Roots decompose faster in clay loam than in sand and clay soils (Silver & Miya, 2001).

Root chemical and morphological characteristics also contribute to their decomposition patterns and rates (Silver & Miya, 2001; Puttaso et al., 2011; Aulen et al., 2012). Labile compounds in root tissues, such as simple sugars, organic acids, small-chain fatty acids, and proteins, can be easily taken up by microorganisms and metabolized, and their concentrations in the plant tissue tend to decline quickly as the decomposition process progresses (Rasse et al., 2005; Berg & McClaugherty, 2008). Conversely, cellulose is a relatively recalcitrant cell wall component compared to water-soluble compounds and the decomposition of plant-derived C typically decreases as cellulose concentration increases (Birouste *et al.*, 2012). However, research has also demonstrated no correlation between cellulose concentration and root decomposition rates (Aulen et al., 2012). Lignin, suberin, cutin, and polyphenols are considered recalcitrant components and generally retard root decomposition (Rasse et al., 2005; von Luetzow et al., 2006; Hättenschwiler & Jørgensen, 2010). In addition, herbaceous species root morphological attributes such as specific root length (SRL) have been positively correlated with root decomposition rate (Aulen et al., 2012). In a 4-year experiment with temperate tree species, root decomposition rates differed among root diameter classes (e.g., <0.5 and 0.5–2 mm) (Sun et al., 2013). In contrast, Birouste et al. (2012) found no correlation between initial SRL or diameter and decomposition rates. Consequently, there are conflicting reports in the literature on the effects of root chemistry and morphology on root decomposition rates, and the complexity is probably due to the interaction of root characteristics and environmental factors.

In general, Poaceae roots decompose more slowly than those of herbaceous species from other taxonomic groups (Aulen et al., 2012; Birouste et al., 2012). The slow decomposition rates in Poaceae roots were often associated with their unique root morphological and chemical traits compared with other species, such as low N and P concentrations, high C/N, and high tissue density (Birouste et al., 2012; Siqueira da Silva *et al.*, 2015). Within the Poaceae family, species also differed in their root morphological and chemical traits and thus decomposition rates (Fornara & Tilman, 2008; Aulen et al., 2012). In addition, perennial warm-season grasses generally exhibit abundant root biomass and deep rooting systems (Monti & Zatta, 2009; Somerville et al., 2010). For instance, root dry weight in a 0- to 120-cm soil profile was 14 Mg ha^{-1} in giant reed (*Arundo donax*), 7.6 Mg ha^{-1} in giant miscanthus (*Miscanthus × giganteus*), and 8.5 Mg ha^{-1} in switchgrass (*Panicum virgatum*) in north Italy (Monti & Zatta, 2009). The large differences in root biomass production and root characteristics among perennial grass species are expected to result in differences in the amount and characteristics of C inputs to the soil.

Despite relatively similar aboveground biomass production (Erickson *et al.*, 2012; Knoll et al., 2012; Fedenko *et al.*, 2013), our understanding regarding the potential impacts of root morphological characteristics and chemical composition of perennial bioenergy crops on belowground C fluxes remains limited, particularly in the southeastern USA, where the warm temperatures associated with abundant rainfall are expected to create favorable conditions for root decomposition. Therefore, the objectives of this research were to characterize root morphological and chemical differences in six bioenergy grass species and to evaluate their impacts on root decomposition over a 12-month period under field conditions. We hypothesized that variability in species root chemistry and morphology would lead to differences in decomposition that will help to better understand C cycling in bioenergy grass cropping systems.

Materials and methods

Plant materials, growth conditions, and treatments

Giant reed (*Arundo donax*), elephantgrass (*Pennisetum purpureum* 'Merkeron'), energycane (*Saccharum* spp. 'L79-1002'), sugarcane (*Saccharum* spp. 'CP89-2143'), sweetcane (*Saccharum arundinaceum* 'IK76-110'), and giant miscanthus (*Miscanthus × giganteus*) were established from vegetative propagules from November 2008 to January 2009 at the University of Florida Plant Science Research and Education Unit (29°24′N 82°10′W) in Citra, Florida. The six species were arranged in a randomized complete block design with four replicates, giving 24 plots in total. Plots consisted of six 6-m-long rows initially planted with 1-m row spacing (i.e., each plot was approximately 36 m^2). All plots were fertilized at a rate of 280 kg N ha^{-1} yr^{-1} split into applications of 90 kg N ha^{-1} in mid-April and 190 kg N ha^{-1} in June. Plots also received ~ 70 and 140 kg ha^{-1} of P_2O_5 and K_2O in each year, respectively, along with micronutrients. Limited irrigation was applied via overhead irrigation during establishment and with appearance of visible water stress (i.e., leaf rolling). Soil temperature at 10 cm was monitored during the experiment and is summarized in Fig. 1.

For aboveground dry biomass yield, plots were harvested once annually in 2009 and 2010 (November) and twice annually (July and November) in 2011 and 2012. A 4-m section (4 m^2) from one of the inner two middle rows in each plot was harvested and weighed in the field. A representative subsample of

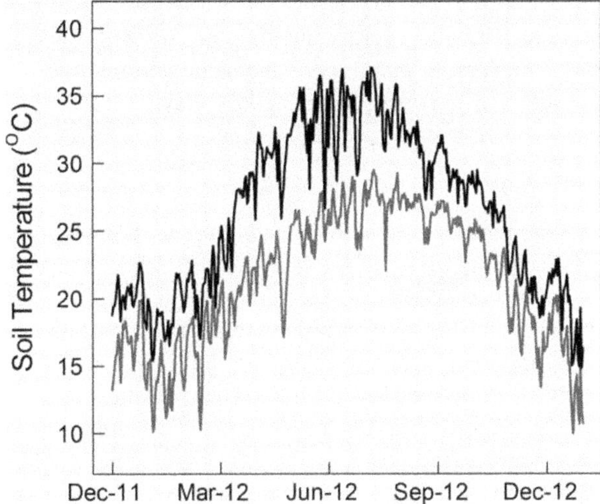

Fig. 1 Daily maximum (black line) and minimum (gray line) soil temperature (°C) at a 100-mm soil depth over the duration of the field root decomposition study.

biomass was then collected and oven-dried at 50 °C to a constant weight to determine dry matter concentration and dry biomass yield.

Root morphology, chemistry, and decomposition

To obtain a representative root biomass sample, at least four soil cores (11 cm diameter × 20 cm depth) were collected from each of the six perennial grass species plots after the above-ground biomass harvest in December 2011. Soil cores from the same plot were pooled into a composited sample and then taken back to the laboratory. Roots were separated from soil using a 2-mm sieve, gently washed with deionized water, and dried in the oven at 30 °C for 1–2 h to approximately 40% moisture. The roots for a given plot were then cut into 5-cm pieces, thoroughly mixed, and homogenized, and 3.45 ± 0.05 g [mean (n = 92) sample weight ±1 SE at 40% moisture] was put into 15 × 20 cm, 250-μm mesh nylon litter bags (Castillo *et al.*, 2010). A subsample of cut roots was placed in the oven at 50 °C and dried to a constant weight to determine the root moisture content and root dry weights for each bag. An additional fresh root subsample from each plot for each species (n = 4) was scanned and analyzed with a digital image analysis system (WinRHIZO, Regent Instrument, Quebec, CA, USA) to determine root length, surface area, and volume. The subsample was then placed in the oven at 50 °C and dried to a constant weight to determine root dry weight. Specific root length (SRL), specific root area (SRA), and specific root volume (SRV) were calculated as the ratio of root length, surface area, and volume divided by the corresponding root dry weight, respectively.

Four root decomposition bags were buried horizontally at a 7.5-cm depth in the soil in the row middles (50 cm away from each row) in their corresponding field plots on December 8, 2011. For species without obvious plant rows after 2 years since planting, for example, giant reed, decomposition bags were buried arbitrarily in the center of the plots. One bag from each plot was collected at 1, 3, 6, and 12 months after installation. Root biomass remaining in the bags after decomposition was oven-dried at 50 °C to a constant weight and then was ashed at 500 °C for at least 6 h to calculate the root biomass on an ash-free basis. The percentage of the initial root biomass remaining after decomposition (M_t, %) was calculated as follows:

$$M_t = \frac{\text{Root}_{\text{mass},t} - \text{Ash}_{\text{mass},t}}{\text{Root}_{\text{mass},i} - \text{Ash}_{\text{mass},i}} \times 100, \qquad (1)$$

where $\text{Root}_{\text{mass},i}$ and $\text{Root}_{\text{mass},t}$ are the initial dry root biomass before decomposition and remaining at each harvest, respectively, and $\text{Ash}_{\text{mass},i}$ and $\text{Ash}_{\text{mass},t}$ are the ash concentrations in the initial and remaining root biomass, respectively.

To estimate the decomposition rate for each species, the proportion of the initial biomass remaining over time (t) was fit with a single-pool negative exponential model (Adair *et al.*, 2010; Birouste et al., 2012):

$$M_t = 100 \times e^{-Kt}, \qquad (2)$$

where K is the decomposition rate constant and is expressed in g kg^{-1} day^{-1}.

Root biomass from each plot before and after the 12-month field incubation was ground using a Restsch Mixer Mill (MM400, Verder Scientific, Inc., Newtown, PA, USA). A subsample of 30–50 mg of ground root was wrapped in tin capsules (9 × 10 mm, Costech Analytic Technology, Inc., Valencia, CA) and then the total C and N concentrations were measured by dry combustion using an elemental analyzer (FLASHEA 112 Series, Thermo Fisher Scientific, Inc., Waltham, MA, USA). On another subsample, nonstructural extractives, structural carbohydrates, and lignin in root tissue before and after 12 months of decomposition were determined according to the procedures fully described in Fedenko et al. (2013). Briefly, to remove nonstructural extractives, 0.5 g of ground root tissue was autoclaved with 100 ml of deionized water in 140-ml sealed pressure tubes (ACE Glass, Inc., Vineland, NJ, USA) at 121 °C and 103 kPa for 1 h. Autoclaved samples were then vacuum-filtered through coarse-porosity filter paper (> 25 μm, Whatman 113, GE Healthcare UK Limited, UK) to capture all structural fiber. Captured structural biomass was then dried at 50 °C to a constant weight for subsequent fiber carbohydrate and lignin analysis. A subsample of 0.3 g of structural fiber was incubated at 30 °C in 3 ml of 72% sulfuric acid for 1 h followed by digestion in 87 ml of 4% sulfuric acid (by adding 84 ml deionized water) in an autoclave at 121 °C and 103 kPa for 1 h. Hydrolyzed samples were vacuum-filtered through a medium-porosity filtering crucible (Coors #60531, CoorsTex, Golden, CO, USA), and the filtrate was collected and analyzed for acid-soluble lignin using a UV–vis spectrophotometer (StellarNet, Inc., Tampa, FL, USA) at a wavelength of 240 nm. The filtered solids were dried at 105 °C to constant weight and then were ashed at 500 °C for at least 6 h. The residuals were weighed for ash concentration, and acid-insoluble lignin was calculated as the difference between dried filtered solids and ash. Total lignin (TL) was calculated as the sum of acid-soluble and acid-insoluble lignin. The remaining filtrate was adjusted to pH to 5–7 with calcium carbonate and filtered through a 0.22-μm syringe filter (Fisher Scientific, Pittsburgh, PA, USA) for fiber carbohydrate analysis using high-performance liquid chromatography (HPLC). The filtrate samples were analyzed by HPLC (Perkin-Elmer Flexar system, Waltham, MA) with an Aminex HPX-87H column (Bio-Rad, Hercules, CA) maintained at 50 °C with HPLC-grade 4 mM sulfuric acid as the mobile phase at 0.4 ml min^{-1} with a 10-μl injection and 40-min run time. Concentrations of total fiber glucose (TFG), total fiber xylose (TFX), total fiber arabinose (TFA), total fiber carbohydrates (TFC) (the sum of TFG, TFX, AND TFA), and TL were all expressed on an ash-free basis (mg g^{-1} DM).

The remaining chemical components after the 12-month field incubation, including total lignin ($TL_{remaining}$), total fiber glucose ($TFG_{remaining}$), total fiber xylose ($TFX_{remaining}$), total fiber arabinose ($TFA_{remaining}$), total fiber carbohydrate ($TFC_{remaining}$), nitrogen ($N_{remaining}$), and carbon ($C_{remaining}$), were calculated using the following equation (Fioretto *et al.*, 2005):

$$R_c = \frac{c_f \times \text{Root}_{\text{mass},f}}{c_i \times \text{Root}_{\text{mass},i}} \times 100, \quad (3)$$

where R_C is the remaining chemical component after the 12-month incubation in the field. C_i and C_f are initial and final concentrations of each chemical component, respectively. $\text{Root}_{\text{mass},f}$ is the final weight of root biomass after the 12-month incubation.

Data analysis

Root chemical (e.g., concentrations of TL, TFG, TFX, TFA, TFC, and total C and N) and morphological (e.g., SRL, SRA, and SRV) characteristics prior to decomposition were analyzed using the generalized linear mixed model (glimmix) procedure of SAS (ver. 9.3, SAS institute, Cary, NC, USA) with species as a fixed effect and block as a random effect. Remaining chemical components and *K* were also analyzed using the glimmix procedure. Pairwise comparisons were made using the lsmeans statement with the Tukey's method with a significance level of $P < 0.05$. Relationships between *K* and root traits were analyzed with Pearson's correlation, principal component analysis (PCA), and factor analysis using SAS procedures of corr, princual, and factor, respectively.

Results

Decomposition rate

The decomposition rate constant, *K*, of elephantgrass was higher at 3.64 g kg^{-1} day^{-1} than all other species except for sweetcane (Table 1). This was associated with 23% of root biomass remaining after the 12-month incubation (Fig. 2), which was also among the lowest of all species in this study. Sweetcane decomposed at a rate constant of 2.77 g kg^{-1} day^{-1}, which was significantly higher than sugarcane and energycane, but did not differ from elephantgrass, giant reed, or giant miscanthus. After the 12-month decomposition study, 40% of sweetcane root biomass remained, which was lower than that in giant reed and energycane (data not shown). Energycane, sugarcane, giant reed, and giant miscanthus did not differ in *K*, averaging 1.72 g kg^{-1} day^{-1}.

Root morphology

Specific root length varied almost three-fold among species, ranging from 9.0 for giant reed to 25.3 m g^{-1} for elephantgrass (Table 1). Elephantgrass roots possessed the highest SRL, whereas giant reed, sugarcane, sweetcane, and energycane were among the lowest in SRL. In contrast, there was less than two-fold variation in SRA, which ranged from 199 to 337 cm^2 g^{-1}. Elephantgrass and giant miscanthus roots had the highest SRA, while giant reed, sweetcane, and sugarcane were among the lowest in SRA. Specific root volume ranged from 3.06 to 4.76 cm^3 g^{-1} across all species. Giant miscanthus, energycane, and sugarcane roots were among the highest in SRV, while elephantgrass, sweetcane, and giant reed were among the lowest for SRV.

Root chemical composition

Elephantgrass roots exhibited a lower TE concentration prior to the 12-month incubation than giant miscanthus, which was not different from other species (Table 2). It also had lower TL concentration than giant reed. Root concentrations of TFG and TFX in elephantgrass roots were higher than almost all other species, in contrast to the relatively lower concentrations in sugarcane. Root concentrations of TFA were relatively lower in giant reed, sweetcane, and elephantgrass. Overall, root TFC concentration of elephantgrass was higher than all other species except sweetcane. Root total C varied little

Table 1 Rate constant (*K*) for decomposition, specific root length (SRL), specific root surface area (SRA), and specific root volume (SRV) of perennial grass fine roots before decomposition

Species	*K* g kg^{-1} day^{-1}	SRL m g^{-1}	SRA cm^2 g^{-1}	SRV cm^3 g^{-1}
Giant reed	1.80 BC*	9.0 C	204 C	3.59 BC
Sugarcane	1.62 C	10.5 C	231 BC	3.97 AB
Sweetcane	2.77 AB	10.1 C	199 C	3.06 C
Energycane	1.48 C	13.1 BC	259 B	4.28 AB
Elephantgrass	3.64 A	25.3 A	311 A	3.09 C
Giant miscanthus	1.99 BC	18.9 B	337 A	4.76 A
P-value	<0.001	<0.001	<0.001	<0.001

*Means (n = 4) followed by different letters within a column differ significantly ($P \leq 0.05$).

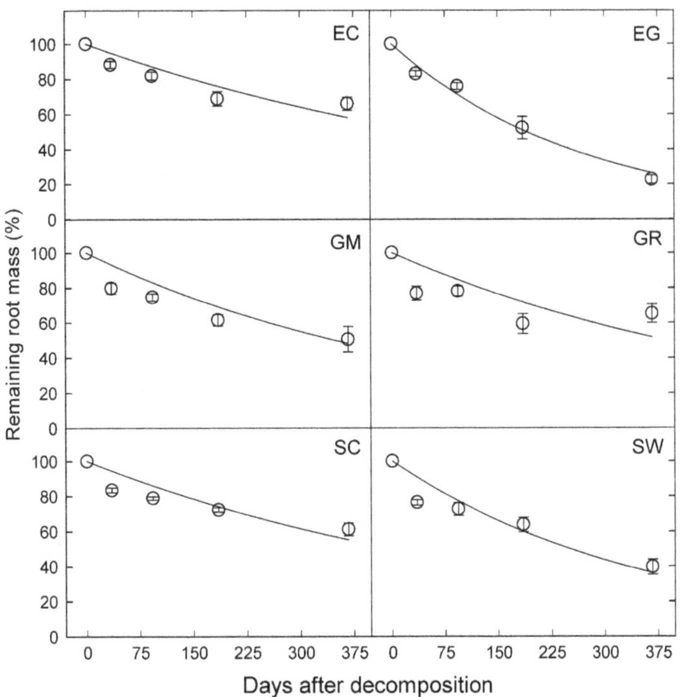

Fig. 2 Root decomposition (% remaining root mass) over the 12-month decomposition period. Species include giant reed (GR), sugarcane (SC), sweetcane (SW), energycane (EC), elephantgrass (EG), and giant miscanthus (GM). Circles represent treatment mean (n = 4) fraction of root biomass on an ash-free basis remaining at each removal date, and the error bars represent ±1 SD. Solid lines represent the fitted lines from the single-pool negative exponential model, $M_t = 100 \times e^{(-Kt)}$, for each of the species, where K is the decomposition rate in g kg^{-1} day^{-1} and t is time in day.

Table 2 Concentrations of total extractives (TE), total lignin (TL), total fiber glucose (TFG), total fiber xylose (TFX), total fiber arabinose (TFA), total fiber carbohydrate (TFC), total C, total N, C:N, and TL:N ratios of root tissue on an ash-free dry matter basis prior to decomposition

Species	TE	TL	TFG	TFX	TFA	TFC	Total C	Total N	C:N	TL:N
	mg g^{-1} DM									
Giant reed	152 AB*	327 A	269 BC	168 BC	19.4 C	455 BC	408 B	6.1 C	67.2 AB	53.8 A
Sugarcane	146 AB	316 AB	248 C	162 C	32.3 B	442 C	439 AB	8.0 AB	54.9 BC	39.6 BC
Sweetcane	144 AB	299 AB	301 B	195 AB	23.6 C	519 AB	428 AB	5.6 C	76.6 A	53.6 A
Energycane	155 AB	301 AB	267 BC	176 BC	38.7 A	481 BC	432 AB	6.8 BC	63.7 AB	44.3 AB
Elephantgrass	126 B	293 B	342 A	207 A	20.6 C	570 A	456 A	7.8 AB	58.7 BC	37.7 BC
Giant miscanthus	174 A	291 B	253 C	172 BC	35.2 AB	459 BC	416 AB	8.8 A	48.6 C	34.0 C
P-value	0.058	0.003	<0.001	0.001	0.001	<0.001	0.048	<0.001	<0.001	<0.001

*Means (n = 4) followed by different letters within a column differ significantly ($P \leq 0.05$).

among species (408–456 mg g^{-1} DM); however, total C of elephantgrass was higher than that of giant reed. Root total N concentration ranged from 5.6 mg g^{-1} in sweetcane to 8.8 mg g^{-1} in giant miscanthus. Giant miscanthus, sugarcane, and elephantgrass were among the species with the highest root N concentrations. Root C:N ratio ranged from about 49 in giant miscanthus to about 77 in sweetcane. Sweetcane, energycane, and giant reed were among the species with the highest root C:N ratios. The high TL:N ratio in giant reed resulted from its high TL concentration and low total N concentration. However, the relatively rapidly decomposing elephantgrass did not show lower C:N or TL:N ratio than most other species.

Root chemical traits were analyzed again after their 12-month incubations and expressed as a fraction of their initial amounts to better understand relevant differences in decomposition (Table 3). Elephantgrass roots exhibited relatively lower remaining TFG, TFX, TFA, TFC, TL, total C, and N (Table 3), which was consistent with its higher K (Table 1). In contrast, roots of giant reed, sugarcane, and energycane showed higher remaining amount of all chemical characteristics (Table 3), which was consistent with their lower K (Table 1).

Among all the six species, initial SRV was positively correlated with remaining TL, TFC, C, and N (Table 4). Remaining TL, TFC, C, and N were also positively correlated with each other.

Relationships between root decomposition rate and root traits

K was positively correlated with a number of root chemical traits prior to decomposition, including TFG, TFX, TFC, and total C, but it was negatively correlated with TL and TFA concentrations and SRV (Table 5). Root TL was negatively correlated with total N concentration, SRL, and SRA. However, TFC concentration was positively correlated with SRL and negatively correlated with SRV.

The first two axes of the PCA performed with 12 root traits and K accounted for 78% of the variance (Fig. 3). The first PCA axis (Component 1) accounted for 42% of the variance and was defined by root chemical and morphological traits. The second PCA axis (Component 2) accounted for 36% of the variance and was defined by K. Concentrations of TFG, TFX, TFC, and total C were grouped together with K, indicating a high correlation among these traits. Additionally, SRL was slightly positively correlated with K. However, SRV, TFA, and TL in the root tissue predecomposition were negatively related to K. Root total N concentration, C:N, TL:N, and SRA prior to decomposition were independent of K, as indicated by the near-90° angle among the directional vectors.

Discussion

Root chemistry plays a dominant role in controlling patterns of decomposition rates at a global scale (Silver & Miya, 2001). Previous studies have commonly focused on plant tissue C:N ratios, which were often negatively correlated with decomposition rate as described in Zhang et al. (2008). However, decomposition is more complex, and rates of decomposition are not necessarily related to C:N ratio, but they can be influenced by a number of factors such as environmental conditions, decomposer composition, other root chemical components, and physical structure (Table 5) (Johnson *et al.*, 2007; Birouste et al., 2012; Smith et al., 2014).

Although there are a number of factors that have been shown to influence decomposition, there is a grow-

Table 3 Amount (% of initial) of total fiber glucose ($TFG_{remaining}$), total fiber xylose ($TFX_{remaining}$), total fiber arabinose ($TFA_{remaining}$), total fiber carbohydrate ($TFC_{remaining}$), total lignin ($TL_{remaining}$), C ($C_{remaining}$), and N ($N_{remaining}$) remaining on an ash-free dry matter basis after a 12-month decomposition period in the field for the six perennial grass species

Species	$TFG_{remaining}$	$TFX_{remaining}$	$TFA_{remaining}$	$TFC_{remaining}$	$TL_{remaining}$	$C_{remaining}$	$N_{remaining}$
Giant reed	50 AB*	48 AB	58 A	50 AB	74 AB	58 A	76 AB
Sugarcane	53 AB	51 AB	60 A	53 AB	70 AB	55 AB	83 AB
Sweetcane	28 BC	28 BC	34 B	28 BC	47 BC	33 BC	57 BC
Energycane	61 A	54 A	66 A	59 A	76 A	63 A	91 A
Elephantgrass	16 C	16 C	17 B	16 C	26 C	17 C	28 C
Giant miscanthus	34 BC	34 BC	34 B	34 BC	54 A-C	42 AB	60 B
P-value	0.0004	0.0007	<0.0001	0.0004	0.0010	0.0001	0.0001

*Means (n = 4) followed by different letters within a column differ significantly ($P \leq 0.05$).

Table 4 Pearson's correlation coefficients between remaining total lignin ($TL_{remaining}$), fiber carbohydrate ($TFC_{remaining}$), C ($C_{remaining}$), and N ($N_{remaining}$) after 12-month decomposition, and specific root length (SRL), surface area (SRA), and volume (SRV) before decomposition of all perennial grass species

	$TL_{remaining}$	$TFC_{remaining}$	$C_{remaining}$	$N_{remaining}$	SRL	SRA
$TFC_{remaining}$	0.84***,†					
$C_{remaining}$	0.97***	0.83***				
$N_{remaining}$	0.93***	0.82***	0.94***			
SRL	−0.34	−0.40	−0.26	−0.41*		
SRA	−0.05	−0.14	−0.02	−0.15	0.87***	
SRV	0.52**	0.47*	0.52**	0.44*	−0.00	0.48**

†$P \leq$ 0.01, 0.05, and 0.1 represented as ***, **, and *, respectively.

Table 5 Pearson's correlation coefficients between root decomposition rate constant (K) and root morphological and chemical traits among all perennial grass species. Root traits prior to decomposition include concentrations of total lignin (TL), total fiber glucose (TFG), total fiber xylose (TFX),total fiber arabinose (TFA), total fiber carbohydrate (TFC), total C and N in root tissue, ratios of TL:N and C:N, and morphological traits of specific root length (SRL), specific root surface area (SRA), and specific root volume (SRV)

	K	TL	TFG	TFX	TFA	TFC	Total C	Total N	C:N	TL:N	SRL	SRA
TL	−0.45**†											
TFG	0.75***	−0.17										
TFX	0.63***	−0.22	0.94***									
TFA	−0.64***	−0.19	−0.51**	−0.29								
TFC	0.67***	−0.23	0.97***	0.99***	−0.31							
Total C	0.40*	−0.15	0.45**	0.40*	−0.02	0.46**						
Total N	−0.05	−0.39*	−0.21	−0.17	0.38*	−0.16	0.13					
C:N	0.10	0.32	0.34	0.31	−0.38*	0.30	0.17	−0.94***				
L:N	−0.15	0.60***	0.13	0.09	−0.40*	0.06	−0.18	−0.95***	0.90***			
SRL	0.38	−0.48**	0.45**	0.47**	0.00	0.49**	0.13	0.52**	−0.51**	−0.61***		
SRA	0.08	−0.44*	0.05	0.12	0.36	0.13	−0.06	0.63***	−0.67***	−0.69***	0.87***	
SRV	−0.58**	0.03	−0.75***	−0.66***	0.75***	−0.66***	−0.36	0.41*	−0.52**	−0.35	−0.00	0.48**

†$P \le 0.01$, 0.05, and 0.1 are represented as ***, **, and *, respectively, which is the same in the following tables for correlation analysis.

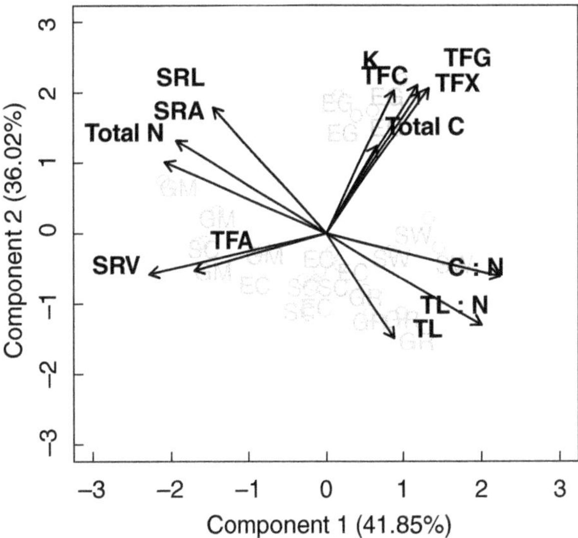

Fig. 3 Principal component analysis for root traits and decomposition rate constant (*K*). Root traits prior to decomposition include specific root length (SRL), specific root surface area (SRA), specific root volume (SRV), and concentrations of total fiber glucose (TFG), total fiber xylose (TFX), total fiber arabinose (TFA), total lignin (TL), total C and N in root tissue, and ratios of TL:N and C:N. Species include giant reed (GR), sugarcane (SC), sweetcane (SW), energycane (EC), elephantgrass (EG), and giant miscanthus (GM).

ing body of literature indicating that tissue carbohydrate concentration and composition are key determinants of root decomposition rates. For example, high fiber carbohydrate concentrations were associated with rapid root decomposition among tree and herbaceous species (Silver & Miya, 2001; Aulen et al., 2012). Results from the present study confirm these findings, as root *K* was positively correlated with TFG, TFX, and TFC prior to decomposition among botanically similar (i.e., Poaceae family) perennial grass species (Table 5 and Fig. 3). The positive correlation between TFC and *K*, and in particular, TFG and *K*, could be explained by energy dominance. Decomposer activity during bio-decomposition is mainly controlled by the energy that can be supplied by substrates contained in the litter (Fioretto et al., 2005; Hättenschwiler & Jørgensen, 2010). Compared with lignin, cellulose is an energy-rich compound. Thus, the faster decomposition of elephantgrass roots in the present study could be because of their relatively higher cellulose concentration and cellulose:lignin ratio compared with the other species.

On the other hand, the energy supplied by more recalcitrant compounds is generally low and the energy required to break them down is high (Fioretto et al., 2005; Hättenschwiler & Jørgensen, 2010; Sun et al., 2013). The energy is therefore not enough to support the level of microbial activity that is seen with less recalcitrant compounds, as indicated by the negative correlation between K and initial TL in the present study (Table 5). Thus, the decomposition of recalcitrant compounds was relatively slow, as indicated by the generally high remaining TL after the 12-month field incubation, with the exception of elephantgrass (Table 3). In the present study, elephantgrass possessed the highest initial TFC concentration among all the six perennial grass species, but its initial TL was similar to other species except for giant reed (Table 2). Furthermore, consistent with the root decomposition study in the field, remaining root biomass was also the lowest in elephantgrass among the six species in another decomposition study performed in pots without plants inside a greenhouse over a 12-month period (data not shown). Root K of elephantgrass from the present study is very close to that from another study with K values ranging from 2.65 to 3.18 g kg^{-1} day^{-1} (Siqueira da Silva et al., 2015). Taken together, rapid decomposition in elephantgrass root tissue, including TL, in the present study could be explained, at least in part, by a relatively high TFC and TFG to TL ratio. The TFC and TFG provided sufficient energy to decompose more recalcitrant compounds similar to the priming effect seen with root exudates and fine root turnover (van der Krift *et al.*, 2002; Fioretto et al., 2005; Talbot & Treseder, 2012).

As a result, the decomposition of labile and recalcitrant compounds was correlated in the study, as indicated by the positive correlation between $TFC_{remaining}$ and $TL_{remaining}$ (Table 3). In the present study, elephantgrass possessed the highest initial TFC concentration among all the grasses, but its initial TL was similar to other species except for giant reed (Table 2). Its high initial TFC concentration probably stimulated or primed lignin decomposition, which resulted in relatively low amounts of both $TFC_{remaining}$ (16%) and $TL_{remaining}$ (26%) after the 12-month field incubation. On the other hand, giant reed had at least 50% of both TFC and TL remaining after the incubation. Consequently, a relatively high TL and/or high TL:TFC reduced K by slowing the loss of hemicellulose and cellulose as well as lignin, implying that lignin may have protected the cell wall from degradation (Fioretto et al., 2005; Talbot & Treseder, 2012).

Besides the physical protection of more easily decomposable components (i.e., cellulose and hemicellulose), the recalcitrance of lignin is related to its chemical characteristics, such as molecule size, polarity, three-dimensional structure, and functional groups (aromatic ring structures) (von Luetzow et al., 2006; Puttaso et al., 2011; Gul & Whalen, 2013). Once plant tissues start to decompose, lignin begins to incorporate N, and condensation reactions take place (Berg & McClaugherty, 2008). These chemical transformations could cause changes in structures that are resistant to degradation by soil microbes and also act as barriers limiting their access to the more labile compounds. High N concentrations also suppress the formation of ligninase (Berg & McClaugherty, 2008) and may help to explain why a positive correlation was found between $TL_{remaining}$ and N, which was consistent with the previous findings, especially in lignin-rich plant tissues (Hobbie, 2000; Perakis *et al.*, 2012). Also, because of the protection of lignin, degradation of TFC could have been retarded, leading to the positive correlation between remaining TFC and N (Table 4).

Root morphological traits represent the economics of root investment: carbon input for root growth vs. the capacity of resource acquisition (Donovan et al., 2014; Reich, 2014). Their correlations with root decomposition rate reflect their potential in C and nutrient cycling across ecosystems (Donovan *et al.*, 2014). In the present study, morphological traits associated with fast root decomposition were high SRL and, even more so, low SRV roots (Fig. 3; Table 5). These findings were in agreement with other studies that have shown SRL to be positively correlated with decomposition rate because a high SRL has the potential to facilitate decomposition through maximizing root surface area and exposure for bio-decomposition (Aulen *et al.*, 2012; Birouste *et al.*, 2012; Donovan et al., 2014; Smith *et al.*, 2014). In addition, the present study identified correlations between root morphological and chemical traits, such as the positive correlation between $TL_{remaining}$ and SRV (Table 4), and the negative correlations between TL, SRL, and SRA (Table 5). Root diameter has also been shown to affect root decomposition rates among switchgrass cultivars (de Graaff *et al.*, 2013), which could be related to different concentrations of soluble carbohydrates and lignin in different root diameter size classes (Fan & Guo, 2010). However, interactive effects of root morphological and chemical traits on root decomposition have not been investigated thoroughly in the previous research. Varied morphological and chemical traits of plant species are determined by both genetic and environmental components. The interaction of root morphological and chemical characteristics, thus, might play an important role in root decomposition under different environments. For instance, after similar time periods of decomposition in the field, the remaining root biomass of the same species can differ greatly under varied environments (Harmon *et al.*, 2009).

Beyond the intrinsic characteristics of roots, environmental factors were also likely to influence root decomposition. For instance, soil temperature has been shown to be closely correlated with root decomposition rates, as decomposition rates increased up to five-fold with a temperature increase from 20 to 30 °C (Solly *et al.*, 2014;

Stewart *et al.*, 2015). In the present study, root mass loss was accelerated at 185 days after decomposition (Fig. 2), which was consistent with the high soil temperature in June (Fig. 1). Additionally, species or environmental factors (e.g., soil temperature, moisture, and fertility) and their interactions could have contributed to differences in microbial community diversity and/or functional activity (van der Heijden *et al.*, 2008; Brzostek *et al.*, 2015) that could have contributed, at least in part, to the observed differences in root decomposition among the species as well.

Overall, perennial grass species, many of which possess similar aboveground annual biomass production, differed substantially in root morphological and chemical traits, and these traits were closely correlated with their decomposition rates. In the context of C flux across ecosystems, root biomass (i.e., quantity and quality) and its decomposition rate are among the most important factors governing soil C dynamics (Aerts *et al.*, 2003; von Luetzow *et al.*, 2006). Although the processes controlling belowground C translocation and allocation of root-derived C into soil are complex, more organic matter input from roots combined with greater microbial activity are presumed to lead to more C transfer and storage in the soil (Kuzyakov & Domanski, 2000; Puttaso *et al.*, 2011). However, the quality of root-derived C inputs can also play a dominant role in the transfer of root biomass to soil organic C. Despite a lack of consensus in the literature, most recent studies suggest that high-lignin materials cannot be used efficiently by the soil microbial community, and high-lignin materials thus contribute less to soil organic C than materials with low lignin concentrations (Hancock *et al.*, 2007; Cotrufo *et al.*, 2013; Stewart et al., 2015). Additionally, introducing readily available organic matter into soils initially brings up priming effects, which causes soil organic C loss through promoted microbial metabolisms (Brzostek et al., 2015; Stewart et al., 2015). However, the negative effects of root-derived C inputs on soil C stocks are generally transient and more pronounced in the rhizosphere (Haichar *et al.*, 2014). As evidenced in previous studies, the extent that root-derived C affects soil organic C in bioenergy production systems is expected to vary considerably depending on soil, plant, and environmental factors (Bandaru *et al.*, 2013; Bonin & Lal, 2014). Thus, changes in soil organic C can be expected to differ between bioenergy grasses. For example, after conversion of native land to miscanthus, soil organic C was building up slowly over time, whereas the conversion to sugarcane caused a large initial loss of soil organic C in the top soil, and the loss could last for a few decades before soil organic carbon rebuilds (Anderson-Teixeira *et al.*, 2009). Such interspecies differences in soil organic C could be related to the characteristics of root chemistry and morphology indicated in the present study. For instance, giant miscanthus and giant reed showed slow root decomposition rates, implying a slow increase in soil organic C. In contrast, elephantgrass roots exhibited all the root traits favorable for fast decomposition: high TFG, high TFC, high SRL, and a low SRV, and elephantgrass could thus be expected to contribute to soil C accumulation in a rapid way.

Acknowledgements

The authors would like to acknowledge Andy Schreffler, Carley Fuller, Rezzy Manning, Jeffery Fedenko, and Cameron Preston for their contributions to experimental setting and data collection. This project was supported by grant no. 2008-34606-19522 from the United States Department of Agriculture and by competitive grant no. 2012-67009-19596 from USDA-NIFA.

References

Adair EC, Hobbie SE, Hobbie RK (2010) Single-pool exponential decomposition models: potential pitfalls in their use in ecological studies. *Ecology*, **91**, 1225–1236.

Aerts R, De Caluwe H, Beltman B (2003) Plant community mediated vs. nutritional controls on litter decomposition rates in grasslands. *Ecology*, **84**, 3198–3208.

Amougou N, Bertrand I, Machet J, Recous S (2011) Quality and decomposition in soil of rhizome, root and senescent leaf from *Miscanthus* × *giganteus*, as affected by harvest date and N fertilization. *Plant and Soil*, **338**, 83–97.

Anderson-Teixeira KJ, Davis SC, Masters MD, Delucia EH (2009) Changes in soil organic carbon under biofuel crops. *Global Change Biology Bioenergy*, **1**, 75–96.

Aulen M, Shipley B, Bradley R (2012) Prediction of *in situ* root decomposition rates in an interspecific context from chemical and morphological traits. *Annals of Botany*, **109**, 287–297.

Bandaru V, Izaurralde RC, Manowitz D, Link R, Zhang X, Post WM (2013) Soil carbon change and net energy associated with biofuel production on marginal lands: a regional modeling perspective. *Journal of Environmental Quality*, **42**, 1802–1814.

Berg B, McClaugherty C (2008) *Plant Litter: Decomposition, Humus Formation, Carbon Sequestration*. Springer-Verlag Berlin Heidelberg, Berlin, Germany.

Birouste M, Kazakou E, Blanchard A, Roumet C (2012) Plant traits and decomposition: are the relationships for roots comparable to those for leaves? *Annals of Botany*, **109**, 463–472.

Bonin CL, Lal R (2014) Aboveground productivity and soil carbon storage of biofuel crops in Ohio. *Global Change Biology Bioenergy*, **6**, 67–75.

Brzostek ER, Dragoni D, Brown ZA, Phillips RP (2015) Mycorrhizal type determines the magnitude and direction of root-induced changes in decomposition in a temperate forest. *New Phytologist*, **206**, 1274–1282.

Carroll A, Somerville C (2009) Cellulosic biofuels. *Annual Review of Plant Biology*, **60**, 165–182.

Castillo MS, Sollenberger LE, Vendramini JMB *et al.* (2010) Municipal biosolids as an alternative nutrient source for bioenergy crops: II. Decomposition and organic nitrogen mineralization. *Agronomy Journal*, **102**, 1314–1320.

Cotrufo MF, Wallenstein MD, Boot CM, Denef K, Paul E (2013) The microbial efficiency-matrix stabilization (MEMS) framework integrates plant litter decomposition with soil organic matter stabilization: do labile plant inputs form stable soil organic matter? *Global Change Biology*, **19**, 988–995.

Davis SC, Parton WJ, Dohleman FG, Smith CM, Del Grosso S, Kent AD, DeLucia EH (2010) Comparative biogeochemical cycles of bioenergy crops reveal nitrogen-fixation and low greenhouse gas emissions in a *Miscanthus* × *giganteus* agro-ecosystem. *Ecosystems*, **13**, 144–156.

Don A, Osborne B, Hastings A *et al.* (2012) Land-use change to bioenergy production in Europe: implications for the greenhouse gas balance and soil carbon. *Global Change Biology Bioenergy*, **4**, 372–391.

Donovan LA, Mason CM, Bowsher AW, Goolsby EW, Ishibashi CDA (2014) Ecological and evolutionary lability of plant traits affecting carbon and nutrient cycling. *Journal of Ecology*, **102**, 302–314.

Drewer J, Finch JW, Lloyd CR, Baggs EM, Skiba U (2012) How do soil emissions of N_2O, CH_4 and CO_2 from perennial bioenergy crops differ from arable annual crops? *Global Change Biology Bioenergy*, **4**, 408–419.

Erickson JE, Soikaew A, Sollenberger LE, Bennett JM (2012) Water use and water-use efficiency of three perennial bioenergy grass crops in Florida. *Agriculture*, **2**, 325–338.

Fan P, Guo D (2010) Slow decomposition of lower order roots: a key mechanism of root carbon and nutrient retention in the soil. *Oecologia*, **163**, 509–515.

Fedenko JR, Erickson JE, Woodard KR *et al.* (2013) Biomass production and composition of perennial grasses grown for bioenergy in a subtropical climate across Florida, USA. *Bioenergy Research*, **2**, 1082–1093.

Fioretto A, Di Nardo C, Papa S, Fuggi A (2005) Lignin and cellulose degradation and nitrogen dynamics during decomposition of three leaf litter species in a Mediterranean ecosystem. *Soil Biology and Biochemistry*, **37**, 1083–1091.

Fontaine S, Barot S, Barre P, Bdioui N, Mary B, Rumpel C (2007) Stability of organic carbon in deep soil layers controlled by fresh carbon supply. *Nature*, **450**, 277–280.

Fornara DA, Tilman D (2008) Plant functional composition influences rates of soil carbon and nitrogen accumulation. *Journal of Ecology*, **96**, 314–322.

de Graaff MA, Six J, Jastrow JD, Schadt CW, Wullschleger SD (2013) Variation in root architecture among switchgrass cultivars impacts root decomposition rates. *Soil Biology and Biochemistry*, **58**, 198–206.

Gul S, Whalen J (2013) Plant life history and residue chemistry influences emissions of CO_2 and N_2O from soil perspectives for genetically modified cell wall mutants. *Critical Reviews in Plant Sciences*, **32**, 344–368.

von Haden AC, Dornbush ME (2014) Patterns of root decomposition in response to soil moisture best explain high soil organic carbon heterogeneity within a mesic, restored prairie. *Agriculture Ecosystems & Environment*, **185**, 188–196.

Haichar FZ, Santaella C, Heulin T, Achouak W (2014) Root exudates mediated interactions belowground. *Soil Biology and Biochemistry*, **77**, 69–80.

Hancock JE, Loya WM, Giardina CP, Li LG, Chiang VL, Pregitzer KS (2007) Plant growth, biomass partitioning and soil carbon formation in response to altered lignin biosynthesis in *Populus tremuloides*. *New Phytologist*, **173**, 732–742.

Harmon ME, Silver WL, Fasth B *et al.* (2009) Long-term patterns of mass loss during the decomposition of leaf and fine root litter: an intersite comparison. *Global Change Biology*, **15**, 1320–1338.

Hättenschwiler S, Jørgensen HB (2010) Carbon quality rather than stoichiometry controls litter decomposition in a tropical rain forest. *Journal of Ecology*, **98**, 754–763.

van der Heijden MGA, Bardgett RD, van Straalen NM (2008) The unseen majority: soil microbes as drivers of plant diversity and productivity in terrestrial ecosystems. *Ecology Letters*, **11**, 296–310.

Hobbie SE (2000) Interactions between litter lignin and soil nitrogen availability during leaf litter decomposition in a Hawaiian montane forest. *Ecosystems*, **3**, 484–494.

Johnson JMF, Barbour NW, Weyers SL (2007) Chemical composition of crop biomass impacts its decomposition. *Soil Science Society of America Journal*, **71**, 155–162.

Jones MB, Donnelly A (2004) Carbon sequestration in temperate grassland ecosystems and the influence of management, climate, and elevated CO_2. *New Phytologist*, **164**, 423–439.

Knoll JE, Anderson WF, Strickland TC, Hubbard RK, Malik R (2012) Low-input production of biomass from perennial grasses in the Coastal Plain of Georgia, USA. *Bioenergy Research*, **5**, 206–214.

van der Krift T, Kuikman P, Berendse F (2002) The effect of living plants on root decomposition of four grass species. *Oikos*, **96**, 36–45.

Kuzyakov Y, Domanski G (2000) Carbon input by plants into soil. Review. *Journal of Plant Nutrition and Soil Science*, **163**, 421–431.

Lee H, Fitzgerald J, Hewins DB, McCulley RL, Archer SR, Rahn T, Throop HL (2014) Soil moisture and soil-litter mixing effects on surface litter decomposition: a controlled environment assessment. *Soil Biology & Biochemistry*, **72**, 123–132.

Lemus R, Lal R (2005) Bioenergy crops and carbon sequestration. *Critical Reviews in Plant Sciences*, **24**, 1–21.

von Luetzow M, Koegel-Knabner I, Ekschmitt K *et al.* (2006) Stabilization of organic matter in temperate soils: mechanisms and their relevance under different soil conditions – a review. *European Journal of Soil Science*, **57**, 426–445.

Monti A, Zatta A (2009) Root distribution and soil moisture retrieval in perennial and annual energy crops in Northern Italy. *Agriculture Ecosystems & Environment*, **132**, 252–259.

Perakis SS, Matkins JJ, Hibbs DE (2012) Interactions of tissue and fertilizer nitrogen on decomposition dynamics of lignin-rich conifer litter. *Ecosphere* **3**, art54.

Prescott CE (2010) Litter decomposition: what controls it and how can we alter it to sequester more carbon in forest soils? *Biogeochemistry*, **101**, 133–149.

Puttaso A, Vityakon P, Saenjan P, Trelo-ges V, Cadisch G (2011) Relationship between residue quality, decomposition patterns, and soil organic matter accumulation in a tropical sandy soil after 13 years. *Nutrient Cycling in Agroecosystems*, **89**, 159–174.

Rasse DP, Rumpel C, Dignac M-F (2005) Is soil carbon mostly root carbon? Mechanisms for a specific stabilisation. *Plant and Soil*, **269**, 341–356.

Reich PB (2014) The world-wide 'fast-slow' plant economics spectrum: a traits manifesto. *Journal of Ecology*, **102**, 275–301.

Silver WL, Miya RK (2001) Global patterns in root decomposition: comparisons of climate and litter quality effects. *Oecologia*, **129**, 407–419.

Siqueira da Silva HM, Batista Dubeux JC, Silveira ML, Viana de Freitas E, Ferreira dos Santos MV, de Andrade Lira M (2015) Stocking rate and nitrogen fertilization affect root decomposition of elephantgrass. *Agronomy Journal*, **107**, 1331–1338.

Six J, Elliott ET, Paustian K (2000) Soil macroaggregate turnover and microaggregate formation: a mechanism for C sequestration under no-tillage agriculture. *Soil Biology & Biochemistry*, **32**, 2099–2103.

Smith SW, Woodin SJ, Pakeman RJ, Johnson D, van der Wal R (2014) Root traits predict decomposition across a landscape-scale grazing experiment. *New Phytologist*, **203**, 851–862.

Solly EF, Schoening I, Boch S *et al.* (2014) Factors controlling decomposition rates of fine root litter in temperate forests and grasslands. *Plant and Soil*, **382**, 203–218.

Somerville C, Youngs H, Taylor C, Davis SC, Long SP (2010) Feedstocks for lignocellulosic biofuels. *Science*, **329**, 790–792.

Stewart CE, Moturi P, Follett RF, Halvorson AD (2015) Lignin biochemistry and soil N determine crop residue decomposition and soil priming. *Biogeochemistry* **124**: 335–351.

Sun T, Mao Z, Han Y (2013) Slow decomposition of very fine roots and some factors controlling the process: a 4-year experiment in four temperate tree species. *Plant and Soil* **372**: 445–458.

Talbot JM, Treseder KK (2012) Interactions among lignin, cellulose, and nitrogen drive litter chemistry-decay relationships. *Ecology*, **93**, 345–354.

Zhang D, Hui D, Luo Y, Zhou G (2008) Rates of litter decomposition in terrestrial ecosystems: global patterns and controlling factors. *Journal of Plant Ecology*, **1**, 85–93.

8

Genotypic diversity effects on biomass production in native perennial bioenergy cropping systems

GEOFFREY P. MORRIS[1], ZHENBIN HU[1], PAUL P. GRABOWSKI[2], JUSTIN O. BOREVITZ[3], MARIE-ANNE DE GRAAFF[4], R. MICHAEL MILLER[5] and JULIE D. JASTROW[5]

[1]*Department of Agronomy, Kansas State University, Manhattan, KS 66506, USA,* [2]*USDA-ARS Dairy Forage Research Center, Madison, WI 53706, USA,* [3]*Research School of Biology, Australian National University, Acton, ACT 2601, Australia,* [4]*Department of Biological Sciences, Boise State University, Boise, ID 83725, USA,* [5]*Biosciences Division, Argonne National Laboratory, Argonne, IL 60439, USA*

Abstract

The perennial grass species that are being developed as biomass feedstock crops harbor extensive genotypic diversity, but the effects of this diversity on biomass production are not well understood. We investigated the effects of genotypic diversity in switchgrass (*Panicum virgatum*) and big bluestem (*Andropogon gerardii*) on perennial biomass cropping systems in two experiments conducted over 2008–2014 at a 5.4-ha fertile field site in northeastern Illinois, USA. We varied levels of switchgrass and big bluestem genotypic diversity using various local and nonlocal cultivars – under low or high species diversity, with or without nitrogen inputs – and quantified establishment, biomass yield, and biomass composition. In one experiment ('agronomic trial'), we compared three switchgrass cultivars in monoculture to a switchgrass cultivar mixture and three different species mixtures, with or without N fertilization. In another experiment ('diversity gradient'), we varied diversity levels in switchgrass and big bluestem (1, 2, 4, or 6 cultivars per plot), with one or two species per plot. In both experiments, cultivar mixtures produced yields equivalent to or greater than the best cultivars. In the agronomic trial, the three switchgrass mixture showed the highest production overall, though not significantly different than best cultivar monoculture. In the diversity gradient, genotypic mixtures had one-third higher biomass production than the average monoculture, and none of the monocultures were significantly higher yielding than the average mixture. Year-to-year variation in yields was lowest in the three-cultivar switchgrass mixtures and Cave-In-Rock (the southern Illinois cultivar) and also reduced in the mixture of switchgrass and big bluestem relative to the species monocultures. The effects of genotypic diversity on biomass composition were modest relative to the differences among species and genotypes. Our findings suggest that local genotypes can be included in biomass cropping systems without compromising yields and that genotypic mixtures could help provide high, stable yields of high-quality biomass feedstocks.

Keywords: big bluestem, biomass feedstock, cultivars, ecotype, fertilization, low-input high-diversity, polymorphism, switchgrass, tallgrass prairie, yield

Introduction

In recent years, there has been great worldwide interest in the development of biomass cropping systems that could provide bioenergy feedstocks and reduce greenhouse gas emissions associated with fossil fuels. In the United States, Congress has mandated a major transition to lignocellulosic biofuels, which will require development of multiple new regionally appropriate biomass cropping systems (Downing *et al.*, 2011). Among the top candidates for biomass energy crops are perennial grass species native to North American tallgrass prairies, including switchgrass (*Panicum virgatum*), big bluestem (*Andropogon gerardii*), and indiangrass (*Sorghastrum nutans*). These grasses share several characteristics that could be valuable for bioenergy production systems. Most notably they are capable of producing substantial yields of dry biomass with limited inputs and on land not suited to row cropping (Schmer *et al.*, 2008; Griffith *et al.*, 2011). In addition, bioenergy cropping systems based on native grasses may provide additional ecosystem services, such as carbon (C) sequestration (Tilman *et al.*, 2006a; Gelfand *et al.*, 2013) and wildlife habitat (Robertson *et al.*, 2011).

Native perennial grasses have undergone some selection for forage, habitat restoration, and, increasingly, bioenergy uses, but most cultivars of these species

*Correspondence: Geoffrey P. Morris
e-mail: gpmorris@ksu.edu

remain essentially samples of the wild gene pool (Casler, 2012). As widespread wind-pollinated species, prairie grasses harbor abundant genetic diversity. The ecotypic diversity and regional gene pools of switchgrass have been particularly well studied. In addition to classical work on continental clines (McMillan, 1959) and upland and lowland ecotypes (Porter, 1966), recent studies have identified phenotypic clinal variation across multiple axes (Casler, 2005; Casler *et al.*, 2007) and many genetic subpopulations (Zalapa *et al.*, 2010; Morris *et al.*, 2011; Zhang *et al.*, 2011; Grabowski *et al.*, 2014). Though less well studied, big bluestem exhibits similar patterns of genotypic diversity at the phenotypic and genomic level (McMillan, 1956, 1961; Gray *et al.*, 2014). As predominantly outcrossing species, switchgrass and big bluestem are both highly heterozygous. This abundant genotypic diversity provides useful genetic variance for crop improvement, but high heterozygosity and limited ability to inbreed can hinder many breeding approaches (Liu & Wu, 2012). Other candidate perennial biomass crops that originated in Asia, such as *Miscanthus* and *Saccharum* spp., exhibit similar patterns of diversity in their native ranges (Dillon *et al.*, 2007; Kim *et al.*, 2012).

In most crops, and most cropping systems, genotypic diversity has decreased over time, either directly due to selection for uniformity or indirectly due to genetic bottlenecks created by selection for other traits (Kingsbury, 2011). Direct selection on uniformity has been particularly central to the modern improvement of row crops, where uniform height and maturity is critical for efficient harvest and consistent quality. Accordingly, most commercial row crops are inbred varieties (e.g., wheat, soybean) or F1 hybrids of inbred lines (e.g., corn, sorghum) (Acquaah, 2012). By contrast, many perennial forage crops retain extensive genotypic diversity, either because uniformity is not necessary or desirable, or because the mating system precludes repeated inbreeding (e.g., obligate outcrossers). Given that the development of bioenergy cropping systems is at an early stage, it raises the question of what level of genotypic diversity should be targeted to increase yields in new perennial biomass crops. Moreover, as perennial biomass cropping systems are expected to provide a number of additional ecosystem services, genotypic diversity may influence trade-offs or synergies among system outputs, such as feedstock yields, feedstock quality, soil C storage, or habitat for biodiversity conservation. While potential trade-offs and synergies of species mixtures in perennial biomass cropping systems have been investigated (Tilman *et al.*, 2006a; Adler *et al.*, 2009; Griffith *et al.*, 2011; Mangan *et al.*, 2011; Jarchow & Liebman, 2012, 2013), the effects of genotype diversity mixtures (i.e., intraspecific diversity) are not well understood.

There are a number of reasons why genotypic diversity may be beneficial in perennial biomass cropping systems. Genetically, heterogeneous crop populations may buffer or hedge against temporal and spatial variability in the production environment (Allard & Bradshaw, 1964). This effect is thought to be important to smallholder production on marginal lands (Haussmann *et al.*, 2012) and may be important for biomass cropping systems if production occurs on marginal lands, as expected (Downing *et al.*, 2011; Uden *et al.*, 2013). The most widespread use of designed genotypic mixtures is in multiline cultivars of cool-season cereals, which can reduce the severity and spread of plant disease (Mundt, 2002). In native perennial systems, diverse mixtures often have higher productivity due to a sampling effect, the tendency of higher diversity systems to include at least one highly productive type (Fargione *et al.*, 2007). Intraspecific diversity may also increase niche complementarity (Cook-Patton *et al.*, 2011), as is commonly seen with wild or cultivated interspecific mixtures (e.g., binary legume–grass mixtures) (Fargione *et al.*, 2007; Nyfeler *et al.*, 2009).

Many cultivars, species, and species mixtures have been investigated as potential lignocellulosic biomass crops because the choice of plant material has multiple impacts on the feasibility and sustainability of the biomass feedstock production system. For producers of biomass feedstock, high yield potential and efficient use of inputs will be critical for profitability (Boyer *et al.*, 2012). Biomass cropping systems that have high net energy ratio (bioenergy output per unit input) yet produce low yields (biomass per unit area) may not be adopted because producer profitability is highly dependent on yields. Another concern for producers will be the opportunity cost of moving from a flexible annual cropping system to a perennial system that would require additional time to profitability (with little or no yields for several years following planting) (Uden *et al.*, 2013). If native perennial bioenergy cropping systems are to be adopted, therefore, it is essential that yield potential be increased substantially, especially during the early years following planting.

For the biomass conversion industry, yields will also be important because ready access to low-cost feedstock is required for plant profitability (Gan & Smith, 2011). Beyond feedstock access and cost, feedstock composition is important for many conversion technologies. For example, high-digestibility low-lignin feedstocks may be preferred for enzyme-based conversion, while energy-dense high-lignin feedstock may be preferred for combustion-based conversion. While ash is generally undesirable, particular elements (e.g., K, Na, Ca, S, and Cl) may be especially harmful for particular conversion technologies (Sims, 2003). The abundant natural variation of biomass composition among and within native perennial species (Sarath *et al.*, 2008; Vogel *et al.*, 2010; Zhang *et al.*, 2015) provides an opportunity to optimize plant mixtures for a given fuelshed (e.g., 10s of km surrounding the bioenergy production facility) based on the conversion technology in use.

To investigate the effects of genotypic diversity on stand establishment, biomass yields, and biomass composition, we conducted two field experiments in northeastern Illinois over 2008–2014, where we varied switchgrass and big bluestem genotypic diversity, as well as other management variables. In the first experiment, we compared a switchgrass cultivar mixture to switchgrass cultivars grown in monoculture and multispecies mixtures, with and without nitrogen inputs. In the second experiment, we created a genotypic diversity gradient by varying cultivar numbers within switchgrass monocultures, big bluestem monocultures, and mixtures of switchgrass and big bluestem. Here, we describe the effects of varying genotypic diversity levels on stand establishment, biomass yield, and biomass composition and discuss the implications of these findings for the development of native perennial biomass cropping systems.

Material and methods

Site description and preparation

The experiments were conducted at the Fermilab National Environmental Research Park in Batavia, IL (N 41.8414, W 88.2297). Prior to the experiment (1971–2007), the 5.4 ha site was maintained as an old field dominated by cool-season grasses – primarily smooth brome (*Bromus inermis*), quackgrass (*Agropyron repens*), and *Poa* species (O'Brien *et al.*, 2013). The soil at the site is Grays silt loam (fine-silty, mixed, superactive, mesic Mollic Oxyaquic Hapludalf), which is rated as prime farmland by US Department of Agriculture. Mean annual precipitation in the area is 920 mm with a mean temperature of 9.5°C. Monthly precipitation totals and monthly mean temperatures during the experiment were obtained from the National Climatic Data Center (http://www.ncdc.noaa.gov/) for Chicago West DuPage Airport weather station (USW00094892), located approximately 8 km north of our field site. In fall 2007, standing vegetation at the site was removed by application of the broad-spectrum herbicide glyphosate followed by burning of the dead vegetation. Subsequent regrowth in spring 2008 was treated again with glyphosate twice before planting. Switchgrass monocultures were seeded at 6.7 kg pure live seed (PLS) ha^{-1} (~575 seeds m^{-2}), which is within the seeding rate range (5.6–7.3 kg PLS ha^{-1}) recommended by the US Department of Agriculture (USDA) Natural Resources Conservation Service (NRCS) for switchgrass feedstock production in the Midwest. Seeding rates in the plots that included other species were adjusted to achieve the same total seeding rate as the switchgrass monocultures (~575 seeds m^{-2}), with the proportion of each species and ecotype given in Table 1. The monocultures and mixtures are described in detail below. All sown species are perennials and native to the region (Swink & Wilhelm, 1994).

Switchgrass and big bluestem germplasm

The switchgrass and big bluestem cultivars were chosen to represent a wide range of diversity in each species. The switchgrass germplasm used consisted of the following, listed in order of most northern to most southern origin. Dacotah is derived from progeny of a single plant from North Dakota and originates from the most northern and driest location (380 mm) among the selected cultivars (Barker *et al.*, 1990). Forestburg is a composite of four accessions from eastern South Dakota (Barker *et al.*, 1988). Sunburst is derived from multiple plants from one county in southeastern South Dakota, which were subjected to three cycles of selection for vigor, leafiness, and seed weight (Boe & Ross, 1998). Southlow switchgrass is an ecopool from southern Lower Michigan, a composite of germplasm from 11 native stands crossed and increased with no purposeful selection (Durling *et al.*, 2008). Cave-In-Rock was developed from germplasm originating at a native stand in southern Illinois (Hanson, 1972). Blackwell originates from a single plant collected at an upland site in northern Oklahoma and was tested in northeastern Kansas prior to release (Hanson, 1972). Kanlow originates from germplasm collected at a lowland site in central Oklahoma and was subject to selection for leafiness, vigor, and late season greenness in northeastern Kansas prior to release (Hanson, 1972).

The big bluestem germplasm used was the following. Southlow big bluestem is an ecopool from southern Lower Michigan, a composite of germplasm from 22 native stands crossed and increased with no purposeful selection (Durling *et al.*, 2007). Champ is a hybrid of sand bluestem (*Andropogon hallii*) from Nebraska and big bluestem from multiple sites in Nebraska and Iowa (Newell, 1968a). Pawnee was developed from germplasm originating from a county in southeastern Nebraska and was selected for several generations in Nebraska (Newell, 1968b). Bonanza is a derivative of Pawnee, selected for three generations for forage yield and digestibility across three sites in Nebraska (Vogel *et al.*, 2006). Rountree originates from a native stand in western Iowa and was selected and increased in eastern Missouri (Alderson & Sharp, 1994). Epic originates from a site in western Arkansas and was selected and increased in eastern Missouri (USDA-NRCS, 2013). Suther originates from a native stand in central North Carolina and was tested and increased in New Jersey (Davis *et al.*, 2002).

Experimental design for agronomic trial

The agronomic trial has seven plant treatments in three randomized complete blocks with two split-plot fertilization treatments, for a total of 42 plots. Each plot is 36 m × 20 m. The plant treatments (Table 1) consist of three switchgrass monocultures [the lowland cultivar Kanlow (KA), upland cultivar Cave-In-Rock (CR), and regional ecopool Southlow (SL)]; a switchgrass mixture with all three varieties (SG); binary

Table 1 Plant materials and sown plant treatments for the agronomic trial. Values in parenthesis indicate the sown cultivar composition (based on % of sown seeds) for switchgrass and big bluestem, or sown species composition for the forbs. The plant treatments are Kanlow switchgrass (KA), Cave-In-Rock switchgrass (CR), Southlow switchgrass (SL), a three switchgrass cultivar mixture (SG), a big bluestem plus switchgrass mixture (BB), a Canada wildrye plus switchgrass mixture (CW), and a prairie mixture (PR)

				Plant mixture treatments (sown %)						
Species	Common name	Type	Cultivar	KA	CR	SL	SG	BB	CW	PR
Panicum virgatum	Switchgrass	C4 grass		100%	100%	100%	100%	50%	60%	20%
			Kanlow	(100%)			(33%)	(17%)	(20%)	(7%)
			Cave-In-Rock		(100%)		(33%)	(17%)	(20%)	(7%)
			Southlow			(100%)	(33%)	(17%)	(20%)	(7%)
Andropogon gerardii	Big bluestem	C4 grass						50%		20%
			Rountree					(17%)		(7%)
			Epic					(17%)		(7%)
			Southlow					(17%)		(7%)
Sorghastrum nutans	Indiangrass	C4 grass								20%
Elymus canadensis	Canada wildrye	C3 grass							40%	20%
		Forbs								20%
Desmodium canadense	Showy tick trefoil	Legume								(2.50%)
Lespedeza capitata	Round-headed bush clover	Legume								(2.50%)
Dalea purpurea	Purple prairie clover	Legume								(2.50%)
Aster nova-angliae	New England aster	Composite								(2.50%)
Coreopsis tripteris	Tall tickseed	Composite								(2.50%)
Heliopsis helianthoides	Smooth oxeye	Composite								(2.50%)
Ratibida pinnata	Yellow coneflower	Composite								(2.50%)
Veronicastrum virginicum	Culver's root	Other forb								(2.50%)

mixtures of switchgrass with big bluestem (BB), or switchgrass with Canada wildrye (CW); and a 12-species prairie mixture that includes the preceding grasses along with other grasses and forbs native to the region (PR). Plots are separated by 2 m (east–west) or 4 m (north–south) alleys sown with a low-stature fescue mix (*Festuca* spp.) and replicate plot treatments are assigned to northern, central, and southern blocks. This experiment was managed using conventional agricultural techniques. Plant treatments were sown in June 2008, by no-till drill-seeding using a native seed drill at a depth of ~0.5 cm in ~20 cm rows. In the fertilized plots, granular urea (46-0-0) was applied annually during the first week of June (beginning in 2009) with a hand broadcast spreader at a rate of 67 kg N ha^{-1}, within the recommended range for switchgrass feedstock production (50–100 kg N ha^{-1}) (USDA-Natural Resources Conservation Service, 2009). The weeds were controlled by broadcast application of Milestone (aminopyralid) and Garlon (triclopyr) broad-leaf herbicides in 2009 (except the PR mixture, which includes broad-leaf species). In 2010, weed control included spot application of Milestone and Garlon to several patches of crown vetch (*Securigera varia*), dogwood (*Cornus* sp.), and oxeye daisy (*Leucanthemum vulgare*) and spot application of Round-Up (glyphosate) to reed canary grass (*Phalaris arundinacea*).

Experimental design for diversity gradient

The diversity gradient consists of four levels of cultivar diversity (1, 2, 4, 6) and three species treatments (switchgrass, big bluestem, or switchgrass and big bluestem mixture) for a total of 164 plots (3 m × 2 m each). Each of seven cultivars of switchgrass and seven cultivars of big bluestem was grown in four replicate monoculture plots (2 species × 7 cultivars × 4 replicates = 56 plots). For the single-species cultivar mixtures ('Swi', 'Big'), there are 12 replicates at each cultivar diversity level (2, 4, 6), with cultivars sampled randomly and seeded at equal rates (2 species × 3 diversity levels × 12 replicates = 72 plots). For the mixed-species cultivar mixtures ('Mix'), there are 12 replicates at each cultivar diversity level with cultivars sampled randomly at a 1 : 1 ratio of switchgrass and big bluestem (3 diversity levels × 12 replicates = 36 plots). Alternating rows are separated by 1 m east–west alleys, and the experiment is surrounded and bisected by larger alleys from the Agronomic trial. Replicates are blocked according to exposure to alleys. Seeds were hand-sown into ~0.5-cm-deep furrows in 20 cm rows. Two months postsowing, all plots showed germination of the sown species (i.e., switchgrass, big bluestem, or both). The plots were hand-weeded in the spring from 2008 to 2010, and treated with broad-leaf herbicide (Milestone and Garlon) in 2009.

Biomass yield measurements

To account for the time it takes for native perennial systems to establish, we define the first 2 years of the experiment as establishment years (2008–2010) and the subsequent 5 years (2009–2014) as production years (Fig. 1). In the first establishment

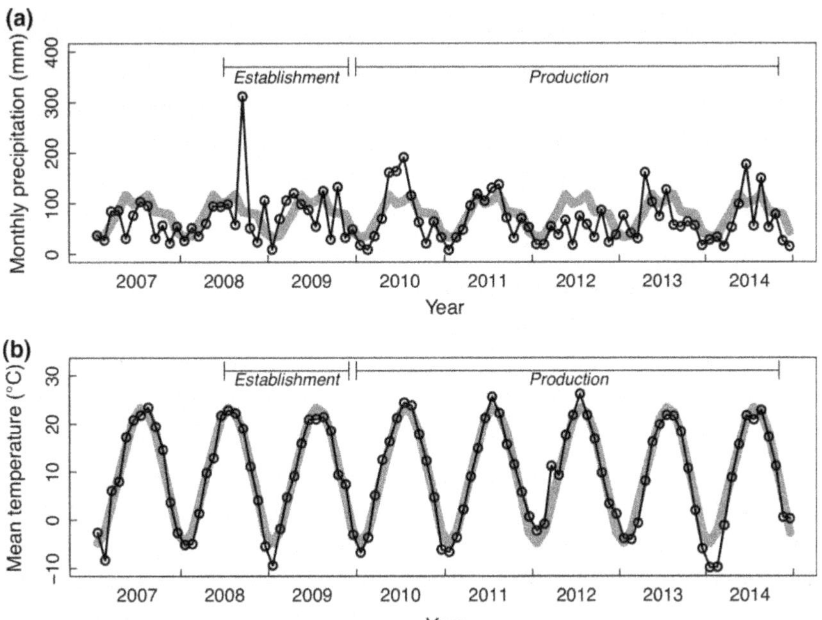

Fig. 1 Monthly precipitation (a) and mean monthly temperature conditions (b) during the experiment. The 20-year averages (1981–2001) are shown with the gray lines. Data are from 8 km north of the field site.

year (November 2008) and first production year (September 2010), the standing crop of aboveground biomass in the agronomic trial was estimated for sown species and weeds using 0.5-m^2 circular quadrats (Kennedy, 1972). Four quadrats were randomly placed in each plot, and all stems within the area were clipped to 2–4 cm aboveground level. In years 2009–2014, aboveground biomass was harvested at ~15 cm aboveground level (USDA-Natural Resources Conservation Service, 2009) after a killing frost (early- to mid-November) using standard commercial hay machinery. Each plot was harvested individually and produced one or more round bales (of variable size) per plot, which were weighed with a hanging scale. For each diversity gradient plot, in years 2009–2014, the entire plot was harvested with a string trimmer just aboveground level, and the biomass was collected, so that essentially all aboveground biomass was removed. Moisture content of biomass from each bale/plot was estimated from a subsample taken immediately prior to baling or weighing, and all reported yields have been adjusted to a dry-weight basis (65°C). The effects of plant treatment and nitrogen fertilization on biomass yields, stand establishment, and plant composition in the agronomic trial were determined by repeated measures split-plot ANOVA implemented with aov in the R statistical computing environment (R Core Team, 2014). The *P*-values given in the text for significant factors are from minimal adequate models, obtained after sequentially dropping nonsignificant ($P > 0.05$) factors from the model (Crawley, 2012).

Genetic diversity estimates

Published Illumina short-read sequence data for 123 switchgrass genotypes along with barcode multiplexing information (Grabowski *et al.*, 2014) were downloaded from the NCBI Sequence Read Archive (Accession: PRJNA252891) and Dryad Digital Repository (doi:10.5061/dryad.k77nh), respectively. The raw data were demultiplexed with ea-utils.1.1.2-537 (Aronesty, 2013), and adapters were removed with the FASTX-Toolkit (http://hannonlab.cshl.edu/fastx_toolkit/). The individual fastq files were aligned on the reference genome with BWA (Li & Durbin, 2009), and single nucleotide polymorphisms (SNPs) were called with SAMtools (Li *et al.*, 2009) with default settings. The genetic distance of individuals and the neighbor-joining (NJ) tree was calculated with TASSEL 4 (Bradbury *et al.*, 2007), and the NJ tree was visualized with MEGA 6 (Tamura *et al.*, 2013). Nucleotide diversity was calculated with $2p(1-p)$, where p represents the allele frequency. The allele frequencies were calculated with VCFtools (Danecek *et al.*, 2011), and the figure was drawn with R (R Core Team, 2014).

Biomass composition analysis

Subsamples of the whole-plot harvested biomass were taken for all plots (both agronomic trial and diversity gradient) in 2012. The samples were dried at 65°C and sent to a commercial plant testing laboratory, Dairyland Laboratories Inc. (Arcadia, WI, USA), for grinding in a cyclone mill and compositional analysis using wet chemistry and near-infrared reflectance spectroscopy (NIRS). Mineral analysis (Ca, P, K, Mg, and S) was carried out using wet chemistry methods (AOAC Official Method 953.01). NIRS estimates of lignin, hemicellulose, cellulose, fat, sugar, ash, and nitrogen were performed on a FOSS 5000 using calibrations developed in WinISI (FOSS North America, Eden Prairie, MN). Calibrations were based on wet chemistry (AOAC Official Methods 973.18, 2002.04, 920.39, 942.05, 990.03) from a worldwide panel of mixed hay samples

(N ranging from 355 to 7140), with an r^2 based on validation samples ranging from 0.73 to 0.99. All values are provided on a dry matter basis.

Results

Precipitation and temperature conditions

Stand establishment and biomass production are expected to be influenced by precipitation and temperature. The monthly precipitation and average monthly temperatures during the experiment are presented along with 20-year averages for comparison (Fig. 1). The first establishment year (2008) and first production year (2010) were unusually wet, with 2008 being the wettest year on record (1871–2014) for the Chicago area (US Department of Commerce, 2015). The third production year coincided with the historic drought of 2012 and total precipitation in the 6 months preceding harvest was just 56% of the 20-year average. The 2012 drought year also had unusually high temperatures in July, 3°C above the 20-year average for the month. The first winter postplanting was unusually cold, with January 2009 temperatures 5°C below the 20-year average. The winter prior to the 2014 production year was also unusually cold with January and February 2014 temperatures 5°C and 7°C below the 20-year averages, respectively.

Stand establishment

To characterize any differences in stand establishment among the treatments in the agronomic trial, we compared the sown plant composition (Fig. 2a) to the observed plant composition in the first establishment year (2008) (Fig. 2b) and first production year (2010) (Fig. 2c,d). In 2008, the average aboveground standing crop was 0.4 Mg ha^{-1}, of which only 25% was accounted for by sown plant species. No significant differences were observed among the plant treatments

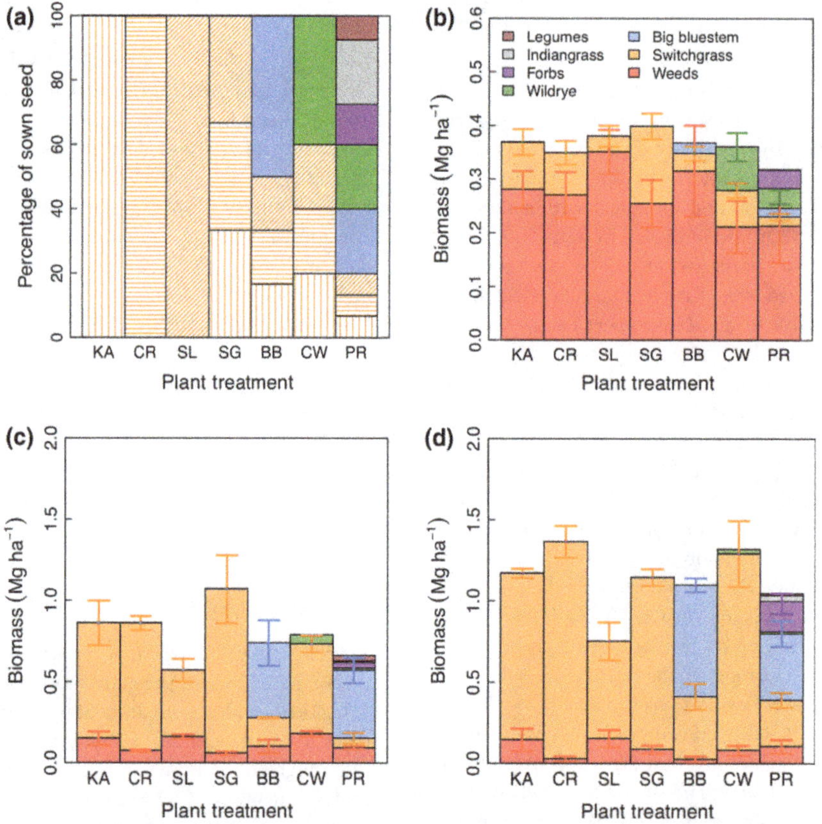

Fig. 2 Plant establishment and plot composition based on peak aboveground biomass in the agronomic trial. Shown are the percentage of sown seed in each plant treatment (a), the observed biomass composition (mean ± standard error) in the establishment year 2008 (b), and the first production year 2010, unfertilized (c), and fertilized (d). The plant treatments are Kanlow switchgrass (KA), Cave-In-Rock switchgrass (CR), Southlow switchgrass (SL), a three switchgrass cultivar mixture (SG), a big bluestem plus switchgrass mixture (BB), a Canada wildrye plus switchgrass mixture (CW), and a prairie mixture (PR). Shading in (a) designates the three switchgrass cultivars, Kanlow (vertical), Cave-In-Rock (horizontal), and Southlow (diagonal). For 2008, (b) the data are averaged over all plots in a given plant treatment because no fertilizer had been applied at that point. 'Forbs' refers to all nonlegume sown forbs.

either for the total amount of biomass ($P = 0.11$) or for the proportion of weed biomass ($P = 0.24$). By 2010, the average biomass was 9.6 Mg ha^{-1} and the proportion of biomass accounted for by sown plant species increased to 90%, on average. Here, there were significant differences in the proportion of sown vs. weed biomass among plant treatments ($P < 0.01$), with Southlow switchgrass showing poor establishment compared to most other plant treatments (~65% of biomass; Tukey's HSD $P < 0.03$). In addition, the unfertilized plots showed a significantly higher proportion of weed biomass compared to the fertilized plots (20% vs. 10%; $P < 0.01$). The total biomass (sown species plus weeds) was 30% less in the unfertilized plots compared to the fertilized plots ($P < 0.0001$). Overall, total biomass was approximately equal in all plant treatments except Southlow switchgrass, which had 34% less biomass than the others ($P < 0.01$). No significant plant treatment by fertilizer interactions on plant composition or total biomass was observed during stand establishment or the first production year.

Biomass yields in genotypic and species mixtures in the agronomic trial

Biomass yields were determined in 2009–2014 by harvesting entire plots after plant senescence (Fig. 3). By the third year of the experiment (2010), the total yield across all plots was statistically indistinguishable from the total yield in later years, supporting the definition of 2008–2009 as establishment years and 2010–2014 as production years. Averaging over the five production years, plant treatment and fertilization effects were both highly significant ($P < 0.0001$). Block effects were not observed ($P = 0.44$) and are not considered further here. The mean yield increase due to nitrogen fertilization (67 kg N ha^{-1}) was 2.2 Mg ha^{-1}, and no interaction was observed between fertilization and plant treatment ($P = 0.55$). The

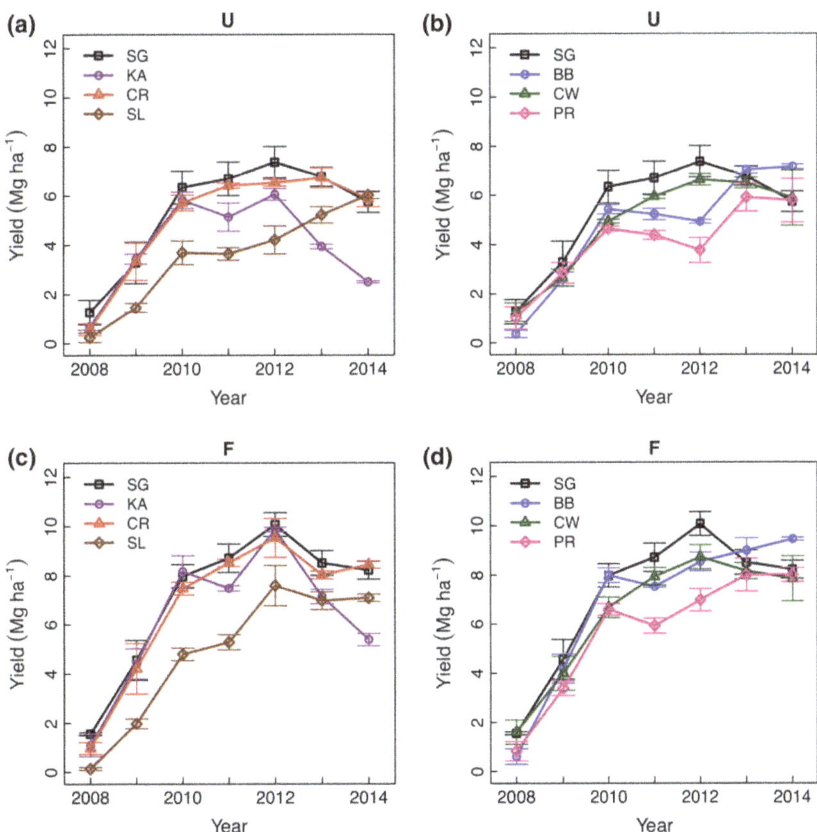

Fig. 3 Biomass yields over 7 years from 14 perennial cropping systems. Plotted are the means ± standard error ($n = 3$) of dry-weight yields for whole-plot harvestable biomass (or for 2008, an estimate based on quadrat data). The plant treatments are Kanlow switchgrass (KA), Cave-In-Rock switchgrass (CR), Southlow switchgrass (SL), a three switchgrass cultivar mixture (SG), a big bluestem plus switchgrass mixture (BB), a Canada wildrye plus switchgrass mixture (CW), and a prairie mixture (PR). The top panels (a and b) are results from unfertilized ('U') plots, while the bottom panels (c and d) are from fertilized ('F') plots. Averaged over the five production years (2010–2014) and across two fertilization levels, the mixture of three switchgrass cultivars (SG) had the highest yield (though not significantly higher than BB, CR, and CW).

switchgrass cultivar mixture (SG) produced the greatest yield overall, averaging 8.7 Mg ha^{-1} across 5 years when fertilized and 6.6 Mg ha^{-1} when unfertilized. Average yields were similar for the Cave-In-Rock cultivar monoculture (CR) and the binary species mixtures of switchgrass with big bluestem (BB) and switchgrass with Canada wildrye (CW) (Tukey's HSD test $P > 0.05$). In contrast, we observed lower average yields for the Kanlow and Southlow cultivar monocultures (KA and SL) and the prairie mixture (PR). Overall, the yields from the switchgrass-only treatments (SG, KA, CR, and SL) were no greater or less than the mixtures with other species (CW, BB, and PR) ($P = 0.79$).

Across the production years, the yield trends were significantly different among plant treatments (ANCOVA $P < 0.0001$). Comparing each plant treatment to Cave-In-Rock, the switchgrass cultivar most likely *a priori* to be well adapted and high yielding in this region, some plant treatments showed significant yield increases over the production years, while others showed yield decreases. An increasing yield over the production years was observed in the Southlow cultivar (SL, 0.6 Mg ha^{-1} yr^{-1}, $P = 0.02$) and the binary mixture of switchgrass and big bluestem (BB, 0.5 Mg ha^{-1} yr^{-1}, $P = 0.02$). In contrast, a trend of decreasing yield was observed in the Kanlow cultivar (KA, −0.7 Mg ha^{-1} yr^{-1}, $P < 0.0001$). There was no evidence of fertilization by year interaction or a three-way interaction of fertilization with plant treatment over years ($P = 0.98$).

Biomass yields over the genotypic diversity gradient

To better understand genotypic diversity effects on biomass production, we carried out a parallel experiment on a diversity gradient with four levels of cultivar richness (1, 2, 4, or 6) and three species treatments (switchgrass, big bluestem, or a binary mixture of the two). The seven switchgrass and seven big bluestem cultivars used in this experiment were chosen to represent a broad sample of the geographic range and genetic variation for each species (see Materials and Methods). For the switchgrass treatments, we estimated the sown genotypic diversity based on nucleotide diversity from genotyping-by-sequencing data and found significantly higher polymorphism in the cultivar mixtures ($P < 0.0001$, $r^2 = 0.33$; Fig. 4a,b). There were large differences in biomass yield among the 14 cultivars when grown in monoculture (Fig. 4c,d). (Note that biomass yields from the diversity gradient plots are presented as Mg ha^{-1} yr^{-1} for comparison with agronomic trial results, not to provide absolute yield estimates). When grown in cultivar monocultures, the big bluestem cultivars yielded somewhat more than the switchgrass cultivars, averaging +1.7 Mg ha^{-1} yr^{-1} over the production years of 2010–2014 ($P < 0.01$). Among the cultivar monocultures, the species effect (switchgrass vs. big bluestem) explains 12% of the variation, while cultivar (nested within species) explains 43% of the variation ($P_{sp} < 0.001$; $P_{sp:cv} < 0.0001$).

To test the effects of genotypic diversity on biomass production, we compared the cultivar monocultures described above ($n = 56$) to randomly chosen mixtures of the same cultivars (2, 4, or 6 cultivars per plot; $n = 108$). Among the cultivar mixture plots, two-thirds were sown with only switchgrass or only big bluestem, while one-third were sown with 1 : 1 mixtures of the two species (e.g., three switchgrass cultivars and three big bluestem cultivars in a six cultivar plot). Averaging yields over the production years (2010–2014), we observed a significant positive relationship between yield and number of cultivars (Fig. 4e,f; $P = 0.02$). Comparing yields of the cultivar monocultures to the genotypic/species mixtures, we see a higher average yield (31% for switchgrass, 9.5% for big bluestem, 34% for the species mixture) in the mixtures (2, 4, or 6 cultivars) than the average yield of the cultivars grown in monoculture ($P < 0.01$). Only one switchgrass cultivar monoculture, Blackwell, had a higher mean yield than the switchgrass cultivar mixtures (8.1 vs. 7.5 Mg ha^{-1}) although this was not significant ($P = 0.15$). Two big bluestem cultivar monocultures, Rountree and Suther, had higher mean yield than the big bluestem cultivar mixtures (9.5 and 9.4 vs. 8.2 Mg ha^{-1}), but the differences were again not significant ($P = 0.1$).

Year-to-year stability of biomass yields

In addition to mean biomass yields, we also considered the differences in year-to-year stability of yields, estimated as the coefficient of variation for yield across years. In the agronomic trial, the plant treatments differ significantly in the year-to-year variation in yield over the production years (Fig. 5a; $P < 0.0001$). The fertilized and unfertilized treatments did not have significantly different coefficient of variation for yield ($P = 0.29$). Kanlow and Southlow had the highest coefficients of variation while Cave-in-Rock and the three-cultivar switchgrass mixture had lower variation (Fig. 5a). Overall, the year-to-year coefficient of variation was slightly higher in the switchgrass cultivar monocultures than the switchgrass cultivar mixture (0.26 vs. 0.1; $P < 0.05$). In the diversity gradient experiment, no significant reduction in year-to-year variation was observed in the plots with higher genotypic diversity Fig. 5b; $P = 0.11$). However, reduced year-to-year variation was observed in the plots with both switchgrass and big bluestem as compared to the plots with only one of the species (Fig. 5c; $P < 0.0001$).

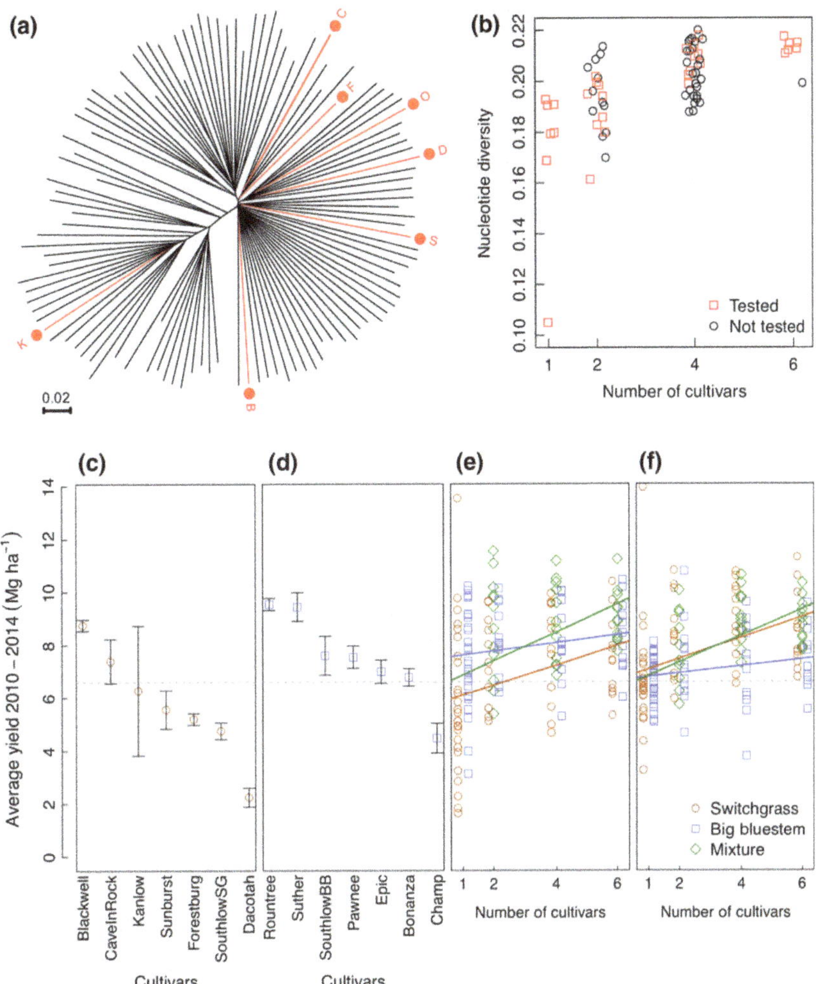

Fig. 4 Biomass yields along a genotypic diversity gradient for switchgrass and big bluestem. (a) Genetic variation of the seven switchgrass cultivars used in the experiment relative to the continent-wide diversity of switchgrass, estimated from single nucleotide polymorphism, and plotted on a neighbor-joining tree (K = Kanlow, B = Blackwell, S = Sunburst, D = Dacotah, O = Southlow, F = Forestburg, C = Cave-In-Rock). (b) The nucleotide diversity (average pairwise difference) in switchgrass monocultures and cultivar mixtures used in the field experiment (red, 'Tested') and other possible mixtures of the seven cultivars (black, 'Not Tested'). (c–f) The mean yields over five production years (± standard error) of the switchgrass (c) and big bluestem (d) cultivar monocultures are compared to those of mixtures (e and f), with the observed mixture yields (e) or the 'scaled' mixture yields (f), which are corrected for the expected yield of each mixture based on the average monoculture yields of the component cultivars. The cultivar monocultures (c and d) are sorted in order of decreasing mean yield. The dotted line indicates the global mean across plant treatments.

Composition of harvested biomass

The final aspect of the cropping systems we evaluated was the composition of the harvested biomass. To identify differences among treatments, we estimated the content of the structural components (lignin, cellulose, hemicellulose), nonstructural organic compounds (sugar, nonfiber carbohydrates, fats), and minerals (N, P, K, S, Mg, Na, Cl) from all plots in both experiments in 2012, midway through the production years. Here, we highlight a few of these compositional effects from each experiment. In the agronomic trial, we observed significant plant treatment differences and fertilization effects for several constituents, and plant treatment by fertilization interaction in a few cases (Table 2). The differences in lignin content among plant treatments were highly significant ($P < 0.0001$), with higher lignin content in the prairie mixture (PR; Tukey's $P < 0.01$) and lower lignin content in Kanlow switchgrass (KA; Tukey's $P < 0.01$). Cellulose content was significantly lower in the big bluestem–switchgrass mixture (BB) and higher in switchgrass biomass (Tukey's $P < 0.0001$). N fertilization leads to a significant, but small, increase in biomass N content from 0.59% to 0.66% ($P < 0.001$). There was no evidence of differing N content among plant treatments, but there was significant interaction

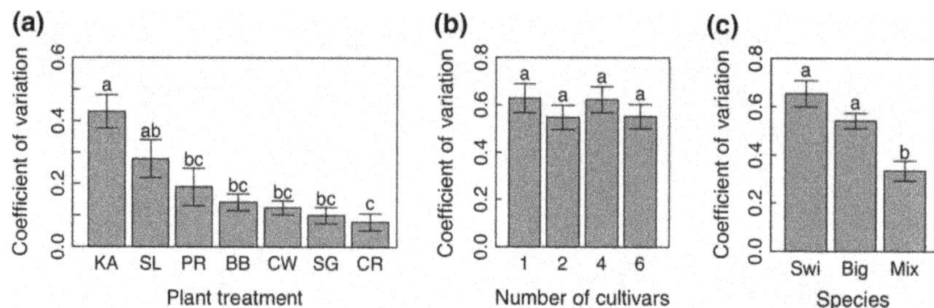

Fig. 5 Differences in year-to-year variation for yield among plant treatments across the production years (2010–2014). Significant differences among plant treatments are observed for the agronomic trial (a). The plant treatments are Kanlow switchgrass (KA), Cave-In-Rock switchgrass (CR), Southlow switchgrass (SL), a three switchgrass cultivar mixture (SG), a big bluestem plus switchgrass mixture (BB), a Canada wildrye plus switchgrass mixture (CW), and a prairie mixture (PR). For the diversity gradient, there is no significant difference by number of cultivars (b), but there is a difference by species treatment (c) ('Swi' = switchgrass, 'Big' = big bluestem, and 'Mix' = mixtures of the two species).

Table 2 Biomass composition by plant treatment (top; $n = 6$) or N fertilization treatment (bottom; $n = 21$). Values that are significantly different (Tukey's HSD, $\alpha = 0.05$) among cultivars or diversity levels are indicated by a different letter

	Lignin	Cell	Hemi	Sugar	NFC	Fat	Ash	N	P	K	Mg	S	Ca	Na	Cl
Plant treatment															
KA	7.5c	47.7a	30.2a	5.0a	3.5c	1.6a	5.4b	0.62a	0.037a	0.16a	0.17a	0.058a	0.21d	103.0a	0.058bc
CR	8.1b	46.5a	28.6ab	4.6ab	5.7abc	1.4ab	5.9ab	0.58a	0.048a	0.19a	0.16a	0.047a	0.33c	78.8a	0.083a
SL	8.3ab	46.8a	28.4b	3.4c	4.0bc	1.3b	7.0a	0.63a	0.060a	0.17a	0.15a	0.050a	0.42ab	83.8a	0.073ab
SG	8.1ab	46.5a	28.4b	4.4ab	5.1abc	1.4ab	6.4ab	0.63a	0.045a	0.20a	0.16a	0.057a	0.36bc	82.0a	0.070abc
BB	7.9bc	44.4b	29.2ab	4.1b	6.7a	1.5ab	6.0ab	0.67a	0.048a	0.15a	0.15a	0.043a	0.36bc	77.8a	0.052c
CW	8.0bc	45.9ab	28.7ab	4.6ab	5.8abc	1.5ab	6.1ab	0.60a	0.047a	0.19a	0.16a	0.047a	0.34c	88.2a	0.068abc
PR	8.7a	46.4ab	26.4c	3.9bc	6.2ab	1.5ab	6.6ab	0.65a	0.045a	0.18a	0.15a	0.045a	0.44a	81.2a	0.058bc
N Fert.															
Unfert.	7.9b	45.8b	28.7a	4.5a	5.8a	1.5a	6.4a	0.59b	0.057a	0.18a	0.15b	0.050a	0.35a	81.7a	0.069a
Fertilized	8.3a	46.8a	28.4a	4.1a	4.8a	1.4b	6.0a	0.66a	0.037b	0.17a	0.17a	0.050a	0.36a	88.2a	0.064a

Cell, cellulose; Hemi, hemicellulose; NFC, nonfiber carbohydrate.

with fertilization ($P = 0.047$) due to higher N levels in the fertilized big bluestem-switchgrass mixture (BB). The biomass P content is significantly lower with N fertilization ($P < 0.0001$), but there were no differences in K content among treatments. The only significant difference in ash content was the higher ash content of Southlow vs. Kanlow switchgrass (7.0% vs. 5.4%; $P = 0.02$).

In the genotypic diversity gradient, switchgrass biomass had lower estimated cellulose content (44.6%) than big bluestem (46.8%; Tukey's $P < 0.0001$) and the mixtures of big bluestem and switchgrass (45.5%; Tukey's $P < 0.001$). Conversely, switchgrass had higher estimated sugar content (4.9%) than big bluestem (4.1%; Tukey's $P < 0.0001$) and the two species mixture (4.5%; Tukey's $P = 0.001$). The level of cultivar diversity had no significant effects on the organic compound content, and little effect on mineral composition.

Discussion

In this study, we characterized the effects of plant genotypic and species diversity on establishment, yields, and biomass composition and consider some implications for production, conversion and environmental impacts. Overall, we found genotype mixtures performed well in terms of yield (Figs 3, 4 and 5). Given that this was true even when some of the constituent cultivars performed poorly in monoculture, this suggests that there is little risk of low yields in using genotypic mixtures as long as some adapted material is included. We see that use of genotypic diversity can stabilize yields (as seen with SG), but so can the choice of a high-yielding and well-adapted cultivar monoculture (in this case, Cave-In-Rock) (Fig. 5a). These observations are consistent with a sampling effect because the yields of the genotypic mixtures are not transgressive (Fargione *et al.*, 2007). In more marginal production environments, where temporal and spatial heterogeneity is proportionally more important, there may be greater benefits of genotypic diversity (Haussmann *et al.*, 2012). Further studies of genotypic mixtures of biomass species grown in marginal production environments (e.g., drought-prone or nutrient deficient soils) will be needed to evaluate this hypothesis. In addition, future studies of targeted genotypic mixtures (i.e., selected for complementary traits or high yield potential) might reveal benefits of genotypic diversity not observed in the random genotype mixtures we studied here.

We did not estimate the abundances of the three switchgrass cultivars separately in the cultivar mixture, but given biomass composition estimates, it appears that all three cultivars were well represented. For example, the higher ash content of SG and SL treatments than KA and CR (Table 2) suggests that Southlow was abundant in the three-cultivar mixture by 2012 despite its slow establishment in monoculture. These findings show that a local ecopool cultivar with high genotypic diversity can meet or exceed the biomass yields from cultivars with high expected yields when production is considered over a longer time scale.

While high average yields are important, the stability of yields may be equally important in marginal production systems with localized value chains. For instance, in drier years, crops with high yield potential but drought sensitivity could fail to meet the year-round feedstock supplies required by the bioconversion plants (Uden *et al.*, 2013). In row cropping and native grassland systems, mixtures of genotypes and species have been shown to stabilize productivity (Smithson & Lenné, 1996; Tilman *et al.*, 2006b) and thus may be expected to limit yield losses to environmental stresses, such as cold spells or drought. In fact, in this experiment, the highest overall biomass yields were observed in the three switchgrass mixture (SG) during the drought year of 2012 (true for both fertilized and unfertilized treatments; Fig. 3). In contrast, the 2012 corn yields in the county were about 20% below the 2011 yields and the long-term yield trend (USDA-NASS, 2013). The ability of diverse native perennial biomass cropping systems to produce high yields when row crop (e.g., corn or sorghum) stover yields are low may be especially valuable for ensuring stable feedstock supplies in a fuel shed despite annual variability (Uden *et al.*, 2013).

Diverse species mixtures are commonly used in forage systems – especially mixtures of legumes and cool-season grasses – and these polycultures often substantially outperform forage monocultures (Picasso *et al.*, 2008). As most of the weed species at our field site were cool-season species, we hypothesized that the plant treatments that included cool-season species (the wildrye with switchgrass and the 12-species prairie mixture) could reduce weed biomass during establishment. However, there was no evidence that the cool-season species we included were able to out compete the cool-season weeds. Instead, vigorous growth in the establishment year by improved cultivars seemed to be most associated with weed suppression. Further studies with other cool-season native species will be needed to determine whether other warm + cool-season native grass mixtures are able to provide improved yields.

While interactions between species diversity and fertilization have been investigated, interactions between genotypic diversity and fertilization have been less studied (Jarchow & Liebman, 2013). In this study, there was no evidence that switchgrass monocultures responded more to N fertilization than the species mixtures or that different switchgrass cultivars or cultivar mixtures responded differently (Fig. 3). This suggests that choosing a higher genotypic and/or species diversity will neither help or hinder production under moderate N fertilization, which is likely to be employed for perennial bioenergy crops given that producer profits are maximized at intermediate N inputs (Boyer *et al.*, 2012). This finding is consistent with studies on other native perennial cropping systems in fertile soil (Jarchow & Liebman, 2013).

A meta-analysis of switchgrass monocultures and switchgrass-dominated mixtures found that mixtures of switchgrass and legumes generally performed well (Wang *et al.*, 2010), presumably due to fertilization effects from N_2 fixation. In our study, however, the treatment that included legumes (PR) had lower yields compared to switchgrass cultivar monocultures and lower diversity mixtures, whether unfertilized or fertilized, suggesting that there was no N fertilization effect due to the legumes (Fig. 3). This may have been due the low abundance of legumes in our mixture (7.5% of sown seed) and the poor establishment of these legumes (Fig. 2). Also, prairie mixture (PR) yields might have been lower because the broadleaf plants (e.g., the large composites) take up a large area when they are green but when they senesce (and are cut and baled), the leaves fall to the ground and are not harvested. Future studies on combined genotypic/species mixtures for biomass production might take advantage of a better understanding of optimal legume composition in biomass cropping systems that has emerged from recent work (Picasso *et al.*, 2008; Johnson *et al.*, 2010; Wang *et al.*, 2010; Griffith *et al.*, 2011).

Another potential impact of native biomass crops is gene flow from cultivars to native stands of the same species. The probability of such gene flow and its effects on native populations and associated communities are not yet understood (Kwit & Stewart, 2011). The inclusion of local ecopools in cultivar mixtures for biomass cropping systems could help conserve local genotypes. One aspect we considered here was whether local germplasm could be used effectively in biomass cropping systems. The results were mixed. The local switchgrass ecopool Southlow did not establish well, contributing little to the biomass in the early years of the experiment

(Fig. 2) and yielding poorly in the early production years (Fig. 3). This is likely because Southlow, unlike Cave-In-Rock and Kanlow, was not intentionally selected for reduced dormancy and vigorous emergence. However, Southlow switchgrass yields continued to rise over the production years compared to Cave-In-Rock (which plateaued) while Kanlow decreased. This is consistent with the expectation that Southlow, from nearby southern Michigan, is well adapted to northeast Illinois (Casler *et al.*, 2007; Grabowski *et al.*, 2014). By the last production years (2013–2014), Southlow yields matched those of Cave-In-Rock and exceeded in those of Kanlow in the unfertilized treatment.

In many photoperiod sensitive species, including switchgrass, the use of low latitude-adapted germplasm in high latitudes is known to dramatically increase biomass yields through delayed flowering (Casler *et al.*, 2007; Rooney *et al.*, 2007). Under favorable conditions, wet-adapted germplasm also produce more biomass than dry-adapted germplasm. Consistent with this expectation, we saw the lowest yields in the most northern and dry-adapted cultivars, Dacotah switchgrass and Champ big bluestem (Fig. 4). The yields of Kanlow, the most southern and wet-adapted switchgrass cultivar in our study, did achieve high yields in some years, but declined dramatically in 2013 and 2014. Cultivars from southern lowland gene pool would be poorly adapted to the northern Midwest climate (Casler *et al.*, 2007; Grabowski *et al.*, 2014). The unusually cold winter of 2013/2014 may have contributed to the decline of the Kanlow stands, as Kanlow stands have shown little survival just 175 km north of our field site (Casler, 2014). Overall, the balance between photoperiod effects and local adaptation effects seems to favor cultivars or mixtures from within the hardiness zone or one zone south of our site in the upper Midwest (e.g., Cave-In-Rock and Blackwell for switchgrass, and Rountree and Suther for big bluestem).

The composition of harvested biomass has important implications for crop management (e.g., balancing N-P-K removal and fertilization), biomass utilization (e.g., lignin/cellulose ratios or slagging and fouling) (Sims, 2003), and the environmental sustainability of the system (e.g., net energy inputs and GHG balance) (Downing *et al.*, 2011; Singh *et al.*, 2012; Bhandari *et al.*, 2013). A previous study comparing four switchgrass and three big bluestem genotypes found the former having significantly higher lignin content, as determined by HPLC (Zhang *et al.*, 2015). However, we saw no evidence of higher lignin in switchgrass overall in our comparison of seven switchgrass vs. seven big bluestem genotypes (Table 3). Given that we did observe among-cultivar variation in lignin (Table 2 and 3), the lack of between-species differences in lignin is not likely due to a lack of precision in the NIRS measurements, but the broader sampling of switchgrass and big bluestem germplasm in our study. Ash and N levels were similar for big bluestem and switchgrass, and big bluestem may have lower levels of minerals involved in corrosion/fouling (e.g., Cl, Ca, S; Table 2). Together with the evidence of high yield potential for big bluestem, these data suggest that big bluestem mixtures and monocultures warrant more consideration for biomass production systems.

The variation in feedstock composition introduced by higher diversity biomass cropping systems has been raised as a potential limitation of these systems (Adler *et al.*, 2009). In our study, the prairie mixture (PR) did have significantly different composition for a number of variables, including notably higher lignin and ash than the Cave-In-Rock (CR) and Kanlow (KA) switchgrass monocultures (Table 2). In contrast, the switchgrass cultivar mixture (SG) and the big bluestem + switchgrass mixture (BB) had lignin levels similar to the Cave-In-Rock monoculture. Using genotypic mixtures of a single species, or simple mixtures of grasses, may be a way to exploit beneficial diversity effects while avoiding undesirable compositional effects of high species diversity. The low N composition of switchgrass had a role in its prioritization over leguminous crops for bioenergy research (Wright, 2007). Here, we observed no difference in biomass N among the plant treatments (Table 2), but this may be due to the small proportion of legumes in the harvested biomass and translocation of nutrients to belowground biomass prior to harvest. The nonlegume forbs, which did contribute substantially to harvested biomass, had no effect on N levels (Table 2). Biomass from N fertilized plots had significantly higher lignin and N than biomass from unfertilized plots (5% greater in both cases), indicating that the increase in yield upon N fertilization may go along with a reduction in feedstock quality.

Given the long breeding cycle for perennial outcrossing crops, the development of regionally optimized genotypes for perennial biomass production may take decades. Effectively implementing and interpreting multi-environment and multiyear field trials in plant breeding programs remain a serious challenge even in major annual row crops (Cooper *et al.*, 2014). Taken together, our findings on biomass yield and composition suggest that genotypic mixtures could be a useful strategy to increase and stabilize yields in biomass feedstock production systems, in parallel with crop improvement efforts. Further studies, especially in marginal production environments and on targeted genotypic mixtures, will help determine the potential value of genotypic mixtures to increase the sustainability and profitability of native perennial bioenergy cropping systems.

Table 3 Biomass composition for switchgrass and bluestem cultivar monocultures (top) or different diversity levels (bottom). Values that are significantly different ($\alpha = 0.05$) among cultivars or diversity levels are indicated by a different letter. For the cultivars, monocultures $n = 4$ and for the diversity levels $n = 42$ (1 cultivar) or $n = 12$ (2, 4, and 6 cultivars)

Species	Cultivar	Lignin	Cell	Hemi	Sugar	NFC	Fat	Ash	N	P	K	Mg	S	Ca	Na	Cl
Switchgrass	Blackwell	7.9ab	45.0ab	29.8ab	4.9bc	6.9b	1.4bc	5.3ab	0.54b	0.045b	0.15bc	0.17a	0.043b	0.29 cd	83.5a	0.095b
	CaveInRock	7.7ab	44.9ab	29.8ab	5.1b	7.0b	1.5bc	5.9ab	0.49b	0.060ab	0.15bc	0.16a	0.043b	0.31bcd	79.5a	0.065b
	Dacotah	7.5ab	39.7c	25.1c	6.5a	11.6a	2.3a	7.9a	1.03a	0.100a	0.30a	0.16a	0.075a	0.52a	123.0a	0.155a
	Forestburg	8.1a	44.5b	29.9ab	4.8bcd	5.9bc	1.4bc	6.4ab	0.57b	0.060ab	0.13c	0.16a	0.045b	0.40bc	92.5a	0.068b
	Kanlow	7.3ab	46.8ab	31.0a	4.9bc	4.5bc	1.5bc	6.5ab	0.55b	0.035b	0.14c	0.17a	0.043b	0.21d	92.2a	0.062b
	SouthlowSG	7.8ab	45.2ab	28.9ab	4.0 cde	6.3b	1.4bc	6.8ab	0.53b	0.053b	0.13c	0.14ab	0.045b	0.40bc	76.5a	0.092b
	Sunburst	8.1a	45.0ab	30.0ab	4.5bcd	5.9bc	1.3c	6.1ab	0.53b	0.055b	0.12c	0.17a	0.062ab	0.42ab	89.0a	0.070b
Big Bluestem	Bonanza	7.2ab	47.1ab	31.2a	3.3e	1.7c	1.6bc	7.3ab	0.58b	0.048b	0.17bc	0.13ab	0.040b	0.29d	107.2a	0.057b
	Champ	7.7ab	47.2ab	28.9ab	4.0cde	4.4bc	1.7b	6.5ab	0.56b	0.058b	0.19abc	0.12ab	0.043b	0.29d	83.0a	0.070b
	Epic	7.0b	46.1ab	29.8ab	4.0cde	4.9bc	1.7bc	7.0ab	0.53b	0.058b	0.28ab	0.10b	0.043b	0.26d	88.5a	0.083b
	Pawnee	7.9ab	45.3ab	30.2ab	4.5bcd	6.0bc	1.6bc	5.0b	0.64b	0.048b	0.18abc	0.14ab	0.043b	0.30cd	78.0a	0.087b
	Rountree	8.1a	48.0a	29.3ab	3.9de	3.4bc	1.5bc	5.8ab	0.58b	0.045b	0.21abc	0.10b	0.037b	0.24d	89.5a	0.068b
	SouthlowBB	7.8ab	48.2a	29.3ab	3.3e	2.5bc	1.4bc	7.1ab	0.54b	0.053b	0.16bc	0.11b	0.040b	0.27d	120.2a	0.058b
	Suther	8.0a	46.7ab	28.5b	5.3b	5.7bc	1.5bc	5.9ab	0.57b	0.045b	0.23abc	0.12ab	0.045b	0.25d	78.0a	0.068b

Species	No. of cultivars	Lignin	Cell	Hemi	Sugar	NFC	Fat	Ash	N	P	K	Mg	S	Ca	Na	Cl
Switchgrass	1	7.8ab	44.5d	29.2a	4.9a	7.0a	1.5a	6.4a	0.61a	0.058a	0.16b	0.16a	0.051a	0.37a	90.9a	0.087a
	2	8.0ab	45.2 cd	29.6a	4.8ab	5.7ab	1.4a	6.3a	0.56a	0.055ab	0.14b	0.16a	0.048ab	0.36ab	87.0a	0.083ab
	4	8.0ab	46.3abcd	29.6a	4.8ab	5.0ab	1.5a	5.8a	0.58a	0.048ab	0.14b	0.17a	0.045ab	0.32abc	119.3a	0.079ab
	6	7.9ab	45.8bcd	29.4a	4.8ab	5.7ab	1.4a	6.2a	0.56a	0.041b	0.14b	0.16a	0.043ab	0.34abc	88.8a	0.084ab
Big Bluestem	1	7.7b	47.0abc	29.6a	4.0 cd	4.1b	1.6a	6.4a	0.57a	0.050ab	0.20a	0.12c	0.041b	0.27c	92.1a	0.070ab
	2	8.2a	48.0a	28.9a	4.1bcd	3.6b	1.5a	6.0a	0.59a	0.055ab	0.19ab	0.12bc	0.040b	0.26c	91.4a	0.058b
	4	7.9ab	47.2abc	29.6a	4.1bcd	3.7b	1.6a	6.3a	0.56a	0.048ab	0.18ab	0.12c	0.040b	0.28bc	98.0a	0.065ab
	6	7.9ab	47.6ab	30.2a	3.8d	3.7b	1.5a	6.2a	0.57a	0.046ab	0.18ab	0.12bc	0.039b	0.27c	83.7a	0.057b
Mixture	2	7.8ab	47.0abc	29.6a	4.4abcd	4.3b	1.5a	7.0a	0.52a	0.041b	0.16b	0.14b	0.043ab	0.30bc	91.2a	0.075ab
	4	7.8ab	46.3abcd	30.3a	4.2bcd	4.8ab	1.4a	6.5a	0.58a	0.047ab	0.15b	0.14ab	0.043ab	0.32abc	85.8a	0.068ab
	6	8.0ab	46.4abc	29.4a	4.7abc	4.9ab	1.4a	6.0a	0.58a	0.043ab	0.14b	0.13bc	0.044ab	0.28bc	80.2a	0.069ab

Cell, cellulose; Hemi, hemicellulose; NFC, nonfiber carbohydrate.

Acknowledgements

Funding for this research was provided by the US Department of Energy, Office of Science, Office of Biological and Environmental Research under contract DE-AC02-06CH11357 to Argonne National Laboratory (RMM and JDJ). Additional support was provided by the Argonne/UChicago Energy Initiative to RMM and JOB, USDA-NIFA grant 2010-03894 to RMM, and a USDA-AFRI grant 2012-67010-20069 to M-AG, JDJ, and GPM. PPG was partially supported by National Institutes of Health Training Grant T32 GM007197. We thank Timothy Vugteveen, Whitney Panneton, Nina Noah, Jeremy Lederhouse, Scott Hofmann, Susan Kirt Alterio, Kelly Moran Sturner, and Cheryl Martin for technical assistance, and two anonymous reviewers for helpful comments and suggestions.

References

Acquaah G (2012) *Principles of Plant Genetics and Breeding* (2 edn). Wiley-Blackwell, Hoboken, NJ. 758 pp.

Adler PR, Sanderson MA, Weimer PJ, Vogel KP (2009) Plant species composition and biofuel yields of conservation grasslands. *Ecological Applications*, **19**, 2202–2209.

Alderson J, Sharp WC, United States. Department of Agriculture (1994) *Grass Varieties in the United States*. U.S. Department of Agriculture, Washington, DC, 310 pp.

Allard RW, Bradshaw AD (1964) Implications of genotype-environmental interactions in applied plant breeding. *Crop Science*, **4**, 503–508.

Aronesty E (2013) Comparison of sequencing utility programs. *The Open Bioinformatics Journal*, **7**, 1–8.

Barker RE, Haas RJ, Jacobson ET, Berdahl JD (1988) Registration of "Forestburg" switchgrass. *Crop Science*, **28**, 192–193.

Barker RE, Haas RJ, Berdahl JD, Jacobson ET (1990) Registration of "Dacotah" switchgrass. *Crop Science*, **30**, 1158.

Bhandari HS, Walker DW, Bouton JH, Saha MC (2013) Effects of ecotypes and morphotypes in feedstock composition of switchgrass (*Panicum virgatum* L.). *GCB Bioenergy*, **6**, 26–34.

Boe A, Ross JG (1998) Registration of "Sunburst" switchgrass. *Crop Science*, **38**, 540.

Boyer CN, Tyler DD, Roberts RK, English BC, Larson JA (2012) Switchgrass yield response functions and profit-maximizing nitrogen rates on four landscapes in Tennessee. *Agronomy Journal*, **104**, 1579–1588.

Bradbury PJ, Zhang Z, Kroon DE, Casstevens TM, Ramdoss Y, Buckler ES (2007) TASSEL: software for association mapping of complex traits in diverse samples. *Bioinformatics*, **23**, 2633–2635.

Casler MD (2005) Ecotypic variation among switchgrass populations from the Northern USA. *Crop Science*, **45**, 388–398.

Casler MD (2012) Switchgrass breeding, genetics, and genomics. In: *Switchgrass* (ed. Monti A), pp. 29–53. Springer, London.

Casler MD (2014) Heterosis and reciprocal-cross effects in tetraploid switchgrass. *Crop Science*, **54**, 2063–2069.

Casler MD, Vogel KP, Taliaferro CM *et al.* (2007) Latitudinal and longitudinal adaptation of switchgrass populations. *Crop Science*, **47**, 2249–2260.

Cook-Patton SC, McArt SH, Parachnowitsch AL, Thaler JS, Agrawal AA (2011) A direct comparison of the consequences of plant genotypic and species diversity on communities and ecosystem function. *Ecology*, **92**, 915–923.

Cooper M, Messina CD, Podlich D *et al.* (2014) Predicting the future of plant breeding: complementing empirical evaluation with genetic prediction. *Crop and Pasture Science*, **65**, 311–336.

R Core Team (2014) *R: A Language and Environment for Statistical Computing*. R Core Team, Vienna, Austria.

Crawley MJ (2012) *The R Book* (2nd edn). Wiley, Chichester, West Sussex, UK. 1076 pp.

Danecek P, Auton A, Abecasis G *et al.* (2011) The variant call format and VCFtools. *Bioinformatics*, **27**, 2156–2158.

Davis KM, Englert JM, Kujawski JL (2002) Improved conservation plant materials released by NRCS and cooperators through September 2002.

Dillon SL, Shapter FM, Henry RJ, Cordeiro G, Izquierdo L, Lee LS (2007) Domestication to crop improvement: genetic resources for Sorghum and Saccharum (Andropogoneae). *Annals of Botany*, **100**, 975–989.

Downing M, Eaton LM, Graham RL *et al.* (2011) *US Billion-ton Update: Biomass Supply for a Bioenergy and Bioproducts Industry*. Oak Ridge National Laboratory (ORNL), Oak Ridge, TN.

Durling JC, Leif JW, Burgdorf DW (2007) Registration of southlow Michigan germplasm big bluestem. *Crop Science*, **47**, 455.

Durling JC, Leif JW, Burgdorf DW (2008) Registration of southlow Michigan germplasm switchgrass. *Journal of Plant Registrations*, **2**, 60.

Fargione J, Tilman D, Dybzinski R *et al.* (2007) From selection to complementarity: shifts in the causes of biodiversity–productivity relationships in a long-term biodiversity experiment. *Proceedings of the Royal Society B: Biological Sciences*, **274**, 871–876.

Gan J, Smith CT (2011) Optimal plant size and feedstock supply radius: a modeling approach to minimize bioenergy production costs. *Biomass and Bioenergy*, **35**, 3350–3359.

Gelfand I, Sahajpal R, Zhang X, Izaurralde RC, Gross KL, Robertson GP (2013) Sustainable bioenergy production from marginal lands in the US Midwest. *Nature*, **493**, 514–517.

Grabowski PP, Morris GP, Casler MD, Borevitz JO (2014) Population genomic variation reveals roles of history, adaptation and ploidy in switchgrass. *Molecular Ecology*, **23**, 4059–4073.

Gray MM, St. Amand P, Bello NM *et al.* (2014) Ecotypes of an ecologically dominant prairie grass (Andropogon gerardii) exhibit genetic divergence across the U.S. Midwest grasslands' environmental gradient. *Molecular Ecology*, **23**, 6011–6028.

Griffith AP, Epplin FM, Fuhlendorf SD, Gillen R (2011) A comparison of perennial polycultures and monocultures for producing biomass for biorefinery feedstock. *Agronomy Journal*, **103**, 617–627.

Hanson AA (Angus A (1972) *Grass Varieties in the United States*. Agricultural Research Service, U.S. Department of Agriculture, Washington, DC, 134 pp.

Haussmann BIG, Fred Rattunde H, Weltzien-Rattunde E, Traoré PSC, Vom Brocke K, Parzies HK (2012) Breeding strategies for adaptation of pearl millet and sorghum to climate variability and change in West Africa. *Journal of Agronomy and Crop Science*, **198**, 327–339.

Jarchow ME, Liebman M (2012) Tradeoffs in biomass and nutrient allocation in prairies and corn managed for bioenergy production. *Crop Science*, **52**, 1330–1342.

Jarchow ME, Liebman M (2013) Nitrogen fertilization increases diversity and productivity of prairie communities used for bioenergy. *GCB Bioenergy*, **5**, 281–289.

Johnson M-VV, Kiniry JR, Sanchez H, Polley HW, Fay PA (2010) Comparing biomass yields of low-input high-diversity communities with managed monocultures across the Central United States. *BioEnergy Research*, **3**, 353–361.

Kennedy RK (1972) The sickledrat: a circular quadrat modification useful in grassland studies. *Journal of Range Management Archives*, **25**, 312–313.

Kim C, Zhang D, Auckland S *et al.* (2012) SSR-based genetic maps of *Miscanthus sinensis* and *M. sacchariflorus* and their comparison to sorghum. *TAG Theoretical and Applied Genetics*, **124**, 1325–1338.

Kingsbury N (2011) *Hybrid: The History and Science of Plant Breeding*. University Of Chicago Press, Chicago, IL; Bristol. 512 pp.

Kwit C, Stewart CN (2011) Gene flow matters in switchgrass (*Panicum virgatum* L.), a potential widespread biofuel feedstock. *Ecological Applications*, **22**, 3–7.

Li H, Durbin R (2009) Fast and accurate short read alignment with Burrows-Wheeler transform. *Bioinformatics*, **25**, 1754–1760.

Li H, Handsaker B, Wysoker A *et al.* (2009) The sequence alignment/map format and SAMtools. *Bioinformatics*, **25**, 2078–2079.

Liu L, Wu Y (2012) Identification of a selfing compatible genotype and mode of inheritance in switchgrass. *BioEnergy Research*, **5**, 662–668.

Mangan ME, Sheaffer C, Wyse DL, Ehlke NJ, Reich PB (2011) Native perennial grassland species for bioenergy: establishment and biomass productivity. *Agronomy Journal*, **103**, 509–519.

McMillan C (1956) Nature of the plant community. I. Uniform garden and light period studies of five grass taxa in Nebraska. *Ecology*, **37**, 330–340.

McMillan C (1959) The role of ecotypic variation in the distribution of the Central Grassland of North America. *Ecological Monographs*, **29**, 286–308.

McMillan C (1961) Nature of the plant community. VI. Texas Grassland Communities under transplanted conditions. *American Journal of Botany*, **48**, 778–785.

Morris GP, Grabowski PP, Borevitz JO (2011) Genomic diversity in switchgrass (*Panicum virgatum*): from the continental scale to a dune landscape. *Molecular Ecology*, **20**, 4938–4952.

Mundt CC (2002) Use of multiline cultivars and cultivar mixtures for disease management. *Annual Review of Phytopathology*, **40**, 381–410.

Newell LC (1968a) Registration of Champ Bluestem (Reg. No. 2). *Crop Science*, **8**, 515.

Newell LC (1968b) Registration of Pawnee Big Bluestem (Reg. No. 1). *Crop Science*, **8**, 514–515.

Nyfeler D, Huguenin-Elie O, Suter M, Frossard E, Connolly J, Lüscher A (2009) Strong mixture effects among four species in fertilized agricultural grassland led to persistent and consistent transgressive overyielding. *Journal of Applied Ecology*, **46**, 683–691.

O'Brien SL, Jastrow JD, McFarlane KJ, Guilderson TP, Gonzalez-Meler MA (2013) Decadal cycling within long-lived carbon pools revealed by dual isotopic analysis of mineral-associated soil organic matter. *Biogeochemistry*, **112**, 111–125.

Picasso VD, Brummer EC, Liebman M, Dixon PM, Wilsey BJ (2008) Crop species diversity affects productivity and weed suppression in perennial polycultures under two management strategies. *Crop Science*, **48**, 331–342.

Porter CL (1966) An analysis of variation between upland and lowland switchgrass, *Panicum virgatum* L., in Central Oklahoma. *Ecology*, **47**, 980–992.

Robertson BA, Doran PJ, Loomis LR, Robertson JR, Schemske DW (2011) Perennial biomass feedstocks enhance avian diversity. *GCB Bioenergy*, **3**, 235–246.

Rooney WL, Blumenthal J, Bean B, Mullet JE (2007) Designing sorghum as a dedicated bioenergy feedstock. *Biofuels, Bioproducts and Biorefining*, **1**, 147–157.

Sarath G, Akin DE, Mitchell RB, Vogel KP (2008) Cell-wall composition and accessibility to hydrolytic enzymes is differentially altered in divergently bred switchgrass (*Panicum virgatum* L.) genotypes. *Applied Biochemistry and Biotechnology*, **150**, 1–14.

Schmer MR, Vogel KP, Mitchell RB, Perrin RK (2008) Net energy of cellulosic ethanol from switchgrass. *Proceedings of the National Academy of Sciences*, **105**, 464–469.

Sims REH (2003) *Bioenergy Options for a Cleaner Environment: In Developed and Developing Countries: In Developed and Developing Countries*. Elsevier, Amsterdam, 200 pp.

Singh MP, Erickson JE, Sollenberger LE, Woodard KR, Vendramini JMB, Fedenko JR (2012) Mineral composition and biomass partitioning of sweet sorghum grown for bioenergy in the southeastern USA. *Biomass and Bioenergy*, **47**, 1–8.

Smithson JB, Lenné JM (1996) Varietal mixtures: a viable strategy for sustainable productivity in subsistence agriculture. *Annals of Applied Biology*, **128**, 127–158.

Swink F, Wilhelm G (1994) *Plants of the Chicago Region* (4th edn). Indiana Academy of Science, Indianapolis. 936 pp.

Tamura K, Stecher G, Peterson D, Filipski A, Kumar S (2013) MEGA6: Molecular evolutionary genetics analysis version 6.0. *Molecular Biology and Evolution*, **30**, 2725–2729.

Tilman D, Hill J, Lehman C (2006a) Carbon-negative biofuels from low-input high-diversity grassland biomass. *Science*, **314**, 1598–1600.

Tilman D, Reich PB, Knops JMH (2006b) Biodiversity and ecosystem stability in a decade-long grassland experiment. *Nature*, **441**, 629–632.

Uden DR, Mitchell RB, Allen CR, Guan Q, McCoy TD (2013) The feasibility of producing adequate feedstock for year-round cellulosic ethanol production in an intensive agricultural fuelshed. *BioEnergy Research*, **6**, 930–938.

US Department of Commerce N (2015) Official Extreme Weather Records for Chicago, IL.

USDA-NASS (2013) Acreage, yield and production by counties, Illinois, 2012.

USDA-Natural Resources Conservation Service (2009) *Planting and Managing Switchgrass as a Biomass Energy Crop*. Plant Materials Program, Technical Note 3.

USDA-NRCS (2013) Iowa Bulletin No. 190-12-5.

Vogel KP, Mitchell RB, Klopfenstein TJ, Anderson BE (2006) Registration of "Bonanza" big bluestem. *Crop Science*, **46**, 2313–2314.

Vogel KP, Dien BS, Jung HG, Casler MD, Masterson SD, Mitchell RB (2010) Quantifying actual and theoretical ethanol yields for switchgrass strains using NIRS analyses. *BioEnergy Research*, **4**, 96–110.

Wang D, Lebauer DS, Dietze MC (2010) A quantitative review comparing the yield of switchgrass in monocultures and mixtures in relation to climate and management factors. *GCB Bioenergy*, **2**, 16–25.

Wright L (2007) *Historical Perspective on how and why Switchgrass was Selected as a "Model" High-Potential Energy Crop*. Bioenergy Resources and Engineering Systems, ORNL/TM-2007/109 Oak Ridge, TN.

Zalapa JE, Price DL, Kaeppler SM, Tobias CM, Okada M, Casler MD (2010) Hierarchical classification of switchgrass genotypes using SSR and chloroplast sequences: ecotypes, ploidies, gene pools, and cultivars. *Theoretical and Applied Genetics*, **122**, 805–817.

Zhang Y, Zalapa JE, Jakubowski AR *et al.* (2011) Post-glacial evolution of *Panicum virgatum*: centers of diversity and gene pools revealed by SSR markers and cpDNA sequences. *Genetica*, **139**, 933–948.

Zhang K, Johnson L, Prasad PVV, Pei Z, Yuan W, Wang D (2015) Comparison of big bluestem with other native grasses: chemical composition and biofuel yield. *Energy*, **83**, 358–365.

9

Bioenergy crop productivity and potential climate change mitigation from marginal lands in the United States: An ecosystem modeling perspective

ZHANGCAI QIN*, QIANLAI ZHUANG*,† and XIMING CAI‡

Department of Earth, Atmospheric, and Planetary Sciences, Purdue University, West Lafayette, IN 47907, USA, †Department of Agronomy, Purdue University, West Lafayette, IN 47907, USA, ‡Ven Te Chow Hydrosystems Laboratory, Department of Civil and Environmental Engineering, University of Illinois at Urbana–Champaign, Urbana, IL 61801, USA

Abstract

Growing biomass feedstocks from marginal lands is becoming an increasingly attractive choice for producing biofuel as an alternative energy to fossil fuels. Here, we used a biogeochemical model at ecosystem scale to estimate crop productivity and greenhouse gas (GHG) emissions from bioenergy crops grown on marginal lands in the United States. Two broadly tested cellulosic crops, switchgrass, and *Miscanthus*, were assumed to be grown on the abandoned land and mixed crop-vegetation land with marginal productivity. Production of biomass and biofuel as well as net carbon exchange and nitrous oxide emissions were estimated in a spatially explicit manner. We found that, cellulosic crops, especially *Miscanthus* could produce a considerable amount of biomass, and the effective ethanol yield is high on these marginal lands. For every hectare of marginal land, switchgrass and *Miscanthus* could produce 1.0–2.3 kl and 2.9–6.9 kl ethanol, respectively, depending on nitrogen fertilization rate and biofuel conversion efficiency. Nationally, both crop systems act as net GHG sources. Switchgrass has high global warming intensity (100–390 g CO_2eq l^{-1} ethanol), in terms of GHG emissions per unit ethanol produced. *Miscanthus*, however, emits only 21–36 g CO_2eq to produce every liter of ethanol. To reach the mandated cellulosic ethanol target in the United States, growing *Miscanthus* on the marginal lands could potentially save land and reduce GHG emissions in comparison to growing switchgrass. However, the ecosystem modeling is still limited by data availability and model deficiencies, further efforts should be made to classify crop-specific marginal land availability, improve model structure, and better integrate ecosystem modeling into life cycle assessment.

Keywords: biofuel, global warming potential, greenhouse gas emission, land use change, life cycle assessment, *Miscanthus*, nitrous oxide, switchgrass

Introduction

Bioenergy, an important renewable energy produced from biological materials, is becoming an increasingly attractive energy choice in the context of economic development, energy security, and climate change. One hand, with increasing world population and rapidly growing regional and global economy, conventional fossil fuel-based energy alone is not likely to provide essential and sufficient support to the functioning of modern economies, due to its limited supply, high or volatile fossil fuel prices, and concerns about national energy independence (Field *et al.*, 2008; Hill *et al.*, 2009). On the other hand, the society is increasingly aware of the destructive impacts of conventional energy use on the environment and climate change, and looking for alternative sources of energy that are renewable and sustainable (Tilman *et al.*, 2009; Fargione *et al.*, 2010). Biofuels, compared with fossil fuels, could potentially support state energy goals, increase domestic energy supplies to reduce dependence on foreign oil and its potential disruptions, and yet reduce GHG emissions and other air pollutants (USDOE, 2011). In the United States, only about 10% of total primary energy consumption is from renewable energy sources, but biomass-derived energy makes up about half of the total renewable energy (EIA, 2012). Compared with some other renewable energy alternatives (e.g. wind, solar power), bioenergy may be one of the most viable options to adopt in the near term (USEPA, 2009).

To meet the mandate targets for biofuel production (US Congress, 2007), a large amount of land will be

Correspondence: Z Qin
e-mails: qin9@purdue.edu; qin.zhangcai@gmail.com

needed to grow energy crops for biomass feedstocks. Among lands that can be used for production of biofuel feedstocks, marginal lands were often introduced as a promising land option for energy cropping purpose, considering that switching food crops to biofuel crops to produce biomass on currently available croplands may raise concerns about food security, ethic issues, and unsustainable farming practices (Field *et al.*, 2008; Tilman *et al.*, 2009; Fargione *et al.*, 2010; Gramig *et al.*, 2013), while converting lands occupied by natural ecosystems (e.g. forest) to biofuel cropland could inevitably cause environmental and ecological problems such as deforestation, biodiversity loss, habitat fragmentation, and land use change induced GHG emissions (Searchinger *et al.*, 2008; Melillo *et al.*, 2009; Dauber *et al.*, 2010). Marginal land refers to those lands where a cost-effective production is not possible under given environmental conditions, cultivation techniques, agricultural management as well as other economic and legal conditions (Wiegmann *et al.*, 2008; Gopalakrishnan *et al.*, 2011), including lands such as idle or fallow cropland, abandoned or degraded cropland, and abandoned pastureland (Cai *et al.*, 2011; Gopalakrishnan *et al.*, 2011). Compared with cropland, marginal land normally has lower inherent agricultural productivity, due to its less fertile soils and often less favorable water, climate, and possibly other environmental conditions. However, certain energy crops with high resource-use-efficiencies are still capable of growing on these lands where traditional food crops may not thrive (Bandaru *et al.*, 2013; Gelfand *et al.*, 2013). For example, some perennial cellulosic crops, such as switchgrass and *Miscanthus*, could provide abundant biomass but require relatively less nutrient than food crops (Lewandowski *et al.*, 2003; Heaton *et al.*, 2004; Stewart *et al.*, 2009). These crops could therefore be used to grow biomass feedstock and produce cellulosic ethanol by using the less favored lands, and thus avoid competing with food crops for cropland (Bandaru *et al.*, 2013).

Field experiments suggested that, cellulosic energy crops or herbaceous vegetation, once well established, could produce considerable biomass feedstocks and have direct GHG emissions mitigation capacity that rivals that of conventional food crops. Switchgrass and *Miscanthus*, for example, can produce comparable or even higher biomass than traditionally used biofuel crop – maize (Fike *et al.*, 2006; Heaton *et al.*, 2008; Nikièma *et al.*, 2011). These perennial cellulosic crops normally have high conversion efficiency of photosynthetically active radiation and are able to enhance carbon (C) accumulation in a wide range of soil and climate conditions because of C4 metabolism (Heaton *et al.*, 2008). A considerable amount of C is assimilated and stored in the belowground biomass and soils, which fosters benefits for carbon dioxide (CO_2) sequestration (Don *et al.*, 2012; Monti *et al.*, 2012). In addition, cellulosic crops generally require only a very limited amount of nutrients (e.g. nitrogen fertilizer) due to their high nutrient-use efficiency, and therefore could possibly reduce fertilization induced nitrous oxide (N_2O) emissions (Lewandowski *et al.*, 2003; Monti *et al.*, 2012). Soil methane (CH_4) fluxes were negligible in these ecosystems (Drewer *et al.*, 2012). Gelfand *et al.* (2013) recently also reported in their comparative experiments that, if grown on marginal lands, successional herbaceous crops could still produce sizeable amounts of biomass and concurrently mitigate GHG emissions due to significant C sequestration in soils and reduction in N_2O emissions.

However, biomass productivity and GHG emissions regarding large-scale bioenergy expansion on marginal lands are rarely studied (Gelfand *et al.*, 2013). During the past several decades, modeling was used extensively to study regional or global scale C, nitrogen (N) dynamics, and GHG emissions of both natural (e.g. forest, grassland) and managed ecosystems (e.g. cropland) (Raich *et al.*, 1991; Bondeau *et al.*, 2007; Huang *et al.*, 2009). More recently, models were increasingly used to assess agroecosystems related to bioenergy crops, either by incorporating agricultural modules into natural ecosystem models, e.g. Agro-BGC(Di Vittorio *et al.*, 2010) and LPJml (Bondeau *et al.*, 2007), or by developing crop-specific models, e.g. ALMANAC (Kiniry *et al.*, 1992) and MISCANMOD (Clifton-brown *et al.*, 2004). These models can be applied to a large region to estimate biomass production or/and GHG emissions (Thomas *et al.*, 2013). As most previous modeling studies concentrated on the land use change due to conversion of natural ecosystems to agroecosystems, or crop switching from food crops to energy crops on cropland (Fargione *et al.*, 2008; Searchinger *et al.*, 2008; Melillo *et al.*, 2009), another land use scenario of growing energy crops on marginal lands was also important but less studied (Qin *et al.*, 2011; Gelfand *et al.*, 2013). Along with the biomass production, GHG emissions produced from or mitigated by marginal lands could significantly affect the total GHG budget in the lifecycle assessment of biofuel production, and therefore additional effort should be made to study potential C and N dynamics and GHG fluxes of these biofuel ecosystems. Here, we use a modeling approach to conduct such a study assuming switchgrass and *Miscanthus* grown on the marginal lands in the conterminous United States. The spatial estimates are made for biomass production, net carbon balance, nitrous oxide emissions, and therefore the total GHG emissions. Biofuel productivity, land use, and global warming potential are further analyzed at regional scales to meet the United States national biofuel mandate by year 2022.

Materials and methods

Energy crops

Switchgrass and *Miscanthus* were introduced as energy crops for biomass production purpose due to their considerable productivity and stress tolerance to unfavorable environments (McLaughlin & Adams Kszos, 2005; Heaton *et al.*, 2008). Switchgrass is a perennial cellulosic crop native to North America, with biomass productivity of 5–20 Mg (1 Mg = 1 t) dry matter (DM) per hectare land. It was widely tested for biomass production across the conterminous United States (Fike *et al.*, 2006; Heaton *et al.*, 2008; Wright & Turhollow, 2010). *Miscanthus* refers to a genus of several perennial grass species mostly native to the subtropical and tropical areas of Asia (Stewart *et al.*, 2009). Its yield could normally reach 20–30 Mg DM ha^{-1} if well cultivated (Heaton *et al.*, 2008). These two perennial crops could be potential biomass sources for cellulosic ethanol production. In this study, Switchgrass and *Miscanthus* are assumed to be grown on marginal lands in the United States to produce biofuel feedstocks.

Model description

AgTEM is a biogeochemical model designed for agroecosystems, by incorporating ecophysiological, biogeochemical, and management related processes into the framework of the Terrestrial Ecosystem Model (TEM) (Raich *et al.*, 1991; McGuire *et al.*, 1992; Zhuang *et al.*, 2003, 2010). The model can be used to simulate C and N dynamics of agroecosystems (Figure S1) at a daily time step, by using spatially explicit forcing data describing climate, soil, vegetation, and agronomic conditions (Qin *et al.*, 2013a,b).

In AgTEM, all algorithms related to C and N fluxes and pools are governed by five equations describing changes of ecosystem states regarding vegetation and soil (Qin *et al.*, 2013a). C cycling in the agroecosystems is modeled as following [Eqn (1)]: atmospheric CO_2 is preliminarily assimilated by plants through photosynthesis and stored in the vegetation. In the model, net primary production (NPP) is the rate at which the plants produce net useful chemical energy. It is the difference between the rate at which the plant produces useful chemical energy (GPP, gross primary production) and the rate at which some of that energy is used during autotrophic respiration. NPP represents the total available biomass of the ecosystem produced, which is partly harvested as harvestable biomass (HBIO), partly used during heterotrophic respiration and partly allocated to soil organic carbon (SOC) and belowground biomass (as in perennial crops). C of HBIO is eventually released as CO_2 through biofuel production and use. The net C balance in the ecosystem is modeled as net carbon exchange (NCE) which accounts for all C fluxes into or out of the system. A positive NCE indicates net ecosystem CO_2 sink while a negative value indicates a CO_2 source (Qin *et al.*, 2013a).

$$CO_2 \Rightarrow GPP \underbrace{\overset{CO_2\uparrow}{\Longrightarrow} NPP \overset{CO_2\uparrow}{\Longrightarrow} \left\{ \begin{array}{l} HBIO \overset{CO_2\uparrow}{\Longrightarrow} \\ \overset{CO_2\uparrow}{\Longrightarrow} C_{SOC} \end{array} \right.}_{NCE} \tag{1}$$

Modeled N_2O accounts for soil N_2O fluxes from both nitrification and denitrification, as in [Eqn (2)] (Qin *et al.*, 2013a):

$$\left. \begin{array}{l} NH_4^+ \overset{NO}{\Longrightarrow} N_2O_{ntf} \\ NO_3^- \overset{N_2}{\Longrightarrow} N_2O_{dtf} \end{array} \right\} \Longrightarrow N_2O \tag{2}$$

where N_2O_{ntf} is N_2O produced from the nitrification process of the biological oxidation of ammonia (NH_4^+) with oxygen, and N_2O_{dtf} is N_2O produced from soil nitrate (NO_3^-) through denitrification process; N2O is the total N_2O fluxes of N_2O_{ntf} and N_2O_{dtf}. Nitric oxide (NO) and nitrogen (N_2) are also produced from the processes of nitrification and denitrification, respectively, but they are not quantified in this study.

The original version of AgTEM 1.0 was calibrated at site levels and applied at regional scales to assess regional C dynamics (Qin *et al.*, 2011), biomass production (Qin *et al.*, 2012), and water balance (Zhuang *et al.*, 2013). The further developed AgTEM 2.0 incorporated processes such as biomass allocation, N cycling and agricultural management (Qin *et al.*, 2013a). In the model, most parameters describing and constraining generic ecosystem processes were either inherited from TEM or predefined in previous studies (e.g. Zhuang *et al.*, 2003, 2010; Qin *et al.*, 2011, 2012). However, some additional vegetation-specific or soil-specific parameters were defined and calibrated for certain ecosystems or processes not previously included in the model. For example, the temperature threshold parameters were determined separately for switchgrass and *Miscanthus* to describe plant photosynthesis and crop phenology. Many additional variables and parameters were included in AgTEM 2.0 to represent nitrification and denitrification processes (Qin *et al.*, 2013a). The parameterized and calibrated model was then used to estimate site-level biomass and N_2O emissions, and they were validated against field observations. The results suggested that the AgTEM 2.0 well reproduced the observations (Qin *et al.*, 2013a) and can be applied to region-level estimations (Qin *et al.*, 2013b). More information concerning AgTEM can be found in previous studies (e.g. Qin *et al.*, 2011, 2013a,b). In this study, the AgTEM 2.0 was used.

Model simulations and regional analyses

By assuming that switchgrass and *Miscanthus* will be grown on available marginal lands in the conterminous United States (Figure S2), we applied the AgTEM 2.0 separately for these two crop systems, to simulate ecosystem C and N dynamics along with crop growth, using spatially referenced data describing climate, soil, vegetation, atmospheric CO_2, and agricultural management. Model estimates were then used to assess spatial distribution of output variables of interest, including NPP, HBIO, NCE, and N_2O. Spatial analyses were finally conducted to estimate spatial and national biomass/biofuel production, CO_2 mitigation, N_2O emissions, and total GHG emissions.

For spatial simulations, model was run grid-by-grid to estimate C and N dynamics at a daily time step with available forcing data from 1989 to 2008. First, we initialized the model by running AgTEM to equilibrium using the first year data. The model was then spun up for 100 years repeatedly using the first 10 years' data to reach equilibrium state. We then ran the transient simulations continuously from 1989 to 2008 using transient forcing data. Spatial forcing data were organized at a 0.25° latitude × 0.25° longitude resolution for the study region. Specifically, climate data describing temperature, precipitation, cloudiness were obtained from the ECMWF (European Centre for Medium-Range Weather Forecasts) Data Server (www.ecmwf.int) and organized at a temporal resolution of 1 day from 1989 to 2008. Annual atmospheric CO_2 concentrations were collected from the NOAA Mauna Loa CO_2 record (www.esrl.noaa.gov/gmd/ccgg/trends/). The elevation data were derived from the Shuttle Radar Topography Mission (SRTM) (Farr *et al.*, 2007) and soil texture data were based on the Food and Agriculture Organization/Civil Service Reform Committee (FAO/CSRC) digitization of the FAO/UNESCO soil map of the World (1971). N fertilization was set at four input rates as 0 (N0), 50 (N1), 100 (N2), and 150 kg N ha^{-1} (N3) for both switchgrass and *Miscanthus* systems, according to field experiments (Fike *et al.*, 2006; Heaton *et al.*, 2008; Propheter *et al.*, 2010; Nikièma *et al.*, 2011). In Cai *et al.*'s (2011) study, global marginal lands were identified according to marginal agricultural productivity based on land suitability indicators such as topography, climate conditions, and soil productivity. The scenario 1 in Cai *et al.* (2011) includes marginal lands from abandoned land and mixed crop and vegetation land, and yet without sacrificing large amounts of cropland and natural lands (forest and grassland) (Figure S2). This scenario was considered as initial land use condition for the modeling purpose in this study, to represent the spatial distribution of marginal lands in the United States. The data in Cai *et al.* (2011) were reorganized at a 0.25° latitude × 0.25° longitude resolution according to the proportion of marginal lands in each pixel.

Spatial analyses were conducted for each crop ecosystem based on model simulations, using geographic information system techniques. Regional analyses based on grid outputs were presented as average of the 1990s. NPP and HBIO were computed for both spatial and national levels as primary and harvested biomass production, respectively. Using biomass-to-biofuel conversion efficiencies, biofuel production was further calculated from HBIO results. Under current technologies, the efficiency of converting biomass-to-biofuel is estimated to be about 282 l ethanol Mg^{-1} DM (Lynd *et al.*, 2008). The potential efficiency could reach about 399 l ethanol Mg^{-1} DM if advanced technologies would be available (Lynd *et al.*, 2008). Net CO_2 balances (NCE) and total N_2O emissions (N2O) were also computed to estimate spatial and national GHG emissions in terms of global warming potential (GWP). The GWP of N_2O was calculated in units of CO_2 equivalent (CO_2eq) over a 100-year time horizon. In addition, GWP was related to energy production by computing global warming intensity (GWP_i) in terms of total GWP relative to biofuel production (Qin *et al.*, 2013b).

Results

Biomass and biofuel production on marginal lands

With increasing use of N fertilizer, the biomass production at ecosystem scale also increases, in both switchgrass and *Miscanthus* ecosystems (Fig. 1). At N0 level, the switchgrass produces NPP (harvest-area weighted) of less than 400 g C m^{-2} in most areas (Fig. 1a). With N addition, the NPP production increases dramatically, especially in those areas with intense cropping, e.g. Wisconsin (Fig. 1b–d). When the N rate reaches N2 (Fig. 1c) and N3 (Fig. 1d) levels, most of the southern areas have NPP of 400–800 g C m^{-2}. In terms of biomass harvested (Table 1), switchgrass produces a national average of 3.5 Mg DM ha^{-1} each year without N application, with additional 1.4 Mg DM ha^{-1} if applied 50 kg N ha^{-1} (N1). The average HBIO could reach 5.7–5.9 Mg DM ha^{-1} with sufficient N fertilizer. *Miscanthus* generally has higher biomass productivity than corresponding switchgrass at the same N application levels (Fig. 1e–h). Without N application, the NPP reaches over 600 g C m^{-2} in most intense cropping areas (Fig. 1e), with a national average HBIO production of about 10 Mg DM ha^{-1} (Table 1). With each additional kg of N application, the *Miscanthus* HBIO increases about 50 kg DM ha^{-1} each year on average, with highest increase of 64 kg DM ha^{-1} from N0 to N1 level and lowest increase of 28 at DM ha^{-1} from N2 to N3 level. When the N rate reaches N3, *Miscanthus* produces the highest HBIO of 17.2 Mg DM ha^{-1}, which almost triples the switchgrass production (Table 1).

Production of cellulosic ethanol using the harvested biomass is highly dependent on biomass-to-biofuel conversion technologies (Table 1). Under currently available technology, switchgrass could produce about 1.0–1.7 kl ethanol from each hectare of marginal land, depending on N application and biomass production. *Miscanthus*, however, could produce 2.9–4.9 kl ethanol ha^{-1} land due to its high biomass productivity. With advanced technology available, the biofuel conversion efficiency could increase by 41.5%. Switchgrass harvested from marginal lands could therefore produce 1.4–2.3 kl ethanol ha^{-1} land and productive *Miscanthus* could produce 4.1–6.9 kl ethanol ha^{-1} land. Generally, with advanced technology and application of high-rate N fertilizer, cellulosic crops grown on marginal lands could have a considerably higher land use efficiency, in terms of biofuel production on given land, than otherwise with current technology and less use of N. *Miscanthus*, in particular, has a higher land use efficiency than switchgrass at each technology × N application level scenario.

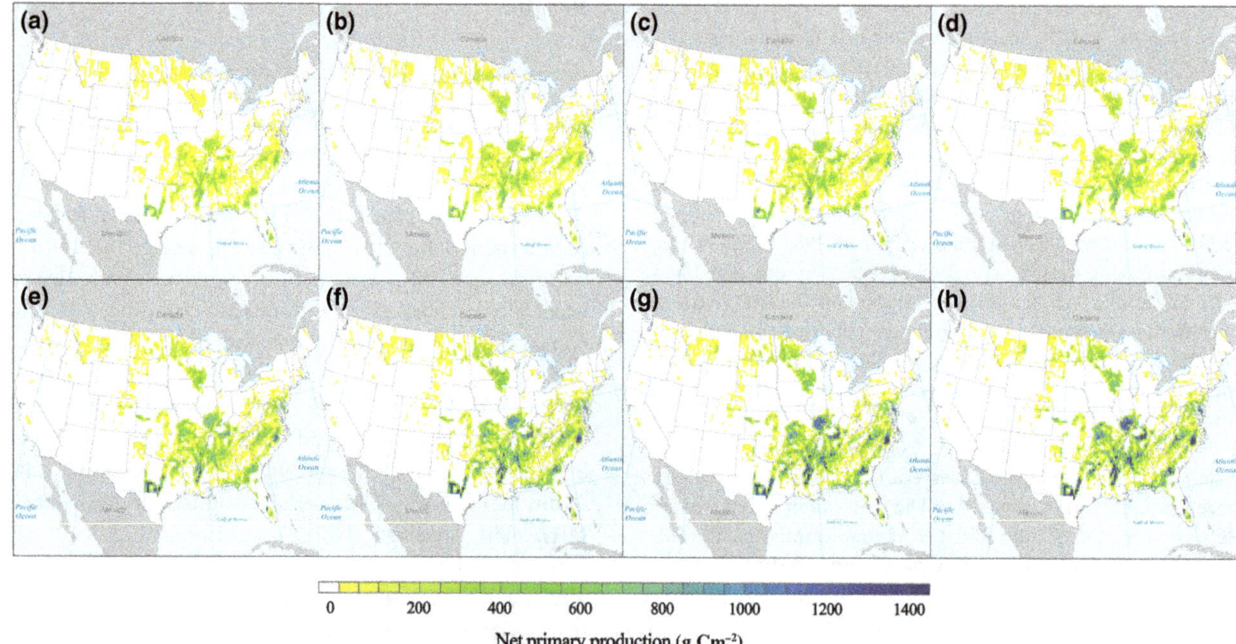

Fig. 1 Modeled net primary production from marginal lands. Area weighted estimates were made for switchgrass grown under nitrogen application levels of (a) N0, (b) N1, (c) N2, and (d) N3, and *Miscanthus* grown under (e) N0, (f) N1, (g) N2, and (h) N3.

Table 1 Estimated harvestable biomass and biofuel production from energy crops grown on marginal lands under different nitrogen application scenarios

Energy crops	Nitrogen application*	Estimated harvestable biomass production (Mg DM ha^{-1} land)	Estimated biofuel production (kl ethanol ha^{-1} land)	
			Current level†	Potential level‡
Swithchgrass	N0	3.5 (0.3)	1.0 (0.1)	1.4 (0.1)
	N1	4.9 (0.5)	1.4 (0.1)	1.9 (0.2)
	N2	5.7 (0.6)	1.6 (0.2)	2.3 (0.2)
	N3	5.9 (0.6)	1.7 (0.2)	2.3 (0.2)
Miscanthus	N0	10.2 (1.0)	2.9 (0.3)	4.1 (0.4)
	N1	13.4 (1.3)	3.8 (0.4)	5.3 (0.5)
	N2	15.8 (1.7)	4.5 (0.5)	6.3 (0.7)
	N3	17.2 (2.0)	4.9 (0.6)	6.9 (0.8)

*Nitrogen fertilization was set at four input rates as 0 (N0), 50 (N1), 100 (N2), and 150 kg N ha^{-1} (N3).

†Current and ‡potential levels of biofuel production are estimated based on current and potential biomass-to-biofuel conversion efficiencies, respectively (Lynd *et al.*, 2008). Values were averaged for the 1990s, with standard deviation in parentheses.

Greenhouse gas emissions in bioenergy ecosystems

GHG emissions (in terms of GWP) are determined by the effects of both ecosystem CO_2 and N_2O emissions. Our model experiments indicate that most of the cropping areas in the southern United States act as net sources of GHG emissions, and the estimated *Miscanthus* GWP (Fig. 2e–h) has a much higher variation than the corresponding switchgrass GWP (Fig. 2a–d) at any specific location. Specifically, in the switchgrass cropping systems, with increasing use of N fertilizer, the GHG emissions increase markedly especially in the intense cropping areas in the middle United States (Fig. 2a–d). For example, after increasing use of N, net GHG sinks in some areas become GHG sources, e.g. Texas (Fig. 2a, b), and some GHG sources become even larger sources, e.g. South Illinois (Fig. 2b, c). In the *Miscanthus* systems, however, the GHG emissions do not necessarily increase with increasing use of N (Fig. 2e–h). It is evident that, for those areas that are already GHG sources without N

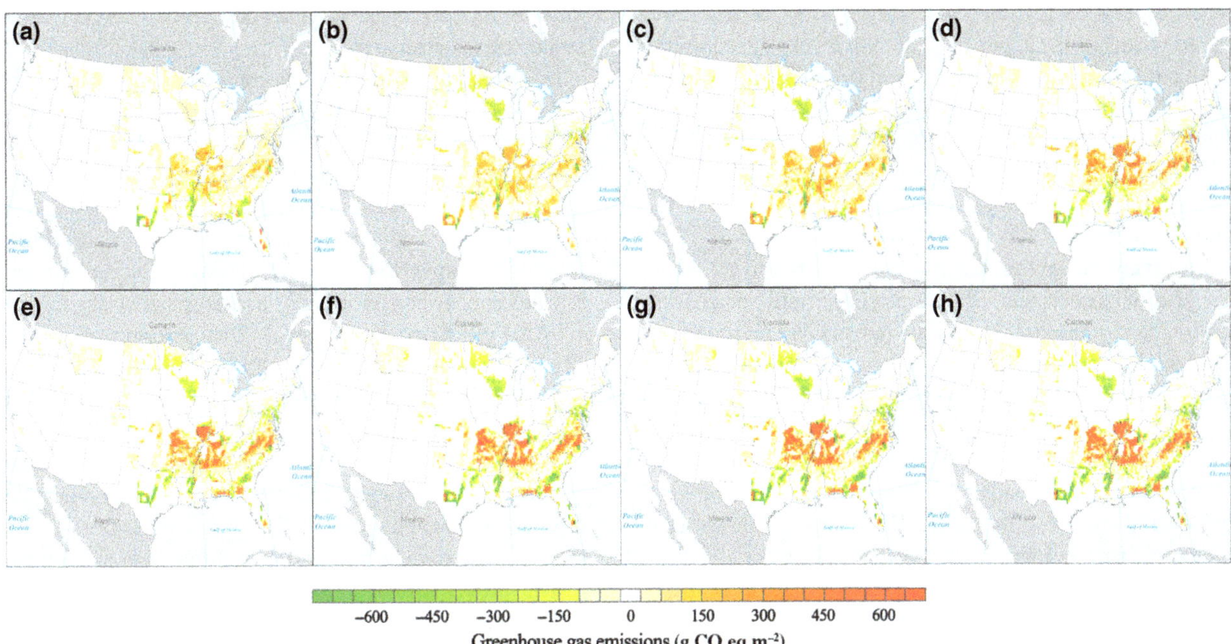

Fig. 2 Modeled GHG emissions from marginal lands. Maps show area weighted total emissions of CO_2 and N_2O (GWP) for switchgrass grown under nitrogen application levels of (a) N0, (b) N1, (c) N2, and (d) N3, and *Miscanthus* grown under (e) N0, (f) N1, (g) N2, and (h) N3. A positive value indicates a net GHG sink while a negative value indicates a net GHG source.

fertilization, e.g. Missouri, Kentucky and Tennessee in the middle of the United States (Fig. 2e), the net GWP tends to be larger after use of N fertilizer (Fig. 2f–h); but for the areas that are originally GHG sinks, e.g. Texas and Louisiana in the South United States (Fig. 2e), their GWP become even smaller, suggesting these areas become even larger GHG sinks.

From the perspective of national average GHG emissions, the changes of net GWP are simply the results of GWP changes in both CO_2 and N_2O. Both ecosystems act as GHG sources at national level and at all N application levels (Fig. 3a). Switchgrass and *Miscanthus* have a similar amount of N_2O emissions at each N application rate, and even similar C sinks at lower N rates (N0, N1). But *Miscanthus* has a much larger C sink than switchgrass at higher N rates (N2, N3). For instance, in the switchgrass systems, with increasing use of N, both N_2O emissions and CO_2 mitigation increase, but the former has a relatively larger value than the latter, resulting in a net source of GHG emissions. This is especially true when the N rate reaches N2 and N3 levels and where the total GHG emissions

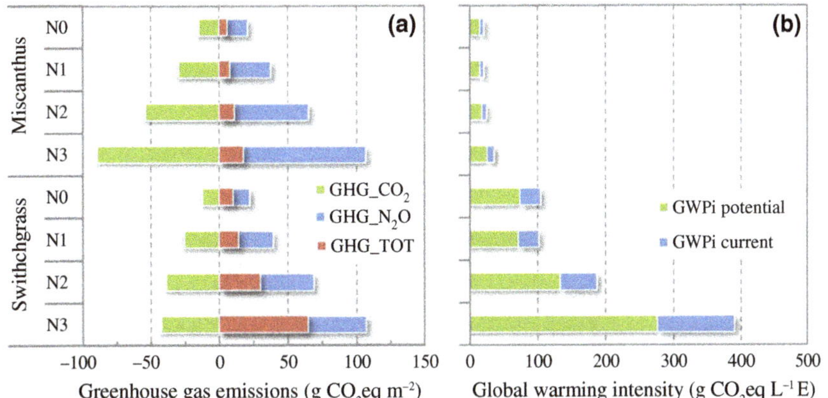

Fig. 3 National average GHG emissions from switchgrass and *Miscanthus* grown on marginal lands. (a) Contributions of CO_2 (GHG_CO2) and N_2O (GHG_N2O) to total GHG emissions (GHG_TOT) under different nitrogen application levels; (b) global warming intensity (GWPi), in terms of GWP relative to ethanol (E) production under current or potential conversion efficiencies.

reach 30 and 65 g CO_2eq m^{-2} respectively, when compared with 10 g CO_2eq m^{-2} at N0 level. By contrast, N_2O emissions and CO_2 mitigation do not change much in *Miscanthus* systems, even when the N rate increases. For example, the GWP (N0) of CO_2 and N_2O are −15 and 20 g CO_2eq m^{-2} respectively, making the net ecosystem GHG emission only 6 g CO_2eq m^{-2}. The GWP of CO_2 and N_2O reaches up to −89 and 107 g CO_2eq m^{-2}, respectively, when the N application gets to the N3 level, but the net ecosystem GHG emission is still only 18 g CO_2eq m^{-2} – about 27% of switchgrass GWP at the same N level. For these cellulosic systems, the emitted N_2O-N accounts for 1.38–1.68% of N applied, which is slightly higher than the IPCC reported default emission factors with a total of 1.325% (IPCC, 2006).

By relating GHG emissions to biofuel production, our model results show that, *Miscanthus* has much smaller global warming intensities than switchgrass, at all N application levels (Fig. 3b). Under currently available technologies, for each liter of ethanol produced, the *Miscanthus* system releases 21–36 g CO_2eq of GHG; with increasing N application, the GWPi also increases. Switchgrass, however, releases much more GHG per unit biofuel than *Miscanthus*, with lowest GWPi of about 100 g CO_2eq l^{-1} at N0 and N1 levels and highest GWPi of 390 g CO_2eq l^{-1} at N3 level. To produce the same amount of ethanol, the switchgrass systems on average release 4–10 times more GHG than *Miscanthus* systems. With advanced conversion technology, the GWPi can be lowered for both systems by reducing about 40% GHG release relative to current GWPi levels. But still, *Miscanthus* ecosystem has significantly lower GWPi than switchgrass ecosystem.

By considering the energy content in ethanol, which is 76 330 British thermal units per gallon (Btu Gal^{-1}) or 21.29 Megajoule per liter (MJ l^{-1}) (GREET, 2012), the global warming intensity can be translated into energy-based GHG emissions which can be further used to calculate carbon intensity in life cycle assessment (LCA) (Table 2). Depending on agricultural management and biofuel conversion efficiency, switchgrass could release 1.4–13.0 g CO_2eq for each MJ of energy produced. *Miscanthus,* however, has a relatively lower GHG intensity, ranging from 0.3 g CO_2eq MJ^{-1} under low N input and high biofuel conversion efficiency to 1.2 g CO_2eq MJ^{-1} under N3 input level and current biofuel conversion efficiency (Table 2). If we assume marginal land to be carbon neutral, the GHG emissions due to cropping of switchgrass and *Miscanthus* would result in a net carbon source in the marginal land. For switchgrass, especially when grown with high N input in the marginal land, the GHG intensity could exceed the earlier estimates for land use changes from cropland, grassland, and forest to cropping switchgrass (Dunn *et al.*, 2013; Elliott *et al.*, 2014) (Table 2).

Discussion

Cellulosic crops as biomass feedstocks

Cellulosic crops, such as switchgrass and *Miscanthus,* normally have higher nutrient-use efficiency

Table 2 Estimated GHG emissions of switchgrass- and *Miscanthus*-based ethanol in ecosystem modeling and life cycle assessment

References	Scenarios	Swtichgrass	*Miscanthus*
		g CO_2eq MJ^{-1}	
Ecosystem modeling			
This study*	N0 input (0 kg N ha^{-1})	1.4~3.5	0.3~0.7
	N1 input (50 kg N ha^{-1})	1.4~3.4	0.3~0.7
	N2 input (100 kg N ha^{-1})	2.6~6.2	0.3~0.8
	N3 input (150 kg N ha^{-1})	5.4~13.0	0.5~1.2
Life cycle assessment†			
Dunn *et al.*, 2013;	Domestic land use change‡	−3.9~13	−12~−3.8
	Total land use change¶	2.7~19	−10~−2.1
Elliott *et al.*, 2014;	Direct land use change‡	−0.13~0.21	0.89~2.35
	Total land use change¶	0.47~3.03	1.28~3.18

*The GHG emissions depend on technology levels, with a lower bound under advanced technology and an upper bound under current technology.

†LCA can estimate both domestic/direct and international/indirect land use change impacts on GHG emissions, and it considers all possible land conversion types.

‡Domestic/Direct land use change refers to conversions due to biofuel cropping within the United States. Cropland, grassland, and forest were normally considered for land use change.

¶Total land use change includes domestic/direct land use change, and international/indirect land use change which refers to land conversions occurred outside the United States because of biofuel cropping in the United States.

(Lewandowski *et al.*, 2003; Fargione *et al.*, 2010) and possibly higher water use efficiency than food crops (Stewart *et al.*, 2009; Zhuang *et al.*, 2013). They could therefore grow on marginal lands instead of competing with food crops for fertile croplands. However, the results here and elsewhere (Gelfand *et al.*, 2013) also show that biomass production from marginal lands may be lower than that from croplands. Our previous studies suggested that, an average of about 5–8 Mg DM ha^{-1} of switchgrass or around 20 Mg DM ha^{-1} of *Miscanthus* could be produced from cropland (Qin *et al.*, 2013b), which is higher than those grown on marginal lands even with high N input (Table 1). This may be partly because that besides nutrient (e.g. N) other factors could also affect biomass production on marginal lands, for example, water availability, climate conditions, and soil fertility (Cai *et al.*, 2011).

N application affects not only biomass production but also the ecosystem GHG emissions. One hand, use of N fertilizer could improve soil nutrient condition and therefore stimulate crop growth. With an increasing rate of N application, for each unit of N use, biomass production increment decreases gradually (Fig. 4a, c). Marginal HBIO production, the change in HBIO arising with each unit change in N input, d(HBIO)/d(N), decreases with N addition (Fig. 4b, d). On the other hand, increasing use of N leads to more N losses through gaseous emissions, leaching, and runoff. With increasing N application, the GHG release also increases (Fig. 4a, c), the marginal GHG emissions i.e. change of GHG arising with each unit change in N input, d(GHG)/d(N) increase with N addition (Fig. 4b, d). It is therefore very important to analyze how N use affects the benefits (e.g. biomass or biofuel production) and costs (e.g. GHG emissions) in marginal lands in our future studies.

Land use and GHG emissions regarding 2022 biofuel target

The Energy Independence and Security Act (US Congress, 2007) established a target of 136 billion liters (36 billion gallons) of renewable fuels in the United States by 2022, including 79 billion liters (21 billion gallons) of cellulosic ethanol. To reach the cellulosic ethanol target, a total of about 280 million tons of cellulosic biomass will be required under current biofuel conversion technology. If switchgrass was to be grown on marginal lands for biofuel feedstocks, a total of 48–81 Mha of land would be required (Fig. 5). According to estimates made by Cai *et al.* (2011), large areas of cropland or natural ecosystems might have to be sacrificed for this purpose. In addition, 8–31 Tg CO_2eq of GHG would be released due to cropping, depending on N input levels (Fig. 5). However, if *Miscanthus* were grown, a large quantity of land could be saved compared with growing switchgrass, only 16–28 Mha of available marginal

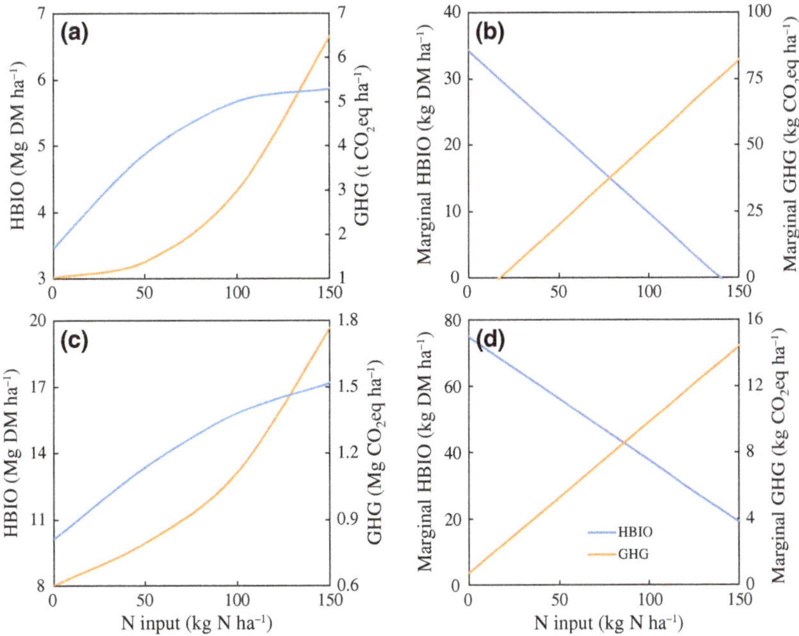

Fig. 4 Modeled change in HBIO and GHG from marginal lands with increasing use of N. Estimates for HBIO and GHG of (a) switchgrass and (b) *Miscanthus* were based on national average results (Table 1); marginal HBIO and marginal GHG of (c) switchgrass and (d) *Miscanthus* were based on polynomial (order 2) relationships between HBIO or GHG and N input.

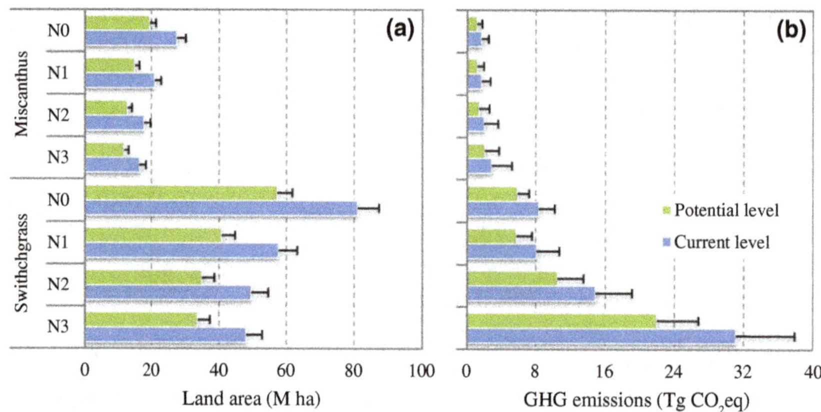

Fig. 5 Estimated demand of marginal lands and GHG emissions to achieve the 2022 biofuel mandate of 79 billion liters of cellulosic ethanol. Model estimates of (a) land demand and (b) GHG emissions were made for switchgrass and *Miscanthus* under current and potential biofuel conversion efficiencies.

lands could be sufficient to produce the required biofuel feedstocks. More importantly, using *Miscanthus* could reduce a considerable amount of GHG emissions; only a total of 1.7–2.9 Tg CO_2eq of GHG would be released from the ecosystem to meet the 2022 target (Fig. 5).

If biofuel conversion efficiency could be improved, i.e. from 282 to 399 l ethanol Mg DM (Lynd *et al.*, 2008), the biomass demand would be dramatically reduced to 200 million ton of dry matter. The land demand and GHG emissions could also be reduced to 71% of those under current technology, for both switchgrass and *Miscanthus* systems. Considering biofuel productivity alone, *Miscanthus* grown under N3 level has the highest land use efficiency. Under this scenario, only 11.6 Mha of marginal lands will serve the purpose of producing 79 million liters of ethanol (Fig. 5). However, if minimizing GHG emissions is the primary concern, then *Miscanthus* grown under the N0 level releases the smallest amount of GHG of just 1.2 Tg CO_2eq, but yet requires 19.6 Mha of land (Fig. 5).

By comparing with previous estimates for biomass produced from cropland (Qin *et al.*, 2012, 2013b), we find that, cellulosic crops have lower productivity grown on marginal lands, and therefore require relatively more land to reach the 2022 target, than if they were grown on fertile cropland. However, compared with maize grown on cropland (Qin *et al.*, 2012, 2013b), marginal land – based *Miscanthus* requires comparable or even less land resources and releases remarkably less amount of GHG, irrespective of N application and technology.

Limitations and future needs

Modeling studies are often limited by data availability and model deficiencies. In this study, data of climate, soil, and vegetation were used to initialize the model and make regional estimates. Most of these data (e.g. temperature, precipitation) are derived or reanalyzed from site/field observations, which inevitably introduce uncertainties into the spatially referenced model simulations due to observation errors, spatial heterogeneity, and possible interpretation biases (Huang *et al.*, 2009; Melillo *et al.*, 2009). In particular, due to lack of spatial data, the fertilization rate was assumed to be constant throughout the whole United States. Even with several different N rates (N0–N3), the fertilization scenario may not necessarily reflect real management practices, mainly because soil fertility is spatially heterogeneous and the fertilization rate can be adjusted accordingly. Also, the N fertilization impacts on N allocation (e.g. Guretzky *et al.*, 2011) and possible N fixation of certain crops (e.g. Davis *et al.*, 2010) are still open to discussion; our modeling experiments did not fully consider these issues due to insufficient mechanism understanding and data unavailability for certain cellulosic crop systems. Another major uncertainty regarding marginal land distribution should be further examined when data are available. This study did not consider crop-specific environmental constraints, such as possible water and temperature limitations, economic profitability, agronomic practicality as well as societal concerns about switchgrass and *Miscanthus*. It is very likely that certain regions may not be suitable for growing switchgrass or *Miscanthus* in the first place. Therefore, it is important to recognize that the land use scenarios considered here did not suggest actual land conversion practices. Crop-specific marginal land distribution and classification data should be developed to assist future modeling decisions with regard to land availability.

In addition, the AgTEM version used in this study did not specifically model the cropping impacts on other environmental factors, especially water quantity (Le *et al.*, 2011) and quality (Ng *et al.*, 2010). As observational and spatial data become available and our understanding regarding bioenergy ecosystems advances, we shall factor these components into AgTEM modeling and regional analysis with higher accuracy. Soil C fluxes were considered as part of the ecosystem C cycling but the SOC was not specifically reported in this study. When more observational data are available to support prediction of SOC under various types of marginal land, we shall further estimate SOC dynamics due to land use change.

As for cost-benefit analysis of energetic, environmental, and economic aspects regarding large-scale bioenergy development (e.g. Hill *et al.*, 2006), LCA will be needed to account for energy system processes along with cellulosic ethanol's life 'from-cradle-to-grave' (e.g. Davis *et al.*, 2009; Scown *et al.*, 2012). The ecosystem analysis in this study is only one segment of the whole LCA chain, and estimates only those processes occurring inside specific ecosystems. As shown in Table 2, the estimated GHG intensity only assesses those GHG emissions during crop growth and harvest, irrespective of previous land use type. The LCA, however, has much broader system boundaries. Besides ecosystem, the LCA also assesses GHG emissions from other system processes, such as transportation, manufacturing, and biofuel use. Even for the assessment of GHG emissions in ecosystem, the LCA analyses could have their specific definitions and boundaries. For example, many LCA studies (e.g. Dunn *et al.*, 2013; Elliott *et al.*, 2014) included GHG emissions induced by land use change (e.g. converting cropland, grassland, or forest to bioenergy crops), which could assess possible C pool changes during land conversion. However, the LCA may not necessarily consider net change in C and/or N pools and fluxes in current (e.g. the estimates in this study) and previous ecosystems. Therefore, caution should be exercised when using the ecosystem modeling results in LCA assessment; especially the system boundary should be clarified to match ecosystem models with LCA processes.

Acknowledgement

The authors thank Dr. Jennifer Dunn and anonymous reviewers for insightful comments on an earlier version of this manuscript. Computing was supported by Rosen Center for Advanced Computing (RCAC) at Purdue University. This study is supported through projects funded by the NASA Land Use and Land Cover Change program (NASA-NNX09AI26G), Department of Energy (DE-FG02-08ER64599), the NSF Division of Information & Intelligent Systems (NSF-1028291), and the NSF Carbon and Water in the Earth Program (NSF-0630319).

References

Bandaru V, Izaurralde RC, Manowitz D, Link R, Zhang X, Post WM (2013) Soil carbon change and net energy associated with biofuel production on marginal lands: a regional modeling perspective. *Journal of Environment Quality*, **42**, 1802.

Bondeau A, Smith PC, Zaehle S *et al.* (2007) Modelling the role of agriculture for the 20th century global terrestrial carbon balance. *Global Change Biology*, **13**, 679–706.

Cai X, Zhang X, Wang D (2011) Land availability for biofuel production. *Environmental Science & Technology*, **45**, 334–339.

Clifton-brown JC, Stampfl PF, Jones MB (2004) *Miscanthus* biomass production for energy in Europe and its potential contribution to decreasing fossil fuel carbon emissions. *Global Change Biology*, **10**, 509–518.

Dauber J, Jones MB, Stout JC (2010) The impact of biomass crop cultivation on temperate biodiversity. *GCB Bioenergy*, **2**, 289–309.

Davis SC, Anderson-Teixeira KJ, DeLucia EH (2009) Life-cycle analysis and the ecology of biofuels. *Trends in Plant Science*, **14**, 140–146.

Davis SC, Parton WJ, Dohleman FG, Smith CM, Del Grosso S, Kent AD, DeLucia EH (2010) Comparative biogeochemical cycles of bioenergy crops reveal nitrogen-fixation and low greenhouse gas emissions in a Miscanthus × giganteus agro-ecosystem. *Ecosystems*, **13**, 144–156.

Di Vittorio AV, Anderson RS, White JD, Miller NL, Running SW (2010) Development and optimization of an Agro-BGC ecosystem model for C4 perennial grasses. *Ecological Modelling*, **221**, 2038–2053.

Don A, Osborne B, Hastings A *et al.* (2012) Land-use change to bioenergy production in Europe: implications for the greenhouse gas balance and soil carbon. *GCB Bioenergy*, **4**, 372–391.

Drewer J, Finch JW, Lloyd CR, Baggs EM, Skiba U (2012) How do soil emissions of N2O, CH4 and CO2 from perennial bioenergy crops differ from arable annual crops? *GCB Bioenergy*, **4**, 408–419.

Dunn JB, Mueller S, Kwon H, Wang MQ (2013) Land-use change and greenhouse gas emissions from corn and cellulosic ethanol. *Biotechnology for Biofuels*, **6**, 51.

EIA (US Energy Information Administration) (2012) Gasoline explained, use of gasoline. Available at: www.eia.gov/energyexplained/index.cfm?page=gasoline_use (accessed January 2013).

Elliott J, Sharma B, Best N *et al.* (2014) A spatial modeling framework to evaluate domestic biofuel-induced potential land use changes and emissions. *Environmental Science & Technology*, **48**, 2488–2496.

Fargione J, Hill J, Tilman D, Polasky S, Hawthorne P (2008) Land clearing and the biofuel carbon debt. *Science*, **319**, 1235–1238.

Fargione JE, Plevin RJ, Hill JD (2010) The ecological impact of biofuels. *Annual Review of Ecology, Evolution, and Systematics*, **41**, 351–377.

Farr TG, Rosen PA, Caro E *et al.* (2007) The shuttle radar topography mission. *Reviews of Geophysics*, **45**, RG2004.

Field CB, Campbell JE, Lobell DB (2008) Biomass energy: the scale of the potential resource. *Trends in Ecology & Evolution*, **23**, 65–72.

Fike JH, Parrish DJ, Wolf DD, Balasko JA, Green JT Jr, Rasnake M, Reynolds JH (2006) Long-term yield potential of switchgrass-for-biofuel systems. *Biomass and Bioenergy*, **30**, 198–206.

Gelfand I, Sahajpal R, Zhang X, Izaurralde RC, Gross KL, Robertson GP (2013) Sustainable bioenergy production from marginal lands in the US Midwest. *Nature*, **493**, 514–517.

Gopalakrishnan G, Cristina Negri M, Snyder SW (2011) A novel framework to classify marginal land for sustainable biomass feedstock production. *Journal of Environment Quality*, **40**, 1593.

Gramig BM, Reeling CJ, Cibin R, Chaubey I (2013) Environmental and economic trade-offs in a watershed when using corn stover for bioenergy. *Environmental Science & Technology*, **47**, 1784–1791.

GREET (2012) The greenhouse gases, regulated emissions, and energy use in transportation model (GREET). Available at: greet.es.anl.gov (accessed January 2014).

Guretzky JA, Biermacher JT, Cook BJ, Kering MK, Mosali J (2011) Switchgrass for forage and bioenergy: harvest and nitrogen rate effects on biomass yields and nutrient composition. *Plant and Soil*, **339**, 69–81.

Heaton E, Voigt T, Long SP (2004) A quantitative review comparing the yields of two candidate C4 perennial biomass crops in relation to nitrogen, temperature and water. *Biomass and Bioenergy*, **27**, 21–30.

Heaton EA, Dohleman FG, Long SP (2008) Meeting US biofuel goals with less land: the potential of *Miscanthus*. *Global Change Biology*, **14**, 2000–2014.

Hill J, Nelson E, Tilman D, Polasky S, Tiffany D (2006) Environmental, economic, and energetic costs and benefits of biodiesel and ethanol biofuels. *Proceedings of the National Academy of Sciences*, **103**, 11206–11210.

Hill J, Polasky S, Nelson E *et al.* (2009) Climate change and health costs of air emissions from biofuels and gasoline. *Proceedings of the National Academy of Sciences*, **106**, 2077–2082.

Huang Y, Yu Y, Zhang W *et al.* (2009) Agro-C: a biogeophysical model for simulating the carbon budget of agroecosystems. *Agricultural and Forest Meteorology*, **149**, 106–129.

IPCC (Intergovernmental Panel on Climate Change) (2006) *2006 IPCC Guidelines for National Greenhouse Gas Inventories*, (eds Eggleston HS, Buendia L, Miwa K, Ngara T, Tanabe K), pp. 11.1–11.54. IGES, Japan.

Kiniry JR, Williams JR, Gassman PW, Debaeke P (1992) A general, process-oriented model for two competing plant species. *Transactions of the ASAE (USA)*, **35**, 801–810.

Le PVV, Kumar P, Drewry DT (2011) Implications for the hydrologic cycle under climate change due to the expansion of bioenergy crops in the midwestern United States. *Proceedings of the National Academy of Sciences*, **108**, 15085–15090.

Lewandowski I, Scurlock JMO, Lindvall E, Christou M (2003) The development and current status of perennial rhizomatous grasses as energy crops in the US and Europe. *Biomass and Bioenergy*, **25**, 335–361.

Lynd LR, Laser MS, Bransby D *et al.* (2008) How biotech can transform biofuels. *Nature Biotechnology*, **26**, 169–172.

McGuire AD, Melillo JM, Joyce LA, Kicklighter DW, Grace AL, Moore B, Vorosmarty CJ (1992) Interactions between carbon and nitrogen dynamics in estimating net primary productivity for potential vegetation in North America. *Global Biogeochemical Cycles*, **6**, 101–124.

McLaughlin SB, Adams Kszos L (2005) Development of switchgrass (Panicum virgatum) as a bioenergy feedstock in the United States. *Biomass and Bioenergy*, **28**, 515–535.

Melillo JM, Reilly JM, Kicklighter DW *et al.* (2009) Indirect emissions from biofuels: how important? *Science*, **326**, 1397–1399.

Monti A, Barbanti L, Zatta A, Zegada-Lizarazu W (2012) The contribution of switchgrass in reducing GHG emissions. *GCB Bioenergy*, **4**, 420–434.

Ng TL, Eheart JW, Cai X, Miguez F (2010) Modeling *Miscanthus* in the soil and water assessment tool (SWAT) to simulate its water quality effects as a bioenergy crop. *Environmental Science & Technology*, **44**, 7138–7144.

Nikièma P, Rothstein DE, Min D-H, Kapp CJ (2011) Nitrogen fertilization of switchgrass increases biomass yield and improves net greenhouse gas balance in northern Michigan, U.S.A. *Biomass and Bioenergy*, **35**, 4356–4367.

Propheter JL, Staggenborg SA, Wu X, Wang D (2010) Performance of annual and perennial biofuel crops: yield during the first 2 years. *Agronomy Journal*, **102**, 806.

Qin Z, Zhuang Q, Zhu X, Cai X, Zhang X (2011) Carbon consequences and agricultural implications of growing biofuel crops on marginal agricultural lands in China. *Environmental Science & Technology*, **45**, 10765–10772.

Qin Z, Zhuang Q, Chen M (2012) Impacts of land use change due to biofuel crops on carbon balance, bioenergy production, and agricultural yield, in the conterminous United States. *GCB Bioenergy*, **4**, 277–288.

Qin Z, Zhuang Q, Zhu X (2013a) Carbon and nitrogen dynamics in bioenergy ecosystems: 1. Model development, validation and sensitivity analysis. *GCB Bioenergy*. doi: 10.1111/gcbb.12107.

USDOE (U.S. Department of Energy) (2011) *US billion-ton update: biomass supply for a bioenergy and bioproducts industry*. Perlack RD and Stokes BJ (Leads), ORNL/TM-2011/224. Oak Ridge National Laboratory, Oak Ridge, TN. 227

USEPA (U.S. Environmental Protection Agency) (2009) State bioenergy primer: information and resources for states on issues, opportunities, and options for advancing bioenergy. Available at: www.epa.gov/statelocalclimate/resources/bioenergy-primer.html (accessed December 2012).

Wiegmann K, Hennenberg KJ, Fritsche UR (2008) Degraded land and sustainable bioenergy feedstock production, Joint international workshop on high nature value criteria and potential for sustainable Use of degraded lands. Paris, June 30–July 1.

Wright L, Turhollow A (2010) Switchgrass selection as a 'model' bioenergy crop: a history of the process. *Biomass and Bioenergy*, **34**, 851–868.

Zhuang Q, McGuire AD, Melillo JM *et al.* (2003) Carbon cycling in extratropical terrestrial ecosystems of the Northern Hemisphere during the 20th century: a modeling analysis of the influences of soil thermal dynamics. *Tellus B*, **55**, 751–776.

Zhuang Q, He J, Lu Y, Ji L, Xiao J, Luo T (2010) Carbon dynamics of terrestrial ecosystems on the Tibetan Plateau during the 20th century: an analysis with a process-based biogeochemical model. *Global Ecology and Biogeography*, **19**, 649–662.

Zhuang Q, Qin Z, Chen M (2013) Biofuel, land and water: maize, switchgrass or *Miscanthus*? *Environmental Research Letters*, **8**, 015020.

Qin Z, Zhuang Q, Zhu X (2013b) Carbon and nitrogen dynamics in bioenergy ecosystems: 2. Potential greenhouse gas emissions and global warming intensity in the conterminous United States. *GCB Bioenergy*. doi: 10.1111/gcbb.12106.

Raich JW, Rastetter EB, Melillo JM *et al.* (1991) Potential net primary productivity in South America: application of a global model. *Ecological Applications*, **1**, 399–429.

Scown CD, Nazaroff WW, Mishra U *et al.* (2012) Lifecycle greenhouse gas implications of US national scenarios for cellulosic ethanol production. *Environmental Research Letters*, **7**, 014011.

Searchinger T, Heimlich R, Houghton RA *et al.* (2008) Use of U.S. croplands for biofuels increases greenhouse gases through emissions from land-use change. *Science*, **319**, 1238–1240.

Stewart JR, Toma Y, Fernández FG, Nishiwaki A, Yamada T, Bollero G (2009) The ecology and agronomy of *Miscanthus* sinensis, a species important to bioenergy crop development, in its native range in Japan: a review. *GCB Bioenergy*, **1**, 126–153.

Thomas ARC, Bond AJ, Hiscock KM (2013) A multi-criteria based review of models that predict environmental impacts of land use-change for perennial energy crops on water, carbon and nitrogen cycling. *GCB Bioenergy*, **5**, 227–242.

Tilman D, Socolow R, Foley JA *et al.* (2009) Beneficial biofuels—the food, energy, and environment trilemma. *Science*, **325**, 270–271.

US Congress (2007) The energy independence and security act of 2007 (H.R. 6). Available at: energy.senate.gov/public/index.cfm?FuseAction=IssueItems.Detail&IssueItem_ID=f10ca3dd-fabd-4900-aa9d-c19de47df2da&Month=12&Year=2007 (accessed May 2011).

10

How willing are landowners to supply land for bioenergy crops in the Northern Great Lakes Region?

SCOTT M. SWINTON[1], SOPHIA TANNER[1], BRADFORD L. BARHAM[2], DANIEL F. MOONEY[2] and THEODOROS SKEVAS[3]

[1]*Department of Agricultural, Food, and Resource Economics, Great Lakes Bioenergy Research Center, Michigan State University, East Lansing, MI, USA,* [2]*Department of Agricultural and Applied Economics, Great Lakes Bioenergy Research Center, University of Wisconsin-Madison, Madison, WI, USA,* [3]*Gulf Coast Research and Education Center, University of Florida, Wimauma, FL, USA*

Abstract

Land to produce biomass is essential if the United States is to expand bioenergy supply. Use of agriculturally marginal land avoids the food vs. fuel problems of food price rises and carbon debt that are associated with crop and forestland. Recent remote sensing studies have identified large areas of US marginal land deemed suitable for bioenergy crops. Yet the sustainability benefits of growing bioenergy crops on marginal land only pertain if land is economically available. Scant attention has been paid to the willingness of landowners to supply land for bioenergy crops. Focusing on the northern tier of the Great Lakes, where grassland transitions to forest and land prices are low, this contingent valuation study reports on the willingness of a representative sample of 1124 private, noncorporate landowners to rent land for three bioenergy crops: corn, switchgrass, and poplar. Of the 11% of land that was agriculturally marginal, they were willing to make available no more than 21% for any bioenergy crop (switchgrass preferred on marginal land) at double the prevailing land rental rate in the region. At the same generous rental rate, of the 28% that is cropland, they would rent up to 23% for bioenergy crops (corn preferred), while of the 55% that is forestland, they would rent up to 15% for bioenergy crops (poplar preferred). Regression results identified deterrents to land rental for bioenergy purposes included appreciation of environmental amenities and concern about rental disamenities. In sum, like landowners in the southern Great Lakes region, landowners in the Northern Tier are reluctant to supply marginal land for bioenergy crops. If rental markets existed, they would rent more crop and forestland for bioenergy crops than they would marginal land, which would generate carbon debt and opportunity costs in wood product and food markets.

Keywords: bioenergy crops, bioenergy supply, contingent valuation, corn, food vs. fuel, land availability, marginal land, poplar, sustainability, switchgrass, Willingness to supply land

Introduction

Research into biofuel and bioelectricity development has been a major focus of scientists, land grant universities, and government agencies in the United States following the passage of the Energy Independence and Security Act of 2007. One guiding assumption of this effort to foster second-generation bioenergy markets has been the notion that significant tracts of marginal agricultural and forestlands could provision biomass without necessarily displacing feed, forage, and timber cultivation. Seminal studies, such as the US Department of Energy (2011) Billion Ton report and the Gelfand *et al.* (2013) article on 'Marginal Lands in the US Midwest', document the stock of rural lands with biophysical conditions that suggest they could be primed to generate large quantities of biomass, in the form of permanent grasses and dedicated fast-growth forests, supplemented by crop residues.

In stark contrast, a growing number of studies probe the critical social and economic questions of whether, and under what market conditions, private landowners of 'marginal lands' would be willing to supply land for biomass production (Jensen *et al.*, 2007; Paulrud & Laitila, 2010; Qualls *et al.*, 2012; Bergtold *et al.*, 2014). Most of these studies survey a representative, random sample of private landowners on willingness to supply a specific type of biomass, such as permanent grasses (Jensen *et al.*, 2007; Bocqueho & Jacquet, 2010; Qualls *et al.*, 2012;), residues from crops (Tyndall *et al.*, 2011; Altman & Sanders, 2012; Altman *et al.*, 2015), or woody biomass

All authors were working in the Great Lakes Bioenergy Research Center at the time this research was conducted.

Correspondence: Scott M. Swinton
e-mail: swintons@msu.edu

(Joshi & Mehmood, 2011; Aguilar *et al.*, 2014). Some studies (Paulrud & Laitila, 2010; Mooney *et al.*, 2015; Skevas *et al.*, 2016) examine multiple biomass sources and attempt to identify the land types that landowners would dedicate to energy biomass production. The advantage of identifying land types is that this information reveals the degree to which 'marginal lands' are likely to play a small or large role in the provisioning of biomass. These findings can then be used to evaluate potential economic and environmental impacts of bioenergy on the landscape.

This study examines land use decisions governing cellulosic biomass on specific land types in the Northern Tier of Michigan and Wisconsin. The region is of interest for several reasons. First, it represents an important 'extensive margin' for biomass provisioning across the north of the United States, one where forests predominate in a cool weather environment, but they are accompanied by both cropland and farmable noncropland that are not a primary source of food from US agriculture. Although all three land types fit an economic definition of 'marginal land' in the sense that rural land prices and rental rates are low (implying low expected profitability of commercial use), only the farmable noncroplands are marginal in the sense that changing their use would affect neither food nor wood markets. The conversion of cropland to bioenergy crops potentially can affect food and feed prices, while the conversion of forestland to bioenergy crops can affect timber markets (NRC, 2011). These conversions also create a 'carbon debt' whereby many years of bioenergy cropping are required to compensate for the carbon released from forest clearing (Fargione *et al.*, 2008).

Second, recent surveys of private landowners in the agricultural areas of southern Michigan and Wisconsin (Mooney *et al.*, 2015; Skevas *et al.*, 2016) reveal that relatively little marginal land would be made available for biomass production even at high rental rates. These findings are explained by: (a) high opportunity costs associated with current land uses, especially the feed and forage demands of integrated livestock operations; (b) uncertainty and sunk investment associated with some land use changes (Song *et al.*, 2011); (c) amenity values associated with current land uses; and (d) other landowner characteristics and preferences.

A third reason for focusing on the Northern Tier of these two states is *prima facie* evidence of lower opportunity costs for biomass provisioning. Land rental rates for cropland and grasslands are considerably lower than in the southern regions of these states. Crop enterprise budget analyses of yields, revenues, and costs associated with cellulosic biomass cultivation in the northern Great Lakes region (Kells & Swinton, 2014) demonstrate that the Northern Tier has a comparative advantage in terms of biomass cultivation (with relatively higher yields of biomass compared to crop and forage production). These comparative advantage conditions are likely to be evident in other Great Lakes states with significant forest cover (e.g., Minnesota and New York). Previous studies of Northern Tier biomass prospects have focused almost exclusively on woody biomass (Joshi & Mehmood, 2011; Aguilar *et al.*, 2014), rather than on the wider range of biomass options afforded by the marginal land types of the region. By contrast, this study probes the full array of land types that landowners could dedicate to biomass production, specifically the choice of using cropland, noncrop marginal land, and forestlands, for any of the three main types of biomass (annual grasses, perennial grasses, and wood). The survey design captures the potential to change current land allocation toward or away from crops, forests, and other uses to biomass provisioning.

The empirical analysis exploits a hurdle model estimation strategy (Cragg, 1971; Ma *et al.*, 2012) that allows us to treat the landowners' problem in two stages: a first-stage probit model of the willingness to participate in each biomass market and a second-stage truncated regression that explores the amount of land dedicated to the activity contingent on participation. This estimation strategy allows a careful examination of the factors shaping the participation and the land quantity decisions. It thus allows for the possibility that factors can shape either, both, or neither of the decisions. The results help to sort out in what ways land use choices are sensitive to land rental rates and to other nonincome-related factors.

The data collection on landowner willingness to rent out land for bioenergy crops relies on contingent valuation methodology (Cameron & James, 1987; Carson & Hanemann, 2005; Mooney *et al.*, 2015; Skevas *et al.*, 2016) to explore the responsiveness of landowners to different rental prices for the biomass types. This type of survey research design randomizes the starting price treatment seen by respondents to probe a wide range of possible rental rates. The rental rate scenarios are preceded by survey questions about current land uses and then succeeded by ones that detail explanatory factors related to landowner wealth and income, preferences for amenities, environmental attitudes, and concerns about rental arrangements. The nonincome factors may be especially relevant to Northern Tier landowners for whom these properties and their use are frequently not significant sources of income but may instead provide major recreational or other nonpecuniary values.

The empirical analysis addresses two main questions related to the supply of biomass in the Northern Tier. First, how much land is available for energy biomass in the Northern Tier of Michigan and Wisconsin?

In particular, how does that availability vary by land type – with specific attention to noncrop marginal land where expansion of bioenergy crops would have minimal effect on food and wood markets and on the level of carbon debt. In order to elicit willingness to supply land for production of bioenergy crops, (a) without requiring the respondent to have the equipment and/or capital to produce their own energy biomass, and (b) without incurring the costs of land clearing, the land supply questions inquire about landowner willingness to rent out land for biomass production, rather than asking whether the landowner would produce energy biomass himself or herself. Second, what factors affect supply of land for renting for bioenergy crops in this region? Specifically, (a) What is the relative importance for landowners in this region of profitability (e.g., rental rate) as compared to amenities (e.g., environmental quality and rental process issues)? (b) How does the relative importance of these attributes compare with findings from agricultural zones, such as the southern parts of these same states? (c) How do the determinants of willingness to supply land for bioenergy crops vary between the decision on *whether* to supply any land at all and *how much* land to offer to rent?

The article is structured as follows: The next section presents our conceptual model of the landowner decisions about land use. The third section develops the empirical methods in three parts: the sample frame, the survey design, and the estimation strategy. Section four presents the main empirical findings. The final section discusses the implications of the findings for bioenergy policy and for future research related to land use decisions by private landowners and other types of landowners.

Conceptual model

Prior research suggests that landowners who own more than one type of land think of land types distinctly (Skevas *et al.*, 2016). They are more inclined to devote land to a closely related use (e.g., change to a different grass crop on cropland) than to undertake a major land use change (e.g., replace an annual grass crop with a perennial tree crop). Hence, we disaggregate the land use decision among three land types: cropland, farmable noncropland, and forestland. We assume that landowners maximize utility from each type of land type (*i*) by choosing the area devoted to a given crop (*j*). We assume that landowner utility comes in part from consumption of marketed goods purchased with money income. That income may be generated as net returns from land-based activities (e.g., crop production, timber harvest) or from nonland income sources. We further assume that landowners derive utility from environmental amenities. Finally, because we elicit willingness to supply land to grow bioenergy crops by hypothesizing a rental market, we assume that the utility function may include disamenities associated with renting land (such as noise and loss of privacy).

Let $\pi^i = \sum_j p_j^i A_j^i$ denote land revenue generated by renting land type i with A_j^i acres in crop j at rental rate p_j^i up to the total area available of land type i, $\bar{A}^i$. Landowners gain utility from consuming goods and services purchased with income that is the sum of land revenue (π) and nonland income (NLI); consumption is denoted $c(\pi^i + \text{NLI})$.

Then, the utility maximization problem on land type i is defined as:

$$\max_{A_j^i} u(c(\pi^i + \text{NLI}), \text{env}^i, \text{rent}^i) \quad s.t. \sum_j A_j^i \leq \bar{A}^i \tag{1}$$

Utility is a function of consumption (c), environmental amenities (env), and rental disamenities (rent) from renting land for bioenergy crops $j = 1,\ldots,J$. Landowners maximize their utility by choosing the area of land to devote to each crop, recognizing that their choice may affect the level of amenities received from the land. The optimal solution to the maximization problem is given by the bioenergy land supply equation:

$$A_j^{i\,*} = A(p_j^i, \text{env}^i, \text{rent}^i | \bar{A}^i, \text{NLI}) \qquad 2$$

For convenience in stating hypotheses, we assume the function *A(.)* to be differentiable in each of its arguments.

The arguments in the bioenergy land supply equation represent theoretical expectations that can be subjected to empirical hypothesis tests that would lead to rejection of the null hypotheses listed below for the reasons indicated:

- H1: Rental rate (p) has no effect on willingness to rent land or amount of land supplied. But if landowners are market oriented, we expect land area to increase in response to higher rental rate offers ($A'(p) > 0$).
- H2: Environmental amenities (env) have no effect on willingness to rent land or amount of land supplied. But if landowners enjoy land-based environmental amenities that might be curtailed by shifting land to bioenergy uses, we expect enjoyment of environmental amenities to reduce land area offered for bioenergy uses ($A'(\text{env}) < 0$).
- H3: Rental disamenities (rent) have no effect on willingness to rent land or amount of land supplied. But if landowners dislike dealing with renters, then we

expect rental disamenities to reduce the land area offered for bioenergy crops ($A'(\text{rent}) < 0$).

- H4: Land available $\bar{A}$ has no effect on willingness to rent land or amount of land supplied. But if owners of larger tracts of land are either more prone to choose to rent out land, or else once they choose to rent they tend to rent out more land, then we expect larger scale landowners to supply more land for bioenergy crops ($A'(\bar{A}) > 0$).

Data and empirical methods

Landowner sampling and survey methods

To study land supply for bioenergy crops at the extensive margin where the cold and short growing season limits agricultural land use, we selected the Northern Tier of Wisconsin and Michigan. This region is primarily composed of forest but includes significant percentages of cropland and other non-forestland, some of it farmable. The region was chosen for its relatively lower agricultural productivity and the associated lower opportunity cost of conversion to bioenergy crops as compared to more agriculturally productive lands to the south (Kells & Swinton, 2014). Figure 1 illustrates the geographical extent of the Northern Tier region in these two states. It is comprised of a 76 county area with boundaries corresponding to the Northern Lake States Forest and Forage Region as defined by the USDA Major Land Resource Area land classification taxonomy (USDA-NRCS, 2006).

The data for our study come from a mail survey of Northern Tier landowners gathered during October 2014 to April 2015. The survey was conducted following Dillman *et al.*'s (2008) total design method. Four mailings were sent out during 2014 as follows: (1) presurvey postcard to alert recipients (October 10), (2) first questionnaire mailing (October 22), (3) reminder postcard (November 3), and (4) second questionnaire mailing to nonrespondents from the first round (November 13). Although nearly all responses were received by the end of February 2015, the survey continued to accept late questionnaire returns until April 30, 2015. A two-page summary of results was mailed to respondents on October 29, 2015.

The landowners contacted were drawn from a list frame consisted of private landowners, farms, and clubs that owned ten or more acres of rural land. The two-stage sampling process used to develop the list frame first entailed selecting a stratified random sample of 18 counties and then continued with secondary stratification within each county. Stratification at the county level involved the designation of land cover classifications for high (≥20%) and low (<20%) levels of crop and grassland cover, respectively (Fig. 1). This ensured an adequate representation of counties with relatively higher levels of cropland and grassland, where planting bioenergy crops is likely to be more viable as compared to more highly forested counties. Data on land cover in cropland or grassland came from the USDA-NASS Cropland Data Layer (2014). In total, six counties were selected at random in Wisconsin (three per stratum) and twelve in Michigan (six per stratum). Twice as many counties were sampled in Michigan because they are roughly half the size of Wisconsin counties. Sampled counties are denoted by stars in Fig. 1.

The second-stage stratification occurred within counties, dividing potential respondents who own at least ten acres of rural land into four strata. The goal was to assure that responses represented (1) landowners who did and did not participate in forest-management programs that could constrain biomass supply possibilities, and (2) landowners with large- and small-scale landholdings. The identification of landowners with land in state forest programs relied on property tax records obtained from county assessor offices. In Michigan, the relevant programs include the Qualified Forest

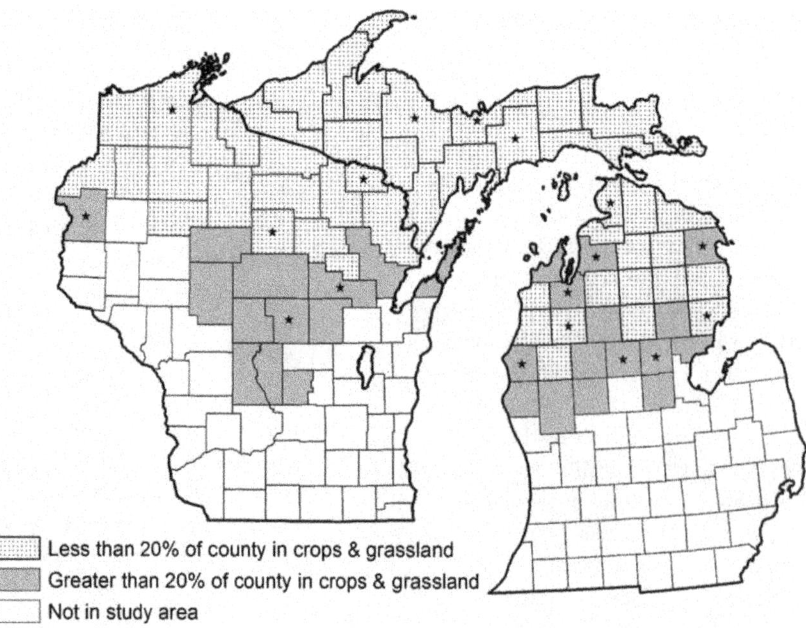

Fig. 1 Northern Tier study zone with county-level land cover stratification categories based on percent of land in crop and grassland from the USDA-NASS Cropland Data Layer (2014). Starred counties were included in the sample.

and Commercial Forest programs, with parcels zoned as 'timber cutover' also counted. In Wisconsin, the Managed Forest Law is the relevant program. The same records identified landowners with 10–100 acres and those with over 100 acres of rural land. The final sample included four strata per county: small-scale landowners not in forest-management programs, small-scale landowners in forest-management programs, large-scale landowners not in forest-management programs, and large-scale landowners in forest-management programs.

To complete the sample selection process, we drew 24 addresses at random from each stratum in (the smaller) Michigan counties and 48 from each stratum in (the larger) Wisconsin counties.[1] This resulted in a balanced final sample frame with 1152 landowners per state. Upon mailing out the study questionnaire to the selected landowner mailing addresses, 134 were returned as undeliverable or otherwise invalid. The final sampled population therefore consisted of 2170 valid addresses, 1124 of which returned completed questionnaires. The final response rate of 51.8% was high enough to provide good assurance of representing the underlying population of private, noncorporate landowners.

Survey experimental design

The questionnaire included sections on current land use and management practices, willingness to rent land for bioenergy crops, opinions about bioenergy, concerns about renting, and demographics. In the land use and management section, respondents were asked how many acres of land they owned of each of the following types – agricultural cropland, farmable noncropland, forestland, and other (e.g., wetlands, lawn). They were further asked whether they used each land type for income or recreation.

The contingent valuation portion was framed as a land rental decision. For each of three potential bioenergy crops – corn, switchgrass, and poplar (*BC* in question example below) – respondents were asked whether they would be willing to rent out land at a given rental rate ($*Y* in question example below, ranging from $15 to $90) on each of three specified land types: cropland, farmable noncropland, and forested land[2] (*LT* in question example below). The question was phrased as follows, 'If somebody offered to rent your *LT* land to grow *BC* for $*Y* an acre per year, would you rent any of it out?' If the answer was yes, the respondent was asked to state the number of acres they would be willing to rent.

Rental rates were varied across surveys. Each crop was given one of four rental rates: 15, 30, 60, or 90 dollars per acre. The average rental rate in the region during 2011–2013 was $30/acre ($46/acre for cropland; $18/acre for pasture) according to the USDA's Cash Rents Survey (2014). In each questionnaire, the two grassy crops (corn and switchgrass) were assigned the same rental rate, and the two sources of woody biomass (poplar and slash) also had the same rate in dollars per acre, for a complete factorial design of sixteen (4×4) rental rate treatments that corresponded to 16 different questionnaire versions.

Attitudes toward bioenergy issues were elicited using a series of statements to which respondents were asked to rate their level of agreement on a scale of 1 (strongly disagree) to 5 (strongly agree). Attitudes toward renting were elicited in a similar manner with respondents rating the degree to which they were concerned about noise, potential legal costs, and having people on their land, along with other potential disamenities from renting.

A complete list of variables used in the econometric models is given in Table 1. The table includes constructed variables from factor analysis of the bioenergy attitude and rental concern variables that are described below and in Tables 2 and 3.

Econometric model

The econometric model is designed to capture a two-stage decision process in which the first decision is whether to rent land for bioenergy crops and, if yes, the second decision is how much land to rent. This class of hurdle model, introduced by Cragg (1971), makes it possible to identify whether the same variables differ in their effects on the first- and second-stage decisions. The first stage, the decision on whether to participate in land rental markets for the bioenergy crops corn, switchgrass, and poplar, was estimated as a binary probit model. For those willing to participate, the second stage on how much land they are willing to rent was estimated using a truncated regression to estimate the number of acres made available (conditional on agreement to rent more than zero acres).

Explanatory variables in both the probit and truncated regressions include current land use, acres of each land type, bioenergy attitudes, rental concerns, and socioeconomic characteristics. There were eleven statements regarding bioenergy attitudes and twelve regarding rental concerns in the questionnaire. Because these variables were measured on a 5-point Likert scale, some were highly correlated. Factor analysis is a method of reducing large numbers of variables by searching for joint variation in response to unobserved factors. Using factor analysis, the eleven attitude variables and twelve concern variables were reduced to four factors each.

For each of the raw variables related to bioenergy attitudes and concerns about land rental, we present the factors and the associated factor loadings after orthogonal varimax factor rotation in Tables 2 and 3. The bioenergy attitude factors are labeled and their loadings of the original Likert-scaled variables are as follows:

- 'Antifossil fuels' factor has high loadings on statements about the need to replace fossil fuels;
- 'Pro-bioenergy' factor has high loadings on bioenergy as superior to other renewable energy sources and liquid biofuels as a promising technology;
- 'Antibioenergy' factor has high loadings on bioenergy crops competing with food needs and leading to loss of forest;
- 'Bioenergy skeptic' factor has positive loadings on the importance of renewable energy and the need to protect biodiversity, with negative loadings on prioritizing bioenergy over other forms of renewable energy.

[1]In several counties, there were fewer than 24 landowners in the forest management program. In those cases, we surveyed all landowners in the stratum and increased the sample non-program participants to maintain equal sized county samples within each state.

[2]In addition to the three crops, landowners were also asked whether they would be willing to contract for woody biomass removal the next time they had timber harvested or thinned. Results from these questions are analyzed separately from this article.

Table 1 Variables included in econometric models of willingness to rent land for bioenergy crops (northern Michigan and Wisconsin landowners, October 2014–April 2015) (n = 1077)

Variable name	Definition	Units	Mean	SD
Current land use and management				
rent_out	rented out land in 2013	(0/1)	0.25	0.04
rent_in	rented in land in 2013	(0/1)	0.06	0.02
farm_land	landowner farmed land in 2013	(0/1)	0.25	0.05
grew_corn	landowner has grown corn	(0/1)	0.38	0.05
timber_harvest	landowner has had timber harvested	(0/1)	0.54	0.05
acres_cropland	acres cropland	acres	34.21	9.94
acres_noncrop	acres farmable noncropland	acres	14.71	2.96
acres_mx_forest	acres mixed natural forest	acres	63.53	6.34
acres_single_spec	acres in single species tree plantations	acres	10.64	6.79
acres_other	acres other rural land	acres	4.16	0.96
forest_program	land enrolled in a state forest program	(0/1)	0.25	0.04
ag_income	cropland used for income	(0/1)	0.22	0.04
ag_personal	cropland used for personal recreation	(0/1)	0.56	0.05
noncrop_income	farmable noncropland used for income	(0/1)	0.06	0.02
noncrop_personal	farmable noncropland – personal recreation	(0/1)	0.54	0.05
forest_income	forestland used for income	(0/1)	0.03	0.01
forest_personal	forestland used for personal recreation	(0/1)	0.88	0.03
Bioenergy attitudes				
BA-1	antifossil fuels		0.09	0.04
BA-2	pro-bioenergy		−0.01	0.03
BA-3	antibioenergy		0.01	0.03
BA-4	bioenergy skeptic		−0.03	0.02
Rental concerns				
RC-1	environmental concern		0.12	0.05
RC-2	rental process		0.02	0.05
RC-3	smell and noise		0.09	0.05
RC-4	unwanted land use change		0.03	0.03
Background information				
age	age	years	57.94	1.22
gender	male gender	(0/1)	0.81	0.04
h_size	household members	count	2.58	0.10
farmer	farmer	(0/1)	0.21	0.04
income	household income	$1000	91.42	4.50
educ	education	1–6*	3.46	0.14
own_duration	duration of land ownership	years	22.95	1.41
family_land	land previously owned by family relative	(0/1)	0.44	0.05
residence	residence on rural land	(0/1)	0.71	0.04

*Education scaled from 1 (less than 12 years) to 6 (graduate degree).

Land rental concerns were similarly reduced to four factors, as follows:

- 'Environmental impact' factor has heavy loadings on increased use of pesticides and fertilizers, loss of biodiversity, reduced soil and water quality, and negative land use changes;
- 'Rental process' factor loads heavily on potential legal costs, contract length, and need for insurance;
- 'Smell and noise' factor loads heavily on potential smell and noise from machinery, with lesser loading from potential legal costs;
- 'Unwanted land use change' factor loads chiefly on the concern about land changing in undesirable ways.

Results

The Northern Tier of Wisconsin and Michigan is dominated by forest. Survey respondents reported owning 299 000 acres of land. Extrapolating from the survey stratum sampling probabilities, forest cover accounts for 55% of rural land cover (50% mixed species; 5% single species) (Fig. 2). Agricultural cropland is the second most important land type, with 28% of area. Farmable noncropland represents the category of agriculturally marginal land that is not currently in crops but could easily be converted to agricultural use. This land type

Table 2 Bioenergy attitude factor analysis: Rotated factor loadings (pattern matrix), northern Michigan and Wisconsin, 2014

Bioenergy attitude variables	BA1-Antifossil fuels	BA2-Pro-bioenergy	BA3-Antibioenergy	BA4-Bioenergy skeptic
Developing renewable energy (e.g., wind, solar, bioenergy, hydro-electrical) is important to our nation's future.	0.546	0.052	−0.112	0.192
Bioenergy should be prioritized over other forms of renewable energy such as wind or solar power.	0.032	0.392	−0.009	−0.170
Burning bioenergy feedstocks to generate electricity instead of burning coal is worth the extra cost.	0.645	0.223	−0.042	−0.085
Substituting bioenergy feedstocks for fossil fuels will help mitigate climate change.	0.707	0.147	−0.009	−0.061
Growing bioenergy feedstocks on cropland will increase competition with food needs.	−0.030	0.022	0.416	0.080
Increased bioenergy feedstock production will result in significant forest loss.	0.003	−0.177	0.462	−0.033
Government should allow regular harvesting of public forestland and CRP land for bioenergy purposes.	0.050	0.418	−0.198	0.093
Biodiversity should be maintained when land use is changed.	0.357	0.083	0.058	0.317
Liquid biofuels are a promising alternative energy technology that will be successful in the future.	0.236	0.356	−0.100	0.108
The use of fossil fuels can be harmful to human health and the environment.	0.641	−0.188	0.046	0.202
The world will run out of fossil fuels (e.g., oil, natural gas) in the next 50 to 120 years.	0.588	−0.143	0.019	0.118

Table 3 Concerns with renting land factor analysis: Rotated factor loadings (pattern matrix) and unique variances, northern Michigan and Wisconsin, 2014

Concerns with renting land variables	RC1-Environ impact	RC2-Rental process	RC3-Smell and noise	RC4-Unwanted land use change
The potential smell	0.156	0.202	0.712	0.068
Noise from harvesting, planting, or other activities	0.238	0.189	0.745	0.025
Potential legal costs of contracting	0.183	0.600	0.452	0.021
The length of the contract	0.219	0.689	0.175	0.074
The possible need for insurance	0.184	0.722	0.194	0.093
Having other people on my land	0.371	0.350	0.285	0.258
The land changing in a way that I can no longer use it as I want	0.463	0.162	0.206	0.375
How profitable it will be	0.056	0.336	−0.043	0.320
A lack of information about the potential feedstocks	0.233	0.354	0.157	0.297
The use of pesticide and fertilizer on my land	0.601	0.233	0.142	0.030
The loss of biodiversity on my land (e.g., insects, birds, mammals, plants)	0.768	0.137	0.176	0.039
The risk of lower soil and water quality	0.734	0.188	0.199	0.077

constituted 11% of the total, with the remaining 5% described as 'other' noncropland (chiefly wetlands).

The overarching finding is that less than 30% of landowners are willing to rent out their land for any bioenergy crop at the rental rates offered (Figs 3–5). Given that these rates ranged up to three times the prevailing $30/acre cash rental rate, landowners are clearly quite reluctant to make their land available for this purpose. Among those who are willing to rent cropland, they generally prefer to do so for corn (Fig. 3), while those willing to rent out farmable noncropland prefer to do so for switchgrass (Fig. 4). Landowners are especially reluctant to rent out forestland for any bioenergy crop (Fig. 5). But if they do, poplar trees are the preferred bioenergy crop (still with fewer than 20% willing to do so). Extremely few (under 10%) are willing to rent out forestland for planting of grassy crops.

The determinants of willingness to supply land to grow bioenergy crops depend importantly on the interaction among land type (three categories) and crops (3),

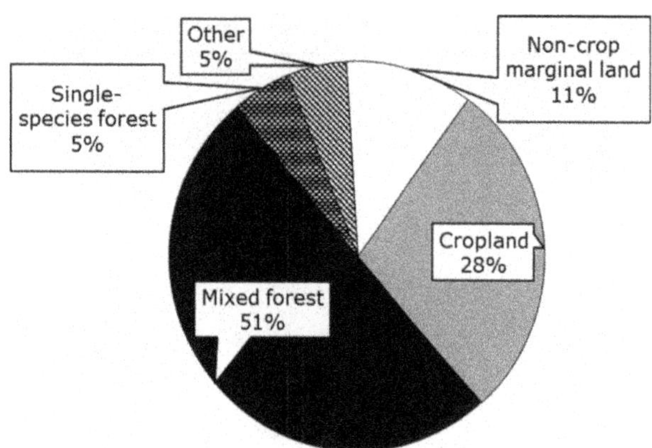

Fig. 2 Land cover type shares, adjusted with survey sampling probability weights, 1077 respondents, northern tier of Michigan and Wisconsin, 2014.

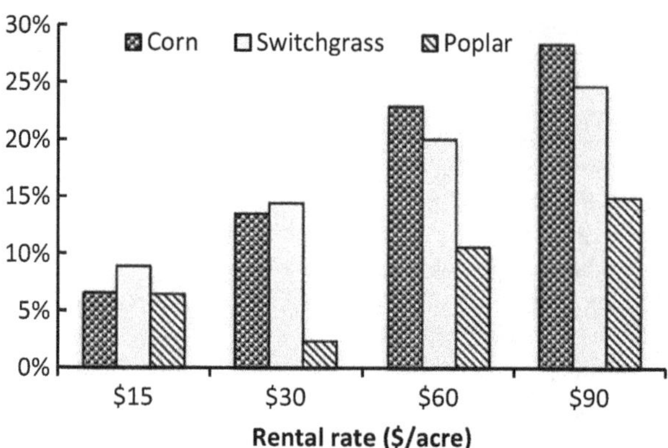

Fig. 3 Willingness to rent out cropland to grow bioenergy crops, adjusted with survey sampling probability weights (n = 690–698), northern tier of Michigan and Wisconsin, 2014.

as well as the two hurdle model stages. In reporting results of the 18 econometric models (Tables 4–9), we describe the consistently influential drivers of land supply before parsing them more carefully at the level of land type, bioenergy crop, or hypothesis area.

Several explanatory variables favored land rental for bioenergy use in nearly all of the nine probit models (Tables 4, 6, and 8). Owners who already rented out land (6 of 9 probit models) and who held pro-bioenergy attitudes (8/9 probits) were more likely to be willing to rent any type of land for bioenergy crops. Likewise, those who had more land, whether cropland (6/9), farmable noncropland (5/9), or mixed forest (7/9), were willing to rent more acres, contingent on being willing to rent out land for bioenergy crops in the first place (Tables 5, 7, and 9). Landowners who held concerns about the rental process were less willing to rent out land for grassy bioenergy crops (corn and switchgrass; Tables 4 and 6).

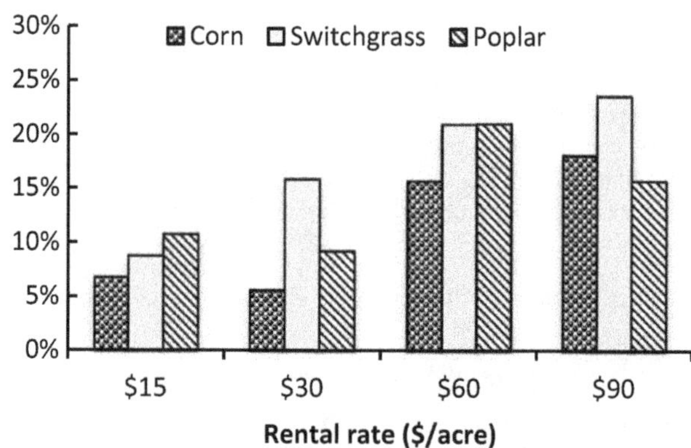

Fig. 4 Willingness to rent out farmable noncrop marginal land to grow bioenergy crops, adjusted with sampling probability weights (n = 732–745), northern tier of Michigan and Wisconsin, 2014.

Results by land type

On cropland (Table 4), as more generally, factors favoring willingness to devote the land to bioenergy crops included having rented out land previously and holding pro-bioenergy views (making, respectively, a typical grower 29% and 6% more likely to rent land). It appears that landowners perceive a connection between their cropland and their farmable noncropland. Landowners who use farmable noncropland for income are 13% less willing to rent out land for corn or switchgrass on cropland, whereas those who use farmable noncropland for personal use are 12–18% more willing to rent out land to grow these bioenergy crops on cropland. In addition, owners who hold concerns about the rental process are also 4–5% less likely to rent out land for bioenergy crops.

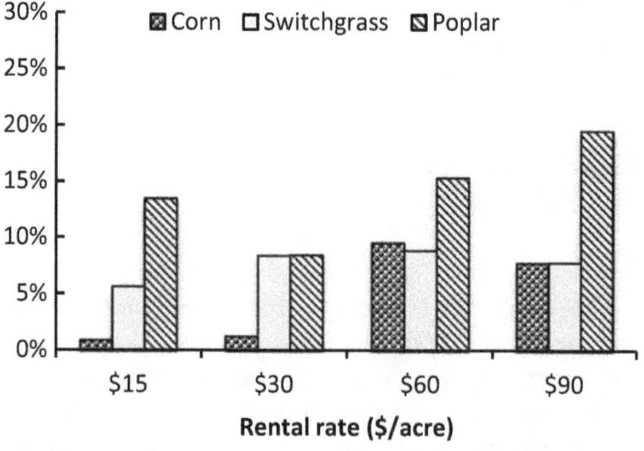

Fig. 5 Willingness to rent out forestland to grow bioenergy crops, adjusted with sampling probability weights (n = 740–748), northern tier of Michigan and Wisconsin, 2014.

Table 4 Marginal effects of determinants of willingness to rent (survey-weighted probit) for 3 bioenergy crops on cropland, northern Michigan and Wisconsin, 2014

	Corn $n = 698$	Switchgrass $n = 692$	Poplar $n = 690$
Rental rate	0.0017***	0.0009	0.0009
Rented out land	0.2939***	0.1931***	0.0467
Rented in land	0.0028	−0.019	−0.3166***
Grew corn	−0.0364	−0.1070**	−0.0603
Had timber harvested	0.1007**	0.0200	0.0193
Acres cropland	−0.0006***	0.0001	0.0002**
Use cropland for income	−0.0374	0.0271	0.0292
Use farmable noncropland for income	−0.1334**	−0.1250**	−0.0615
Use forestland for income	0.0526	0.0779	0.0838**
Personal use for cropland	−0.0177	−0.0445	0.0140
Personal use for farmable noncropland	0.1248**	0.1790***	0.0218
Personal use for forestland	−0.1260**	−0.0729	0.0540
Age	0.0011	−0.0013	−0.0024
Income	−0.0001	0.0001	0.0004*
BA2-pro-bioenergy	0.0637**	0.0807**	0.0678*
RC2-rental process	0.0112	−0.0137	−0.0412**
RC3-smell and noise	−0.0400*	−0.0509*	−0.0240

P-value of chi-square from likelihood ratio test = zero for all models.
Significance (*t*-test probability > 0): ***1%; **5%; *10%.

The determinants of willingness to rent out cropland for bioenergy poplar (Table 4) differed somewhat from those for bioenergy corn or switchgrass. Those with higher income and more cropland were more willing to rent it out for poplar, especially if they already used forestland for income.

The area of cropland offered by those landowners who were willing to rent it out (Table 5) was increased among the ones who owned more land (of any type), who used cropland for income, who were farmers, whose land was previously in the family, who earned more income, and who held concerns about the rental process (suggesting that those concerns had been assuaged when they decided to rent out land at all). Less land was offered for rental among landowners willing to rent for bioenergy crops who were more educated, who had a personal use for cropland, and who were bioenergy skeptics.

On farmable noncropland, as with cropland, willingness to rent the land out for bioenergy crops (Table 6) was 13–18% greater among owners who already rented out land, while it was 5–8% less among those with rental process concerns. It was also 8–20% less among those who grew corn – at least to make the noncropland available for corn or switchgrass. Those who favor bioenergy were 5–8% more willing to rent out land for the grass crops.

Table 5 Determinants of area rented of cropland for 3 bioenergy crops among willing renters (survey-weighted truncated regression), northern Michigan and Wisconsin, 2014

	Corn $n = 143$	Switchgrass $n = 128$	Poplar $n = 68$
Rental rate	−0.1814	0.4494	−0.0038
Rented in land	−34.6315	−70.4944	−2286.642**
Farmed land	14.4131	43.8403	18.0394
Acres cropland	0.7204***	0.2174***	1.3107***
Acres noncropland	0.4905*	1.0911***	0.4080
Acres mixed forest	0.2491***	0.3866***	0.9206***
Acres single species	0.3966	0.0468	1.0528***
Acres other	0.4824	2.2534***	1.1808
Enrolled in forest program	−19.9057	−18.9259	21.601
Use cropland for income	46.9705**	70.0677***	25.9531
Use farmable noncropland for income	−14.9797	23.8311	−81.4256*
Use forestland for income	2.4199	40.5554	−84.7840*
Personal use for cropland	−29.8684*	−70.4109***	−152.5169***
Farmer	49.3676*	56.2523*	58.8810
Income	0.2712*	0.5113***	−0.1669
Education	−5.4484***	−12.5929***	−7.3265**
Duration of land ownership	0.5233*	0.2620	−1.3221
Land previously in family	25.4351	44.3883**	56.9597
BA1-antifossil fuels	3.2794	−28.4319	−51.7005*
BA4-bioenergy skeptic	−55.6504**	−103.2028***	−30.5390
RC1-environmental impact	−17.1396	−6.2123	−53.3789***
RC3-smell and noise	23.5600*	46.9646**	63.8777***

P-value of chi-square from likelihood ratio test = zero for all models.
Significance (*t*-test probability > 0): ***1%; **5%; *10%.

Table 6 Marginal effects of determinants of willingness to rent (survey-weighted probit) for 3 bioenergy crops on farmable noncrop marginal land, northern Michigan and Wisconsin, 2014

	Corn n = 745	Switchgrass n = 738	Poplar n = 732
Rental rate	0.0012***	0.0010	0.0010
Rented out land	0.1342***	0.1817***	0.1523***
Grew corn	−0.0813**	−0.1967***	−0.0996*
Acres cropland	−0.0005**	0.0002*	0.0003*
Acres farmable noncropland	0.0004**	0.0005	0.0000
Enrolled in forest program	−0.0613*	−0.0532	0.0237
Use noncropland for income	−0.0939**	−0.0591	0.0663
Use forestland for income	0.1336***	0.0926	−0.0193
Personal use for noncropland	0.0370	0.0880**	0.0730
Income	0.0002	0.0003	0.0002
BA1-antifossil fuels	−0.0157	0.0252	0.0394
BA2-pro-bioenergy	0.0470**	0.0710*	0.0367
BA3-antibioenergy	−0.0189	−0.0386	−0.0861**
RC3-smell and noise	−0.0456**	−0.0812***	−0.0619**

P-value of chi-square from likelihood ratio test = zero for all models.
Significance (*t*-test probability > 0): ***1%; **5%; *10%.

The area of farmable, noncropland that willing owners would avail for bioenergy crops echoes results from cropland (Table 7). Those with more land were willing to rent more land for bioenergy crops, as were those with rental process concerns (assuming those concerns could be addressed). Factors that reduced the area that respondents were willing to rent for bioenergy crops were personal use of noncropland and environmental concerns about bioenergy crop production. Several additional factors worked against renting out land for corn production.

On forestland, owners were much less willing to rent for corn and switchgrass production (Table 8, Fig. 5) – fewer than 10%, even at the highest rental rate (which was over double prevailing cash rents). Those who held pro-bioenergy attitudes, who were offered higher rental rates, and who used cropland for income were relatively more willing. Those who grew corn, held antibioenergy attitudes, and who were concerned about change in land use or loss of profitability tended to offer less land to rent out for bioenergy crops. Surprisingly, those who used forestland for personal use were more willing to rent it out for poplar production.

As for the area of forestland that owners would be willing to rent out for bioenergy crop production (Table 9), renting land for poplar was much preferred to the grass crops. Those with more forestland would rent out more of it for poplar (and those with more cropland would rent out less forestland for poplar). Environmental concerns linked to bioenergy crop production also detracted from the area that owners would

Table 7 Determinants of area rented of farmable noncrop marginal land for three bioenergy crops among willing renters (survey-weighted truncated regression), northern Michigan and Wisconsin, 2014

	Corn n = 86	Switchgrass n = 106	Poplar n = 101
Rental rate	0.2217	0.4764	−0.9638
Farmed land	16.6768	27.7282	321.3955**
Has had timber harvested	37.3231**	23.5186	124.103
Acres cropland	0.2133**	0.0996***	0.2166*
Acres noncropland	0.0888	0.3920***	1.8026**
Acres mixed forest	0.2884***	0.0517	0.7016***
Acres single species forest	−1.0029	−0.3494	2.9794***
Enrolled in forest program	−61.3598**	−26.1975	−34.0651
Use cropland for income	55.6611**	5.9788	91.6095
Use noncropland for income	−72.7822**	6.9606	−74.6996
Personal use for cropland	−28.5437**	−20.8304	−164.5243
Personal use for noncropland	−31.8586**	−40.0688**	−169.7366
Age	−0.2413	0.9791	0.4235
Male	−3.2353	−52.3486**	−212.5848
Farmer	3.9624	−32.2367	−205.6984
Income	0.0165	0.3477**	−0.0068
Duration of land ownership	−0.5742*	−0.2128	−0.8408
BA1-antifossil fuels	9.4527	−16.8962*	−80.4048
BA4-bioenergy skeptic	−20.4121	57.1869**	181.4276*
RC1-environmental concern	−27.0000***	−39.6014***	−218.4342***
RC2-rental process	21.0484**	−0.5186	−96.9972*
RC3-smell and noise	30.1475**	44.7237**	93.5486
RC4-unwanted land change	2.6550	−16.2695	3.9270

P-value of chi-square from likelihood ratio test = zero for all models.
Significance (*t*-test probability > 0): ***1%; **5%; *10%.

Table 8 Marginal effects of determinants of willingness to rent (survey-weighted probit) for 3 bioenergy crops on forestland, northern Michigan and Wisconsin, 2014

	Corn n = 748	Switchgrass n = 740	Poplar n = 742
Rental rate	0.0011***	0.0003	0.0012*
Rented out land	0.0875***	0.0021	0.0822
Grew corn	−0.0977***	−0.1756***	−0.1944***
Has had timber harvested	0.0672**	0.0466	0.0170
Acres noncropland	−0.0001	−0.0005*	−0.0007
Acres mixed forest	−0.0000	0.0000	−0.0000
Acres single species forest	−0.0003	−0.0008	−0.0000
Acres other	0.0001	0.0006**	0.0000
Enrolled in forest program	−0.0220	−0.0514*	0.0480
Use cropland for income	−0.0148	0.0817**	0.1458**
Use forest for income	0.0059	−0.0403	−0.0096
Personal use for noncropland	−0.0162	0.0193	−0.0715*
Personal use for forest	0.0102	−0.0676	0.1325***
Income	−0.0000	−0.0005**	0.0001
BA1-antifossil fuels	−0.0238***	−0.0011	0.0072
BA2-pro-bioenergy	0.0631***	0.0924***	0.1438***
BA3-antibioenergy	−0.0385**	−0.0449*	−0.0233
RC2-rental process	0.0148	0.0607***	0.0694***
RC4-unwanted land change	−0.0258	−0.0856***	−0.1169***

P-value of chi-square from likelihood ratio test = zero for all models.
Significance (*t*-test probability > 0): ***1%; **5%; *10%.

rent for poplar. Higher rental rate favored offering more forestland, a notable difference from other land types where the rental rate did not affect the area that would be rented for bioenergy crops.

Results by bioenergy crop

Although land type was a dominant factor shaping the willingness of landowners to rent out land for bioenergy crops, there were also clear differences by proposed crop. Corn tended to be the preferred bioenergy crop for rental of cropland. For corn on all land types, a $10/acre increase in rental rate tended to increase the probability of renting out land for bioenergy corn by 1.1–1.7% (on forestland and cropland, respectively). Prior land rental favored willingness to rent out land for corn by 9–29%. Neither rental rate nor prior land rental affected the decision on how many acres to rent. Landowners who already grew corn were disinclined to rent out land to grow more of it on forest or noncropland. As for area of land to rent, the use of cropland or noncropland for income led to more land rented out. By contrast, use of any land type for recreation reduced willingness to rent and/or the area that owners were willing to rent out for corn.

Switchgrass was the preferred bioenergy crop on farmable noncropland. Rental rate did not significantly affect the decision to rent out land for switchgrass, although having rented out land in the past and having a positive attitude toward bioenergy favored doing so. Concerns about land rental and environmental effects of bioenergy crops detracted from willingness to rent land out for switchgrass, as did the use of noncropland or cropland for recreation.

Renting land out for poplar was strongly favored by rental rate, past land rental, and acres of land available (especially forestland). On forestland, those who had had recently harvested timber were more willing to rent land for poplar, whereas those with more single species forest and with environmental concerns about bioenergy were not.

Hypothesis tests

Our conceptual model motivated four hypotheses about determinants of willingness to rent land for bioenergy crops. The null hypotheses turn out to have different effects on the two sides of the econometric hurdle model: the participation probit vs. the area commitment truncated regression.

Rental rate (the price variable in these models) turned out to affect the probability of renting land for corn. More formally, we reject null hypothesis H1 of no rental rate effect in the probit models for corn on all land types, as well as for poplar on forestland. Rental rate did not significantly affect the decision to plant switchgrass on any land type or to plant poplar on cropland or farmable noncropland. Rental rate also affected the area of land rented (at 10% probability of Type I Error) for three cases: switchgrass on cropland, corn on noncropland, and poplar on forestland. Of these, the last is most meaningful, as it implies that rental rate affects both the decision to rent and the area rented for poplar on the land type that is by far the most common.

Environmental amenities tended to have little effect on the decision to rent land out for bioenergy crops, but more effect on the area offered, leading to rejection of H2 for the area offered models. Based on the factor analysis, the 'environmental concerns' factor had positive loadings on three Likert-scaled questions regarding concern about the use of pesticide and fertilizer, the loss of biodiversity, and the risk of lower soil and water quality. Environmental concerns reduced the area of land rented out for both corn and switchgrass on

Table 9 Determinants of area rented of forestland for 3 bioenergy crops among willing renters (survey-weighted truncated regression), northern Michigan and Wisconsin, 2014

	Corn n = 42	Switchgrass n = 47	Poplar n = 126
Rental rate	0.0890	−0.7482**	1.0592***
Rented out land	−15.8086	−98.1590***	−95.7849
Farmed land	−102.9061***	−80.3244**	−7.8465
Has had timber harvested	84.0684***	52.2233*	15.1255
Acres cropland	0.0553	0.0414	−0.2069**
Acres noncropland	0.2677	−0.7829*	−0.8205
Acres mixed forest	0.0977***	0.0725	0.7422***
Acres single species forest	−0.6079***	−0.2546	1.4862***
Acres other	−0.1081	0.5861*	0.9125***
Enrolled in forest program	46.1883**	56.4403*	25.0419
Use cropland for income	−4.8840	42.0171	99.4685
Use noncropland for income	35.7989	12.9629	0.1143
Use forest for income	−50.7284***	−12.4012	45.9190
Personal use for cropland	120.6784***	−84.0853**	−45.3676*
Personal use for noncropland	−87.6434***	106.2819**	10.4705
Personal use for forest	−34.3131***	−94.9490**	−46.6516
Age	−2.4832***	1.3602	−0.9723
Male	−37.3785***	−125.5723**	17.2692
Farmer	−7.8139	72.0467**	42.8510
Income	0.2277***	0.4232**	−0.19614
Education	2.5208**	−3.3241	−7.1597**
Duration of ownership	0.37854	0.1197	0.7592
Residence on land	−58.4296***	81.3367***	33.1348
BA1-antifossil fuels	23.5741***	24.5973**	−15.5943
BA2-pro-bioenergy	3.1775	−0.4279	−5.9331
BA3-antibioenergy	79.0707***	26.9548*	0.0228
BA4-bioenergy skeptic	37.1143***	9.3018	39.9936
RC1-environmental impact	0.5209	−50.4690**	−74.7638***
RC2-rental process	2.2534	−17.6661	49.9649***
RC3-smell and noise	78.3871***	−14.9422	16.5576
RC4-unwanted land change	−87.2139***	−8.5021	−7.8919

P-value of chi-square from likelihood ratio test = zero for all models.
Significance (t-test probability > 0): ***1%; **5%; *10%.

farmable noncropland as well as on forestland. They had the same effect for area of land rented for poplar on cropland and forestland.

Concerns about the rental process had a surprising contrapuntal effect: Rental process concerns reduced the probability of renting land to grow bioenergy crops, but among those willing to rent land, rental process 'concerns' had apparently been dealt with, as this factor was associated with renting more land. More formally, hypothesis H3 that rental concern would have no effect was rejected for five of the participation probits. Rental process concerns reduced the probability of renting cropland for switchgrass or poplar, as well as renting farmable noncropland for any of the three bioenergy crops. A related concern – that of irreversible land use change – detracted from the probability of renting forestland for any of the three bioenergy crops.

The rental process 'concerns' factor had a positive effect on the second-stage area commitment truncated model in six instances. This was true for all three bioenergy crops on cropland, corn and switchgrass on farmable noncropland, and corn on forestland. Presumably this result follows because the landowners who were willing to rent out land for bioenergy crops were those who had resolved any rental process issues.

The land resource constraint clearly affected how much land area was supplied by willing landowners. A robust result for almost all bioenergy crops on all land types was that more land area owned increased the area of land that the owner was willing to make available, implying rejection of H4 for the truncated models. However, in certain instances, land area owned also affected the decision of *whether* to rent land for bioenergy crops. In particular, owners with more cropland were more willing to rent land for switchgrass on noncropland and for poplar on cropland or noncropland, but, oddly, less willing to rent out land for corn on cropland. Land area owned had no effect on the decision to rent out forestland.

Discussion

Land supply for bioenergy crops

Agriculturally marginal regions, including those at the frost-limited northern extensive margin, are potentially attractive for bioenergy crops both because such crops tend not to replace food crops and because the opportunity cost of land is lower. However, this study of the Northern Tier zone of the Great Lakes region in Michigan and Wisconsin finds that the private, noncorporate landowners are willing to supply relatively little land at foreseeable rents more than double current agricultural cash rents in the region. Moreover, most of the land they would supply either has forest cover or crop cover, meaning that bioenergy crops would displace desirable current land covers.

More specifically, the most widespread land cover among the private, noncorporate landowners in the Northern Tier is forest, accounting for 55% of land cover reported.[3] Landowners in the region are reluctant to replace forest with bioenergy crops. Even at rental rates of $90/acre (2–5 times average rental rates for cropland and pasture, respectively), less than 10% of landowners would rent out land for corn or switchgrass, and less than 20% would do so for poplar. Over the range of rental rates reviewed, only 6–14% of landowners would rent forestland for a bioenergy crop (with poplar the preferred choice on forestland). Not only is this a limited land supply, but removing timber to plant bioenergy crops would create a 'carbon debt' that would significantly lengthen the time period before bioenergy crops would make a net reduction in greenhouse gas emissions (Fargione *et al.*, 2008).

Cropland is the second most common land use, at 28% of area managed by respondents. At double the prevailing $45/acre rental rate for cropland in the region, 28% of landowners expressed willingness to rent out land for corn as a bioenergy crop. But this still amounts to just 8% of the aggregate land area, and it carries the opportunity cost of reduced crop output, particularly of livestock feed.

[3]This percentage is lower than the mean for forest land cover generally, because private, non-corporate landowners use a smaller share of their lands for forest than corporate land owners and state/federal forest services.

Farmable noncrop marginal land is the category of greatest interest, due to its low opportunity cost. However, it is the least common type of land, accounting for only 11% of the land held among private, noncorporate landowners in the Northern Tier. Such land typically rents for $15–20/acre, at which rates only 11% of owners would rent out the land. At $90/acre, roughly 5 times the norm, 23% would rent out noncrop marginal land for switchgrass, for an area supply range of 1–2% of total area from these Northern Tier lands with the lowest opportunity cost.

The willingness of landowners to supply noncrop marginal land for bioenergy crops turns out to be similar to the agricultural zones just south of the Northern Tier region. In southern Michigan at rents that range from one-half to three times the $100/acre norm for cropland ($50–300/acre), landowners were willing to supply 20–40% of their noncrop marginal land for bioenergy crops (with corn preferred) (Skevas *et al.*, 2016). Focusing on comparable rents in the $15–90/acre range (when average is $30/acre), Northern Tier landowners were willing to supply an estimated 10–25% of their noncrop marginal land (based on figure 3 of Skevas *et al.*, 2016). In southern Wisconsin, farm landowners with marginal agricultural land would provide less than 5% of their land for bioenergy crops at prices providing similar income (Mooney *et al.*, 2015).

In terms of the overall supply of marginal lands for bioenergy crops, the Northern Tier does not appear any more attractive than more southerly agriculturally dominated regions. In both areas, potential bioenergy supply is quite limited and geographically fragmented, which would in turn increase costs of collection for demand points such as biorefineries or power plants.

Landowner preferences among land types and bioenergy crops

The determinants of land use decisions for cropland and farmable noncropland from this study in the Northern Tier are comparable to those for similar land use categories (cropland, pasture, and other marginal lands) in related studies conducted in southern Michigan and Wisconsin (Mooney *et al.*, 2015; Skevas *et al.*, 2016).

A common finding is that current land cover tends to dictate the preferred bioenergy crops. Respondents preferred not to convert their land from one broad type of cover to another. On agricultural land and farmable noncropland, owners preferred to grow grassy bioenergy crops. On cropland, they tended to favor corn, while on noncrop marginal land, they tended to favor switchgrass. On forestland, they strongly preferred not to convert to bioenergy crops, but the few who were willing to do so strongly preferred to grow poplar, a tree crop, rather than corn or switchgrass.

Land use decisions among Northern Tier landowners appear less motivated by income generation and more motivated by nonmonetary amenities than in the more agricultural zones of southern Michigan and Wisconsin. Evidence of less income orientation in the Northern Tier comes from the coefficients on the rental rate variable in the probit models for both studies. In the cropland and marginal land use categories, rental rate mattered only in 2 of 6 probit models for the Northern Tier (both times for corn). Yet, rental rate mattered in all 6 probit models for southern Michigan, while in southern Wisconsin, biomass price was a significant driver of farm landowners' initial decision of whether to supply land for bioenergy crops.

By contrast, environmental amenities and bioenergy attitudes were stronger drivers of land use decisions in the Northern Tier. In this region, pro-bioenergy views affected willingness to rent land for bioenergy crops in 5/6 probits, with two other bioenergy attitudes also significant. By contrast, in the southern Michigan study, only 1 of 6 probit models had an influential environmental attitude variable. The same pattern is true of the truncated regression models that predict the area of land supplied by willing renters. In the Northern Tier, the pro-environment 'bioenergy skeptic' attitude factor figured in 3 of 6 models (with environmental impact in one other), while in the southern region, environmental or bioenergy attitudes mattered in only 1 of 6 of truncated models (Skevas *et al.*, 2016). In southern Wisconsin, favorable views toward renewable energy and concern for environmental quality boosted the supply of land for bioenergy crops, but the magnitude of these effects was relatively small (Mooney *et al.*, 2015).

Rental concerns, generally disamenities, also played a bigger role in land use decisions in the Northern Tier than in the south. In the Northern Tier, the smell and noise factor mattered in 5 of 6 of probits and 6 of 6 truncated models (with rental process also figuring in 2 of 6 truncated regressions). By contrast, in southern Michigan, land rental concerns mattered in just 2 of 6 of probits, with agricultural production concerns mattering in 1 of 6 of probits and 2 of 6 truncated models.

Conclusion

In conclusion, private, noncorporate landowners in the Northern Tier of the Great Lakes are largely unwilling to supply land for production of bioenergy crops, even at land rental rates 2–5 times prevailing values in 2014. Their reluctance appears to stem in part from caring more for environmental amenities and renting disamenities than for income generation on these lands. Hence, even though the economic opportunity costs of rural land in this region appear lower than in agriculturally dominated lands to the south, the potential supply of land for bioenergy crops is limited in this landowner population. While some biomass could come from timber residues associated with thinning or harvesting commercial forests, such supply is likely to be too dispersed to cost-effectively meet the needs of medium- to large-sized biorefineries or bioenergy-powered electrical generating plants (Epplin *et al.*, 2007).

There remain two potentially attractive avenues for bioenergy crop production in this region that deserve future research. The first is to examine the current data with greater spatial discrimination. Although the percentage of bioenergy-available land in aggregate is small, future research can use spatial analysis to determine whether there exist geographic clusters of landowners who are more willing to supply their land.

The second avenue is to look beyond private, noncorporate landowners. Apart from this group, there exist two other major types of landowners in the Northern Tier: governments and corporations (McDonough *et al.*, 1999; Leefers *et al.*, 2003; Vasievich & Leefers, 2006). Most government forest managers are required to target 'mixed use' criteria, but revenue generation is one important objective. Likewise, corporate land (including real estate investment trusts) is typically managed for income generation. Future research into the availability of land for bioenergy crops in the Northern Tier should examine the potential supply from these institutional and corporate landowners.

Acknowledgements

This work was funded in part by the DOE Great Lakes Bioenergy Research Center (DOE BER Office of Science DE-FC02-07ER64494) and DOE OBP Office of Energy Efficiency and Renewable Energy (DE-AC05-76RL01830), as well as by MSU AgBioResearch and the USDA National Institute of Food and Agriculture. For data collection and input, we thank Daniel Prager, Matthew Kaplan, Michaela Palmer, and Zhuli Stoyanova. For helpful comments, we thank Sarah Klammer, Conner Bailey, and two anonymous reviewers.

References

Aguilar FX, Cai Z, D'Amato AW (2014) Non-industrial private forest owner's willingness-to-harvest: How higher timber prices influence woody biomass supply. *Biomass and Bioenergy*, **71**, 202–215.

Altman I, Sanders D (2012) Producer willingness and ability to supply biomass: evidence from the U.S. Midwest. *Biomass and Bioenergy*, **36**, 176–181.

Altman I, Bergtold J, Sanders D, Johnson T (2015) Willingness to supply biomass for bioenergy production: a random parameter truncated analysis. *Energy Economics*, **47**, 1–10.

Bergtold J, Fewell J, Williams J (2014) Farmers' willingness to produce alternative cellulosic biofuel feedstocks under contract in Kansas using stated choice experiments. *BioEnergy Research*, **7** , 876–884.

Bocqueho G, Jacquet F (2010) The adoption of switchgrass and miscanthus by farmers: impact of liquidity constraints and risk preferences. *Energy Policy*, **38** , 2598–2607.

Cameron T, James M (1987) Estimating willingness to pay from survey data: an alternative pre-test-market evaluation procedure. *Journal of Marketing Research*, **24**, 389–395.

Carson R, Hanemann W (2005) Contingent valuation. *Handbook of Environmental Economics*, **2**, 821–936.

Cragg JG (1971) Some statistical models for limited dependent variables with application to the demand for durable goods. *Econometrica*, **39** , 829–844.

Dillman DA, Smyth JD, Christian LM (2008) *Internet, Mail, and Mixed-Mode Surveys: The Tailored Design Method*, 3rd edn. Wiley, Hoboken, NJ.

Epplin FM, Clark CD, Roberts RK, Hwang S (2007) Challenges to the development of a dedicated energy crop. *American Journal of Agricultural Economics*, **89**, 1296–1302.

Fargione J, Hill J, Tilman D, Polasky S, Hawthorne P (2008) Land clearing and the biofuel carbon debt. *Science*, **319** , 1235–1238.

Gelfand I, Sahajpal R, Zhang X, Izaurralde RC, Gross KL, Robertson GP (2013). Sustainable bioenergy production from marginal lands in the US Midwest. *Nature*, **493**, 7433, 514–517.

Jensen K, Clark CD, Ellis P, English B, Menard J, Walsh M, de la Torre Ugarte D (2007) Farmer willingness to grow switchgrass for energy production. *Biomass and Bioenergy*, **31**, 773–781.

Joshi O, Mehmood SR (2011) Factors affecting nonindustrial private forest landowners' willingness to supply woody biomass for bioenergy. *Biomass and Bioenergy*, **35**, 186–192.

Kells BJ, Swinton SM (2014) Profitability of cellulosic biomass production in the Northern Great Lakes Region. *Agronomy Journal*, **106** , 397–406.

Leefers LA, Potter-Witter K, McDonough M (2003) *Social and Economic Assessment for the Michigan National Forests.* Department of Forestry, Michigan State University, East Lansing, MI.

Ma S, Swinton SM, Lupi F, Jolejole-Foreman C (2012) Farmers' willingness to participate in payment-for-environmental-services programmes. *Journal of Agricultural Economics*, **63** , 604–626.

McDonough M, Fried J, Potter-Witter K *et al.* (1999) *The Role of Natural Resources in Community and Regional Economic Stability in the Eastern Upper Peninsula.* Michigan Agricultural Experiment Station, East Lansing, MI.

Mooney DF, Barham BL, Lian C (2015) Inelastic and fragmented farm supply response to second-generation bioenergy feedstocks: ex ante survey evidence from Wisconsin. *Applied Economics Perspectives and Policy*, **37**, 287–310.

National Research Council (NRC) (2011) *Renewable Fuel Standard: Potential Economic and Environmental Effects of U.S. Biofuel Policy.* National Academies Press, Washington, DC.

Paulrud S, Laitila T (2010) Farmers' attitudes about growing energy crops: a choice experiment approach. *Biomass and Bioenergy*, **34**, 1770–1779.

Qualls D, Jensen K, Clark C, English B, Larson J, Yen S (2012) Analysis of factors affecting willingness to produce switchgrass in the Southeastern United States. *Biomass and Bioenergy*, **39**, 159–167.

Skevas T, Hayden NJ, Swinton SM, Lupi F (2016) Landowner willingness to supply marginal land for bioenergy production. *Land Use Policy*, **50**, 507–517.

Song F, Zhao J, Swinton SM (2011) Switching to perennial energy crops under uncertainty and costly reversibility. *American Journal of Agricultural Economics*, **93** , 768–783.

Tyndall J, Berg E, Colletti J (2011) Corn stover as a biofuel feedstock in Iowa's bio-economy: an Iowa farmer survey. *Biomass and Bioenergy*, **35**, 1485–1495.

U.S. Department of Energy (2011) *US Billion-ton Update: Biomass Supply for a Bioenergy and Bioproducts Industry.* Oak Ridge National Laboratory, Oak Ridge, TN.

USDA-NASS Cash Rents Survey (2014) QuickStats 2.0: Pre-Defined Query for County Estimates. [Online]. Available at: www.nass.usda.gov/Surveys/Guide_to_NASS_Surveys/Cash_Rents_by_County (Accessed 11 Aug 2015, verified 5 Nov 2015). U.S. Department of Agriculture, Washington, DC.

USDA-NASS Cropland Data Layer (2014) Published crop-specific data layer [Online]. Available at http://nassgeodata.gmu.edu/CropScape/ (accessed 22 Sept 2014, verified 22 Oct 2015). USDA-NASS, Washington, DC.

USDA-NRCS (2006) Land Resource Regions and Major Land Resource Areas of the United States, the Caribbean, and the Pacific Basin [Online]. Available at http://www.nrcs.usda.gov/Internet/FSE_DOCUMENTS/nrcs142p2_050898.pdf (Accessed 22 Sept 2014, verified 22 Oct 2015). U.S. Department of Agriculture Handbook 296.

Vasievich JM, Leefers LA (2006) *Social and Economic Assessment for Michigan's State Forests.* Tessa Systems, LLC, East Lansing, MI. Prepared for: Michigan Department of Natural Resources – Forest, Mineral and Fire Management Division.

11

Perennial rhizomatous grasses as bioenergy feedstock in SWAT: parameter development and model improvement

ELIZABETH M. TRYBULA[1,2,‡], RAJ CIBIN[1,‡], JENNIFER L. BURKS[2], INDRAJEET CHAUBEY[1,3], SYLVIE M. BROUDER[2] and JEFFREY J. VOLENEC[2]

[1]*Department of Agricultural and Biological Engineering, Purdue University, West Lafayette, IN, USA,* [2]*Department of Agronomy, Purdue University, West Lafayette, IN, USA,* [3]*Department of Earth, Atmospheric and Planetary Sciences, Purdue University, West Lafayette, IN, USA*

Abstract

The Soil and Water Assessment Tool (SWAT) is increasingly used to quantify hydrologic and water quality impacts of bioenergy production, but crop-growth parameters for candidate perennial rhizomatous grasses (PRG) *Miscanthus* × *giganteus* and upland ecotypes of *Panicum virgatum* (switchgrass) are limited by the availability of field data. Crop-growth parameter ranges and suggested values were developed in this study using agronomic and weather data collected at the Purdue University Water Quality Field Station in northwestern Indiana. During the process of parameterization, the comparison of measured data with conceptual representation of PRG growth in the model led to three changes in the SWAT 2009 code: the harvest algorithm was modified to maintain belowground biomass over winter, plant respiration was extended via modified-DLAI to better reflect maturity and leaf senescence, and nutrient uptake algorithms were revised to respond to temperature, water, and nutrient stress. Parameter values and changes to the model resulted in simulated biomass yield and leaf area index consistent with reported values for the region. Code changes in the SWAT model improved nutrient storage during dormancy period and nitrogen and phosphorus uptake by both switchgrass and *Miscanthus*.

Abbreviations

ACRE = agronomy center for research and education
BIO_E = radiation use efficiency × 10
BLAI = maximum leaf area index
CMN = rate of humus mineralization
CYLD = nutrient fraction at harvest
DLAI = point of the growing season when senescence begins
HEFF = harvest efficiency
HI = harvest index
HU = heat unit
LAI = leaf area index
OAT = one-at-a-time method
PAR = photosynthetically active radiation
PLTFR = plant nutrient fraction
PLTNFR = plant nitrogen fraction
PLTPFR = plant phosphorus fraction
PRG = perennial rhizomatous grasses
RUE = radiation use efficiency
SWAT = soil and water assessment tool
T_BASE = base temperature
WQFS = water quality field station.

Keywords: bioenergy feedstock, hydrologic model, *Miscanthus*, model parameterization, perennial rhizomatous grass, soil and water assessment tool, switchgrass

[‡]These authors contributed equally.

Correspondence: Indrajeet Chaubey
e-mail: ichaubey@purdue.edu

Introduction

Anticipated increases in bioenergy feedstock production as a result of cellulosic ethanol production mandates have instigated research to quantify the environmental impacts of candidate perennial cropping systems such as switchgrass (*Panicum virgatum*) and *Miscanthus* (*Miscanthus* × *giganteus*) (National Research Council, 2011). Production of these dedicated energy crops in existing agricultural fields, either intensively managed or currently under conservation reserve programs, has been hypothesized to impact both hydrologic and water quality responses (Stephens *et al.*, 2001; Simpson *et al.*, 2008; Vanloocke *et al.*, 2010). These impacts are unknown or poorly characterized and depend upon crop selection, associated field management decisions such as tillage and fertilizer application, as well as soil and topographic conditions (Engel *et al.*, 2010). Understanding these impacts is critical to developing crop production strategies that will not only help meet biomass production demand, but will also mitigate the impacts of production on environmental quality. Given the fact that large-scale production of these energy crops has yet to be realized, models such as the Soil and Water Assessment Tool (SWAT) are being used to simulate anticipated hydrologic and water quality response to land use conversion into perennial bioenergy feedstock production at the watershed scale (Ng *et al.*, 2010; Love & Nejadhashemi, 2011; Cibin *et al.*, 2012; Wu & Liu, 2012; Wu *et al.*, 2012). However, for model results to be considered credible and useful, it is important that model parameters and processes are utilized effectively to reduce uncertainty and increase reproducibility of model outputs, in this case, via realistic estimation of biomass production, runoff, and losses of sediment and nutrients to receiving water bodies. For this reason, input parameters must accurately capture the species-specific ontological and growth attributes influencing radiation, water and/or nutrient use efficiency.

The SWAT 2009 crop database currently has 110 crops, including some perennial energy crops such as switchgrass, for which parameters have been developed. However, the crop database in SWAT is parameterized for a lowland switchgrass cultivar Alamo, using data from field experiments across Texas (Kiniry *et al.*, 1996; Arnold *et al.*, 2011). Growth, development, yield, and stress tolerance of lowland ecotypes of switchgrass differ in important ways from upland ecotypes (Parrish & Fike, 2005). While research has been conducted for cultivars across northern sites for simulation of upland ecotypes in the ALMANAC model (Kiniry *et al.*, 2008a,c), these data were not published as parameter values in the SWAT crop database. To our knowledge, there are no publications that parameterize upland switchgrass cultivars in SWAT, though measured values and simulations across the United States identify consistent ecotype differences in yield as a response to regional characteristics (Kiniry *et al.*, 2008c; Jager *et al.*, 2010; Wullschleger *et al.*, 2010).

Similarly, there have been attempts to simulate *Miscanthus* growth in SWAT; however, many of the model parameters were not directly measured. For example, Ng *et al.* (2010) proposed parameter values for *Miscanthus* in SWAT using simulated growth from a second model, BioCro (Miguez *et al.*, 2009, 2012), to estimate biomass and leaf area development values, while adapting seasonal plant tissue concentrations of nitrogen (N) and phosphorus (P) from the work of Beale and Long in the United Kingdom (Beale & Long, 1997) to develop nutrient uptake parameters. Love & Nejadhashemi (2011) adapted crop database values for Alamo switchgrass to represent *Miscanthus* in large-scale bioenergy simulations based on 'literature (values) and consultation with bioenergy crop experts'. If the impacts of large-scale production of switchgrass and *Miscanthus* are to be accurately and consistently simulated in SWAT, it is critical that crop parameters are improved and documented using evidence-based values that represent production scenarios (Kiniry *et al.*, 2008b,c). It should be noted that parallel field studies across the Midwest continue to gather crop-growth measurements that can be used to further improve model representation at the field and watershed scales (Heaton *et al.*, 2004; Dohleman & Long, 2009; Kiniry *et al.*, 2011; Thompson *et al.*, 2012; Burks, 2013). As agronomists and modelers increasingly work to improve model representation, opportunity exists not only for field data to inform models, but also for models to inform designs for empirical experiments and their sampling protocols.

SWAT model crop growth, parameter description, and representation of perennial cropping systems

SWAT (http://swat.tamu.edu) is an international, open-source model capable of simulating hydrologic and water quality response at varying spatial (e.g., field to large river basins) and temporal (e.g., sub-daily to decadal) scales (Arnold *et al.*, 1998; Gassman *et al.*, 2007). Its accuracy, flexibility, broad use and extensive documentation make it a strong candidate for quantifying impacts of bioenergy feedstock production on hydrology and water quality (Engel *et al.*, 2010). SWAT models crop growth using algorithms adapted from the Environmental Policy Integrated Climate (EPIC) model (Neitsch *et al.*, 2011), employing over 35 crop-growth parameters to simulate leaf area development, biomass production, water uptake, and nutrient uptake. These parameters generate values for cropping system characteristics that have direct and indirect effects on hydrologic and water quality processes including evapotranspiration, canopy interception, nutrient pooling, and surface runoff. While this simulation produces outputs at user-defined spatial scales, the basic spatial simulation unit in the SWAT model is called a hydrologic response unit (HRU). The HRU is a unique combination of land use, soil, and slope characteristics. The model estimates crop growth, hydrology and water quality attributes on a daily time scale at the HRU level. These elements are aggregated at or routed to the watershed outlet to obtain watershed-scale results. A thorough description of all crop-growth parameters is available in the SWAT input–output manual as well as the theoretical documentation (http://swat.tamu.edu/documentation/) (Arnold *et al.*, 2011; Neitsch *et al.*, 2011). A short summary is also included in Data S1.

While the process of developing growth parameters for upland switchgrass and *Miscanthus* requires measured values from field study, it should also include analysis of observed physiological characteristics of perennial rhizomatous grass (PRG) monocultures and the impacts of those characteristics on model outputs (Kiniry *et al.*, 2008b). To this point, it is important to consider how SWAT represents differences between annual and perennial life cycle elements that influence nutrient pooling, biomass development, and management scenario determination. Annual crop simulation is driven by scheduled planting, timing of physiological maturity, grain yield, and harvest dates. PRG simulation is driven by emergence after dormancy, peak biomass, timing of physiological maturity, harvest dates, and overwintering. When compared to an annual, grain crop, simulating the multi-year survival strategy of PRGs (with specific emphasis on biomass production) changes key aspects of growth simulation, maturity definition, and management options.

SWAT initiates annual crop growth via scheduled planting, while perennial crop growth is initiated when mean daily temperature reaches a base threshold (T_BASE parameter). Annual crops start each growing season with a new rooting system; the root network from the prior year is mineralized to nutrient pools. Conversely, PRGs maintain a network of nutrient and carbohydrate storage structures (roots, rhizomes, and stem bases) throughout the year. The belowground biomass for an established stand of *Miscanthus*, is reported to not fluctuate significantly from year-to-year (Dohleman *et al.*, 2012; Burks, 2013), creating a consistent storage reserve from year-to-year. Perennial systems have long been known to use these storage organs to supply nutrients for winter survival and spring regrowth (Volenec *et al.*, 1996). The plant's ability to accumulate nutrients and carbohydrates during late-growing season and translocate these resources from storage organs to rapidly growing structures during early crop regrowth period after dormancy offers these perennial systems a competitive advantage over annual cropping systems, including early spring growth, reduction in weed competition, and reduced reliance on external fertilizer application (Beale & Long, 1997; Dohleman *et al.*, 2012). This seasonal nutrient mobilization from vegetative to storage structures and *vice versa* is currently represented in nutrient fraction parameters within SWAT (PLTNFR for nitrogen and PLTPFR for phosphorus), while the variation in root-to-shoot ratio throughout the season is represented at emergence and maturity via the root fraction (RFR1C and RFR2C).

SWAT definition of maturity ($_{fr}PHU = 1$) coincides with physiological maturity in annual cropping systems (grain has formed), whereas maturity definition for PRGs may coincide with peak biomass accumulation depending on management decisions and photoperiod sensitivity. *Miscanthus* specifically exhibits indeterminate growth as well as unique cold tolerance for a C_4 grass (Beale *et al.*, 1996; Lewandowski *et al.*, 2003; Naidu *et al.*, 2003), allowing the crop to begin growth early in the spring, as well as to survive beyond the traditional harvest dates of annual crops. Currently, SWAT model is not capable of simulating nutrient translocation at senescence, storage during dormancy period, and availability during the following year. Beyond year-to-year management, inherent differences between annual and perennial cropping systems are further complicated by the 3–4 year periods necessary for aboveground canopy establishment of *Miscanthus* and switchgrass (Heaton *et al.*, 2010), during which time it is likely that biomass production, overwintering, and nutrient partitioning vary substantially from established systems.

This work is a first effort to inform the most sensitive parameter values with region-specific data and literature value comparisons based on collaborative efforts between agronomists and agricultural engineers in the spirit of cooperative research that has defined SWAT model development (Williams *et al.*, 2008). We present parameter values that reflect representative conditions for growth of *Miscanthu* × *giganteus* (*Miscanthus*) and the upland switchgrass cultivar Shawnee (Vogel *et al.*, 1996) on high-production, tile-drained soils in the upper Midwest. We also propose several modifications to the SWAT crop-growth model algorithms to better reflect PRG development. This was done to address the following questions: (i) Which crop-growth parameters produce a significant response in biomass production, hydrologic, and/or water quality output and does agronomic data exist to establish values for the most sensitive model parameters, (ii) Do empirically derived parameters for *Miscanthus* and Shawnee switchgrass improve simulated crop growth and hydrologic response at the most basic simulation unit (the HRU) compared to existing PRGs (such as Alamo switchgrass) previously used to simulate bioenergy scenarios in SWAT, and (iii) What changes to the source code may improve the conceptual representation of PRGs in SWAT to better represent bioenergy feedstock scenario development?

Materials and methods

Parameter sensitivity analysis to direct data analysis

Relative sensitivity for crop-growth parameters was calculated using the one-at-a-time (OAT) method to evaluate parameter influence on biomass production as well as hydrologic outputs at the watershed outlet, including water yield, nitrate load, and total P load. Outcomes of the sensitivity analysis directed compilation of existing data and additional field measurement used to develop model parameters. This approach allowed model needs to direct data collection efforts to derive an evidence-based range of parameter values.

Experimental site and crop management

Research was conducted at the Water Quality Field Station (WQFS) at the Purdue University Agronomy Center for Research and Education (ACRE), West Lafayette, IN (40° 29′ 55.20″ N; 86° 59′ 53.23″ W; elevation 215 m). The soil series are Drummer silty clay loam (fine-silty, mixed, superactive, mesic Typic Endoaquoll) and Raub silt loam (fine-silty, mixed, superactive, mesic Aquic Argiudoll) with an average soil organic matter level of 45 g kg^{-1}. The WQFS consists of 48 treatment plots (10.8 m wide × 48 m long) that are individually drained with in-ground 24 m × 9 m drainage lysimeters with plastic agricultural tile (0.1 m diameter) placed 0.9 m below the soil surface. Treatments were arranged in a randomized complete-block design consisting of 12 treatments with four replicate plots of each treatment. Two treatments, switchgrass and *Miscanthus*, were used in this study alongside a continuous corn (*Z. mays* L.) control and big-bluestem dominant (*A.gerardii* Vitman), mixed prairie reference plots. Switchgrass plots were seeded at 20 kg ha^{-1} of pure, live seed using a Brillion corrugated seeder on May 15, 2007. *Miscanthus* plots were hand transplanted with greenhouse-grown vegetative propagules in 1-L pots on 1 m^2 centers on May 13 and 19, 2008. In 2008, switchgrass plots were fertilized with 75 kg N ha^{-1} in early-May, and *Miscanthus* plots were fertilized with 84 kg N ha^{-1} in early-August. These were fertilized with Agrotain®-coated urea (Agrotain International, St. Louis, MO, USA) using a Gandy airflow spreader (The Gandy Company, Owatonna, MN, USA). In 2009, 2010, and 2011, the rate for *Miscanthus* and switchgrass was adjusted to current recommendations of 56 kg N ha^{-1}, applied in early- to mid-May. Continuous corn plots were chiseled in late-October and disked in late-May (2009 and 2010) to early June (2011). It was fertilized with 179.3 kg N ha^{-1} preplant and 23.8 kg N ha^{-1} starter in late-April to early-May. Unfertilized, mixed prairie plots were seeded in 1992 and burned annually until 2007, when aboveground biomass was removed annually at harvest to represent a bioenergy feedstock scenario.

Solar radiation, precipitation, temperature, and heat units

Daily precipitation, daily minimum, average, and maximum air temperature, and daily minimum and maximum soil temperature data were collected from the West Lafayette 6 NW station, accessed via the Indiana Climate Network (http://www.iClimate.org). Daily solar radiation was collected at an automated weather station located at ACRE (40° 28′ 1.2″ N, 86° 58′ 58.8″ W), located 0.5 km from the study site. Missing data values were filled using the Throckmorton Purdue Agricultural Center (TPAC, 40° 17′ 52.8″ N, 86° 54′ 10.8″ W) weather station, located 22 km from the study site. Daily photosynthetically active radiation (PAR) was calculated as one-half of measured daily solar radiation (Arnold *et al.*, 2011). Average annual temperature and cumulative precipitation were 11.8 °C and 916 mm in 2009 and 10.7 °C and 920 mm in 2010, respectively. From 1901 to present, average annual temperature was 10.3 °C and cumulative precipitation was 947 mm per year.

Base temperature was estimated using one- and two-week moving averages of average daily air temperature in conjunction with observed PRG emergence dates at the WQFS. The range of reasonable base temperature values was compared to reported values in literature (Hsu *et al.*, 1985; Lewandowski *et al.*, 2003; Wullschleger *et al.*, 2010; Zub & Brancourt-Hulmel, 2010). Estimated base temperature values were further validated using Illinois Climate Network data (http://www.isws.illinois.edu/warm/datatype.asp) coinciding with emergence dates reported by Heaton (2006). The optimal temperature was assumed as 25 °C for both crops based on SWAT theoretical documentation for warm season grasses (Wullschleger *et al.*, 2010; Neitsch *et al.*, 2011). Heat units to maturity for switchgrass and *Miscanthus* were estimated using the SWAT-PHU program (http://swat.tamu.edu/software/potential-heat-unit-program/). The program estimates heat units to maturity using long-term regional climate data, crop-specific base temperature and number of growing days. Estimates were compared with calculated heat unit accumulation from measured mean daily temperature, emergence dates, and measured biomass values.

Harvest yield, crop residue, harvest efficiency, and biomass sampling

Aboveground yield samples were collected by hand monthly from May to September during 2009 and 2010 and in August and October of 2011 as reported by Burks (2013). A quadrat (0.165 m^2) was randomly placed in each switchgrass plot, where biomass was removed with hand clippers and collected. Biomass sampling for switchgrass occurred as a composite of four locations per plot from April 2009 to April 2010 and of three locations per plot from April 2010 onward. *Miscanthus* biomass was collected in May and June 2009 from one individual plant within each plot. From July to September 2009 *Miscanthus* top growth was collected from one-half of an individual plant. An individual plant consists of many individual tillers, the number of which increase as the plant ages from year-to-year. As the *Miscanthus* plants grew older, due to sampling difficulty *Miscanthus* top growth was collected from one-quarter of an individual *Miscanthus* plant (area of 0.25 m^2) during the 2010 and 2011 growing seasons. Harvest yield was obtained using a self-propelled biomass harvester in October of all 3 years. Plot residue was collected within two 1 m^2 quadrats arbitrarily selected within the harvester path immediately following the 2011 harvest (Figure S1). The remaining stem bases were clipped at the soil surface and transect residue was raked into sample bags. All samples were analyzed using standard methods as described in Burks (2013). Harvest efficiency (HEFF) was calculated using yield and residue data [Eqn (1)].

$$\mathrm{HEFF} = 1 - \left(\frac{\mathrm{Residue}}{\mathrm{Residue} + \mathrm{Harvested\ Yield}}\right) \quad (1)$$

Stem base (area of plant spanning 8 cm above the soil surface to 2.5 cm below the soil surface), rhizome, and root biomass was collected monthly from April to September, and December 2, 2009, and monthly from March 31 (roots and rhizomes only) to September and once December 2, 2010, and March 31 (roots and rhizomes only), August, November, and December 12, 2011. Sampling of roots, rhizomes, and stem bases from switchgrass plots was done using turf grass cup cutters (Lever Action Hole Cutter Model No. 1001-1, Par Aide, Lino Lakes, MN, USA) measuring 18 cm deep and 10 cm in diameter. Belowground samples were obtained from the same locations as the aboveground biomass. From April to June 2009, *Miscanthus* belowground samples were obtained to soil depth of 27 cm using a shovel, as this was the length of the shovel blade. From July 2009 onward a battery operated reciprocating saw (Craftsman Model No. 315.115790, Sears Inc.) equipped with a 30.5 cm wood blade was used to collect roots and rhizomes. Tissues were placed in coolers on ice. Tissues were washed free of soil, separated into roots, rhizomes, and stem bases, and placed in paper bags and stored at −80 °C for at least 24 h. Tissues were transferred to −4 °C and held until freeze-dried (FreeZone 12 freeze dryer, Labconco Corporation, Kansas City, MO, USA). Tissues were weighed and ground to pass through a 1 mm screen as described above.

Leaf area index and light extinction coefficient

Photosynthetically active radiation was measured simultaneously above the canopy and below the canopy with a Decagon AccuPAR LP-80 ceptometer (Decagon Devices, Inc., Pullman, WA, USA). Below-canopy PAR was measured 5 cm above the soil surface at random intervals across plot replicates using a modification of the Plant Method outlined by Johnson *et al.* (2010). Measurement occurred at 3-to-4 week intervals between June and October in 2010 and between May and September in 2011; more frequent measurement was not possible due to repeated instrument failure. Five transects were taken across each plot replicate in 2010 and were reduced to three transects per plot in 2011. Five measurements were averaged across each transect. Light extinction coefficient and the fraction of intercepted PAR were calculated for each measurement using measured above- and below-canopy PAR values. To confirm ceptometer calculations, destructive sampling was conducted using portions of the August, 2011 aboveground biomass sample. Samples were defoliated and leaves were passed through a Li-COR LI-3100C leaf area meter (Li-COR, Lincoln, NE, USA) to measure LAI values.

Radiation use efficiency

Radiation use efficiency (RUE) was calculated as the slope of the relationship between cumulative, intercepted PAR and biomass yield distributed throughout the growing season. Cumulative, intercepted PAR was calculated between emergence (4/10/2010, 4/13/2011) and biomass sampling dates in 2010 and 2011. Linear regression models were plotted for the mean, as well as the first and third quartile biomass values to estimate suggested and range of RUE, respectively. BIO_E, the model parameter representing RUE, was calculated as 10 times the slope of the regression model (Arnold *et al.*, 2011).

Nutrient fractions

Tissue total N concentrations were determined for roots, rhizomes, stem bases, and top growth using a flash combustion elemental analyzer (Flash EA 1112 Series, Thermo Fisher Scientific, the Netherlands). P analysis was done using methods outlined by Murphy & Riley (1962). Total plant nutrient fractions were calculated using a weighted average for constituent tissues [Eqn (2)]:

$$\mathrm{PLTXFR} = \frac{\sum\left[\left(\frac{\%\mathrm{D.M.}}{100}\right)(\mathrm{X\ mg\ g^{-1}})\right]}{1000} \quad (2)$$

where PLTXFR is the plant nutrient fraction,% DM is the tissue dry matter weight as a percent of the entire plant, and X is the nutrient concentration (N or P). Plant fractions (PLTFR) for the three growing periods were derived using data from the period of emergence in April (PLTFR1), halfway through the growing season in July (PLTFR2), and peak sampling in August (PLTFR3). Nutrient fractions for N and P in harvested biomass (CYLD) were derived from tissue concentrations in aboveground dry matter sampled during October harvest. Data from 2009 and 2010 growing seasons were used to derive switchgrass ranges and suggested parameter value, while data from 2010 were used exclusively for *Miscanthus* parameter values as 2009 could be considered an establishment year for the crop.

Changes to the SWAT Code

Although the crop-growth model consistently produced reliable simulations of corn production at the study site, preliminary simulations revealed limitations of the crop-growth model for PRG production. Simulations were inconsistent with field observations of PRG leaf area development, biomass yields, and calculated N budgets. Algorithms representing plant nutrient uptake, leaf area development, root decay and harvest index were modified to better represent differences characteristic of PRG growth. Code modification for PRG process representation was verified by SWAT model developers and is included in the published version of SWAT model (revision 612). The list of code modification is included in supplementary information (Table S1). Users can also obtain the FORTRAN source code of SWAT model with the revisions discussed in this manuscript from http://swat.tamu.edu/software/swat-model/.

Crop nutrient uptake is simulated in SWAT as a function of two driving factors: plant tissue nutrient fraction at different growth stages and soil nutrient availability. SWAT 2009 does not limit nutrient uptake on temperature or water stress days. A simple if-then loop was added to the grow.f code to bracket the nutrient uptake subroutine; when drought stress occurred, nutrient uptake subroutines were not used.

Crop physiological activities such as biomass accumulation, nutrient uptake, and transpiration stop at crop maturity (PHU = 1). SWAT 2009 initiates crop senescence at the end of the growing season via the leaf area index parameter DLAI (Fig. 1). During senescence, LAI reduces linearly from Maximum Leaf Area Index (BLAI) until it reaches 0, which correlates with crop maturity (PHU = 1). Using physiological maturity to regulate the timing of development curves proved problematic with bioenergy cropping systems; specifically species that demonstrate indeterminate growth or photoperiod sensitivity, such as *Miscanthus*. In the current parameter estimation process, we defined crop maturity as peak biomass, stopping subsequent biomass accumulation once the peak occurred (often in late-August to early-September at our site). However, *Miscanthus* remained green in the field for an additional month or more after the peak biomass was recorded. Although senescence was observed in the lower canopy at this time, we expect the plant to be physiologically active during this period, using nutrients for seed development and transpiring without biomass accumulation. To accommodate this developmental pattern, we modified the LAI curve (Fig. 1) to allow the DLAI parameter to have a user-defined value greater than 1, while creating an ending heat unit value to be entered as a user-defined parameter. The decrease in LAI to represent senescence (between DLAI and the end heat unit) was calculated as a linear reduction. Using the measured heat unit data and field crop condition parameters, we estimated the DLAI of 1.1 for *Miscanthus* and 1 for switchgrass, while the end heat unit fractions were estimated to be 1.2 for *Miscanthus* and 1.15 for switchgrass.

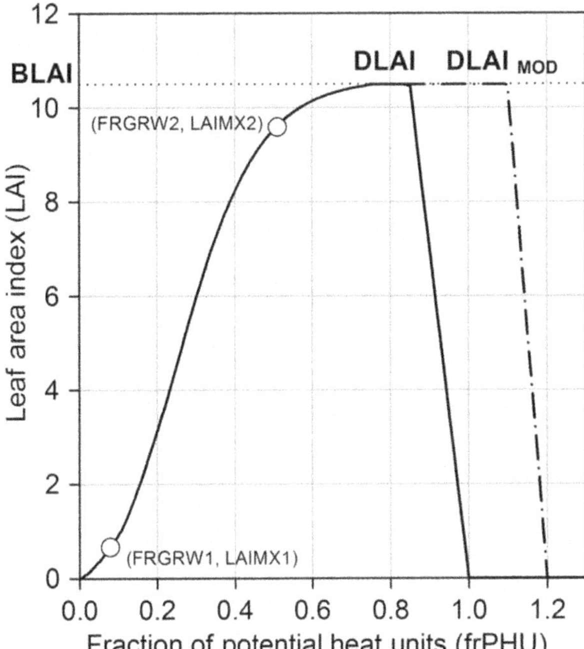

Fig. 1 Modification of the point in the growing season when leaf area index (LAI) declines ($DLAI_{MOD}$) of optimal leaf area development curve delays perennial rhizomatous grass (PRG) senescence. Peak biomass definition of PRG maturity does not coincide with physiological maturity definition for annual grain crops. $DLAI_{MOD}$ improves model representation of field conditions to extend period of photosynthesis and transpiration after PRG maturity. Additional parameter acronyms defined in Table 1.

SWAT 2009 represents harvested yield via the harvest index parameter where yields (both seed and biomass) are reduced in response to water stress. The harvest algorithms (harvestop.f) for perennials were modified to exclude harvest index adjustments since the reduction in biomass is already accounted for once during the stress period.

While SWAT 2009 harvestop.f also allocates about 80% of roots to the annual residue pool postharvest, the absence of overwinter data required that we assume zero loss in belowground biomass due to root and rhizome decay. Measured root biomass at harvest period and a month before emergence in the following year remained similar for both *Miscanthus* and switchgrass. To represent the minimal root decay observed for the PRG, root decay variables (rtres) were turned off. This prevented any loss in calculated biomass (bio_ms) due to root decay.

SWAT model application

As PRGs have yet to be deployed at the watershed scale, we are not aware of any data related to production of PRGs and associated hydrologic and water quality response that could be used to validate the SWAT model at the watershed scale. Consequently, the proposed model changes and parameters developed in this study were tested using a single HRU at the field scale and compared with the measured data. While SWAT was developed primarily as a watershed-scale model, strong model performance at the basic simulation level is an intuitive benchmark from which reasonable aggregate values can be produced at the watershed scale.

SWAT model was developed with WQFS-specific soil and weather characteristics to validate derived parameters and growth algorithm improvements. Soil data for the model was obtained as Drummer soil from SSURGO soil database representative of soil characteristics at the WQFS. Daily weather data inputs included precipitation, solar radiation, minimum and maximum temperature for 7 years (2004–2011)

Table 1 Suggested values and potential parameter range for *Miscanthus × giganteus* and upland switchgrass (*P. virgatum*) cultivar Shawnee compared to current lowland switchgrass (c.v. Alamo) in the SWAT 2009 crop database. Shaded parameters estimated

			Miscanthus × giganteus MISG		Shawnee Switchgrass (*Panicum virgatum*) SWSH		Alamo Switchgrass SWCH
Parameter	Acronym	Unit	Suggested	Range	Suggested	Range	Database value
Optimal Temperature (degrees Celsius)	T_OPT†'‡'**	°C	Existing Alamo value		Existing Alamo value		25
Base Temperature (degrees Celsius) [Potential Heat Units]	T_BASE*,†'‡'§(PHU)	°C	8 (1830)	7–10 (2100–1600)	10 (1400)	8–12 (1600–1200)	12
Radiation Use Efficiency in ambient CO_2	BIO_E*,§¶	$\frac{g}{MJ} \times 10$	41 (39§§)	–	17 (12§§)	–	47
Root fraction at emergence	RFR1C	NA	0.87	0.76–0.96	0.89	0.80–0.97	Default (0.40)
Root fraction at maturity	RFR2C	NA	0.18	0.12–0.22	0.49	0.44–0.57	Default (0.20)
Harvest Index	HVSTI††	NA	1	–	1	–	0.9
Harvest Efficiency	HEFF*	NA	0.7	0.65–0.75	0.75	0.7–0.75	
Lower Limit of Harvest Index due to stress	WSYF	NA	1	–	1	–	0.9
Maximum Leaf Area Index (LAI)	BLAI*	$\frac{m^2}{m^2}$	11	10–13	8	–	6
Fraction of growing season when growth declines	DLAI*, ††	NA	1.1	–	1		0.7
Minimum LAI for plant during dormant period	ALAI_MIN‡‡	$\frac{m^2}{m^2}$	0	–	0	–	0
Light extinction coefficient	EXT_COEFF*	NA	0.55	0.45–0.65	0.5	0.4–0.55	0.33
First point fraction of BLAI for optimum growth curve	LAIMX1*,§	NA	0.1	–	0.1	–	0.2
Second point fraction of BLAI for optimum growth curve	LAIMX2*'§	NA	0.85	–	0.85	–	0.95
Fraction of growing season coinciding with LAIMX1	FRGRW1*,§	NA	0.1	–	0.1	–	0.1
Fraction of growing season coinciding with LAIMX2	FRGRW2*,§	NA	0.45	–	0.4	–	0.2
Plant nitrogen fraction at emergence (whole plant)	PLTNFR(1)*	$\frac{kg\ N}{kg\ DM}$	0.0100	0.0097–0.0104	0.0073	0.0066–0.0081	0.035
Plant nitrogen fraction at 50% maturity (whole plant)	PLTNFR(2)*	$\frac{kg\ N}{kg\ DM}$	0.0065	0.0062–0.0070	0.0068	0.0067–0.0072	0.015
Plant nitrogen fraction at maturity (whole plant)	PLTNFR(3)*	$\frac{kg\ N}{kg\ DM}$	0.0057	0.0053–0.0060	0.0053	0.0051–0.0055	0.0038
Plant nitrogen fraction in harvested (aboveground) mass	CNYLD*	$\frac{kg\ N}{kgyield}$	0.0035	0.0034–0.0035	0.0054	0.0053–0.0058	0.0160
Plant phosphorus fraction at emergence (whole plant)	PLTPFR(1)*	$\frac{kg\ P}{kg\ DM}$	0.0016	0.0016–0.0017	0.0011	0.0010–0.0012	0.0014

Table 1 (continued)

Plant phosphorus fraction at 50% maturity (whole plant)	PLTPFR(2)*	$\frac{\text{kg P}}{\text{kg DM}}$	0.0012	0.0010–0.0014	0.0014	0.0013–0.0016	0.001
Plant phosphorus fraction at maturity (whole plant)	PLTPFR(3)*	$\frac{\text{kg P}}{\text{kg DM}}$	0.0009	0.0007–0.0011	0.0012	0.0011–0.0012	0.0007
Plant phosphorus fraction in harvested (aboveground) mass	CPYLD*	$\frac{\text{kg P}}{\text{kg yield}}$	0.0003	0.0003–0.0004	0.0010	0.0010–0.0011	0.0022
Max. Canopy Height (m)	CHTMX*,**	*m*	3.5	–	2	–	2.5
Max. Rooting Depth (m)	RDMX**	*m*	3	2–4	3	2–4	2.2
Min. Crop Factor for Water Erosion	USLE_C‡‡	NA	Existing Alamo Value		Existing Alamo Value		0.003
Vapor pressure deficit	VPDFR‡‡	kPa	Existing Alamo Value		Existing Alamo Value		4
Stomatal conductance	GSI‡‡	$\frac{m}{s}$	Existing Alamo Value		Existing Alamo Value		0.005
GSI fraction corresponding to the second point on the stomatal conductance curve	FRGMAX‡‡	NA	Existing Alamo Value		Existing Alamo Value		0.75
Rate of decline in RUE due to increase in vapor pressure deficit	WAVP‡‡	NA	Existing Alamo Value		Existing Alamo Value		8.5

*Original data from the Purdue University Water Quality Field Station.
†Daily minimum, maximum, and mean temperature Indiana State Climate Office.
‡Daily minimum, maximum, and mean temperature Illinois Climate Network.
§Heaton EM, 2006.
¶Kiniry *et al.*, 2011.
**Zub & Brancourt-Hulmel, 2010.
††Modified parameter for perennial rhizomatous grass representation.
‡‡Assumed.
§§Preliminary value using top growth data, replaced by value using total biomass data.

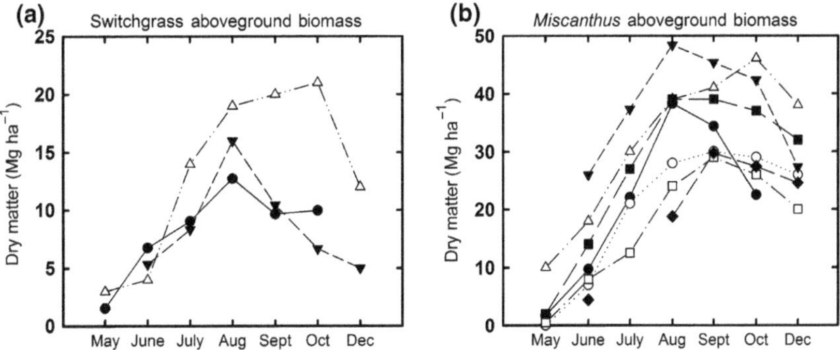

Fig. 2 Aboveground biomass comparison of current study values with published values for switchgrass (a) and *Miscanthus* (b) from this study (closed circle), Beale and Long (open circle), Heaton *et al.* (closed triangle), Dohleman *et al.* (open triangle), Nassi *et al.* (closed square), Himken *et al.* (open square), and Schwarz *et al.* (closed diamond).

from data sources discussed in study site description. A model warm up period of 4 years (2004–2007) was included to stabilize the initial conditions in the model. Although there is a recognized period of establishment in the first 3-to-4 years of PRG development, simulations modeled established PRG growth only. The HRU was tile drained with a depth to tile set at 1 m.

Miscanthus and switchgrass were represented in the model as multi-year crop rotations with planting on year one and base temperature guided emergence from dormancy from year two, which often occurred in March/April. Crop management practices included urea application of 56 kg N ha^{-1} in early-May and harvest on October 31. The parameter regulating the rate of humus mineralization (CMN) was input as 0.003 to account for high rate of mineralization expected in WQFS soils. Average organic matter content of soils at the study site tested at 4.5%. Rapid mineralization was also implied by measured annual N uptake for *Miscanthus* during the study period, which was greater than N application rate. Total N content for aboveground biomass was 150 kg N ha^{-1} for samples collected in August (averaged over 2-years). The difference in N was assumed to come from mineralization of soil organic N and crop residue. Runoff curve numbers for switchgrass and *Miscanthus* were assumed to be the same as other PRGs in the crop database: 31, 59, 72, and 79 for soil hydrologic groups A, B, C, and D, respectively. Model simulations for *Miscanthus*, switchgrass, and continuous corn were compared with the measured data for biomass, plant N uptake and plant P uptake. A Monte Carlo simulation was conducted ($N = 1000$) using suggested parameter ranges to quantify their effects on model outputs, specifically influencing crop growth and nutrient uptake.

Results

Sensitivity analysis

Radiation use efficiency (BIO_E), base temperature (T_BASE), optimum temperature (T_OPT), the point in the growing season when LAI declines (DLAI), and the N fraction of harvested biomass (CNYLD) were the most sensitive parameters for multiple model outputs (Table S2). Biomass yield (harvested) was most sensitive to BIO_E, T_BASE, and T_OPT parameters, whereas water yield was more sensitive to T_BASE and maximum leaf area index (BLAI). Emergence threshold, baseline heat unit accumulation, and temperature stress regulation are all controlled by T_BASE, which also affects the timing of peak biomass and plant maturity. Simulated watershed nitrate load, unlike simulated total P, was also sensitive to T_BASE. Simulated nitrate load was also highly sensitive to DLAI, as well as CNYLD, BIO_E, and T_OPT. Simulated total P load was most sensitive to BIO_E and DLAI.

Base Temperature (T_BASE) and heat units to maturity (PHU)

One- and two-week moving averages of measured, daily average air temperature leading up to and during the period of emergence (early- to mid-April) supported estimates of T_BASE parameter values of 8 °C for *Miscanthus* and 10 °C for switchgrass (Figure S2). This is within the range of reported values between 7 and 10 °C for *Miscanthus* and 8 to 12 °C for switchgrass (Beale *et al.*, 1996; Beale & Long, 1997; Lewandowski *et al.*, 2003). SWAT-PHU program estimated heat units to maturity (peak biomass) for *Miscanthus* and switchgrass as 1830 and 1400, respectively, with 140 and 120 days as number of annual growing days. These estimates fell within the observed range between plant emergence and peak biomass (maturity) from 2009 to 2011 (Figure S2).

Biomass development and radiation use efficiencies (BIO_E)

Although observed yields were lower than regionally similar studies with similar stand age (Fig. 2), aboveground biomass measured in this study for switchgrass and *Miscanthus* demonstrated comparable trends to the work of Heaton *et al.* (2008) in Illinois. Belowground development (roots and rhizomes) for *Miscanthus* was notably low: two-year means were 5.4 Mg ha^{-1} in May, 6.9 Mg ha^{-1} in September, and 14.6 Mg ha^{-1} in December. Belowground biomass for switchgrass (two-year means) was 9.3 Mg ha^{-1}in May, 11.3 Mg ha^{-1} in September, and 8.7 Mg ha^{-1} in December. These values are consistent with similar belowground studies that also remark on the low root mass of *Miscanthus*, which is offset only by its prolific rhizome development (Dohleman *et al.*, 2012). The majority of the *Miscanthus* belowground roots and rhizomes at the WQFS were found within the top 27 cm of the soil profile (Burks, 2013). Core sampling of *Miscanthus* roots to a depth of 60 cm revealed only modest root production (0.2 Mg ha^{-1}) in the 20 to 60 cm soil depth when compared to prairie root production in an adjacent plot at that same depth (0.5 Mg ha^{-1}). Low root biomass for *Miscanthus* resulted in a similarly low belowground fraction with a two-year mean of 0.87 in April/May, and 0.18 in August. Switchgrass belowground biomass production was more pronounced; the mean of belowground fractions was 0.89 in April/May, and 0.49 in August. To avoid unintentionally limiting growth simulation, maximum root depth (RDMX) was estimated at 3 m for both crops (Table 1).

Slope calculations using mean aboveground biomass values resulted in suggested BIO_E parameter values of 39 for *Miscanthus* and 12 for switchgrass (Fig. 3). However, preliminary simulations using these values resulted in consistent under-prediction of switchgrass yields. Use of aboveground biomass data limited the influence of belowground production on RUE (Zhang *et al.*, 2011). While biomass partitioning is included as a parameter input in the newest version of the SWAT model, it also has a clear impact on RUE. When total biomass values were used in RUE calculations, mean *Miscanthus* RUE was marginally affected, while mean switchgrass RUE increased notably (Fig. 3). RUE calculations using total

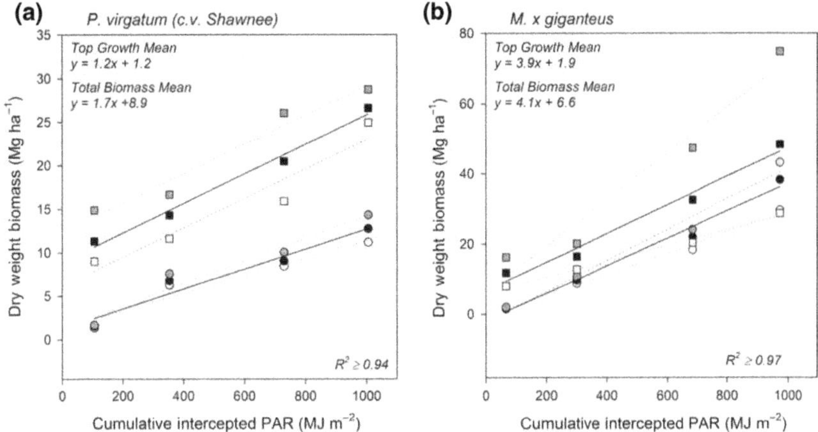

Fig. 3 Radiation use efficiency (RUE) values calculated using mean, first, and third quartiles of total biomass and top growth sampled in 2010 for switchgrass (a) and *Miscanthus* (b). Shaded squares: third quartile values for total biomass; black squares: mean values for total biomass; white squares: first quartile values for total biomass; Shaded circles: third quartile values for top growth; black circles: mean values for top growth; white circles: first quartile values for top growth. Quartile value regression lines are dashed; mean value regression lines are solid. SWAT RUE parameter (BIO_E) value is ten times the calculated RUE.

biomass resulted in suggested BIO_E parameter values of 41 for *Miscanthus* and 17 for switchgrass (Table 1).

Miscanthus values for RUE were consistent with studies in central Illinois (Heaton *et al.*, 2008) and Elsberry, Missouri, USA (Kiniry *et al.*, 2011), which had similar climates to this study and included approximately equivalent mean BIO_E values of 41 and 37, respectively. While the upper quartile range of *Miscanthus* BIO_E is consistent with what Zhang *et al.* (2011) anticipated, the modest change in mean (suggested) BIO_E value is likely due to the low root biomass and subsequent impacts on belowground contribution to total biomass. There are no clear explanations for how *Miscanthus* is able to produce such substantial top growth with a proportionally smaller root system, which merits additional investigation. However, while low, the values reported in this study are consistent with root fractions and root-to-shoot ratios reported both in Illinois and Western Europe (Beale & Long, 1997; Amougou *et al.*, 2011; Dohleman *et al.*, 2012).

The central Illinois study using a Cave-in-Rock (upland) cultivar observed a BIO_E mean value of 12 (Heaton *et al.*, 2008), identical to the suggested value using aboveground biomass values. Switchgrass RUE values were notably low compared to documented lowland varieties (Kiniry *et al.*, 1999), which highlights the importance of ecotype variation and site-specific conditions in parameter development.

Harvest index (HVSTI) and harvest efficiency (HEFF)

Inefficiencies in harvest operations impact yield and residue outcomes in the field. Residue sampling after harvest operation revealed that only 70% of aboveground biomass is typically removed by the forage harvester for *Miscanthus* stands, while mechanical switchgrass harvests removed 75% of available aboveground biomass. The suggested HEFF parameter value is the replicate mean of this study; the range includes the first and third quartiles of measured replicates in a single season (Table 1). A harvest index of one was used in conjunction with suggested HEFF value to ensure that no aboveground biomass was transferred as live material to the following year.

Optimal leaf area development and light extinction coefficient

Maximum LAI (BLAI) recorded during the study period (2010–2011) was 11.2 for *Miscanthus* stands and 11.7 for switchgrass (Figure S3). Light extinction coefficient calculations using the same data resulted in values from 0.45 to 0.65 for *Miscanthus*, and from 0.40 to 0.55 for switchgrass replicates (Table 1). Destructive sampling conducted during peak biomass indicated that the AccuPAR LP80 slightly underpredicted stand LAI across *Miscanthus* replicates, though meter results were consistent with reported values from central Illinois using the same method (Heaton *et al.*, 2008). In contrast, destructive sampling suggested that the AccuPAR LP80 consistently overpredicted switchgrass LAI after July. This was likely due to the influence of observed mid-season lodging on canopy structure. The impacts of equipment failure, drought, and lodging can be observed in reported values (Figure S3). To address this discrepancy, BLAI range for *Miscanthus* includes maximum value of 13 observed in destructive sampling, and suggested switchgrass BLAI was estimated as eight using data collected prior to lodging and comparative literature review of other upland cultivars within the region (Table 1) (Heaton *et al.*, 2008).

Optimal leaf area development fractions were estimated using fitted curves of mean LAI values represented as a fraction of BLAI over the fraction of PHU (Figure S3). Fractions greater than 1 can be observed in switchgrass LAI fraction due to use of estimated BLAI described above (Table 1). Estimates were overlaid with values derived using published data from Illinois (Heaton, 2006), where fractions of LAI and PHU were consistently within one-tenth of study values for each parameter with the exception of the switchgrass second point fraction of BLAI (LAIMX2), which was within two-tenths.

Nutrient fractioning

The fraction of *Miscanthus* N and P tissue concentrations was highest in April and declined throughout the growing season. While the fraction of switchgrass N followed a similar trend, fraction of switchgrass P between April and July did not demonstrate a consistent trend from year-to-year (Fig. 4). SWAT plant nutrient fractions represent plant demand, regulating nutrient uptake and associated nutrient mass balance. For PRGs, nutrient fractions can also characterize nutrient mobilization between aboveground tissue and plant storage structures such as stem bases and rhizomes at varying points in crop development and in harvested tissues. SWAT is able to model end-of-season nutrient mobilization via a mass balance approach that uses the difference between pre- and postharvest plant nutrient content as quantified by PLTFR3 and CNYLD or CPYLD. Nutrients that are not mobilized out of aboveground tissues are removed at harvest; a smaller value of kg N ha^{-1} or kg P ha^{-1} removed during harvest is desirable to ensure adequate amounts of N and P available to the plant during spring regrowth.

SWAT model validation for crop simulation

Changes in model code and parameter values simulated biomass accumulation, leaf area development, and yield outputs for PRGs and corn consistent with observed growth in this study and reported values for the Midwest (Heaton *et al.*, 2004, 2008; Dohleman & Long, 2009; Kiniry *et al.*, 2011; Dohleman *et al.*, 2012). Peak total biomass production for *Miscanthus* simulations ranged from 50 to 60 Mg ha^{-1} with mean annual harvested yield approximately 30 Mg ha^{-1} (Figure S4a.). Total biomass production for switchgrass simulations peaked between 29 to 33 Mg ha^{-1} with a mean yield of approximately 12 Mg ha^{-1} across the 3-year simulation period. Peak total biomass from continuous corn simulations ranged from 23 to 30 Mg ha^{-1}. Simulated values for corn were consistent with reported values by Hernandez-Ramirez *et al.* (2011) from 2005–2006 for the study site where aboveground biomass for continuous corn averaged 21.5 Mg ha^{-1}, which is roughly equivalent to 32.7 Mg ha^{-1} total biomass based on accepted root-to-shoot ratios for corn (Bray, 1963 as cited by Kramer & Boyer, 1995).

Simulated LAI development curves (Figure S4b.) consistently mirrored the extended growth periods of PRGs compared to continuous corn while maintaining a reasonable range of LAI values throughout the growing season compared to observed values (Figure S3). *Miscanthus* and switchgrass simulations reflected a longer growing season than the continuous corn due to early-growing-season emergence prior to corn planting as well as delayed onset of senescence at the end of the growing season. Additionally, the effects of different base temperature values were observed between *Miscanthus* and switchgrass simulations (Figure S4a, b); switchgrass emergence was delayed briefly in the first and third years, though it corresponded closely to *Miscanthus* emergence in the second year. These patterns are consistent with field observations.

Simulation outputs also revealed differences in mean annual uptake of plant N and P between PRGs and continuous corn (Figure S4c., d.). Year-to-year nutrient storage was the most pronounced difference. *Miscanthus* and switchgrass maintained between 10 and 100 kg N ha^{-1} and between 1 and 35 kg P ha^{-1} from one season into the next. There appeared to be a gradual decline in simulated plant N and P from 2009 to 2011. Nutrients in the continuous corn system cannot be stored, but instead are carried forward as residue. The retained nutrients in simulated PRG systems reduced simulated uptake in early stages of shoot emergence and development in spring of the following year. Mean annual N uptake within a given growing season was simulated as approximately 250 kg N ha^{-1} for continuous corn, but only as 150 kg N ha^{-1} for *Miscanthus*.

Changes in the model code

Removing root decay variables from harvestop.f code maintained belowground biomass through the dormant period for PRGs into the following growing season, averaging a 9 Mg ha^{-1} difference from the SWAT 2009 model (Figure S5). Similarly, belowground tissue N and P persisted through dormancy as nutrient storage pools (Figure S5b, d, f); postharvest belowground tissue N and P corresponded to early-growing season plant N and P, which provided a nutrient resource for

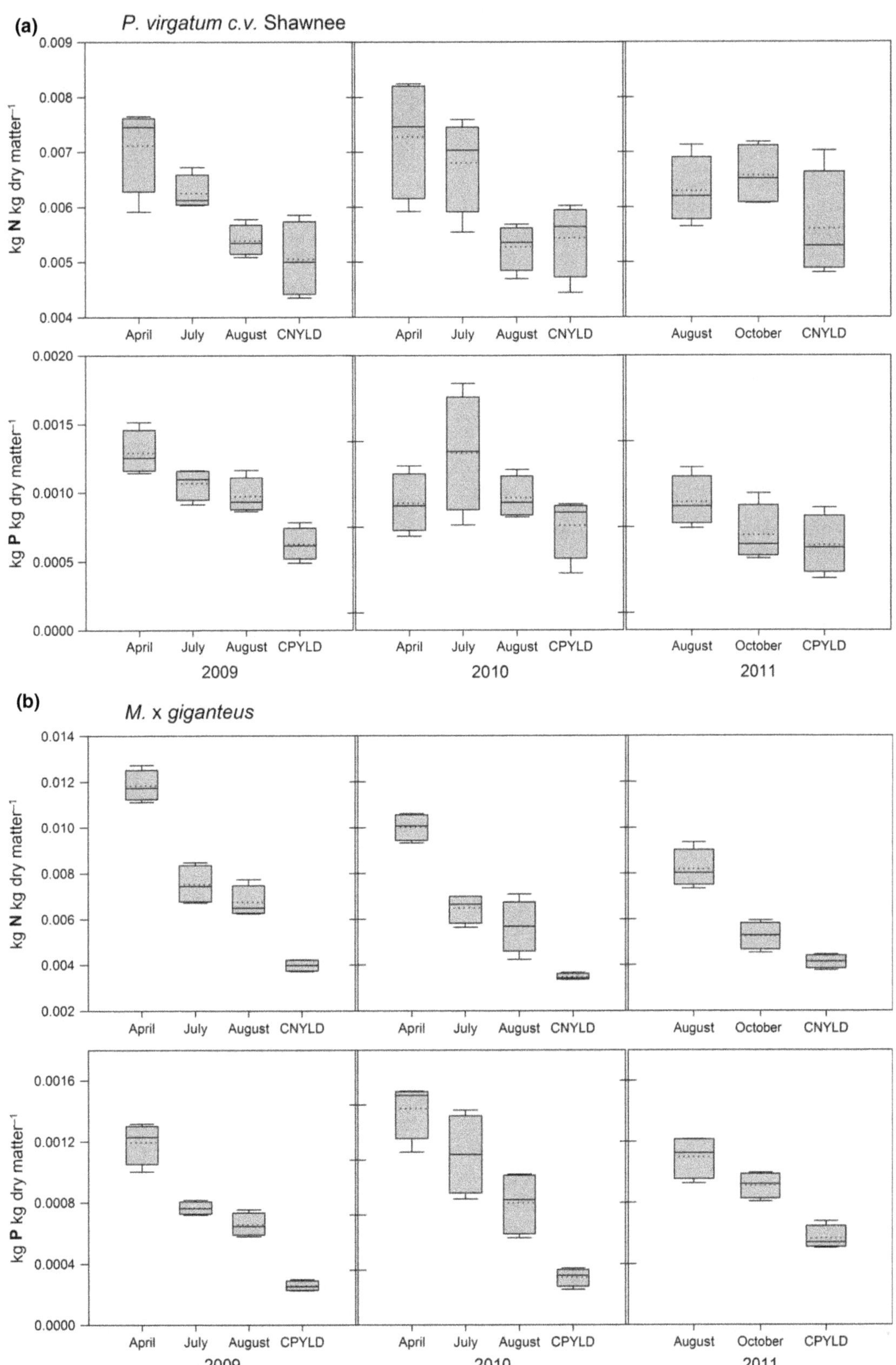

Fig. 4 Measured switchgrass (a) and *Miscanthus* (b) nitrogen (N) and phosphorus (P) nutrient fractions. PLTNFR and PLTPFR are whole-plant nutrient fractions (kg nutrient kg DM^{-1}), while CNYLD and CPYLD are nutrient fractions in harvested (October) biomass. Dotted line in box plot is the mean, solid line is the median. Nutrient fraction acronyms defined in Table 1.

early-growth plant requirements. While this improved simulated plant N compared to measured plant N (Fig. 6c), simulated plant P did not capture the lower, early-season, measured plant P content (Fig. 6d). The approximately 100 kg ha^{-1} N contribution from belowground organs decreased annual plant nutrient uptake requirements of the perennial crops, which affected nutrient pools and subsequent water quality impacts.

DLAI algorithm improvement (Fig. 1) successfully extended the period of plant respiration and subsequent transpiration by maintaining LAI beyond peak biomass, while eliminating the postharvest LAI value currently simulated by SWAT 2009 (Figure S5c). Maintaining LAI past peak biomass increased late-growing season evapotranspiration (ET) compared to SWAT 2009 simulations (Figure S5e).

Range analysis and output comparison

Random selection of parameters from suggested ranges produced an array of alternative scenarios for which simulated total biomass and LAI development varied to a lesser extent than plant nutrient content and uptake (Figs 5; 6). Although factorial analysis of interactive effects was beyond the scope of this study, a few key observations were made. The 2011 simulated LAI captured the range and shape of measured LAI (Figs 5b; 6b), but a temporal shift was observed when comparing simulated and observed LAI in 2010; simulated LAI increased at a faster rate than measured LAI. This may be due to the influence of the long-term estimation method of PHU. In 2010, heat units accumulated more quickly than they did in 2011, resulting in earlier maturity. At the same time, 2010 experienced a very wet spring season, for which flooding may have delayed in-field development that the model was unable to simulate. This may also reaffirm the importance of future research in frequent early- and mid-season LAI measurements for precise determination of shape coefficient parameters.

While the suggested parameter values for plant N fraction resulted in simulated plant N content for *Miscanthus* consistent with observed plant N throughout the growing season (Fig. 5c), simulated plant N values for switchgrass parameter values consistently overpredicted winter and early–growing-season plant N (75 kg ha^{-1}) compared to observed values (50 kg ha^{-1}) (Fig. 6c) Similarly, both *Miscanthus* and switchgrass suggested parameter values consistently overpredicted simulated plant P content in winter and early-growing season (Figs 5d; 6d) compared to observed plant P content by almost 25 kg ha^{-1} (*Miscanthus*) and 5 kg ha^{-1} (switchgrass). However, when the full range of parameters was used in the range analysis, the plant N and P content in the winter and early-growing season were adequately simulated. The associated shift in nutrient uptake response extended and decreased uptake timing depending on the specific parameter combination. Nutrient fraction variability also affected the timing of nutrient uptake (Fig. 5e, f; Fig. 6e, f).

Hydrologic and water quality response in the Single-HRU model

Hydrologic and water quality response of *Miscanthus* and Shawnee switchgrass was considerably different from Alamo switchgrass (default switchgrass available in the SWAT model) and corn (Table 2). Results from a single-HRU simulation indicate that annual evapotranspiration losses are potentially higher resulting in lower water yield than Alamo switchgrass and corn. Surface runoff values were comparable for the PRGs (while remaining lower than corn), and erosion values reflected the reduction in runoff. Organic and mineral N and P values for *Miscanthus* and Shawnee switchgrass were consistent, but lower, than Alamo, suggesting that larger-scale application of revised parameters may indicate greater reduction in erosion, N, and P losses in the Corn Belt compared to simulations done with Alamo switchgrass. Watershed-scale impacts are outside the scope of this paper as we focus on the most basic unit of SWAT simulations – HRU scale. Interested readers can refer to Cibin (2013) for detailed discussions on watershed-scale impacts of Shawnee switchgrass and *Miscanthus* production.

Discussion

We have compared the results of this study with existing regional data and published values to improve estimates, which should be considered in the use of suggested parameter values. For the most sensitive attributes of crop development, regionally based parameters are anticipated to improve simulation results and subsequent interpretation of bioenergy cropping system impacts at the regional scale. For example, the United States Department of Energy anticipates switchgrass and *Miscanthus* to be among the leading cellulosic feedstocks in the Great Plains and Midwest USA (U.S. Department of Energy, 2011). In the absence of any large-scale measured data, it is likely that decisions on biomass production and siting of bio-refineries will be based on model simulations. Similarly, quantification of unintended consequences on hydrology and water quality will also be based on model simulations. Under these scenarios, it is important that models are parameterized with evidence-based values that represent regional site, climate, and cultivar characteristics.

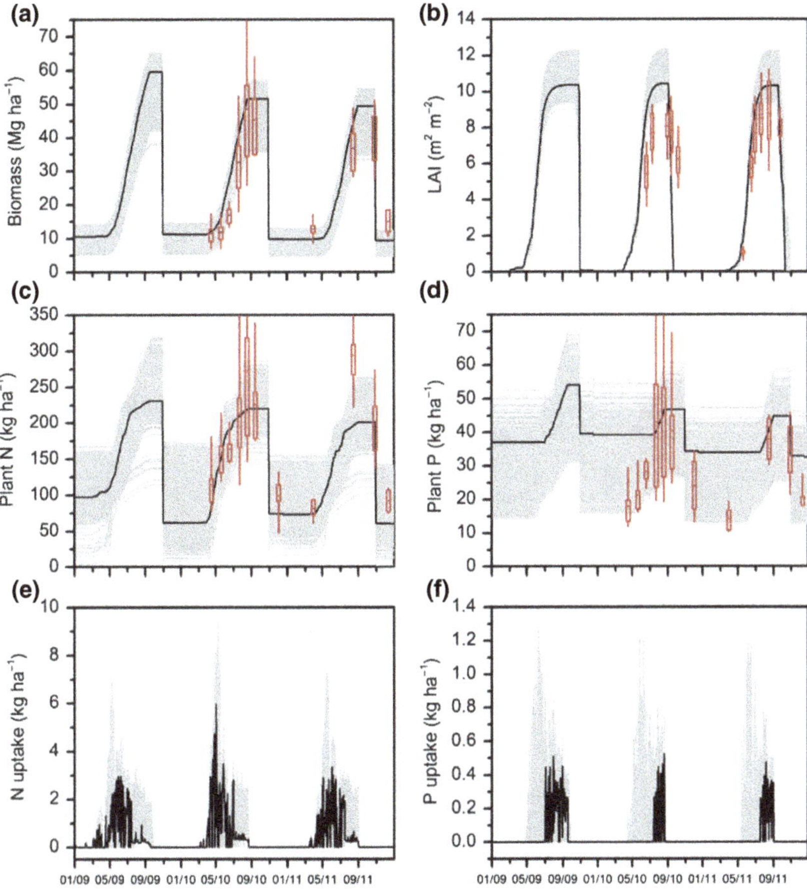

Fig. 5 Analysis of suggested parameter ranges for *Miscanthus*; the gray lines demonstrate 1000-sample Monte Carlo simulation of randomly chosen parameter values from suggested ranges. Suggested parameter values represented by black line; boxplots show measured data for (a) total biomass, (b) leaf area index, (c) total plant nitrogen, (d) total plant phosphorus, (e) plant nitrogen uptake, and (f) plant phosphorus uptake.

At the same time, regional studies indicate where opportunities for changes to the model may be appropriate. This study points to the need for transparency and provenance in parameter selection, consideration of nutrient fractionation and availability, as well as the representation of PRGs for bioenergy production using traditional definitions of physiological maturity as the framework for optimal leaf area development. We anticipate that improvements in the model parameterization along with crop-growth process representation will enhance our ability to accurately simulate large-scale production of PRGs and associated responses. Similarly, landscape optimization studies to determine biomass production strategies under single- (e.g., maximization of biomass produced) or multiple objective functions (e.g., maximization of biomass produced and minimization or erosion) will benefit from improvements in the SWAT model at HRU scale as it forms the basic unit of spatial optimization (Maringanti *et al.*, 2011).

Generating regional parameters for improved model representation of crop response to regional conditions is a process that has tradeoffs. As we work to integrate measured values into model development it is critical to understand limitations of the methods employed. The number of parameters required to model crop growth in SWAT makes consistent field measurements that represent optimal crop growth and development incredibly challenging under realistic data collection constraints, while evidence-based parameters appear to generate crop growth that reflects what is actually occurring in the field. Site-specific variation must be considered as the parameter values contribute to regional assessment. We present parameter ranges to emphasize this point, while encouraging thoughtful selection and interpretation of crop-growth parameters to specific study conditions. While this uncertainty provides limitations to an otherwise pragmatically sound approach, datasets continue to expand and regional characteristics will likely emerge with the compilation and analysis of additional data from multiple field studies (Miguez *et al.*, 2012).

The results of this study demonstrate hydrologic and water quality impacts attributed to distinct physiologi-

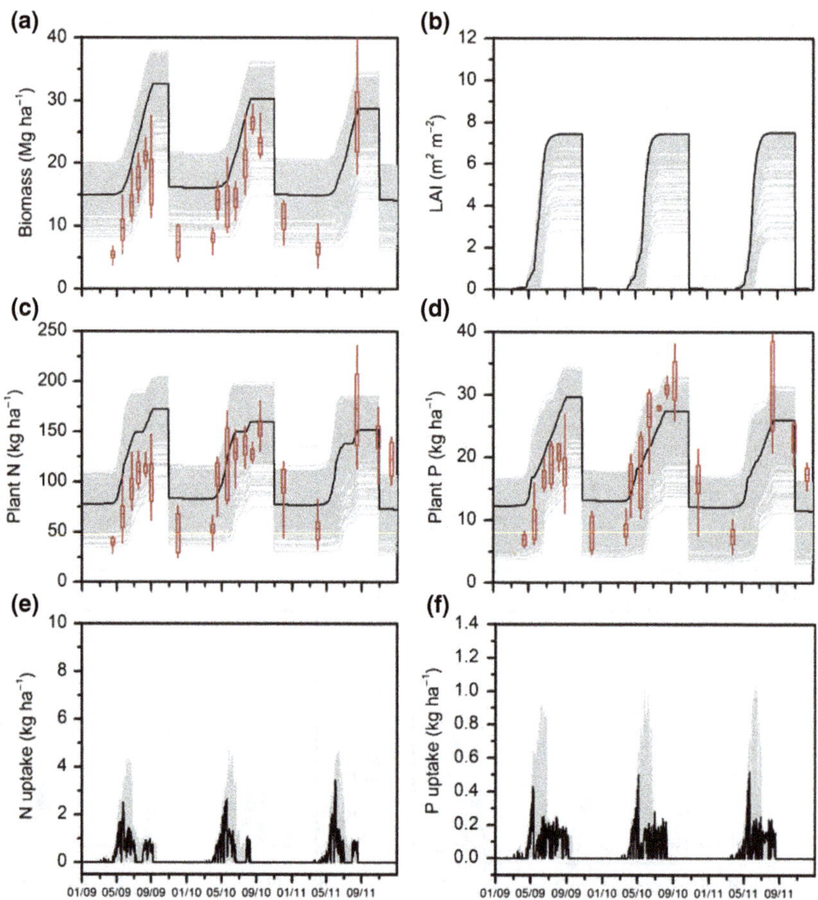

Fig. 6 Analysis of suggested parameter ranges for *Panicum virgaum* (c.v. *Shawnee*); the gray lines demonstrate 1000-sample Monte Carlo simulation of randomly chosen parameter values from suggested ranges. Suggested parameter values represented by black line; boxplots show measured data. for (a) total biomass, (b) leaf area index, (c) total plant nitrogen, (d) total plant phosphorus, (e) plant nitrogen uptake, and (f) plant phosphorus uptake.

Table 2 Single-HRU watershed outlet values for single-crop scenarios using revised model code for all perennial crops and presented parameters for *Miscanthus* and Shawnee switchgrass. Corn and Alamo switchgrass simulations used default crop-growth database parameter values

	ET (mm)	Surface Runoff (mm)	Water yield (mm)	Erosion (Mg ha^{-1})	Org N (kg ha^{-1})	Org P (kg ha^{-1})	Nitrate (kg ha^{-1})	Min P (kg ha^{-1})
Corn	702	202	343	4.454	27.96	3.435	30.46	1.141
Alamo switchgrass	610	61	435	0.021	0.14	0.017	18.39	0.028
Shawnee switchgrass	786	39	259	0.010	0.07	0.009	14.59	0.020
Miscanthus	845	33	199	0.009	0.06	0.007	8.20	0.022

cal attributes of PRGs as well as crop-specific parameters. For example, the extended growing season observed in *Miscanthus* due to early spring emergence and indeterminate growth into September is combined with substantial vegetative biomass production throughout the growing season, resulting in an increase in transpiration losses (Figure S5). The modification of the optimal leaf area development curve (Fig. 1) provided the conceptual basis for these improvements. However, unanticipated challenges with LAI measurement occurring both in this study as well as others (Heaton, 2006) offer opportunity for future work to generate more robust LAI values at key times in plant development to develop shape coefficients with greater precision.

Additionally, new input options such as the root fraction provide increasing flexibility of the model to simulate in-field conditions, and reasonable outputs are possible with the use of existing parameters. At the same time, the current model inherently attributes structures as above- or belowground. While this has practical elements, the biomass partitioning of PRGs may be more effectively categorized by role, such as vegetative, uptake, or storage structure as opposed to above- and belowground. The allocation of stem base biomass is a useful example. There are logical arguments for allocating stem bases in either the above- or belowground pools: though they are predominantly an aboveground structure, their role in nutrient and carbohydrate storage over winter is very similar to rhizomes. We allocated half of stem base mass into the belowground biomass pool to represent their contribution to overwinter survival and total biomass at emergence, and half the mass into aboveground pools. However, this means some storage tissues become residue due to harvest inefficiencies. Ideally, the user would be able to distinguish the mass and nutrient contributions of this pool independent of vegetative or root structures.

Consistent increases in total biomass production of *Miscanthus* stands suggest that belowground biomass is still developing as the perennial stands mature. It is likely that model outputs are sensitive to differences between parameter values for PRGs during the period of establishment compared to established PRGs (Kiniry *et al.*, 2008a). Above- and belowground biomass partitioning, associated ET, nutrient fractions, and nutrient uptake are all parameters that may benefit from additional investigation (Strullu *et al.*, 2011; Cadoux *et al.*, 2012; Dohleman *et al.*, 2012). While there are challenges to modeling a landscape mosaic of establishing vs. established PRG monocultures in HRUs, modifying existing algorithms that allow for partitioning of juvenile and adult phases of tree growth may provide a mechanism to implement phases of PRG life cycles. Similarly, there are few measured data to compare changes in belowground biomass over winter. More detailed studies with belowground biomass data during dormancy period are required to model root decay in PRGs during dormancy.

Simulated impacts on watershed hydrology and water quality outcomes in single-HRU and watershed-scale scenarios (Cibin 2013) support the importance of region- and crop-specific parameter accuracy to effectively model environmental impacts with SWAT. Future collaborative efforts between modelers and agronomists provide the opportunity to improve model representation of bioenergy cropping systems as research continues to document potential impacts of these crops on hydrologic processes.

Acknowledgement

We would acknowledge Niki DeArmond and Suzanne Cunningham for supporting field data collection and laboratory analysis. This work was made possible due to funding from the United States Sun Grant Initiative, United States Department of Energy (based upon work supported by Award Number DE-EE0004396), CenUSA Bioenergy supported by Agriculture and Food Research Initiative (2011-68005-30411), and the Purdue University C. Andrews Graduate Fellowship.

Conflict of interest: The authors declare no competing financial interest.

Author Contributions

The manuscript was written through contributions of all authors. All authors have given approval to the final version of the manuscript

References

Amougou N, Bertrand I, Machet JM, Recous S (2011) Quality and decomposition in soil of rhizome, root and senescent leaf from *Miscanthus × giganteus*, as affected by harvest date and N fertilization. *Plant and Soil*, **338**, 83–97.

Arnold JG, Srinivasan R, Williams JR (1998) Large area hydrologic modeling and assessment part I: model development. *Journal of the American Water Resources Association*, **34**, 73–89.

Arnold JG, Kiniry JR, Srinivasan R, Williams JR, Haney EB, Neitsch SL (2011) Soil and water assessment tool input/output file documentation version 2009. Texas Water Resources Institute Technical Report No. 365. College Station, TX, USA 1–643. (accessed: April 1 2011).

Beale CV, Long SP (1997) Seasonal dynamics of nutrient accumulation and partitioning in the perennial C4-grasses *Miscanthus × giganteus* and *Spartina cynosuroides*. *Biomass and Bionergy*, **12**, 419–428.

Beale CV, Bint DA, Long SP (1996) Leaf photosynthesis in the C4-grass *Miscanthus × giganteus*, growing in the cool temperate climate of southern England. *Journal of Experimental Botany*, **47**, 267–273.

Bray JR (1963) Root production and the estimation of net productivity. *Canadian Journal of Botany*, **41**, 65–72.

Burks JL (2013) Eco-physiology of three perennial bioenergy systems. PhD Dissertation. Purdue University, West Lafayette, IN, USA, 237.

Cadoux S, Riche AB, Yates NE, Machet J (2012) Nutrient requirements of *Miscanthus × giganteus*: conclusions from a review of published studies. *Biomass and Bioenergy*, **38**, 14–22.

Cibin R, Chaubey I, Engel B (2012) Simulated watershed scale impacts of corn stover removal for biofuel on hydrology and water quality. *Hydrological Processes*, **26**, 1629–1641.

Cibin R (2013) *Optimal Land use Planning on Selection and Placement of Energy Crops for Sustainable Biofuel Production*. PhD dissertation, Department of Agricultural and Biological Engineering, Purdue University, ProQuest, UMI Dissertations Publishing.

Dohleman FG, Long SP (2009) More productive than maize in the Midwest: how does *Miscanthus* do it? *Plant Physiology*, **150**, 2104–2115.

Dohleman FG, Heaton EA, Arundale RA, Long SP (2012) Seasonal dynamics of above- and below-ground biomass and nitrogen partitioning in *Miscanthus × giganteus* and *Panicum virgatum* across three growing seasons. *Global Change Biology Bioenergy*, **4**, 543–544.

Engel B, Chaubey I, Thomas M, Saraswat D, Murphy P, Bhaduri B (2010) Biofuels and water quality: challenges and opportunities for simulation modeling. *Biofuels*, **1**, 463–477.

Gassman PW, Reyes MR, Green CH, Arnold JG (2007) The soil and water assessment tool: historical development, applications and future research directions. *Transactions of the American Society of Agricultural and Biological Engineering*, **50**, 1211–1250.

Heaton EA (2006) The comparative agronomic potential of *Miscanthus × giganteus* and *Panicum virgatum* as energy crops in Illinois. PhD Dissertation. Urbana-Champaign, IL

Heaton EA, Voigt TB, Long SP (2004) A quantitative review comparing the yields of two candidate C4 perennial biomass crops in relation to nitrogen, temperature, and water. *Biomass and Bioenergy*, **27**, 21–30.

Heaton EA, Dohleman FG, Long SP (2008) Meeting US biofuel goals with less land: the potential of *Miscanthus*. *Global Change Biology*, **14**, 2000–2014.

Heaton EA, Dohleman FG, Miguez AF *et al.* (2010) *Miscanthus*: a promising bioenergy crop. *Advances in Botanical Research*, **56**, 75–137.

Hernandez-Ramirez G, Brouder SM, Smith DR, Van Scoyoc GE (2011) Nitrogen partitioning and utilization in corn cropping systems: rotation, N source, and N timing. *European Journal of Agronomy*, **34**, 190–195.

Hsu FH, Nelson CJ, Matches AG (1985) Temperature effects on germination of perennial warm-season forage grasses. *Crop Science*, **25**, 215–220.

Jager HI, Baskaran LM, Brandt CC, Davis EB, Gunderson CA, Wullshleger SD (2010) Empirical geographic modeling of switchgrass yields in the United States. *Global Change Biology Bioenergy*, **2**, 248–257.

Johnson MVV, Kiniry JR, Burson BL (2010) Ceptometer deployment method affects measurement of fraction of intercepted photosynthetically active radiation. *Agronomy Journal*, **102**, 1132–1137.

Kiniry JR, Sanderson MA, Williams JR *et al.* (1996) Simulating Alamo switchgrass with the ALMANAC model. *Agronomy Journal*, **88**, 602–606.

Kiniry JR, Tischler CR, Van Esbroeck GA (1999) Radiation use efficiency and leaf CO_2 exchange for diverse C4 grasses. *Biomass and Bioenergy*, **17**, 95–112.

Kiniry JR, Lynd L, Greene N, Johnson MV, Casler M, Laser MS (2008a) Biofuels and water use: comparison of maize and switchgrass and general perspectives. In: *New Research on Biofuels* (eds Wright JH, Evans DA), pp. 17–30. Nova Science Publishers, New York, NY, USA.

Kiniry JR, MacDonald JD, Armen RK, Watson B, Putz G, Ellie EP (2008b) Plant growth simulation for landscape-scale hydrological modelling. *Hydrological Sciences Journal*, **53**, 1030–1042.

Kiniry JR, Schmer MR, Vogel KP, Mitchell RB (2008c) Switchgrass biomass simulation at diverse sites in the Northern Great Plains of the U.S. *BioEnergy Research*, **1**, 259–264.

Kiniry JR, Johnson MV, Bruckerhoff SB, Kaiser JU, Cordsiemon RL, Harmel RD (2011) Clash of the titans: comparing productivity via radiation use efficiency for two grass giants of the biofuel field. *BioEnergy Research*, **5**, 41–48.

Kramer PJ, Boyer JS (1995) Roots and root systems. In: *Water Relations of Plants and Soils*, pp. 115–166. Academic Press, San Diego, CA, USA. Available at: http://books.google.com/books?hl=en&lr=&id=H6aHAwAAQBAJ&oi=fnd&pg=PP2&dq=kramer+boyer+water&ots=BWEe5ZQ50G&sig=1DHHV2KfdIUBEqcYzr3_RqvAjXs#v=onepage&q=kramer%20boyer%20water&f=false (accessed 15 December 2013).

Lewandowski I, Scurlock JMO, Lindvall E, Christou M (2003) The development and current status of perennial rhizomatous grasses as energy crops in the US and Europe. *Biomass and Bioenergy*, **25**, 335–361.

Love BJ, Nejadhashemi AP (2011) Water quality impact assessment of large-scale biofuel crops expansion in agricultural regions of Michigan. *Biomass and Bioenergy*, **35**, 2200–2216.

Maringanti C, Chaubey I, Arabi M, Engel B (2011) Application of a multi-objective optimization method to provide least cost alternatives for NPS pollution control. *Environmental Management*, **48**, 448–461.

Miguez FE, Zhu X, Humphries S, Bollero GA, Long SP (2009) A semimechanistic model predicting the growth and production of the bioenergy crop *Miscanthus*× *giganteus*: description, parameterization and validation. *Global Change Biology Bioenergy*, **1**, 282–296.

Miguez FE, Maughan M, Bollero GA, Long SP (2012) Modeling spatial and dynamic variation in growth, yield, and yield stability of the bioenergy crops *Miscanthus* × *giganteus* and *Panicum virgatum* across the conterminous United States. *Global Change Biology Bioenergy*, **4**, 509–520.

Murphy J, Riley JP (1962) A single solution method for the determination of phosphate in natural waters. *Analytical Chimica Acta*, **27**, 31–36.

Naidu SL, Moose SP, Al-Shoaibi AK, Raines CA, Long SP (2003) Cold tolerance of C4 photosynthesis in *Miscanthus* × *giganteus*: adaptations in amounts and sequence of C4 photosynthetic enzymes. *Plant Physiology*, **132**, 1688–1697.

National Research Council (2011) *Renewable Fuel Standard: Potential Economic and Environmental Effects of U.S. Biofuel Policy*. The National Academies Press, Washington, D.C, USA.

Neitsch SL, Arnold JG, Kiniry JR, Williams JR (2011) Soil and Water Assessment Tool theoretical documentation version 2009. Texas Water Resources Institute Technical Report No. 406. College Station, TX, USA. 1–618. (accessed April 1 2011).

Ng TL, Eheart JW, Cai X, Miguez F (2010) Modeling *Miscanthus* in the soil and water assessment tool (SWAT) to simulate its water quality effects as a bioenergy crop. *Environmental Science and Technology*, **44**, 7138–7144.

Parrish DJ, Fike JH (2005) The biology and agronomy of switchgrass for biofuels. *Critical Reviews in Plant Sciences*, **24**, 423–459.

Simpson TW, Sharpley AN, Howarth RW, Paerl HW, Mankin KR (2008) The new gold rush: fueling ethanol production while protecting water quality. *Journal of Environmental Quality*, **37**, 318–324.

Stephens W, Hess T, Knox J (2001) Review of the effects of energy crops on hydrology. NF0416, Cranfield University, MAFF.

Strullu L, Cadoux S, Preudhomme M, Jeuffroy M-H, Beaudoin N (2011) Biomass production and nitrogen accumulation and remobilization by *Miscanthus* × *giganteus* as influenced by nitrogen stocks in belowground organs. *Field Crops Research*, **121**, 381–391.

Thompson M, Anex R, Bach E *et al.* (2012) Biomass production and ecosystem services in Iowa biofuel cropping systems. Oral presentation at the ASA, CSSA, and SSSA International Annual Meeting. Cincinnati, OH, USA.

U.S. Department of Energy (2011) Chapter 5: Biomass energy crops. In: *US Billion-Ton Update: Biomass Supply for a Biorefinery and Bioproducts Industry* (eds Perlack RD, Stokes BJ), pp. 227, Oak Ridge National Laboratory, Oak Ridge, TN, USA.

Vanloocke A, Bernacchi CJ, Twine TE (2010) The impacts of *Miscanthus*× *giganteus* production on the Midwest US hydrologic cycle. *Global Change Biology Bioenergy*, **2**, 180–191.

Vogel KP, Hopkins AA, Moore KJ, Johnson KD, Carlson IT (1996) Registration of 'Shawnee' switchgrass. *Crop Science*, **36**, 1713.

Volenec JJ, Ourry A, Joern BC (1996) A role for nitrogen reserves in forage regrowth and stress tolerance. *Physiologia Plantarum*, **97**, 185–193.

Williams JR, Arnold JG, Kiniry JR, Gassman PW, Green CH (2008) History of model development at Temple Texas. *Hydrological Sciences Journal*, **53**, 948–960.

Wu Y, Liu S (2012) Impacts of biofuels production alternatives on water quantity and quality in the Iowa River Basin. *Biomass and Bioenergy*, **36**, 182–191.

Wu M, Demissie Y, Yan E (2012) Simulated impact of future biofuel production on water quality and water cycle dynamics in the Upper Mississippi river basin. *Biomass and Bioenergy*, **41**, 44–56.

Wullschleger SD, Davis EB, Borsuk ME, Gunderson CA, Lynd LR (2010) Biomass production in switchgrass across the United States: database description and determinants of yield. *Agronomy Journal*, **102**, 1158–1168.

Zhang X, Izaurralde RC, Arnold JG, Sammons NB, Manowitz DH, Thomson AM, Williams JR (2011) Comment on 'Modeling *Miscanthus* in the soil and water assessment tool (SWAT) to simulate its water quality effects as a bioenergy crop'. *Environmental Science & Technology*, **45**, 6211–6212.

Zub HW, Brancourt-Hulmel M (2010) Agronomic and physiological performances of different species of *Miscanthus*, a major energy crop A review. *Agronomy for Sustainable Development*, **30**, 201–214.

12

Comparative assessment of ecosystem C exchange in *Miscanthus* and reed canary grass during early establishment

ÓRLAITH NÍCHONCUBHAIR[1,2], BRUCE OSBORNE[2,3], JOHN FINNAN[4] and GARY LANIGAN[1]

[1]*Teagasc Environmental Research Centre, Johnstown Castle, Co. Wexford, Ireland,* [2]*UCD School of Biology & Environmental Science, University College Dublin, Dublin 4, Ireland,* [3]*UCD Earth Institute, University College Dublin, Dublin 4, Ireland,* [4]*Teagasc Crops Research Centre, Oak Park, Carlow, Ireland*

Abstract

Land-use change to bioenergy crop production can contribute towards addressing the dual challenges of greenhouse gas mitigation and energy security. Realisation of the mitigation potential of bioenergy crops is, however, dependent on suitable crop selection and full assessment of the carbon (C) emissions associated with land conversion. Using eddy covariance-based estimates, ecosystem C exchange was studied during the early-establishment phase of two perennial crops, C_3 reed canary grass (RCG) and C_4 *Miscanthus*, planted on former grassland in Ireland. Crop development was the main determinant of net carbon exchange in the *Miscanthus* crop, restricting significant net C uptake during the first 2 years of establishment. The *Miscanthus* ecosystem switched from being a net C source in the conversion year to a strong net C sink (-411 ± 63 g C m^{-2}) in the third year, driven by significant above-ground growth and leaf expansion. For RCG, early establishment and rapid canopy development facilitated a net C sink in the first 2 years of growth (-319 ± 57 (post-planting) and -397 ± 114 g C m^{-2}, respectively). Peak seasonal C uptake occurred three months earlier in RCG (May) than *Miscanthus* (August), however *Miscanthus* sustained net C uptake longer into the autumn and was close to C-neutral in winter. Leaf longevity is therefore a key advantage of C_4 *Miscanthus* in temperate climates. Further increases in productivity are projected as *Miscanthus* reaches maturity and are likely to further enhance the C sink potential of *Miscanthus* relative to RCG.

Keywords: bioenergy crops, C_4 photosynthesis, carbon balance, eddy covariance, grassland, land-use change, leaf longevity, perennial rhizomatous grasses, *Phalaris arundinacea*, reed canary grass

Introduction

Global concerns surrounding the impact of anthropogenic greenhouse gas (GHG) emissions on our climate allied with challenges to global energy security are driving interest in renewable energy, including bioenergy. In the European Union, member states have committed to increasing the contribution of renewable energy to 20% of total energy consumption by 2020 (EU, 2009). Furthermore, limitations identified in first-generation liquid biofuels have engendered an increased focus on second-generation alternatives produced from ligno-cellulosic plant materials (Sims *et al.*, 2010). Realisation of the potential of bioenergy in the EU will, however, require significant agricultural land area, estimated at between 18 and 21 million hectares (Özdemir *et al.*, 2009).

Correspondence: Órlaith Ní Choncubhair
e-mail: o.nichoncubhair@teagasc.ie

Perennial rhizomatous grasses (PRGs), such as *Miscanthus* × *giganteus* and reed canary grass (RCG), confer many advantages as potential nonfood bioenergy crops and have received increasing attention in Europe and the USA in recent decades (Landström *et al.*, 1996; Lewandowski *et al.*, 2003). *Miscanthus*, a C_4 plant originating from East Asia, can be highly productive, yielding 10–25 t dry matter ha^{-1} y^{-1} in central and northern Europe (Lewandowski *et al.*, 2000; Finnan & Burke, 2014), while yields of RCG, a C_3 perennial grass indigenous to temperate regions of Europe, Asia and North America, ranged from 5 to 12 t dry matter ha^{-1} y^{-1} in trials in northern Europe (Landström *et al.*, 1996; Saijonkari-Pahkala, 2001; Lewandowski *et al.*, 2003; Kandel *et al.*, 2013). As perennial species, both *Miscanthus* and RCG invest significant resources below ground, thus building reserves for more rapid canopy development in the spring compared with annual crops (Beale & Long, 1995; McLaughlin & Walsh, 1998).

High productivity in *Miscanthus* has historically been attributed to the superior light-, water- and nitrogen-use efficiency afforded by the C_4 photosynthetic pathway (Long, 1983). For example, the maximum efficiency of solar energy conversion in C_4 crops has been estimated to be 40% higher than that of C_3 species (Monteith, 1978). Furthermore, studies have highlighted the crop's exceptional ability to maintain high photosynthetic productivity even in cool temperate climates (Beale & Long, 1995). This is most likely achieved through a combination of reduced susceptibility to photoinhibition and decreased sensitivity to chilling temperatures (Beale *et al.*, 1996; Naidu & Long, 2004; Wang *et al.*, 2008).

More recent studies have, however, identified extended leaf longevity and high leaf area as the driving factors contributing to greater productivity in *Miscanthus*. In a side-by-side comparison of field-scale stands of *Miscanthus* and C_4 maize (*Zea mays*) in the USA, Dohleman & Long (2009) showed that the efficiency of captured sunlight-to-biomass conversion was almost identical in both crops averaged over two growing seasons. However, light interception efficiency was 61% higher in *Miscanthus*, which developed a closed canopy a month earlier than maize and maintained it a month longer in the autumn. This resulted in substantial net carbon gains for the cold-tolerant *Miscanthus* crop due to enhanced leaf area duration (Dohleman & Long, 2009).

The question remains, however, as to whether *Miscanthus* can out-perform native C_3 bioenergy crop candidates in temperate regions in terms of productivity, leaf longevity and net C sequestration. RCG, in particular, grows vigorously after seed establishment and easily out-competes weeds after the first year of establishment (Lewandowski *et al.*, 2003). Indeed, its competitive advantages have allowed RCG to become an invasive species in certain wetland areas of the mid-western and north-western USA (Wrobel *et al.*, 2009). To date, no direct comparisons of *Miscanthus* and RCG under the same environmental conditions have been made to understand the relative dynamics of crop development and C assimilation in these ecosystems. This information would: (1) provide focus for exploiting favourable plant traits and developing superior genotypes for bioenergy production, (2) highlight the relative merits of *Miscanthus* and RCG in different climatic zones with varying growing seasons and (3) reveal the C balance implications of land-use change (LUC) to these crops.

Additional factors, such as the previous land use and the magnitude of LUC-related emissions, will significantly impact the long-term C balance of established bioenergy crops. The initial transition phase of LUC can be associated with substantially increased GHG emissions (Guo & Gifford, 2002; Fargione *et al.*, 2008; Donnelly *et al.*, 2011; Poeplau *et al.*, 2011; Houghton *et al.*, 2012), particularly if bioenergy crop yields are low (Don *et al.*, 2012). The time taken for the crop to achieve maximum productivity is also likely to be a significant determinant of the early establishment C balance.

In mature *Miscanthus* crops, soil C sequestration rates of 0.4–0.66 Mg ha^{-1} y^{-1} have been reported for plantations established on former croplands (Don *et al.*, 2012; Zimmermann *et al.*, 2012; Poeplau & Don, 2014). Less information is available on grasslands converted to *Miscanthus*, but limited meta-analysis data has shown an annual carbon sequestration rate close to zero (no positive or negative effect) (Don *et al.*, 2012; McCalmont *et al.*, 2016; Qin *et al.*, 2016). Measurements of carbon exchange in RCG plantations have provided evidence of net C uptake in these ecosystems (Shurpali *et al.*, 2009; Mander *et al.*, 2012; Lind *et al.*, 2015); however, these studies were largely confined to drained peat extraction areas and there is little information on the carbon balance of RCG crop plantations established on mineral soils.

This study presents the first side-by-side field-scale comparison of *Miscanthus* and RCG to investigate comparative differences in crop development, leaf longevity and ecosystem-scale C fluxes from initial establishment to near maturity. The bioenergy crops were established on land previously under permanent grass. Conversion of grassland to bioenergy crops is of particular relevance in Ireland and in the wider European continent as over 90% and 35% of utilised agricultural area in Ireland and the EU-27, respectively, is currently used for grass production (Central Statistics Office, 2014; Huyghe *et al.*, 2014). The specific aim of the study was to address the following questions: (1) what are the C emissions associated with the initial LUC from grassland, which we define as the 'transitional phase'? (2) What are the longer term C balance implications of establishing *Miscanthus* and RCG crops on permanent grasslands (the 'post-establishment phase')? (3) How does leaf longevity in *Miscanthus* compare to that of an indigenous C_3 bioenergy crop and does it result in higher net C uptake?

Materials & Methods

Site description & management

The study was carried out from late April 2009 to the end of December 2011 at the Teagasc Environmental Research Centre, Johnstown Castle, Co. Wexford in the south-east of Ireland (52.3°N, 6.5°W, 67 m above sea level). This region has a maritime temperate climate with a mean annual rainfall of 1038 mm distributed evenly across the year and a mean annual air temperature of 10.4 °C. The seasonal range in temperature is narrow (Fig. 1), with an average summer air temperature of

14.9 °C and an average winter air temperature of 6.3 °C. The prevailing wind direction is south-westerly.

The *Miscanthus* and RCG crops were established on two former grassland sites (2 ha and 1 ha in area, respectively) (Fig. 2). Most of the experimental area had been maintained as grassland for at least 37 years and was managed organically for beef production since 2006. The *Miscanthus* site had been conventionally tilled and reseeded with perennial ryegrass in 2000 and surface seeded with white clover in 2005. The RCG site had been conventionally tilled and reseeded with perennial ryegrass and red clover in 2005. Otherwise, historical land management activities were comparable at both sites. The swards received organic fertiliser in the form of cattle slurry and farmyard manure. Grazing took place every 3–4 weeks until October 2008. Soils in this area are variable and are classified as imperfectly drained Gleys (FAO classification: Gleyic Cambisol) or moderately to well-drained Brown Earth soils (Cambisol). Selected soil physical and chemical properties of the *Miscanthus*, RCG and adjacent reference grassland sites are summarised in Table 1. The wilting point, field capacity and water content at saturation were calculated using a hydraulic properties model (http://hydrolab.arsusda.gov/soilwater/Index.htm) as 0.17, 0.32 and 0.59, respectively, for all sites.

On the 1st April 2009, both sites were sprayed with glyphosate to eradicate the extant vegetation. The soil was conventionally tilled using a mouldboard plough to a depth of 20 cm on the 27th April 2009 (approximately one tenth of the *Miscanthus* site) and completed on the 29th April 2009, followed by power-harrowing on the 1st and 5th June 2009. *Miscanthus* rhizomes were planted on one site on the 9th and 10th June 2009 and the soil was consolidated using a heavy roller 1 week later. Additional herbicides were applied to the *Miscanthus* site in the early establishment phase to reduce competition from grass and broad-leaf weeds. The selective herbicide MCPA was sprayed to control broad-leaf species (13th August 2009 and 27th July 2010) while glyphosate was applied soon after harvesting (5th March 2010 and 8th March 2011) to control grass weeds. The crop was cut with a conditioner mower on the 4th March 2010 and 4th March 2011 but the limited biomass material that was cut was left on the ground.

The second site lay fallow until April 2010 when RCG was established. For this, the site was power-harrowed and seeded with RCG (*Phalaris arundinacea* L.) at a rate of 30 kg ha^{-1} on the 15th April 2010, and then consolidated with a heavy roller 1 day later. MCPA was applied on the 29th June 2010 to reduce competition from broad-leaf species. The RCG crop was harvested once during the study period, on the 12th October 2010, while harvesting of the 2011 crop was delayed until spring 2012, which is the preferred time for harvesting. No fertilisers were applied over the duration of the study.

Micrometeorological measurements

Ecosystem-scale CO_2 fluxes were measured using an open-path eddy covariance (EC) system commencing on the 28th April 2009 and 15th April 2010 at the *Miscanthus* and RCG sites, respectively. The instrumentation was identical at both sites and consisted of an open-path infrared gas analyser (IRGA) (LI-7500, LI-COR Biosciences, Lincoln, NE, USA) coupled with a 3D sonic anemometer (CSAT3, Campbell Scientific, Logan, UT, USA). EC data were collected at a frequency of 10 Hz and averaged over 30-minute intervals. The flux tower was located in the north-eastern corner of the fields to maximise the fetch in the direction of the prevailing south-westerly wind (Fig. 2). Tower height was increased during periods of active growth to maintain its position above the canopy while restricting the flux footprint to the experimental area for as much time as possible. The maximum tower height was 4 m and 3 m at the *Miscanthus* and RCG sites, respectively, while the minimum sensor to canopy height ratios were 1.47 (*Miscanthus*) and 1.52 (RCG).

Ancillary biometric sensors included an air temperature and relative humidity probe (HMP45C, Campbell Scientific, Logan, UT, USA), a net radiation sensor (NR-Lite, Kipp & Zonen, Delft, The Netherlands) and a down-welling quantum sensor (SKP 215, Skye Instruments Ltd., Llandrindod Wells, UK). Two self-calibrating soil heat flux plates were installed at 8 cm soil depth (HFP01SC, Hukseflux, Delft, The Netherlands) and averaging soil temperature probes were installed at 2 cm and 6 cm depth above the soil heat flux plates. Time domain reflectometers (CS616, Campbell Scientific, Logan, UT, USA) measured soil volumetric water content (VWC) in the upper 15 cm of soil. Daily meteorological data (mean air temperature, total rainfall and global solar radiation) were obtained from a Met Éireann synoptic weather station located 1.7 km from the field site.

Quality assurance and flux analysis

Data quality control procedures included spike removal (Vickers & Mahrt, 1997), time lag compensation using a covariance maximisation procedure and compensation for air density fluctuations using the WPL term (Webb *et al.*, 1980). The double rotation method (Kaimal & Finnigan, 1994) was used to correct for sonic anemometer tilt as the alternative planar fit method requires several weeks of measurement with constant instrumental set-up and is often not recommended for measurements over canopies with dynamic height variation (Moureaux *et al.*, 2012). Spectral attenuation effects were corrected following the analytical methods of Moncrieff *et al.* (1997). Tests on developed turbulence and stationarity were applied to the calculated fluxes (Table 13, Mauder & Foken, 2004) and data of questionable quality (QC-flag = 2) were removed, while data of moderate quality (QC-flag = 1) were retained but not included in the regression analysis performed in the gap-filling procedure. Data screening based on the results of these tests has been shown to result in a less systematic distribution of data gaps compared to the removal of fluxes below a derived friction velocity (u_*) threshold (Ruppert *et al.*, 2006). Plausible limits were also applied to net ecosystem exchange (NEE) ($-50 < NEE < 30$ μmol CO_2 m^{-2} s^{-1}), latent heat ($-20 < LE < 600$ W m^{-2}) and sensible heat ($-100 < H < 300$ W m^{-2}) fluxes.

The peak location of the flux footprint, x_{max}, and the distance from the flux tower which includes 90% of the source area contributing to the measured flux, x_{R90}, were estimated for the prevailing south-westerly wind direction using the Kljun *et al.* (2004) model as 35 m and 95 m, respectively, for the *Miscanthus* site and 32 m and 87 m, respectively, for the

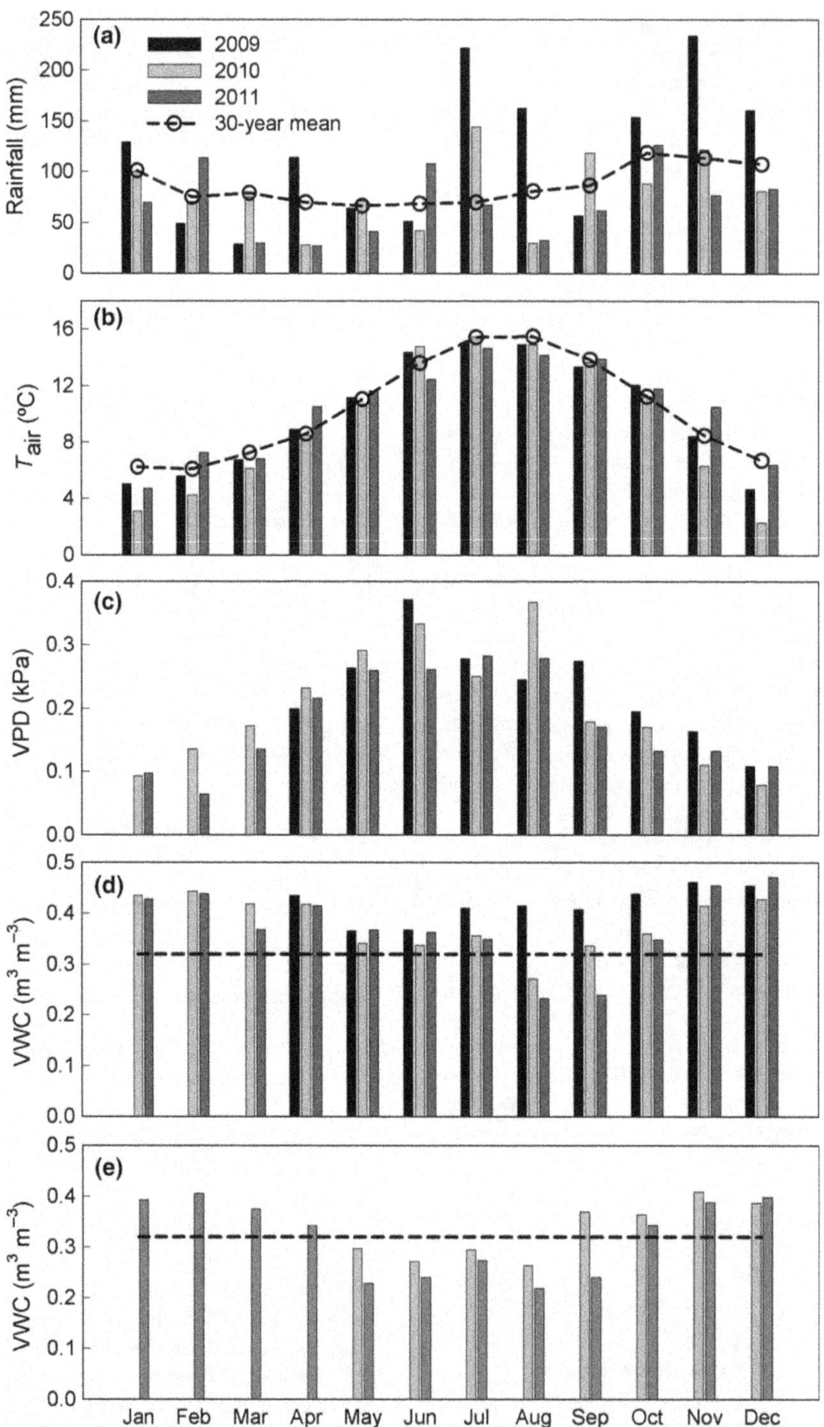

Fig. 1 Mean monthly values for total rainfall (a), air temperature (b), vapour pressure deficit (VPD) (c) and soil volumetric water content (VWC) at the *Miscanthus* (d) and RCG (e) sites for 2009, 2010 and 2011. The 30-year (1981–2010) mean monthly rainfall and air temperature values for Johnstown Castle are also shown. The soil VWC at field capacity is included in (d) and (e) as dashed black lines.

RCG site. Owing to the relatively constrained field sizes, detailed footprint analysis was conducted on half-hourly fluxes based on the analytical Kormann & Meixner (2001) footprint model. Since analytical models tend to overestimate flux footprints in comparison to more complex Lagrangian stochastic models due to the neglect of along-wind velocity fluctuation (Kljun *et al.*, 2003), this approach is likely to be conservative. Sample flux footprints for the two towers at their maximum height under unstable conditions are shown in Fig. 2. At the maximum height, the footprint of the *Mis-*

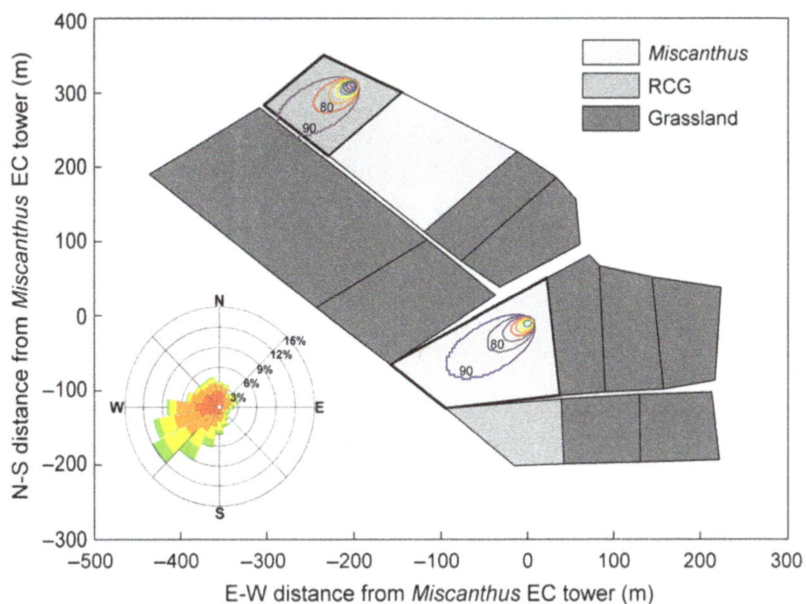

Fig. 2 Experimental site map showing the location and dimensions of the *Miscanthus* and RCG field sites and surrounding grassland fields, as well as a wind rose displaying the distribution of mean wind direction for the site. The wind rose was generated using the freeware WRPLOT View (http://weblakes.com). Contributions from wind speed classes 0.5–2.1 (red), 2.1–3.6 (orange), 3.6–5.7 (yellow), 5.7–8.8 (green) and $\geq$ 8.8 m s^{-1} (dark green) are depicted. Sample footprints are shown for the *Miscanthus* and RCG eddy covariance towers at their maximum height under unstable conditions ($z/L = -0.06$, where z is the height and L is the Monin-Obukhov length). Contour lines represent the crosswind integrated cumulative contribution (%) to the estimated footprint.

canthus tower was well conserved within the *Miscanthus* field while close to 90% of the flux footprint from the RCG tower was derived from the RCG field. At lower heights (April 2009 to August 2011 for the *Miscanthus* tower and April 2010 to June 2011 for the RCG tower), the flux footprints were better confined to the experimental areas. Half-hourly data were rejected if less than 70% of the derived footprint originated from the respective experimental areas, similar to previous studies (Ammann *et al.*, 2007; Baum *et al.*, 2008; Davis *et al.*, 2010; Vanderborght *et al.*, 2010). This resulted in 17.4% and 14.9% of fluxes being rejected overall for the *Miscanthus* and RCG sites, respectively.

The positioning of the EC sensors in relation to the inertial boundary layer was also monitored due to the relatively low measurement heights. The thickness of the inertial boundary layer, calculated according to equation 5 of Munro & Oke (1975), reached maximum values of 4.9 m, 4.9 m and 7.0 m in 2009, 2010 and 2011, respectively, for *Miscanthus* and values of 2.8 m and 3.7 m in 2010 and 2011, respectively, for RCG. The lower limit of the inertial boundary layer was estimated to occur between $(z\text{–}d) = 5z_0$ and $(z\text{–}d) = 10z_0$ (Garratt, 1992), where z is the height above ground, d is the zero-plane displacement height and z_0 is the roughness length of the crop surface. This yielded maximum values in the range 1.1–1.6 m, 1.1–1.5 m and 3.0–4.2 m in 2009, 2010 and 2011, respectively, for *Miscanthus* and values of 0.8–1.1 m and 1.7–2.5 m in 2010 and 2011, respectively, for RCG. These estimates suggested that the sensors were within the appropriate inertial boundary layer for the vast majority of the experiment.

The quality of the CO_2 flux estimates was assessed by examining energy balance closure (EBC) at the site. This routine provides an independent check of the degree to which turbulent

Table 1 Physical and chemical characteristics of the soils at the *Miscanthus*, RCG and adjacent reference grassland sites in autumn 2011. Bulk density (BD), texture, pH and total nitrogen (TN) are reported for 0–15 cm soil depth. The soils at all sites are classified as loam. Total carbon (TC) and total organic carbon (TOC) are shown for three depths: 0–15 cm (A), 15–30 cm (B) and 30–45 cm (C)

							TC (%)			TOC (%)		
	BD (g cm^{-3})	Sand (%)	Silt (%)	Clay (%)	pH	TN (%)	A	B	C	A	B	C
Miscanthus	0.98	50	32	18	6.4	0.3	3.2	2.9	1.5	2.4	2.2	1.1
Grassland reference	0.99	51	32	18	6.5	0.3	2.9	2.9	2.1	2.2	2.2	1.5
RCG	0.97	48	33	19	6.7	0.3	2.8	2.5	1.9	2.2	1.9	1.4
Grassland reference	0.88	52	30	19	6.2	0.3	2.7	2.5	2.1	2.0	1.8	1.5

fluxes are captured in the boundary layer and may highlight significant bias in measurements (Twine *et al.*, 2000). Energy balance closure was tested by comparing half-hourly and daily sums of net radiation with the sum of good quality (QC-flag = 0, 1) *LE* and *H* fluxes and energy storage terms using the equation

$$R_n = LE + H + G + S_s + S_p \quad (1)$$

where R_n is net radiation, *LE* is the latent heat flux, *H* is the sensible heat flux, *G* is the soil heat flux, S_s is soil heat storage above the heat flux plates and S_p is the energy stored in photosynthate. Heat storage in the soil surface layer (of depth Δz) was calculated as

$$S_s = \frac{\Delta T(\theta_v \rho_w c_w + \rho_s c_s)\Delta z}{\Delta t} \quad (2)$$

where ΔT is the change in average soil temperature above the heat flux plates over the time interval Δt, θ_v is the soil volumetric water content, ρ_w is the density of water, c_w is the specific heat capacity of water, ρ_s is the soil bulk density and c_s is the specific heat capacity of soil [a value of 837 J kg^{-1} K^{-1} was used for c_s (Scott, 2000)]. The energy captured during photosynthesis and stored in biomass, S_p, was computed by equating photosynthetic fixation of 2.5 mg CO_2 m^{-2} s^{-1} to an energy flux of 28 W m^{-2} (Meyers & Hollinger, 2004). For this calculation, photosynthetic rates were based on gross primary productivity (GPP) estimates calculated by subtracting modelled total ecosystem respiration (TER) from measured NEE. Heat storage in plant biomass and canopy air was not measured and considered minor relative to the other terms in equation 1, particularly in the case of *Miscanthus* which had a sparse, open canopy for much of the study.

The regression analysis of half-hourly energy fluxes yielded a slope of 0.945 ± 0.004 [95% confidence interval (CI)] and intercept of 5.2 W m^{-2} for the *Miscanthus* site and a slope of 0.905 ± 0.004 and intercept of 6.3 W m^{-2} for the RCG site (Fig. 3). Calculating daily sums of energy fluxes resulted in a small increase in the slope for *Miscanthus* to 0.963 ± 0.013 and a larger increase in the slope for RCG to 0.964 ± 0.016 (Fig. 3). This suggests that there was a small contribution from additional nonestimated storage terms, such as heat storage in plant biomass or canopy air, to the energy balance of these ecosystems. These contributions tend to be negligible on a daily scale (Oncley *et al.*, 2007). Although these slopes were significantly different from one, they compare well with

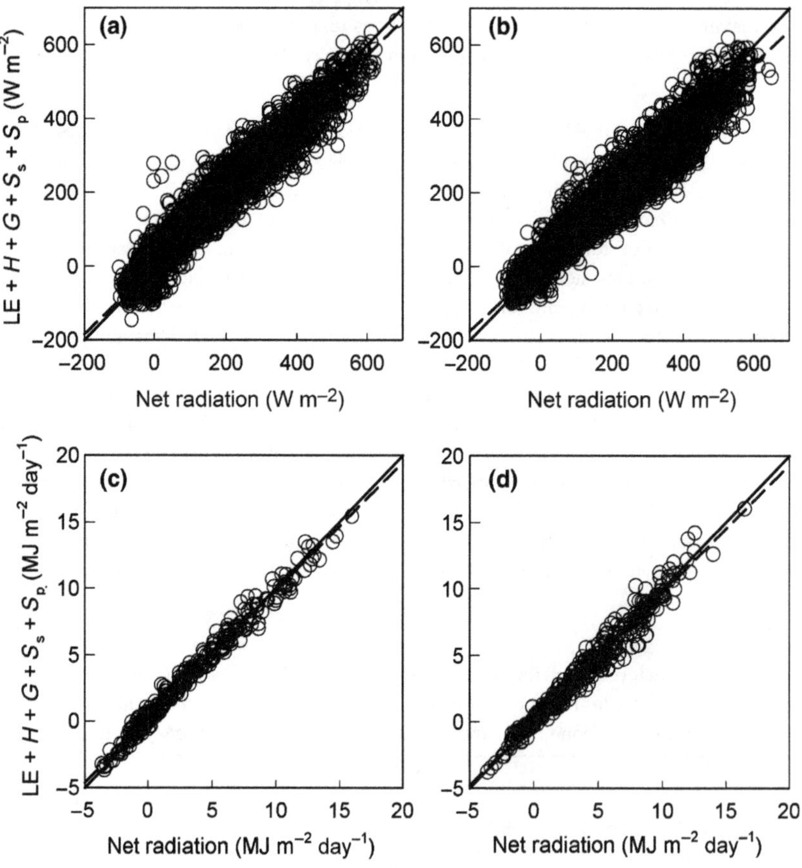

Fig. 3 Energy balance closure on a half-hourly timescale for *Miscanthus* (a) and RCG (b) and a daily timescale for *Miscanthus* (c) and RCG (d) in 2011. The dashed black lines represent the regression lines [$y = 0.95x + 5.24$, $R^2 = 0.96$ for *Miscanthus* (half-hourly); $y = 0.91x + 6.3$, $R^2 = 0.96$ for RCG (half-hourly); $y = 0.96x + 0.2$, $R^2 = 0.99$ for *Miscanthus* (daily); $y = 0.96x + 0.04$, $R^2 = 0.97$ for RCG (daily)]. The solid black lines indicate a 1 : 1 relationship. Daily sums were calculated using the available data in a 24-h period.

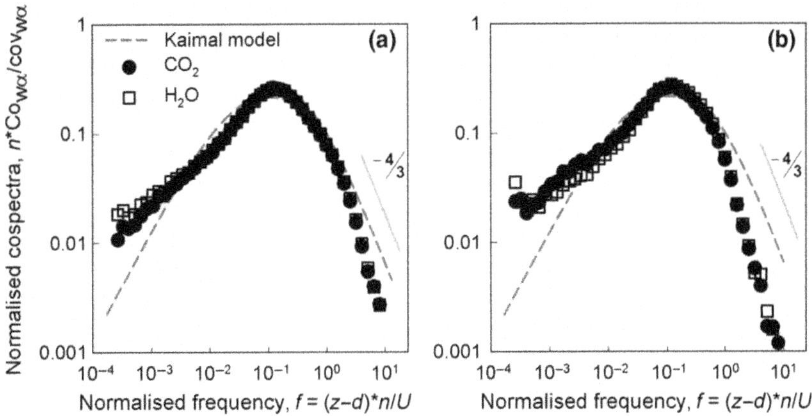

Fig. 4 Normalised ensemble-averaged cospectra of CO_2 and H_2O in unstable conditions plotted against normalised frequency, *f*, for the *Miscanthus* (a) and RCG (b) sites. The dashed grey curve and solid grey line represent the theoretical universal Kaimal cospectra and the −4/3 slope predicted by Kolmogorov's Law, respectively (Kaimal *et al.*, 1972).

reported ranges of EBC of 0.53–0.99 (Wilson *et al.*, 2002) and 0.70–0.94 (Stoy *et al.*, 2013) and suggest that the vast majority of the available energy was accounted for. Consequently, turbulent energy exchange was satisfactorily resolved by the EC measurement system.

In addition, cospectral analysis of the high frequency data was performed to assess the frequency response of the EC system. Fast Fourier transforms were applied to the 10 Hz data to yield full cospectra, which were subsequently reduced into exponentially spaced frequency bins and ensemble-averaged. Normalised, ensemble-averaged and frequency-weighted cospectra for both sites are presented in Fig. 4 for unstable conditions and compared with the universal theoretical Kaimal cospectral function (Moncrieff *et al.*, 1997). The shapes of the measured cospectra are generally consistent with the Kaimal curve, except for small deviations from the theoretical −4/3 slope in the inertial subrange. This result is indicative of some small-scale dampening of the high frequency signal, however, the flux postprocessing routines employed included a correction for high frequency losses. The CO_2 spectral correction factor (SCF) for *Miscanthus* had a median of 1.17 and mean ± 95% CI of 1.19 ± 0.001. For RCG, the median was 1.15 and the mean was 1.16 ± 0.002. These values fall in the expected range of CO_2 correction factors (1.04–1.25; Aubinet *et al.*, 2001) and compare well with an SCF of 1.12 for a similar open-path system (Haslwanter *et al.*, 2009).

Gaps in the flux data created by measurement failures and data quality analysis procedures were filled using semiempirical gap-filling techniques. TER was modelled by relating soil temperature to nocturnal measured NEE using the exponential Lloyd & Taylor (1994) equation

$$\mathrm{TER} = R_{10}e^{E_0(1/(283.15-T_0)-1/(T-T_0))} \tag{3}$$

where R_{10} is ecosystem base respiration at a reference temperature of 10 °C, E_0 is an activity energy parameter, T_0 was set to 227.13 K as in the original study and T is soil temperature (K). Significant seasonality has been demonstrated in the E_0 parameter, particularly in summer active ecosystems (Reichstein *et al.*, 2005), and therefore bimonthly best fit estimates of E_0 were obtained in each year using equation 3. These short-term bimonthly estimates were then averaged over each year to derive more realistic long-term E_0 values. Temporal variability in the R_{10} parameter was also considered. For each available nocturnal NEE value and adopting the relevant long-term E_0 value, equation 3 was rearranged to provide an estimate of R_{10} in that half-hourly period. The weekly running mean, $R_{10\mathrm{Mean}}$, was then computed and employed instead of R_{10} in equation 3 above to incorporate time-varying influences on R_{10} such as soil moisture status, plant phenology etc. The dependence of nocturnal TER on temperature was extended to daytime and estimates of TER were obtained for all 30-minute periods in this way.

GPP was initially computed by subtracting the estimated TER from measured daytime half-hourly NEE. Daytime GPP values were then pooled for bimonthly periods and subdivided into fixed 4 °C temperature bins within each bimonthly period. GPP model parameters were then derived for each subset using a rectangular hyperbola function (Falge *et al.*, 2001)

$$\mathrm{GPP} = \frac{\alpha Q_{\mathrm{PPFD}} A_{\max}}{\alpha Q_{\mathrm{PPFD}} + A_{\max}} \tag{4}$$

where α is the ecosystem quantum (photon) yield (mol CO_2 [mol photon]$^{-1}$), Q_{PPFD} is the photosynthetic photon flux density (μmol [photon] m^{-2} s^{-1}) and $A_{\max}$ is the maximum assimilation rate (μmol CO_2 m^{-2} s^{-1}).

The net ecosystem carbon balance (NECB) is defined as the net rate of C accumulation or loss from an ecosystem, taking account of all physical, biological and anthropogenic sources and sinks of carbon in an ecosystem (Chapin *et al.*, 2006). NECB was calculated as $-\Sigma\mathrm{NEE} - \mathrm{C}_{\mathrm{harvest}}$, where $\mathrm{C}_{\mathrm{harvest}}$ is the carbon removed from the ecosystem at harvest, assuming that contributions from carbon monoxide, methane, volatile organic carbon (VOC), dissolved carbon and particulate carbon were negligible.

An assessment of the uncertainty associated with annual CO_2 flux estimates was performed in a similar way to Black *et al.* (2007). Firstly, the contribution of uniform systematic errors to the measured fluxes was evaluated following Goulden *et al.* (1996). Assuming spectral similarity between latent heat, sensible heat and CO_2 flux, the imbalance between available energy and measured latent and sensible heat, as calculated in the long-term energy balance, was used as an approximation of the underestimation of ecosystem exchange by the EC measurement system. Secondly, sampling uncertainty errors, associated with the imputation of half-hourly missing data during gap-filling procedures, were assessed in a similar way to Falge *et al.* (2001). Artificial datasets were created with 10, 25, 35 and 45% of the data replaced by gaps and gap-filling procedures were followed to fill the artificial gaps. The mean bias error was calculated as the mean difference between measured and calculated values for the complete data series. The final uncertainty estimate was computed for each measured or gap-filled half-hourly flux based on either the uniform systematic error derived from the energy balance closure, expressed as a percentage, or the mean bias error arising from sampling uncertainty (gap-filling).

Soil respiration

Soil respiration was monitored using a closed dynamic chamber coupled with a portable infrared gas analyser (Environmental Gas Monitor EGM-4, PP Systems, Hitchin, UK). Chamber measurements were made on bare soil at 10 sampling points in the *Miscanthus* and RCG sites, yielding an estimate of the flux of CO_2 from the soil surface which combines contributions from autotrophic (plant root) and heterotrophic (microbes and soil fauna) components. Measurements were carried out from early June 2010 to early September 2011 at approximately weekly intervals during the growing season and monthly intervals during the winter. Soil temperature and VWC (0–7 cm) were recorded at each sampling point using a WET sensor (Delta-T Devices, Cambridge, UK).

Crop analysis

Above-ground *Miscanthus* biomass was calculated as the product of average shoot density and total dry biomass (leaf + stem) per shoot sampled prior to harvesting and at least monthly during the growing seasons of 2010 and 2011. Shoot density estimates (shoots m^{-2}) were based on the number of shoots counted in a quadrat of area 2.1 m^2 at 10 random locations in the *Miscanthus* plantation. Total dry biomass per shoot was measured by cutting a minimum of 35 random shoots at ground level during each sampling event and drying the leaf and stem biomass at 70 °C to constant weight. Below-ground biomass was sampled on three occasions (February 2010, January 2011 and September 2011) by excavating all below-ground plant material associated with three randomly located *Miscanthus* plants to a depth of 40 cm. Roots, live rhizomes and dead rhizomes were separated, washed free of soil over a 2 mm sieve and dried to constant weight at 70 °C. Below-ground biomass was up-scaled from the plant scale to a per unit area (m^{-2}) basis using calculated above- to below-ground ratios and average above-ground biomass on each individual sampling date. The contribution of understory vegetation was assessed by clipping all above-ground material in quadrats of area 0.25 m^2 placed inside the large *Miscanthus* quadrats with subsequent determination of dry matter yields after oven drying at 70 °C.

Above-ground RCG biomass was assessed by cutting all vegetation at ground level in quadrats of area 0.25 m^2 randomly positioned at five locations within the crop. Sampling was carried out in September 2010, prior to harvest in October 2010 and in June and September 2011. A root auger of volume 750 cm^3 (Eijkelkamp Agrisearch Equipment, Giesbeek, The Netherlands) was used to sample below-ground biomass to a depth of 45 cm in January and October 2011 at 4 random locations in the crop. The intact soil cores were mixed with water in the laboratory, washed over a 0.2 mm sieve and dried at 70 °C to constant weight.

Miscanthus green leaf area index (GLAI) was determined by firstly establishing an allometric relationship between leaf area and the product of leaf length and width, following Clifton-Brown *et al.* (2000). The length and width of leaves harvested on three different occasions were measured and images of the leaves were captured with a flatbed scanner (CanoScan LiDE 35, Canon, Tokyo, Japan). Leaf area estimates were obtained using the open-source software IMAGEJ (http://imagej.nih.gov/ij/) to generate the linear regression equation

$$\text{leaf area} = 0.76\,(\text{length} \times \text{width}) \qquad (5)$$

where both measured quantities are in cm^2, $R^2 = 0.93$ and $n = 295$. This equation was applied to leaf length × width measurements made during above-ground biomass sampling. *Miscanthus* GLAI was computed as the product of green leaf area per shoot and shoot density. The GLAI of understory vegetation was estimated based on an allometric relationship between dry biomass weight and calculated leaf area.

RCG leaf area index (LAI) was monitored with a Sunscan Canopy Analysis System (Delta-T Devices, Cambridge, UK) at approximately monthly intervals, providing a measure of the total leaf area per unit ground area. Individual measurements made at 40 locations within the crop were averaged on each sampling date. These measurements involved a non-destructive sampling technique based on the fraction of photosynthetically active radiation (PAR) intercepted by the canopy (Campbell & Norman, 1989). Therefore, RCG LAI values included contributions from both live and senescent leaves as both influence the transmission of light through the canopy.

Results

Environmental conditions

Mean annual air temperatures were close to the 30-year mean temperature in 2009 (10.0 °C) and 2011 (10.4 °C) but dipped in 2010 to 9.3 °C. Two prolonged periods of cold weather were experienced during these years (December 2009 to February 2010 and November 2010 to January 2011) (Fig. 1). In addition, summer mean temperature was 7% lower than the long-term mean in 2011. Rainfall totals were highly variable, with higher than average rainfall in 2009 (1427 mm), close to the average in 2010 (972 mm) and below average in 2011 (839 mm). Exceptionally wet summer months characterised 2009, with atypically high rainfall also occurring in November. The mean monthly soil VWC at the *Miscanthus* site was high for much of the study, only reducing below field capacity (0.32 m^3 m^{-3}) in August 2010 and August and September 2011. Soil VWC in the RCG site was below field capacity from May to August of each year but remained above field capacity (0.32 m^3 m^{-3}) outside of the peak growth period. In general, soil VWC was higher in *Miscanthus* than RCG, most obviously from May to July 2011. Vapour pressure deficit (VPD) was always low, ranging from a monthly average of 0.06 kPa in February 2011 to a maximum monthly average of 0.37 kPa in June 2009 and August 2010 (Fig. 1).

Crop development

Establishment and growth of the *Miscanthus* crop was slow in the first 2 years after planting. Prior to the first harvest in March 2010, above-ground dry biomass totalled just 10 ± 4.0 (standard error of the mean, SEM) g m^{-2} while below-ground biomass stocks were estimated at 13.8 ± 7.1 g m^{-2} (Table 2). Following the growing season of 2010, above-ground biomass reached a seasonal maximum of over 100 ± 20.2 g m^{-2} in September and a maximum GLAI of 0.55 ± 0.05 m^2 m^{-2} was recorded in October (Fig. 5). The contribution of understory vegetation to ecosystem productivity was substantial at this time. Above-ground biomass stocks of understory vegetation, dominated mostly by grass species (*Agrostis stolonifera, Poa trivialis* and *Alopecurus geniculatus*), totalled 110 ± 21.7 g m^{-2} in September 2010, with an associated GLAI of 3.56 ± 0.70 m^2 m^{-2}.

Table 2 Above- and below-ground biomass stocks in the *Miscanthus* and RCG crops in g (dry matter) m^{-2}. Total above- and below-ground stocks in *Miscanthus* are divided into their constituent pools. Values in parentheses represent the standard error of the mean

	Above-ground (g m^{-2})			Below-ground (g m^{-2})			
Date	Stem	Leaf	Total	Live Rhizome	Dead Rhizome	Root	Total
Miscanthus							
February 2010	8.2 (3.3)	1.6 (0.7)	9.9 (4.0)	10.7 (6.2)	0	3.0 (0.9)	13.8 (7.1)
September 2010	63.9 (17.3)	36.6 (5.2)	100.5 (20.2)				
January 2011	33.1 (13.9)	6.5 (2.2)	39.6 (16.1)	73.6 (3.8)	9.3 (2.1)	22.3 (2.7)	105.2 (8.5)
March 2011	53.3 (15.2)	4.9 (1.4)	58.2 (16.4)				
June 2011	52.3 (23.3)	31.5 (9.2)	83.8 (32.5)				
September 2011	1001.6 (362.3)	308.9 (102.6)	1310.5 (464.6)	331.1 (138.4)	16.0 (6.3)	63.7 (21.8)	410.8 (166.4)
Reed Canary Grass							
September 2010			653.0 (76.4)				
October 2010			589.3 (28.5)				
January 2011							489.0 (46.3)
June 2011			675.4 (93.9)				
September 2011			706.4 (61.2)				834.5 (72.5)

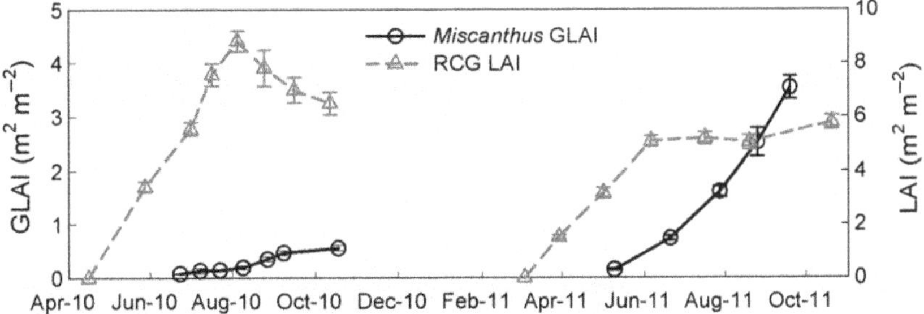

Fig. 5 Temporal pattern of green leaf area index (GLAI) in *Miscanthus* and leaf area index (LAI) in RCG during the 2010 and 2011 growing seasons. Vertical bars represent the standard error of the mean.

Following initial slow development in 2011, growth rates increased exponentially towards the end of the summer, yielding 1311 ± 465 g m^{-2} in late September 2011. GLAI increased steadily over this time (Fig. 5, Table 2). Above-ground crop expansion was facilitated by a four-fold increase in below-ground biomass between January and late September 2011 when total below-ground *Miscanthus* biomass was 410.8 ± 166.4 g m^{-2}, comprising 80.6% live rhizome, 3.9% dead rhizome and 15.5% root (Table 2). Correspondingly, stocks of understory vegetation declined, associated with herbicide application in spring 2011 and increased competition from *Miscanthus* plants. Estimated above-ground biomass from weeds was 5.8 ± 2.2 g m^{-2} and 14.5 ± 6.3 g m^{-2} in late June and late September 2011, respectively, while their GLAI was calculated as 0.19 ± 0.07 m^2 m^{-2} and 0.47 ± 0.20 m^2 m^{-2} at the same times, representing significant reductions from 2010 values.

In contrast with the slow establishment of *Miscanthus*, rapid growth was observed in the new RCG plantation after sowing in mid-April 2010. A dense green canopy developed early in the first growing season resulting in a seasonal maximum LAI of 8.8 ± 0.4 m^2 m^{-2} at the start of August 2010 (Fig. 5). A reduction in LAI was then observed as the crop senesced. At the time of the first harvest in October 2010, 589 ± 29 g m^{-2} had accumulated in above-ground biomass and the LAI was 6.5 ± 0.4 m^2 m^{-2}. In the second year of establishment, above-ground growth began earlier than in the conversion year and, by early June 2011, a growing season maximum LAI of 5.1 ± 0.2 m^2 m^{-2} was recorded and above-ground biomass totalled 675 ± 94 g m^{-2}. Flowering and senescence occurred earlier in 2011 than in 2010 with the result that, by August 2011, the crop had completely senesced. Significant below-ground biomass stocks were recorded in the RCG crop in the second year of establishment (Table 2), with the maximum below-ground stock in September 2011 (835 ± 73 g m^{-2}) exceeding above-ground biomass estimates. Understory vegetation made a minor contribution to above-ground biomass in the RCG ecosystem, accounting for just 3.4% and 0.5% of total above-ground biomass in June and late September 2011, respectively.

Carbon fluxes

Flux measurements at the *Miscanthus* site began 1 day before the majority of the grassland field was tilled. Ecosystem exchange of carbon was dominated by respiratory fluxes in the first month post-disturbance. Half-hourly NEE fluxes ± SEM averaged 4.0 ± 0.1, 3.3 ± 0.1, 3.3 ± 0.1 and 2.7 ± 0.1 μmol CO_2 m^{-2} s^{-1} in the first

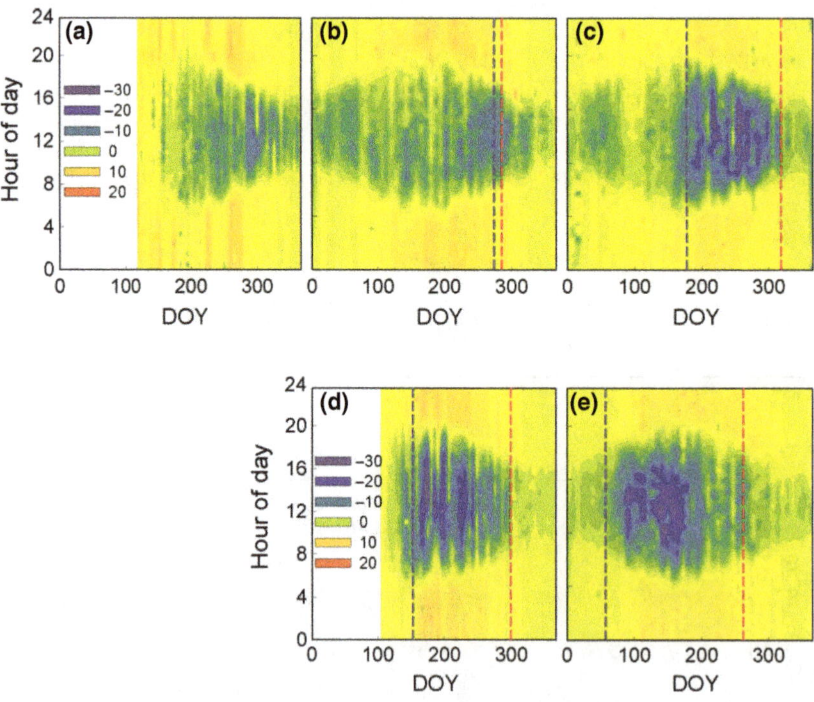

Fig. 6 Diurnal course of half-hourly fluxes of net ecosystem carbon exchange in μmol CO_2 m^{-2} s^{-1} in the *Miscanthus* ecosystem during 2009 (beginning in late April) (a), 2010 (b) and 2011 (c) and in the RCG ecosystem during 2010 (beginning in mid-April) (d) and 2011 (e). Dashed vertical dark blue lines mark the 'break-even point' at which cumulative NEE becomes negative (net C uptake). Dashed vertical red lines indicate the timing of the end of net C uptake.

4 weeks following tillage, respectively, with positive fluxes indicating a net release of C to the atmosphere. Monthly cumulative NEE ± uncertainty amounted to 104.7 ± 3.1, 62.7 ± 2.7 and −12.1 ± 3.1 g C m^{-2} in the first 3 months, respectively, post-tillage, representing significant initial net losses of carbon from the ecosystem.

A small recovery in photosynthetic activity was apparent in subsequent months, concurrent with the slow development of the newly-established *Miscanthus* crop and the emergence of understory vegetative species prompted by a flush of germination after the initial soil disturbance. However, seasonal maximum rates of instantaneous net C assimilation, recorded in mid-October 2009, were no greater than −15 μmol CO_2 m^{-2} s^{-1} (Fig. 6). The magnitude of photosynthetic uptake was similarly modest in 2010, peaking at −17 μmol CO_2 m^{-2} s^{-1} in late September. The seasonal pattern of NEE changed dramatically in 2011, however, concomitant with significant increases in above-ground *Miscanthus* biomass, large-scale leaf expansion and higher GLAI. Maximum instantaneous rates of net C assimilation were −30 to −35 μmol CO_2 m^{-2} s^{-1} between mid-August and late September 2011.

Measurements of NEE in the RCG crop began the same day the site was harrowed and seeded in mid-April 2010. Early half-hourly fluxes of NEE were dominated by ecosystem respiration but C uptake became evident from mid-May onwards (Fig. 6). Weekly-averaged NEE was 2.6 ± 0.1, 1.7 ± 0.1, 0.8 ± 0.1 and −0.2 ± 0.1 μmol CO_2 m^{-2} s^{-1} in the first 4 weeks following planting, respectively. As a result, the ecosystem represented a net source of carbon over the first month (cumulative NEE was 33.3 ± 2.4 g C m^{-2}) but switched to a net sink during the second (−88.0 ± 4.8 g C m^{-2}) and third month (−127.3 ± 13.9 g C m^{-2}).

Rapid establishment and extensive early canopy development in the RCG crop accompanied strong photosynthetic activity early in the first growing season. Just over 2 months after planting (late June 2010), a seasonal maximum NEE of −35 μmol CO_2 m^{-2} s^{-1} was recorded. Net C uptake continued until late September 2010 at a decreasing rate (Fig. 6). Winter NEE fluxes were small as a result of the exceptionally cold temperatures recorded from November 2010 to January 2011. However, photosynthetic activity resumed when daily mean temperatures began to exceed about 5 °C in February 2011. Net ecosystem C accumulation continued until a seasonal maximum NEE of −38 μmol CO_2 m^{-2} s^{-1} was recorded in mid-May 2011, after which rates of C uptake declined.

Daily fluxes of TER and GPP are shown in Fig. 7. Maximal daily values of GPP in the *Miscanthus* crop increased from −7.3 and −9.2 g C m^{-2} d^{-1} in September 2009 and June 2010, respectively, to −14.1 g C m^{-2} d^{-1} in August 2011. Maximal daily TER was 10.0, 9.5 and 7.4 g C m^{-2} d^{-1} in September 2009, June 2010 and July 2011, respectively for *Miscanthus*. In the RCG crop, maximal daily GPP values of −17.6 and −16.3 g C m^{-2} d^{-1} were recorded in late June 2010 and

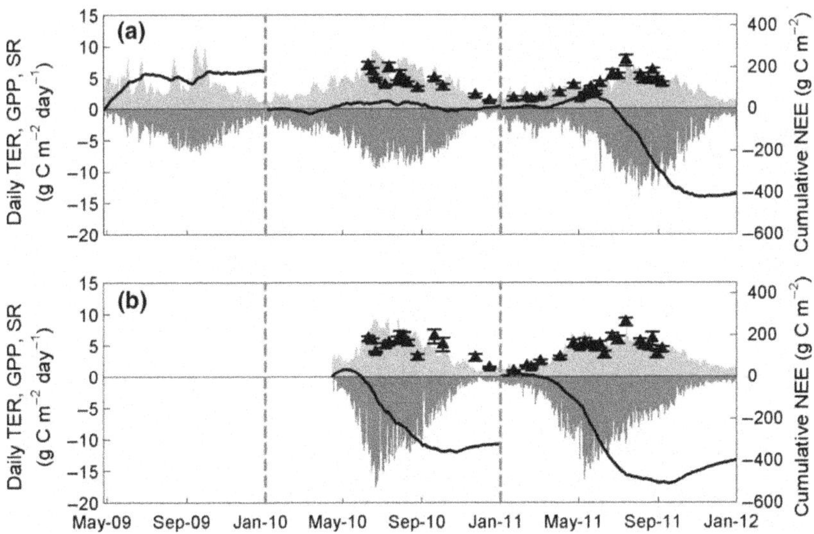

Fig. 7 Daily total ecosystem respiration (TER) (light grey area), daily gross primary productivity (GPP) (dark grey area) and cumulative net ecosystem exchange (NEE) (solid black line) over the duration of the *Miscanthus* (a) and RCG (b) studies. Cumulative NEE was reset to zero at the end of each calendar year (indicated by a dashed vertical grey line). Measured soil respiratory (SR) fluxes are also shown (black triangles), representing below-ground respiration from both autotrophic and heterotrophic sources. Vertical bars represent the standard error of the measurement means.

mid-May 2011, respectively, while maximal daily TER values of 9.2 and 7.6 g C m^{-2} d^{-1} were observed in mid-July 2010 and 2011, respectively. Higher TER values were generally associated with warmer soil temperatures and higher ecosystem productivity.

The cumulative effect of the measured NEE fluxes over the course of the study is also shown in Fig. 7. In the *Miscanthus* crop, cumulative NEE was 183 ± 28 g C m^{-2} from late April to the end of December 2009, 13 ± 45 g C m^{-2} in 2010 and −411 ± 63 g C m^{-2} in 2011 (Table 3). GPP increased from −797 g C m^{-2} in 2009 to −1684 g C m^{-2} in 2011. Cumulative TER peaked in 2010 (1514 g C m^{-2}) and decreased in 2011 to 1273 g C m^{-2}, concomitant with below average summer temperatures. In the RCG crop, cumulative NEE values were −319 ± 57 g C m^{-2} from mid-April to the end of December 2010 and −397 ± 114 g C m^{-2} during 2011. An increase in GPP was observed from the conversion year (−1430 g C m^{-2}, 8½-month period) to the second year of establishment (−1708 g C m^{-2}), while TER also increased over this period from 1112 to 1311 g C m^{-2}.

The 'break-even' point at which cumulative NEE became negative (net C uptake) occurred in early October in 2010 and late June in 2011 for *Miscanthus* (Fig. 6). For RCG, the 'break-even' point occurred much earlier in the year than for *Miscanthus*: in early June in the conversion year (2010) and late February in 2011. With regard to winter fluxes, net C uptake did not cease in *Miscanthus* until mid-October and mid-November in 2010 and 2011, respectively, and the *Miscanthus* ecosystem subsequently remained relatively C-neutral until year end. In contrast, net C loss was observed from late October onwards in the conversion year for RCG and at a higher rate in 2011 from late September onwards.

The net rate of long-term C accumulation in an ecosystem is better described by the NECB. Since no removal of C through harvesting occurred during the *Miscanthus* experiment, the cumulative NEE values quoted for 2009, 2010 and 2011 are representative of the NECB of this ecosystem. For the RCG crop, the NECB was calculated as 66 ± 58 g C m^{-2} (a net C sink), while the NECB for the second year of RCG establishment was 397 ± 114 g C m^{-2}, as no biomass was harvested in this year. These values assume that contributions from leaching, methane, VOC, carbon monoxide and particulate carbon were negligible. Leaching of dissolved organic carbon (DOC) and dissolved inorganic carbon (DIC) are likely to be the most significant of these nonestimated fluxes (Osborne *et al.*, 2010; Smith *et al.*, 2010). Kindler *et al.* (2011) reported average DOC and biogenic DIC losses from European croplands of 4.1 ± 1.3 and 14.6 ± 4.8 g C m^{-2} y^{-1}, respectively. This study included an Irish cropland (Carlow) with well-drained, sandy loam soil which lost 2.6 ± 0.5 and 15.2 ± 4.1 g C m^{-2} y^{-1} as DOC and biogenic DIC, respectively. Following tillage disturbance of a grassland lysimeter at Johnstown Castle with comparable soil to the *Miscanthus* and RCG sites (low to medium drainage capacity), DOC leaching amounted to 1.6 g C m^{-2} over a 33-week period (Ó. Ní Choncubhair, B. Osborne, K. Richards and G. Lanigan, unpublished results). While such C losses comprise a small fraction of the annual NEE estimates for *Miscanthus* and RCG in 2011, their significance for the long-term NECB is far greater.

Based on chamber measurements, soil respiration (below-ground autotrophic and heterotrophic respiration) accounted for the majority of TER, with small contributions from leaf and stem respiration (Fig. 7). This was particularly evident in 2011 when significant below-ground biomass had developed in both crops. Soil respiration in *Miscanthus* varied from 1.3 ± 0.1 (SEM) g C m^{-2} d^{-1} in December 2010 to 7.7 ±

Table 3 Annual cumulative fluxes of gross primary productivity (GPP), total ecosystem respiration (TER), net ecosystem exchange (NEE) and net ecosystem carbon balance (NECB) for the *Miscanthus* and RCG ecosystems

	GPP	TER	NEE* ± uncertainty	NECB† ± uncertainty
Miscanthus				
28th Apr 2009 – 31st Dec 2009 (Ploughing – year end)	−797	979	183 ± 28	−183 ± 28
2010	−1501	1514	13 ± 45	−13 ± 45
2011	−1684	1273	−411 ± 63	411 ± 63
RCG				
15th Apr 2010 – 31st Dec 2010 (Harrowing/planting – year end)	−1430	1112	−319 ± 57	66 ± 58
2011	−1708	1311	−397 ± 114	397 ± 114

*Positive NEE values represent a net release of C to the atmosphere, negative values a net uptake of C by the ecosystem.
†NECB = $-\Sigma$NEE − $C_{harvest}$ where $C_{harvest}$ is the carbon removed from the ecosystem at harvest.

0.9 g C m^{-2} d^{-1} in July 2011. In RCG, fluxes ranged from 0.9 ± 0.1 in January 2011 to 8.7 ± 1.0 g C m^{-2} d^{-1} in July 2011. During intervals of active RCG growth and slow *Miscanthus* development, soil respiration in RCG exceeded that of *Miscanthus*, most notably in autumn 2010 and in April and May 2011. Outside of these periods, soil respiration fluxes were comparable in both ecosystems.

Controls on ecosystem carbon fluxes

Net assimilation of carbon by the *Miscanthus* ecosystem was closely related to crop development during the growing season, as shown in Fig. 8(a). When *Miscanthus* GLAI values were 0.5 m^2 m^{-2} or lower, gross photosynthesis was cancelled out by a comparable ecosystem respiration rate and monthly sums of NEE fluctuated around zero. As leaf expansion progressed and a greater fraction of incident light was intercepted by the *Miscanthus* crop, large increases in monthly NEE were recorded, reaching a peak of −161.8 g C m^{-2} during August 2011.

A strong positive relationship was observed between monthly sums of TER and GPP for much of 2009 and 2010 and the winter months of 2011 (Fig. 8b). Notable deviations from a 1 : 1 linear relationship in May and June 2009 correspond to significant reductions in photosynthetic capacity after conventional inversion tillage. Monthly TER also exceeded monthly GPP in September 2009. This month was characterised by a 3-week period when rainfall was significantly reduced (total 1.2 mm) after a summer of exceptionally wet weather. In addition, respiration outweighed photosynthesis in March and April 2010 and in April 2011 following systemic herbicide application. However, the largest departures from slopes close to unity were observed in the growing season of 2011, when GPP was approximately 1.5 times greater than TER.

The relationship between monthly sums of TER and GPP in the RCG ecosystem is shown in Fig. 9(a) for 2010 and 2011. A strong positive association was observed between the two variables in both years, with respiratory fluxes representing roughly 50% of GPP across the full measurement period. The only months when TER exceeded GPP were April 2010 (after planting), November – December 2010 and October – December 2011. The impact of soil VWC on monthly TER fluxes in the RCG crop can be seen in Fig. 9(b). A negative relationship was evident in 2010 as VWC increased and reached values in excess of 0.4 m^3 m^{-3}, presumably due to oxygen-limitation. Monthly sums of TER in 2011 (drier year) were comparable to 2010 values at high moisture contents, but when soil VWC decreased below a mid-range level (~ 0.26 m^3 m^{-3}), a positive relationship was observed between TER and VWC.

Discussion

In the discussion below, we address the three questions posed, relating to (1) the transitional-phase C fluxes, (2) the postestablishment phase C fluxes associated with the *Miscanthus* and RCG crops and (3) the role of leaf longevity in regulating net C exchange. We discuss how our results compare with previous studies involving land-use transitions and examine the implications of our findings for bioenergy crop selection.

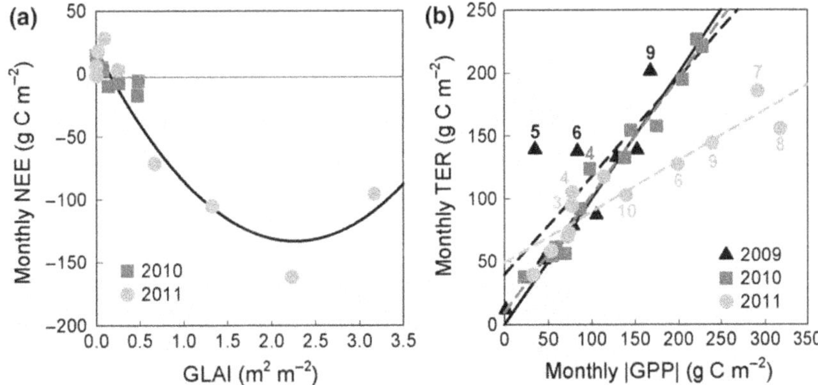

Fig. 8 Monthly sums of net ecosystem exchange (NEE) in the *Miscanthus* ecosystem are plotted against green leaf area index (GLAI) for 2010 and 2011 in (a). Panel (b) shows the correlation for the *Miscanthus* ecosystem of monthly sums of total ecosystem respiration (TER) with monthly sums of gross primary productivity (GPP), plotted in absolute values (|GPP|) to facilitate the comparison. The black solid line indicates a 1 : 1 relationship while the dashed lines represent regression lines: 2009 (black line) $y = 0.78x + 39.6$, $R^2 = 0.58$; 2010 (dark grey line) $y = 0.93x + 9.6$, $R^2 = 0.97$; 2011 (light grey line) $y = 0.40x + 49.3$, $R^2 = 0.85$. Months showing deviations from the 1 : 1 line are marked with their respective month number in the calendar year.

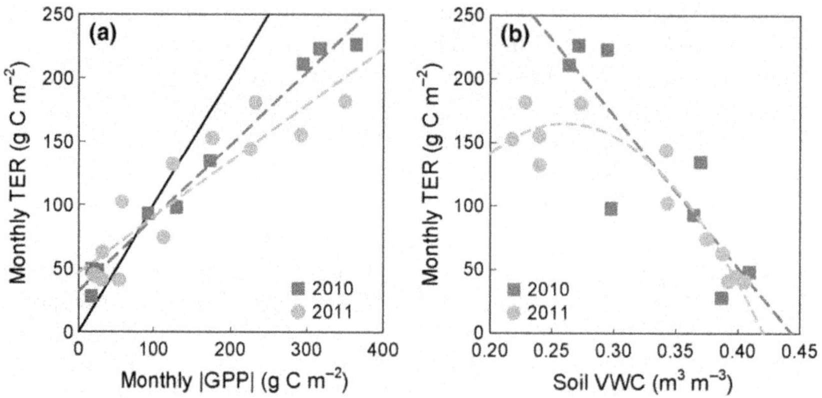

Fig. 9 Correlation of monthly sums of total ecosystem respiration (TER) with monthly sums of gross primary productivity (GPP) for 2010 and 2011 in the RCG ecosystem (a). GPP is plotted in absolute values (|GPP|). The black solid line indicates a 1 : 1 relationship while the dotted lines represent regression lines: 2010 (dark grey line) $y = 0.57x + 32.1$, $R^2 = 0.98$; 2011 (light grey line) $y = 0.44x + 46.5$, $R^2 = 0.81$. Panel (b) shows the relationship between TER and soil volumetric water content (VWC) during 2010 and 2011 in the RCG ecosystem. The dashed lines represent regression lines: 2010 (dark grey line) $y = -1194.2x + 529.4$, $R^2 = 0.72$; 2011 (light grey line) $y = -6451.4x^2 + 3362.2x - 273.1$, $R^2 = 0.91$.

Carbon fluxes during land-use transition to bioenergy crops

Significant carbon emissions and changes in SOC have been attributed to land-use transitions such as deforestation and conversion to cropland (Davidson & Ackerman, 1993; Houghton *et al.*, 2012; Poeplau & Don, 2013). Net C losses in the early stage of LUC can arise as a result of diminished photosynthetic capacity following herbicide application, enhanced mineralisation after tillage as well as reduced C uptake while the new crop develops to full maturity.

Table 4 summarises findings from previous research on C fluxes during land-use transitions. In a reseeding experiment in Johnstown Castle, cumulative TER from a grass sward with similar soil type to the bioenergy crop sites was 278 ± 29 g C m^{-2} in the month following herbicide application (Ó. Ní Choncubhair, B. Osborne, K. Richards and G. Lanigan, unpublished results). Including photosynthetic uptake (albeit at a diminishing rate) would reduce this value significantly. For example, Zenone *et al.* (2011) recorded a cumulative NEE of 19–46 g C m^{-2} in the month after herbicide application to long-term grassland. However, further variable net C losses occur in the posttillage period, dependent on the former land use, environmental conditions and tillage intensity. Cumulative NEE in the first 3 months posttillage in the *Miscanthus* ecosystem was 155.4 ± 8.9 g C m^{-2} and falls in the mid-range of estimated short-term tillage-induced emissions. These losses were primarily driven by minimal photosynthetic activity and slow above-ground development of the rhizome-propagated plants. This slow establishment is typical of Irish and many northern European trials where 3–5 years is generally required for full establishment (Lewandowski *et al.*, 2000; Clifton-Brown *et al.*, 2007).

Following RCG planting in April, net C losses of 33.3 ± 2.4 g C m^{-2} were recorded in the first month. However, the ecosystem became a strong net C sink after this. A number of factors may have contributed to this result. Firstly, the RCG site had been ploughed 1 year previous to crop establishment. Therefore, any increased mineralisation of SOC made available by tillage is most likely to have occurred in the months following the initial soil disturbance (Vellinga *et al.*, 2004; Willems *et al.*, 2011). Secondly, a high seeding rate was employed to guarantee successful establishment. Thirdly, RCG is considered a highly competitive plant that grows rapidly and pre-empts the development of other vegetation early in the growing season (Lavergne & Molofsky, 2004). For example, Adams & Galatowitsch (2005) highlighted a shift from a low root:shoot ratio (< 1) in the first 4 months of development to a higher ratio (> 2) in the remainder of the 2-year study, which enabled RCG to initially monopolise above-ground space and later to spread vegetatively below ground.

Rapid early colonisation and crop development can therefore reduce the early transitional C losses associated with land-use change. However, our results suggest that significant net ecosystem C losses can be generated as a result of reduced photosynthetic assimilation and enhanced decomposition rates, particularly in C-rich grasslands. This loss represents a short-term land conversion carbon debt that must be overcome before net C sequestration is attained in the newly established bioenergy crop.

Table 4 Transitional-phase and post-establishment phase C fluxes associated with different land-use transitions. Annual fluxes from permanent grasslands are included for comparison. Cumulative fluxes refer to net ecosystem exchange (NEE), total ecosystem respiration (TER) or soil respiration (SR). For transitional-phase fluxes, the time period refers to the sampling duration in months (M). For long-term fluxes, the length of establishment is given in years (Y), where Y1 is the conversion year

Original land use	New land use	Location	Time period	Cumulative NEE[A], TER[B] or SR[C] (g C m^{-2})	Research note	Reference
Transitional-phase fluxes						
Post-herbicide						
Grass	Grass	Ireland	1 M	278 ± 29[B]	Summer reseeding	Unpublished results*
Grass	Soybean	USA	1 M	19–46[A]		Zenone *et al.* (2011)
Post-tillage						
Grass	*Miscanthus*	Ireland	3 M	155 ± 9[A]		This study
Wheat	Wheat	USA	2 M	29–41[C]		Dao, 1998
Grass	Spring barley	Denmark	3 M	260[C]		Eriksen & Jensen (2001)
Conversion year (Y1)						
Grass	*Miscanthus*	Ireland	8 M	183 ± 28[A]	From tillage	This study
Arable	*Miscanthus*	USA	7 M	−58[A]	From planting	Zeri *et al.* (2011)
Grass	RCG	Ireland	8.5 M	−319 ± 57[A]	From planting	This study
Grass/Arable	RCG	Finland	5.3 M	−57[A]	From 1.5M post-planting	Lind *et al.* (2015)
Grass/Arable	Poplar	Belgium	7 M	75 ± 4.4[A]	From 2M post-planting	Zenone *et al.* (2015)
Arable	Poplar	Canada	12 M	312[A]	From 4M post-tillage	Cai *et al.* (2011)
Grass/Wheat	Switchgrass	USA	12 M	−31[A]	From 4M post-planting	Skinner & Adler (2010)
Grass	Soybean	USA	12 M	205–262[A]	No-till	Zenone *et al.* (2013)
Post-establishment (long-term) fluxes						
Grass	*Miscanthus*	Ireland	Y2	13 ± 45[A]		This study
Grass	*Miscanthus*	Ireland	Y3	−411 ± 63[A]		This study
Arable	*Miscanthus*	USA	Y3	−554 ± 20[A]		Zeri *et al.* (2011)
Grass	RCG	Ireland	Y2	−397 ± 114[A]		This study
Grass/Arable	RCG	Finland	Y2, 3†	−259[A]	Fertilised	Lind *et al.* (2015)
Drained peat	RCG	Finland	Y4–7	−9 to −211[A]		Shurpali *et al.* (2009)
Drained peat	RCG	Estonia	Y4	−91 to −163[A]‡	± Fertiliser	Mander *et al.* (2012)
Arable	Switchgrass	USA	Y3	−485 ± 20[A]		Zeri *et al.* (2011)
NA	Switchgrass	USA	Y3, 4†	−448[A]	Growing season only	Wagle *et al.*, 2015;
Grass/Arable	Poplar	Belgium	Y2	−96 ± 15[A]		Zona *et al.* (2013)
Arable	Poplar	Canada	Y5	−17[A]		Cai *et al.* (2011)
Permanent grassland						
Grass		Ireland	Annual	−200 to −385[A]	Johnstown Castle; similar soil	Peichl *et al.* (2012)
Grass		Ireland	Annual	−193 to −258[A]	South-west region	Jaksic *et al.* (2006)
Grass		Europe	Annual	−240 ± 70[A]	9 sites	Soussana *et al.* (2007)

NA, not available.
*Órlaith Ní Choncubhair, Bruce Osborne, Karl Richards, Gary Lanigan.
†Mean value given for these years.
‡Chamber measurements.

C balance implications postestablishment

The *Miscanthus* ecosystem switched from a net C source of 183 ± 28 g C m^{-2} in the conversion year (8-month period) to C-neutral in the second year and a strong C sink in the third year (−411 ± 63 g C m^{-2}). In contrast, rapid and early development of the RCG crop resulted in the ecosystem being a net C sink both during the conversion year (−319 ± 57 g C m^{-2}, 8½-month period post-planting) and in the second year of establishment (−397 ± 114 g C m^{-2}). Previous assessments of C fluxes in the conversion year show highly variable results (Table 4), with cumulative NEE ranging from −58 to +312 g C m^{-2} depending on crop type and establishment technique, location and duration of measurements. Only a limited number of studies revealed small C sinks in the conversion year, similar to RCG in this study. Even after harvested biomass losses were incorporated, the NECB of this RCG ecosystem post-planting was still positive, indicating a net C sink of 66 ± 58 g C m^{-2} in the conversion year.

Net C fluxes in 2011 compare favourably with long-term fluxes from previous studies (Table 4), which range from close to C-neutral for poplar and RCG in certain years to in excess of -450 g C m^{-2} y^{-1} for *Miscanthus* and switchgrass. Good agreement exists between our results and the studies on *Miscanthus* in the USA (-554 ± 20 g C m^{-2} y^{-1}, Zeri *et al.*, 2011) and RCG on mineral soil in Finland (-259 g C m^{-2} y^{-1}, Lind *et al.*, 2015), taking account of location and comparative differences in season length and environmental controls. Studies focussed on the LUC transition from permanent grassland to *Miscanthus* or RCG are severely lacking, however, despite the fact that almost one-third of utilised agricultural land in Europe is grassland (Fischer *et al.*, 2010).

In this study, the key driver of ecosystem C fluxes in *Miscanthus* was crop development, constraining monthly-cumulated NEE in the first 2 years close to zero as primary production was counterbalanced by ecosystem respiration. In addition to the inherent limitations of slow establishment in northern Europe, various other factors may have contributed to the poor agronomic performance observed including relatively late planting, possible rhizome failure and competition from understory grass species. However, in the third growing season, exponential above-ground growth and leaf expansion was accompanied by a 4-fold increase in below-ground biomass, resulting in strong net C uptake from June to October of that year.

In the case of RCG, environmental conditions and crop phenology were the main drivers of ecosystem C exchange. Since crop establishment was rapid and early canopy development was uninhibited by competition from other species, physiological activity was dictated by soil temperature, moisture and incident radiation. Net C uptake of -35 μmol CO_2 m^{-2} s^{-1} was observed just over two months after planting, culminating in a strong net C sink of -319 ± 57 g C m^{-2} in the conversion year and an increased net sink in the second year (-397 ± 114 g C m^{-2}). Furthermore, below-ground biomass represented a significant C stock of 3.4 t C ha^{-1}, similar to the results of Xiong & Kätterer (2010).

During periods when RCG productivity exceeded that of *Miscanthus* (autumn of the conversion year and early in its second year), higher soil respiration (below-ground autotrophic and heterotrophic) was recorded in RCG than *Miscanthus*. This may indicate that there was a larger contribution from the autotrophic component to total below-ground respiration in RCG at these times. Soil respiration fluxes were similar in both crops outside of these periods. In a Finnish RCG crop which had 70% of total plant biomass below ground during the peak growth period, autotrophic respiration was the dominant component (about 55%) of TER (Shurpali *et al.*, 2008). Soil respiration constituted the majority of TER in this study, particularly in the final year. While recognising the methodological differences between chamber and EC estimates, this result highlights the significant and increasing contribution of below-ground processes to C cycling in these developing perennial crops. Enhanced below-ground C allocation and root-associated C cycling may favour long-term soil carbon sequestration (Xiong & Kätterer, 2010; Anderson-Teixeira *et al.*, 2013), boosted also by the absence of regular soil disturbance through tillage (Freibauer *et al.*, 2004).

For the early postestablishment phase, our findings suggest that RCG provides a more favourable C balance than *Miscanthus*. However, the outperformance of RCG by *Miscanthus* in the third growth year, both in terms of yield and net C accumulation, indicates that the C sink potential of *Miscanthus* will improve further as the crop matures and above-ground productivity approaches the maximum yield.

Leaf longevity in Miscanthus and RCG

Light-use efficiency is a critical determinant of primary production and crop yield (Monteith, 1977) and is closely related to the crop's ability to maintain a closed canopy during the growing season (Beale & Long, 1995). Greater leaf longevity and higher leaf area have been demonstrated in *Miscanthus* compared with C_4 maize (Dohleman & Long, 2009). However, little information exists on the relative performance of *Miscanthus* when compared to native C_3 plants well-adapted to the temperate conditions of northern Europe.

In this study, strong coupling of GPP and crop phenology was highlighted by a striking 3-month difference in the timing of peak C assimilation in RCG (May) and *Miscanthus* (August) in 2011. RCG emerged early in the spring and achieved net C uptake (the 'break-even' or compensation point) by late February. This is consistent with the temperate climatic conditions of the current study and the associated long growing season. However, the subsequent occurrence of peak RCG productivity in mid-May resulted in the early onset of senescence in August. Crop phenology, therefore, introduced an asynchrony between the timing of maximum leaf area and maximum solar radiation for RCG, which could have implications for the ability of RCG to fully exploit peak irradiances and to maximise yields under some environmental conditions. Furthermore, RCG TER fluxes tended to exceed GPP in the winter months and reduced the cumulative C sink by 113 g C m^{-2} in the second year of establishment.

In contrast with RCG which exhibited early season leaf development, significant increases in leaf area did not occur until June for *Miscanthus*. As a result, the 'break-even' point of net C uptake did not occur until late June in 2011. However, *Miscanthus* maintained growth and substantial leaf area late into the autumn and sustained net ecosystem C accumulation until mid-November. After this, the crop remained C-neutral until the end of the year, similar to the findings of Zeri *et al.* (2011). Indeed, strong coupling of TER and GPP was observed during much of the current study, which meant that significant C losses did not occur.

Dohleman & Long (2009) demonstrated a 59% longer growing season in *Miscanthus* (199 days on average) than C_4 maize (126 days on average). Our results show that the longevity of photosynthetically active leaves in *Miscanthus* is comparable even to native C_3 crops. Indeed, the number of recorded days with net C uptake in 2011 was almost identical for both crops (215 and 212 days for *Miscanthus* and RCG, respectively). This result compares very favourably with an average cropping season length of 149 days in Irish spring barley (Davis *et al.*, 2010) and 212 days of net C uptake (six-year average) in an Irish grassland (Peichl *et al.*, 2011). This further highlights the exceptional performance of C_4 *Miscanthus* in cool temperate climates where C_4 metabolism should be temperature-limited and implies that *Miscanthus* may confer an advantage as a bioenergy crop in the long term. If leaf photosynthetic capacity and longevity is comparable to native C_3 crops and C emissions outside of the growing season are close to zero, the net C balance of *Miscanthus* is likely to be favourable.

Implications of the study

Bioenergy crop cultivation in northern Europe may focus more on grassland conversion to avoid a reduction in the area of croplands dedicated to food and feed production. However, permanent grasslands in northern Europe show strong annual net C uptake, with NEE values ranging from −193 to −385 g C m^{-2} y^{-1} (Table 4). This highlights the potential negative impacts associated with disturbing grasslands that are highly productive and supply substantial amounts of stabilised C to the soil (Jackson *et al.*, 1996; Jones & Donnelly, 2004; Poeplau *et al.*, 2011). Our study demonstrated that a significant C debt can be associated with the early-establishment phase of these bioenergy crops but highlighted the future potential of *Miscanthus* to surpass RCG and possibly long-term grasslands in terms of its C sink strength. Additional measurements of other associated GHGs, such as N_2O and CH_4, will be necessary, however, to assess the full GHG implications of land-use change to this crop.

Furthermore, the duration of the full crop production cycle must be considered. In the case of *Miscanthus*, productive yields may be achievable for up to 15 years or more (Clifton-Brown *et al.*, 2007; Christian *et al.*, 2008; Arundale *et al.*, 2014) but this is likely to be much shorter for RCG (approximately 7–10 years; Saijonkari-Pahkala, 2001; Finnan, 2007). Therefore, more cultivation and replanting will be required in long-term RCG plantations and this has associated C balance implications due to more regular soil disturbance and concomitant reductions in productivity.

Although the theoretical light-use efficiency benefits afforded by the C_4 photosynthetic pathway are often highlighted, the empirical results of this work suggest that high biomass productivity will be controlled more by leaf- or canopy-related factors, both genetic and environmentally derived, rather than the photosynthetic characteristics of individual leaves. Further to this, the distinct difference in the timing of peak C uptake between the two crops is significant and provides information on the suitability of these crops in different climatic zones. While the late season performance of *Miscanthus* may constrain its productivity in regions with a short growing season, RCG may be a better candidate in these regions due to its early emergence in spring and subsequent rapid development. Growing *Miscanthus* under a clear, plastic film, as is common practice for maize production in parts of northern Europe, may also be a valuable tool to encourage earlier emergence in spring and enhance further the duration of net C uptake in the crop (Clifton-Brown *et al.*, 2011).

A final point worth noting is the relative allocation of biomass above and below ground in these two bioenergy crops. At the time of peak biomass yield in 2011, more than 50% of total RCG biomass was below ground compared with 24% in *Miscanthus* and above-ground biomass in *Miscanthus* was almost double that recorded in RCG. Greater investment of resources below ground could enhance long-term C sequestration; however, lower above-ground yields have significant implications for the economic viability and C-offsetting potential of the RCG crop.

Acknowledgements

This research was funded by the Department of Agriculture, Food and the Marine Research Stimulus Fund (Project Ref. 07 527). The authors also gratefully acknowledge the technical assistance of Brendan Swan, Kevin McNamara, Vincent Staples, Carmel O' Connor and Teresa Cowman in Johnstown Castle.

References

Adams CR, Galatowitsch SM (2005) *Phalaris arundinacea* (reed canary grass): rapid growth and growth pattern in conditions approximating newly restored wetlands. *Ecoscience*, **12**, 569–573.

Ammann C, Flechard C, Leifeld J, Neftel A, Fuhrer J (2007) The carbon budget of newly established temperate grassland depends on management intensity. *Agriculture Ecosystems & Environment*, **121**, 5–20.

Anderson-Teixeira K, Masters M, Black C, Zeri M, Hussain M, Bernacchi C, Delucia E (2013) Altered belowground carbon cycling following land-use change to perennial bioenergy crops. *Ecosystems*, **16**, 508–520.

Arundale R, Dohleman F, Heaton E, McGrath J, Voigt T, Long S (2014) Yields of *Miscanthus* × *giganteus* and *Panicum virgatum* decline with stand age in the Midwestern USA. *Global Change Biology Bioenergy*, **6**, 1–13.

Aubinet M, Chermanne B, Vandenhaute M, Longdoz B, Yernaux M, Laitat E (2001) Long term carbon dioxide exchange above a mixed forest in the Belgian Ardennes. *Agricultural and Forest Meteorology*, **108**, 293–315.

Baum K, Ham J, Brunsell N, Coyne P (2008) Surface boundary layer of cattle feedlots: implications for air emissions measurement. *Agricultural and Forest Meteorology*, **148**, 1882–1893.

Beale C, Long S (1995) Can perennial C_4 grasses attain high efficiencies of radiant energy conversion in cool climates? *Plant, Cell and Environment*, **18**, 641–650.

Beale CV, Bint DA, Long SP (1996) Leaf photosynthesis in the C_4-grass *Miscanthus* × *giganteus*, growing in the cool temperate climate of southern England. *Journal of Experimental Botany*, **47**, 267–273.

Black K, Bolger T, Davis P *et al.* (2007) Inventory and eddy covariance-based estimates of annual carbon sequestration in a Sitka spruce (*Picea sitchensis* (Bong.) Carr.) forest ecosystem. *European Journal of Forest Research*, **126**, 167–178.

Cai T, Price D, Orchansky A, Thomas B (2011) Carbon, water, and energy exchanges of a hybrid poplar plantation during the first five years following planting. *Ecosystems*, **14**, 658–671.

Campbell G, Norman J (1989) The description and measurement of plant canopy structure. In: *Plant Canopies: Their Growth, Form and Function* (eds Russell G, Marshall B, Jarvis P), pp. 1–19. Cambridge University Press, Cambridge, UK.

Central Statistics Office (2014) *Crops and Livestock Survey June 2013 - Final Results*. Central Statistics Office, Cork, Ireland.

Chapin F, Woodwell G, Randerson J *et al.* (2006) Reconciling carbon-cycle concepts, terminology, and methods. *Ecosystems*, **9**, 1041–1050.

Christian D, Riche A, Yates N (2008) Growth, yield and mineral content of *Miscanthus* × *giganteus* grown as a biofuel for 14 successive harvests. *Industrial Crops and Products*, **28**, 320–327.

Clifton-Brown J, Neilson B, Lewandowski I, Jones M (2000) The modelled productivity of *Miscanthus* × *giganteus* (GREEF et DEU) in Ireland. *Industrial Crops and Products*, **12**, 97–109.

Clifton-Brown J, Breuer J, Jones M (2007) Carbon mitigation by the energy crop, *Miscanthus*. *Global Change Biology*, **13**, 2296–2307.

Clifton-Brown J, Robson P, Sanderson R, Hastings A, Valentine J, Donnison I (2011) Thermal requirements for seed germination in *Miscanthus* compared with switchgrass (*Panicum virgatum*), reed canary grass (*Phalaris arundinaceae*), maize (*Zea mays*) and perennial ryegrass (*Lolium perenne*). *GCB Bioenergy*, **3**, 375–386.

Dao T (1998) Tillage and crop residue effects on carbon dioxide evolution and carbon storage in a Paleustoll. *Soil Science Society of America Journal*, **62**, 250–256.

Davidson E, Ackerman I (1993) Changes in soil carbon inventories following cultivation of previously untilled soils. *Biogeochemistry*, **20**, 161–193.

Davis P, Brown J, Saunders M *et al.* (2010) Assessing the effects of agricultural management practices on carbon fluxes: spatial variation and the need for replicated estimates of net ecosystem exchange. *Agricultural and Forest Meteorology*, **150**, 564–574.

Dohleman F, Long S (2009) More productive than maize in the Midwest: how does *Miscanthus* do it? *Plant Physiology*, **150**, 2104–2115.

Don A, Osborne B, Hastings A *et al.* (2012) Land-use change to bioenergy production in Europe: implications for the greenhouse gas balance and soil carbon. *Global Change Biology Bioenergy*, **4**, 372–391.

Donnelly A, Styles D, Fitzgerald J, Finnan J (2011) A proposed framework for determining the environmental impact of replacing agricultural grassland with *Miscanthus* in Ireland. *Global Change Biology Bioenergy*, **3**, 247–263.

Eriksen J, Jensen L (2001) Soil respiration, nitrogen mineralization and uptake in barley following cultivation of grazed grasslands. *Biology and Fertility of Soils*, **33**, 139–145.

EU (2009) Directive 2009/28/EC of the European Parliament and of the Council of 23 April 2009 on the promotion of the use of energy from renewable sources and amending and subsequently repealing Directives 2001/77/EC and 2003/30/EC. EU, Brussels.

Falge E, Baldocchi D, Olson R *et al.* (2001) Gap filling strategies for defensible annual sums of net ecosystem exchange. *Agricultural and Forest Meteorology*, **107**, 43–69.

Fargione J, Hill J, Tilman D, Polasky S, Hawthorne P (2008) Land clearing and the biofuel carbon debt. *Science*, **319**, 1235–1238.

Finnan J (2007) Reed canary grass. In: *Fact Sheet Tillage No. 7*. Teagasc, Carlow, Ireland.

Finnan J, Burke B (2014) Nitrogen dynamics in a mature *Miscanthus* × *giganteus* crop fertilized with nitrogen over a five year period. *Irish Journal of Agricultural and Food Research*, **53**, 171–188.

Fischer G, Prieler S, van Velthuizen H, Berndes G, Faaij A, Londo M, de Wit M (2010) Biofuel production potentials in Europe: sustainable use of cultivated land and pastures, Part II: land use scenarios. *Biomass and Bioenergy*, **34**, 173–187.

Freibauer A, Rounsevell M, Smith P, Verhagen J (2004) Carbon sequestration in the agricultural soils of Europe. *Geoderma*, **122**, 1–23.

Garratt J (1992) *The Atmospheric Boundary Layer*. Cambridge University Press, Cambridge, UK.

Goulden M, Munger J, Fan S, Daube B, Wofsy S (1996) Measurements of carbon sequestration by long-term eddy covariance: methods and a critical evaluation of accuracy. *Global Change Biology*, **2**, 169–182.

Guo L, Gifford R (2002) Soil carbon stocks and land use change: a meta analysis. *Global Change Biology*, **8**, 345–360.

Haslwanter A, Hammerle A, Wohlfahrt G (2009) Open-path vs. closed-path eddy covariance measurements of the net ecosystem carbon dioxide and water vapour exchange: a long-term perspective. *Agricultural and Forest Meteorology*, **149**, 291–302.

Houghton R, House J, Pongratz J *et al.* (2012) Carbon emissions from land use and land-cover change. *Biogeosciences*, **9**, 5125–5142.

Huyghe C, De Vliegher A, van Gils B, Peeters A (2014) *Grasslands and Herbivore Production in Europe and Effects of Common Policies*. Éditions Quae, Versailles, France.

Jackson R, Canadell J, Ehleringer J, Mooney H, Sala O, Schulze E (1996) A global analysis of root distributions for terrestrial biomes. *Oecologia*, **108**, 389–411.

Jaksic V, Kiely G, Albertson J, Oren R, Katul G, Leahy P, Byrne KA (2006) Net ecosystem exchange of grassland in contrasting wet and dry years. *Agricultural and Forest Meteorology*, **139**, 323–334.

Jones M, Donnelly A (2004) Carbon sequestration in temperate grassland ecosystems and the influence of management, climate and elevated CO_2. *New Phytologist*, **164**, 423–439.

Kaimal J, Finnigan J (1994) *Atmospheric Boundary Layer Flows: Their Structure and Measurement*. Oxford University Press, Oxford, UK.

Kaimal J, Izumi Y, Wyngaard J, Coté R (1972) Spectral characteristics of surface-layer turbulence. *Quarterly Journal of the Royal Meteorological Society*, **98**, 563–589.

Kandel T, Elsgaard L, Karki S, Laerke P (2013) Biomass yield and greenhouse gas emissions from a drained fen peatland cultivated with reed canary grass under different harvest and fertilizer regimes. *Bioenergy Research*, **6**, 883–895.

Kindler R, Siemens J, Kaiser K *et al.* (2011) Dissolved carbon leaching from soil is a crucial component of the net ecosystem carbon balance. *Global Change Biology*, **17**, 1167–1185.

Kljun N, Kormann R, Rotach M, Meixner F (2003) Comparison of the Langrangian footprint model LPDM-B with an analytical footprint model. *Boundary-Layer Meteorology*, **106**, 349–355.

Kljun N, Calanca P, Rotach M, Schmid H (2004) A simple parameterisation for flux footprint predictions. *Boundary-Layer Meteorology*, **112**, 503–523.

Kormann R, Meixner F (2001) An analytical footprint model for non-neutral stratification. *Boundary-Layer Meteorology*, **99**, 207–224.

Landström S, Lomakka L, Anderson S (1996) Harvest in spring improves yield and quality of reed canary grass as a bioenergy crop. *Biomass and Bioenergy*, **11**, 333–341.

Lavergne S, Molofsky J (2004) Reed canary grass (*Phalaris arundinacea*) as a biological model in the study of plant invasions. *Critical Reviews in Plant Sciences*, **23**, 415–429.

Lewandowski I, Clifton-Brown J, Scurlock JMO, Huisman W (2000) *Miscanthus*: European experience with a novel energy crop. *Biomass and Bioenergy*, **19**, 209–227.

Lewandowski I, Scurlock J, Lindvall E, Christou M (2003) The development and current status of perennial rhizomatous grasses as energy crops in the US and Europe. *Biomass and Bioenergy*, **25**, 335–361.

Lind S, Shurpali N, Peltola O *et al.* (2015) Carbon dioxide exchange of a perennial bioenergy crop cultivation on a mineral soil. *Biogeosciences Discussions*, **12**, 16673–16708.

Lloyd J, Taylor J (1994) On the temperature dependence of soil respiration. *Functional Ecology*, **8**, 315–323.

Long S (1983) C_4 photosynthesis at low temperatures. *Plant Cell and Environment*, **6**, 345–363.

Mander U, Jarveoja J, Maddison M, Soosaar K, Aavola R, Ostonen I, Salm J (2012) Reed canary grass cultivation mitigates greenhouse gas emissions from abandoned peat extraction areas. *Global Change Biology Bioenergy*, **4**, 462–474.

Mauder M, Foken T (2004) Documentation and instruction manual of the eddy covariance software package TK2. Universitat Bayreuth, Abt. Mikrometeologie, Arbeitsergebnisse Nr. 26, Germany.

McCalmont JP, Hastings A, McNamara NP, Richter GM, Robson P, Donnison IS, Clifton-Brown J (2016) Environmental costs and benefits of growing *Miscanthus* for bioenergy in the UK. *GCB Bioenergy*. doi:10.1111/gcbb.12294.

McLaughlin S, Walsh M (1998) Evaluating environmental consequences of producing herbaceous crops for bioenergy. *Biomass and Bioenergy*, **14**, 317–324.

Meyers T, Hollinger S (2004) An assessment of storage terms in the surface energy balance of maize and soybean. *Agricultural and Forest Meteorology*, **125**, 105–115.

Moncrieff J, Massheder J, Debruin H *et al.* (1997) A system to measure surface fluxes of momentum, sensible heat, water vapour and carbon dioxide. *Journal of Hydrology*, **189**, 589–611.

Monteith J (1977) Climate and the efficiency of crop production in Britain. *Philosophical Transactions of the Royal Society B: Biological Sciences*, **281**, 277–294.

Monteith J (1978) Reassessment of maximum growth rates for C_3 and C_4 crops. *Experimental Agriculture*, **14**, 1–5.

Moureaux C, Ceschia E, Arriga N, Béziat P, Eugster W, Kutsch W, Pattey E (2012) Eddy covariance measurements over crops. In: *Eddy Covariance: A Practical Guide to Measurement and Data Analysis* (eds Aubinet M, Vesala T, Papale D), pp. 319–331. Springer Atmospheric Sciences, New York, NY.

Munro D, Oke T (1975) Aerodynamic boundary-layer adjustment over a crop in neutral stability. *Boundary-Layer Meteorology*, **9**, 53–61.

Naidu S, Long S (2004) Potential mechanisms of low-temperature tolerance of C_4 photosynthesis in *Miscanthus* × *giganteus*: an in vivo analysis. *Planta*, **220**, 145–155.

Oncley S, Foken T, Vogt R *et al.* (2007) The Energy Balance Experiment EBEX-2000. Part I: overview and energy balance. *Boundary-Layer Meteorology*, **123**, 1–28.

Osborne B, Saunders M, Walmsley D, Jones M, Smith P (2010) Key questions and uncertainties associated with the assessment of the cropland greenhouse gas balance. *Agriculture Ecosystems & Environment*, **139**, 293–301.

Özdemir E, Härdtlein M, Eltrop L (2009) Land substitution effects of biofuel side products and implications on the land area requirement for EU 2020 biofuel targets. *Energy Policy*, **37**, 2986–2996.

Peichl M, Leahy P, Kiely G (2011) Six-year stable annual uptake of carbon dioxide in intensively managed humid temperate grassland. *Ecosystems*, **14**, 112–126.

Peichl M, Carton O, Kiely G (2012) Management and climate effects on carbon dioxide and energy exchanges in a maritime grassland. *Agriculture Ecosystems & Environment*, **158**, 132–146.

Poeplau C, Don A (2013) Sensitivity of soil organic carbon stocks and fractions to different land-use changes across Europe. *Geoderma*, **192**, 189–201.

Poeplau C, Don A (2014) Soil carbon changes under *Miscanthus* driven by C_4 accumulation and C_3 decompostion - toward a default sequestration function. *Global Change Biology Bioenergy*, **6**, 327–338.

Poeplau C, Don A, Vesterdal L, Leifeld J, van Wesemael B, Schumacher J, Gensior A (2011) Temporal dynamics of soil organic carbon after land-use change in the temperate zone - carbon response functions as a model approach. *Global Change Biology*, **17**, 2415–2427.

Qin Z, Dunn JB, Kwon H, Mueller S, Wander MM (2016) Soil carbon sequestration and land use change associated with biofuel production: empirical evidence. *GCB Bioenergy* **8**, 66–80.

Reichstein M, Falge E, Baldocchi D *et al.* (2005) On the separation of net ecosystem exchange into assimilation and ecosystem respiration: review and improved algorithm. *Global Change Biology*, **11**, 1424–1439.

Ruppert J, Mauder M, Thomas C, Luers J (2006) Innovative gap-filling strategy for annual sums of CO_2 net ecosystem exchange. *Agricultural and Forest Meteorology*, **138**, 5–18.

Saijonkari-Pahkala K (2001) Non-wood plants as raw material for pulp and paper. *Agricultural and Food Science*, **10**, 1–101.

Scott HD (2000) *Soil Physics: Agricultural and Environmental Applications*. Iowa State University Press, Ames, IA.

Shurpali N, Hyvonen N, Huttunen J, Biasi C, Nykanen H, Pekkarinen N, Martikainen P (2008) Bare soil and reed canary grass ecosystem respiration in peat extraction sites in Eastern Finland. *Tellus Series B-Chemical and Physical Meteorology*, **60**, 200–209.

Shurpali N, Hyvonen N, Huttunen J *et al.* (2009) Cultivation of a perennial grass for bioenergy on a boreal organic soil - carbon sink or source? *Global Change Biology Bioenergy*, **1**, 35–50.

Sims R, Mabee W, Saddler J, Taylor M (2010) An overview of second generation biofuel technologies. *Bioresource Technology*, **101**, 1570–1580.

Skinner R, Adler P (2010) Carbon dioxide and water fluxes from switchgrass managed for bioenergy production. *Agriculture Ecosystems & Environment*, **138**, 257–264.

Smith P, Lanigan G, Kutsch W *et al.* (2010) Measurements necessary for assessing the net ecosystem carbon budget of croplands. *Agriculture Ecosystems & Environment*, **139**, 302–315.

Soussana J, Allard V, Pilegaard K *et al.* (2007) Full accounting of the greenhouse gas (CO_2, N_2O, CH_4) budget of nine European grassland sites. *Agriculture Ecosystems & Environment*, **121**, 121–134.

Stoy P, Mauder M, Foken T *et al.* (2013) A data-driven analysis of energy balance closure across FLUXNET research sites: the role of landscape scale heterogeneity. *Agricultural and Forest Meteorology*, **171**, 137–152.

Twine T, Kustas W, Norman J *et al.* (2000) Correcting eddy-covariance flux underestimates over a grassland. *Agricultural and Forest Meteorology*, **103**, 279–300.

Vanderborght J, Graf A, Steenpass C *et al.* (2010) Within-field variability of bare soil evaporation derived from eddy covariance measurements. *Vadose Zone Journal*, **9**, 943–954.

Vellinga T, van den Pol-van Dasselaar A, Kuikman P (2004) The impact of grassland ploughing on CO_2 and N_2O emissions in the Netherlands. *Nutrient Cycling in Agroecosystems*, **70**, 33–45.

Vickers D, Mahrt L (1997) Quality control and flux sampling problems for tower and aircraft data. *Journal of Atmospheric and Oceanic Technology*, **14**, 512–526.

Wagle P, Kakani V, Huhnke R (2015) Net ecosystem carbon dioxide exchange of dedicated bioenergy feedstocks: switchgrass and high biomass sorghum. *Agricultural and Forest Meteorology*, **207**, 107–116.

Wang D, Portis A, Moose S, Long S (2008) Cool C_4 photosynthesis: pyruvate P_i dikinase expression and activity corresponds to the exceptional cold tolerance of carbon assimilation in *Miscanthus* × *giganteus*. *Plant Physiology*, **148**, 557–567.

Webb E, Pearman G, Leuning R (1980) Correction of flux measurements for density effects due to heat and water-vapor transfer. *Quarterly Journal of the Royal Meteorological Society*, **106**, 85–100.

Willems A, Augustenborg C, Hepp S, Lanigan G, Hochstrasser T, Kammann C, Muller C (2011) Carbon dioxide emissions from spring ploughing of grassland in Ireland. *Agriculture Ecosystems & Environment*, **144**, 347–351.

Wilson K, Goldstein A, Falge E *et al.* (2002) Energy balance closure at FLUXNET sites. *Agricultural and Forest Meteorology*, **113**, 223–243.

Wrobel C, Coulman B, Smith D (2009) The potential use of reed canarygrass (*Phalaris arundinacea* L.) as a biofuel crop. *Acta Agriculturae Scandinavica Section B-Soil and Plant Science*, **59**, 1–18.

Xiong S, Kätterer T (2010) Carbon-allocation dynamics in reed canary grass as affected by soil type and fertilization rates in northern Sweden. *Acta Agriculturae Scandinavica Section B-Soil and Plant Science*, **60**, 24–32.

Zenone T, Chen J, Deal M *et al.* (2011) CO_2 fluxes of transitional bioenergy crops: effect of land conversion during the first year of cultivation. *Global Change Biology Bioenergy*, **3**, 401–412.

Zenone T, Gelfand I, Chen J, Hamilton S, Robertson G (2013) From set-aside grassland to annual and perennial cellulosic biofuel crops: effects of land use change on carbon balance. *Agricultural and Forest Meteorology*, **182**, 1–12.

Zenone T, Fischer M, Arriga N *et al.* (2015) Biophysical drivers of the carbon dioxide, water vapor, and energy exchanges of a short-rotation poplar coppice. *Agricultural and Forest Meteorology*, **209**, 22–35.

Zeri M, Anderson-Teixeira K, Hickman G, Masters M, Delucia E, Bernacchi C (2011) Carbon exchange by establishing biofuel crops in Central Illinois. *Agriculture Ecosystems & Environment*, **144**, 319–329.

Zimmermann J, Dauber J, Jones M (2012) Soil carbon sequestration during the establishment phase of Miscanthus × giganteus: a regional-scale study on commercial farms using 13C natural abundance. *GCB Bioenergy*, **4**, 453–461.

Zona D, Janssens I, Aubinet M, Gioli B, Vicca S, Fichot R, Ceulemans R (2013) Fluxes of the greenhouse gases (CO_2, CH_4 and N_2O) above a short-rotation poplar plantation after conversion from agricultural land. *Agricultural and Forest Meteorology*, **169**, 100–110.

13

Productivity and water use efficiency of *Agave americana* in the first field trial as bioenergy feedstock on arid lands

SARAH C. DAVIS[1,2], EMILY R. KUZMICK[1], NICHOLAS NIECHAYEV[1] and DOUGLAS J. HUNSAKER[3]

[1]*Voinovich School of Leadership and Public Affairs, Ohio University, Athens, OH 45701, USA,* [2]*Department of Environmental and Plant Biology, Ohio University, Athens, OH 45701, USA,* [3]*USDA-ARS Arid Lands Agricultural Research Center, Maricopa, AZ 85138, USA*

Abstract

Agave species are high-yielding crassulacean acid metabolism (CAM) plants, some of which are grown commercially and recognized as potential bioenergy species for dry regions of the world. This study is the first field trial of *Agave* species for bioenergy in the United States, and was established to compare the production of *Agave americana* with the production of *Agave tequilana* and *Agave fourcroydes*, which are produced commercially in Mexico for tequila and fiber. The field trial included four experimental irrigation levels to test the response of biomass production to water inputs. After 3 years, annual production of healthy *A. americana* plants reached 9.3 Mg dry mass ha^{-1} yr^{-1} (including pup mass) with 530 mm of annual water inputs, including both rainfall and irrigation. Yields in the most arid conditions tested (300 mm yr^{-1} water input) were 2.0–4.0 Mg dry mass ha^{-1} yr^{-1}. *Agave tequilana* and *Agave fourcroydes* were severely damaged by cold in the first winter, and produced maximum yields of only 0.04 Mg ha^{-1} yr^{-1} and 0.26 Mg ha^{-1} yr^{-1}, respectively. The agave snout weevil (*Scyphophorus acupunctatus*) emerged as an important challenge for *A. americana* cropping, killing a greater number of plants in the higher irrigation treatments. Physiological differences in *A. americana* plants across irrigation treatments were most evident in the warmest season, with gas exchange beginning up to 3 h earlier and water use efficiency declining in treatments with the greatest water input (780 mm yr^{-1} water input). Yields were lower than previous projections for *Agave* species, but results from this study suggest that *A. americana* has potential as a bioenergy crop and would have substantially reduced irrigation requirements relative to conventional crops in the southwestern USA. Challenges for pest management and harvesting must still be addressed before an efficient production system that uses *Agave* can be realized.

Keywords: agriculture, bioethanol, biofuel, CAM, crassulacean acid metabolism, desert, semiarid, southwest USA, WUE

Introduction

Droughts limit agricultural production in arid and semiarid regions, yet plants that exhibit crassulacean acid metabolism (CAM) can improve crop yields in these conditions (Cushman *et al.*, 2015; Davis *et al.*, 2015; Yang *et al.*, 2015). Plants in the *Agave* (L.) genus are obligate CAM plants, some of which are produced commercially, and are being considered for bioenergy crops (Borland *et al.*, 2009; Davis *et al.*, 2011, 2014, 2015; Yang *et al.*, 2015). In regions of the world where *Agave* is grown commercially for beverages or fiber, yields range from 8.5 to 22 Mg ha^{-1} depending on species and location (Davis *et al.*, 2014), but the latitudinal range for these crops is uncertain. Despite the potential advantages and high theoretical yields (Davis *et al.*, 2011, 2014; Owen *et al.*, 2015), there are many dry regions of the world including the southwestern USA where *Agave* has not yet been cultivated commercially or experimentally, making it difficult to assess the realistic potential of this genus for agricultural production. There are no field experiments to date that compare the yields of *Agave* species in arid to semiarid conditions or the response of *Agave* species to variable water inputs. Here, we provide the first field-scale yield estimates of *Agave* in the arid southwestern USA.

Climate projections indicate that many semiarid and arid parts of the world will be subject to more frequent and extreme drought events in the future (IPCC 2014). Agricultural production in many of these areas is already vulnerable to drought, dependent on large volumes of irrigation, and susceptible to soil degradation (Scanlon *et al.*, 2007). The yields of 8.5–22 Mg for *Agave* referenced above (Davis *et al.*, 2014) reflect commercial production in Mexico and Tanzania where the climate

Correspondence: Sarah C. Davis
e-mail: daviss6@ohio.edu

is semiarid due to prolonged periods of drought even though heavy rainfall does occur intermittently. As water use efficiency (WUE) is the primary advantage of CAM plants that has sparked interest in these plants for dryland agricultural production, there is a need to resolve how much precipitation would be needed to achieve commercially viable yields of *Agave*. Commercial production of bioenergy in the USA is primarily supported by *Zea mays* L. (corn) feedstock, which yields an average of 9.4 Mg ha^{-1} yr^{-1} on prime agricultural land, but the dominant commercial cropping systems in Arizona yield far less (*e.g.*, cotton yields 1.5 Mg ha^{-1} yr^{-1}).

The key physiological advantage of CAM, to distinguish from C_3 and C_4, is the temporal separation of carboxylase activities catalyzed by the enzyme PEPC (phosphoenolpyruvate carboxylase) and the enzyme RUBISCO (ribulose *bis*phosphate [RubP] carboxylase/oxygenase) so that PEPC is primarily active during the night when stomata are open (opposite of C_3 and C_4 plants), and RUBISCO is primarily active during the day when light energy is available. With PEPC activation occurring at night when temperatures are cooler, evaporative demand is lessened and the ratio of carbon dioxide to water that diffuses through the stomata is increased relative to what would be expected during daytime conditions. The carbon is stored as malic acid overnight in large vacuoles typical of succulent plants and then released by decarboxylation at dawn. This strategy for growth gives CAM plants a strong advantage over C_3 and C_4 plants in conditions of water stress (Osmond, 1978; Nobel, 1996; Winter & Smith, 1996; Davis *et al.*, 2014). Many CAM plants also have thick waxy cuticles, sunken stomata, and roots that are responsive to water stress; all adaptations that hydraulically isolate the plant in extremely dry conditions. Theoretical projections indicate that high-yielding CAM plants such as *Agave* and *Opuntia* (Mill.) can contribute substantially to the global renewable energy portfolio (Davis *et al.*, 2011, 2014; Cushman *et al.*, 2015; Owen *et al.*, 2015).

There is growing interest in *Agave* species for bioenergy production in the arid and semiarid southwestern region of the USA. Commercial agricultural production has been demonstrated for species native to Mexico (e.g., *Agave tequilana* F.A.C. weber and *Agave fourcroydes* Lemair), but these have not been grown in the USA outside of small horticultural and landscaping plantings. *Agave americana* L., also known as the American century plant, is native to the USA (native range includes Arizona, California, Florida, Louisiana, and Texas) and has been reported to reach sizes similar to its high-yielding Mexican relatives (2 m in height before flowering, >3 m in width) and achieve high photosynthetic rates under a range of environmental conditions (Neales, 1973; Gentry, 1982; Nobel, 2003; Garcia-Moya *et al.*, 2011), but field-scale yields of *A. americana* have not been assessed in previous literature. *A. americana* is produced for a beverage called pulque in Mexico (La Barre, 1938), but there is less information about the agricultural production of this species than what is available from the larger-scale tequila and fiber industries. Tissue composition of *A. americana* has been measured, and the leaves of this species have a low percentage of lignin mass relative to other species of *Agave* as well as high-soluble carbohydrate concentrations relative to other advanced bioenergy crops (Mylsamy & Rajendran, 2010; Li *et al.*, 2012; Corbin *et al.*, 2015), both beneficial traits for biochemical conversion pathways to liquid fuel (Davis *et al.*, 2011).

The first US field trial of *Agave* species for bioenergy was established in 2012 in Maricopa, AZ, a desert agricultural area where annual rainfall averages about 190 mm yr^{-1}. This study reports the results to date for this experiment. The agricultural plantings were designed as a fully factorial two-way experiment to determine differences in survivorship, productivity, and WUE among three species (*A. americana, A. tequilana,* and *A. fourcroydes*) and four irrigation treatments (ranging from arid to well-watered conditions), and to determine whether responses to irrigation are species dependent. The study site is at the edge of the climate tolerance range for *A. tequilana* and *A. fourcroydes* (Lewis *et al.*, 2015), but these are the large-scale commercial species against which a comparison of the US native species, *A. americana*, would be meaningful. Significant mortality and cold damage of *A. tequilana* and *A. fourcroydes* occurred after a rare cold event in the first year, so the focus of the study was narrowed by the third year to resolving differences in *A. americana* production, WUE, and survivorship across irrigation treatments. Early survivorship and production of *A. tequilana*, *A. fourcroydes*, and *A. americana* are reported, but differences in carbon dioxide assimilation, biomass yields, and WUE across four irrigation treatments were resolved only for *A. americana* in year 3.

Materials and methods

Study site and experimental design

The study site is a 1.2-ha field located at the University of Arizona Maricopa Agricultural Center (MAC) in Maricopa, AZ, and is centered at 33°03′32″ N latitude, 111°58′12″ W longitude, and 361 m above sea level. The field was plowed, tilled, and leveled before planting. Soil in the field is classified as Casa Grande sandy clay loam, which is described as fine-loamy, mixed, superactive, hyperthermic Typic Natrargids (Post *et al.*, 1988). The field was subdivided into eight sections (each approximately 32 m by 46 m) separated by berms approximately 50 cm in height to contain water that was allocated to each treatment by way of flood irrigation (Fig. 1). Two replicates of four irrigation treatments were assigned randomly to the eight sections. Irrigation water was provided by a concrete-lined irrigation ditch located on the south end of the field. A 2.5-m long piece of PVC pipe (305 mm, diameter) was hard-plumbed through a hole drilled in the irrigation ditch. Irrigation water to the field was controlled by a swivel valve located inside the ditch at the PVC pipe entry hole. The PVC pipe system was reduced to a 4.6-m long piece of 203-mm-diameter PVC pipe that contained an in-line propeller-driven flow meter. The PVC-valve system was then connected to a 203-mm-diameter plastic polypipe that ran along the center of the field in the north–south direction (Fig. 1). Gates installed along the polypipe at 1.0-m spacing allowed individual sections to be irrigated at desired times. Water was applied in approximately 100-mm increments (the minimum needed to flood-irrigate the section uniformly), but exact amounts varied, as explained below.

Irrigation treatments were initiated in the second year (2013), after planted sections had been uniformly irrigated in 2012 (year one). The control treatment received only precipitation (200 mm annually on average from 2013 to 2014) and the minimal water needed to apply fertilizer. In addition to the native precipitation, the irrigation schedule and levels targeted for other plots were as follows: (i) twice annually (in March and July) to add ~200 mm yr^{-1} (or ~2 Ml ha^{-1} yr^{-1}), (ii) every other month from March to October to add ~400 mm yr^{-1} (or ~4 Ml ha^{-1} yr^{-1}), and (iii) monthly from March to October to add ~800 mm yr^{-1} (or ~8 Ml ha^{-1} yr^{-1}). The actual water inputs achieved differed some from these target quantities due to variations in water flow rate, soil hydraulic and roughness conditions, integrity of berms, which occasionally required rebuilding after being broken by flood water, and practical decisions made by technicians during sporadic heavy rainfall events. Exact amounts of water applied during irrigation were measured (both flow rate and total volume) by the in-line flow meter. The actual 2-year (2013 and 2014) average amounts of irrigation applied for the four treatment levels were 100 mm yr^{-1}, 260 mm yr^{-1}, 330 mm yr^{-1}, and 580 mm yr^{-1}. Including the mean annual rainfall (~200 mm; see detailed climate in Fig. S1), total water inputs in the treatments were 300 mm yr^{-1}, 460 mm yr^{-1}, 530 mm yr^{-1}, and 780 mm yr^{-1}.

The irrigation treatments described above were initiated in the spring of 2013 (year 2) and repeated through 2014 (year 3). The entire field was fertilized annually in June with N:P:K fertilizer in a ratio of 40:80:80 based on recommendations for commercial plantation management used in the tequila industry. Applied rates each year were 56 kg N ha^{-1}, 112 kg P ha^{-1}, and 112 kg K ha^{-1}. Weeding was accomplished manually in the first year, and required regular attention (twice per month). In the second and third years, weeds were controlled with herbicide (Surflan) applied 3–4 time per year at a rate of 4.7–9.5 l ha^{-1}, by mowing between rows where possible, and manual hoeing of weeds near the base of plants.

Within each of the replicated irrigation sections, there were six plots of *Agave* for a total of 48 plots across the field (Fig. 1). Two plots each of *A. tequilana*, *A. fourcroydes*, and *A. americana* were randomly assigned within each replicated irrigation treatment, so that across the whole field there were four replicate plots of each species in each irrigation treatment. Each plot was an experimental unit and contained 49 plants (7 × 7) spaced 2 m apart both within and along rows. The experimental design in the field was purposeful to statistically test the difference in survivorship, yield, and WUE among species and among irrigation treatments. We treated this as a fully factorial design instead of a split-plot design because each combination

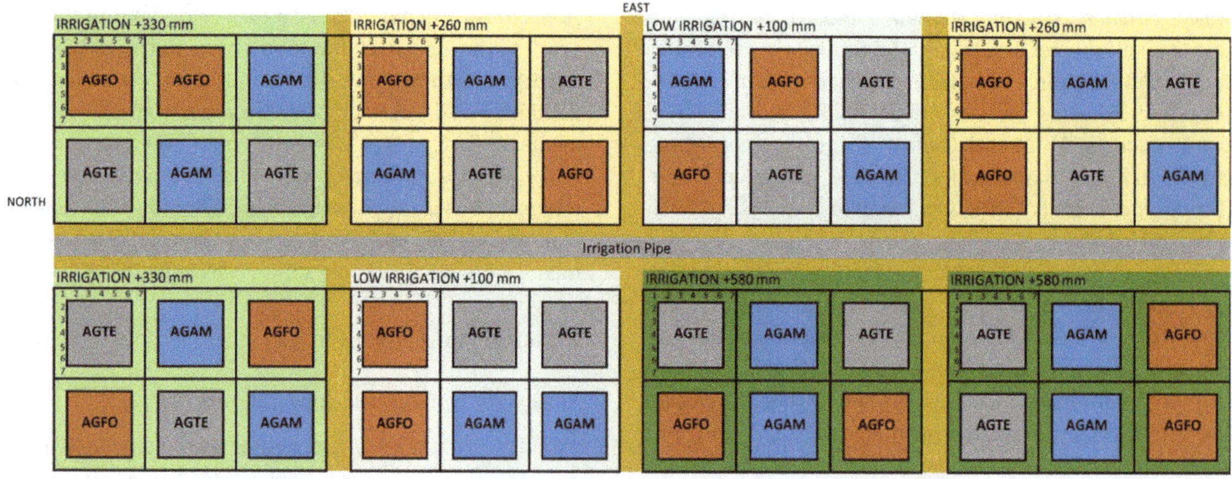

Fig. 1 Field layout of the experimental design with three species (AGAM: *A. americana*, AGFO: *A. fourcroydes*, and AGTE: *A. tequilana*) and four irrigation treatments. Each square plot contained 49 plants spaced 2 m apart. Four replicate plots of each species were randomly assigned to each irrigation treatment with two replicated and randomly located areas of the field assigned to each irrigation level (100, 260, 330, and 580 mm yr^{-1}). Total water inputs including rainfall for each treatment were 300, 460, 530, and 780 mm yr^{-1}. The irrigation pipe (shown in gray) was laid through the center of the field to allow access to each section that would be flood-irrigated at separate times. Berms (shown in brown) were constructed to separate each irrigated section.

of irrigation and species treatments was replicated throughout a single field and the sections delineated in the field for each irrigation were not unique.

Agave tequilana plants were a single genotype of weber var. azul propagated clonally and purchased from growers in Oaxaca, Mexico. *Agave fourcroydes* plants were mixed unidentified genotypes purchased from growers in Yucatan, Mexico. Both *A. tequilana* and *A. fourcroydes* were imported bare rooted with permits from the U.S. Department of Agriculture and were uniform in age of just under 1 year. *Agave americana* plants were from a variety of unidentified genotypes purchased from a grower in California, USA, and shipped to the site in one gallon pots (also approximately 1 year in age).

All species were planted in the spring of 2012 between March 23 and May 5 and irrigated weekly for the first month to ensure establishment, then irrigated monthly until October 2012 (~1000 mm water applied in total). Establishment was successful for all three species, and there was no mortality observed prior to January of 2013. There was an unusually cold period with nine consecutive days of temperatures below freezing in January of 2013, and minimums between −6 °C and −8 °C for five consecutive nights. The *A. tequilana* and *A. fourcroydes* plots suffered severe tissue damage. The monthly mean minimum temperature was 0.6 °C, 67% lower than the mean minimum for January in the other growth years. Although many plants remained, they were severely damaged and the biomass of both *A. fourcroydes* and *A. tequilana* were magnitudes lower than the biomass of *A. americana* as a result. The study is therefore focused on the response of *A. americana* to irrigation, although survivorship and yields were quantified and compared among species after year two. There were an insufficient number of plots of *A. tequilana* remaining across irrigation treatments to statistically assess the response to irrigation for this species.

Measurements

A field-calibrated neutron probe (Model 503, Campbell Pacific Nuclear, CPN, Martinez, CA) was used to measure the volumetric soil water contents (θ) of plots from 0.1 m to 1.9 m in 0.2 m incremental depths. Between mid-April and mid-May, 2013, 33 neutron access tubes were installed vertically to a soil depth of 2.0 m using a tractor-mounted Giddings soil sampler (Model 25-TS; Giddings Machine Company, Windsor, CO, USA). Of the 33 tubes, 16 were located within each of the 16 *A. americana* plots (i.e., four per irrigation level). Tubes were installed at a distance of about 1.0 m away from plants. During installation of the neutron access tubes, soil samples in 0.3-m increments were collected at each location to a depth of 1.8 m (with 6 depth increments). Each separate soil sample was analyzed for soil particle size fraction using the Bouyoucos hydrometer method (Gee & Bauder, 1986). Although sandy clay loam and sandy loam textures were predominate, soil texture analyses revealed approximately 10% of the total samples were classified either as clay loam, loam, or sandy clay. The volumetric soil water content measurements at the 33 access tube locations began on May 21, 2013. In general, soil water contents were measured twice monthly, just before and a few days after a scheduled irrigation. Soil water contents were measured for all access tubes and depths with each irrigation event, even when only a subset of the plots were irrigated.

Evapotranspiration ($ET_{c.}$) for the irrigation treatments of *A. americana* was determined over successive soil water content measurement dates as the residual of the soil water balance of an estimated root zone of 2.0 m. Following Jensen *et al.* (1990), the soil water balance for $ET_{c.}$ occurring between the first measurement date, denoted as day $i = 1$, and the second date $\approx$ 15 days later on day n was calculated as:

$$ET_c = \sum_{i=1}^{10}(S_{i,1} - S_{i,n}) + \sum_{j=1}^{n-1}(R_j + IW_j) \tag{1}$$

where $ET_{c.}$ is the total evapotranspiration that occurred from day 1 to day n, $S_{i,1}$ and $S_{i,\,n}$ are soil water storage measured at each of the 10 neutron probe soil depth layers on day $i = 1$ and on day n, respectively. R_j and IW_j are, respectively, the measured rainfall and applied irrigation depths received on day j. All units are in mm. Rainfall data were provided by a University of Arizona, Meteorological Network (AzMet; ag.arizona.edu/azmet) weather station located at MAC, about 1.2 km from the field site. For each of the 10 soil depth increments, soil water storage was calculated as the measured soil water contents (m^3/m^3) at the layer times the soil depth layer in m (i.e., $\theta \times 0.2$ m). For each soil water measurement location, the volumetric soil water contents at field capacity (FC) and permanent wilting point (PWP) were determined for each layer from the soil texture data using soil water characteristics estimation procedures developed by Saxton & Rawls (2006). Equation 1 does not include a deep percolation component, as there was no evidence of soil water content increasing below a soil depth of 1.5 m. Fig. 2 shows the periodically measured θ for each layer with time for the wettest irrigation treatment (*i.e.*, 580 mm). The figure includes the FC and PWP for each layer determined from soil texture. For 2013 (Fig. 2a), soil water contents measured a few days after irrigations increased to a depth of 1.5 m. However, the soil water contents at lower layers did not increase above FC, indicating negligible drainage from the root zone. Conversely, it can be seen, particularly in 2014 (Fig. 2b), that soil water was depleted somewhat to 1.9 m, suggesting that a root zone of 2.0 m for agave water use was reasonable.

Survival and reproduction rates of plants were recorded and compared across species and irrigation treatments in the second year of growth. Survival rates were recorded in July of 2013 (after all cold damage could be assessed) and biomass productivity was recorded in January of 2014 (after two full growing seasons). The mean percentage of surviving plants in four replicate plots was used to estimate survivorship of each species in each irrigation treatment. No significant cold damage was observed after the spring of 2013. Survivorship and biomass productivity was measured again in February of 2015 (after three full growing seasons). The mean number and condition of reproductive offsets (pups) was also measured in each plot.

The response of dusk-to-dawn carbon dioxide uptake to the different irrigation treatments was determined by measuring gas exchange in *A. americana* using Li-Cor LI-6400XT Portable Photosynthesis Systems (Li-Cor; Lincoln, NE, USA). These infrared gas analyzers were used to record carbon uptake rates

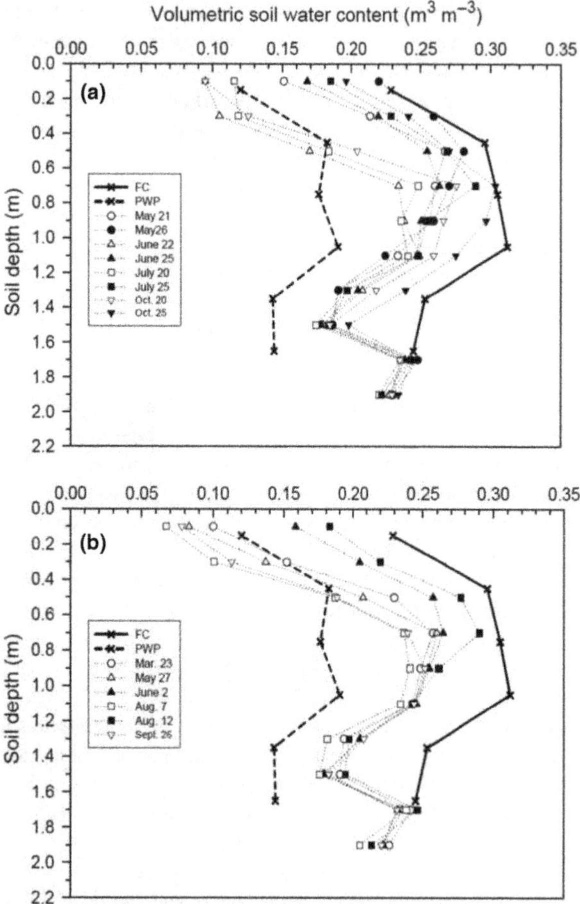

Fig. 2 Soil water content measurements from 0.1 to 1.9 m, in 0.2-m increments (dotted lines), field capacity (FC, solid line), and permanent wilting point (PWP, dashed line) derived from soil texture analyses for 0.3-m soil layers from 0 to 1.8 m for the 800-mm treatment plot in 2013 (a) and 2014 (b). Only a subset of all measurements are shown as examples. Note soil water content measurements made after an irrigation are denoted by dark symbols. The FC and PWP data are plotted at the center of the layer depth (e.g., at 0.15 m for the 0- to 30-m layer).

for plants in each irrigation treatment over a time course starting before sunset and extending until after dawn at three times throughout the year (March, June, and November). Total net carbon assimilation was estimated by integrating the measurements of gas exchange over time (typically a 15-h period). Sampling began in the late afternoon in attempt to include gas exchange during Phase IV, but gas exchange was not observed prior to 6 pm. In November of 2014, a full 24-h time period was measured to verify that the 15-h measurements reflected the full duration of stomatal opening. Dynamics of gas exchange were also compared to determine whether the timing of peak carbon uptake rates and the phases of CAM differed across the four different irrigation treatments.

During each seasonal sampling, gas exchange over the 15-h period from 6:00 pm to 9:00 am was measured at approximately 30-min intervals in each irrigation treatment. Each incremental estimate during the diurnal time period was a mean of the gas exchange rate measured in a mature leaf located midway between the outside whorl and center spike of three randomly selected plants in each irrigation treatment. This approach allowed us to estimate gas exchange levels using a mean of several plants within an irrigation treatment instead of that of a single plant. A square 2 cm × 3 cm cuvette was used with thick gaskets to accommodate the thick succulent leaves, and measurements were made at a point on the leaf approximately two thirds the distance from the base. There were occasional gaps in data (a missing 30-min increment) due to the need to switch batteries in the LiCor 6400 or other maintenance requirements. When calculating the total dusk-to-dawn CO_2 assimilation, measurements from the next closest time interval were used to estimate fluxes during each gap.

Aboveground biomass of a random plant from each of the sixteen *A. americana* plots was measured after destructively harvesting at the end of second and third years of growth (mid-January 2014 and early February 2015). In February of 2015, aboveground biomass of all plants that were killed by the agave snout weevil was also measured (the number of plants killed per plot ranged from 3 to 38). Because most plants harvested in February of 2015 were infested with the snout weevil, an additional harvest occurred in the spring of 2015 to assess the productivity of plants that remained healthy. Subsamples from each plant, consisting of three randomly selected leaves, were dried in an oven for 2 weeks at 65 °C and then weighed to estimate plant tissue water content. The water mass was subtracted from the total wet biomass to estimate dry biomass production in each plot.

Net WUE was estimated in two ways: WUE_1, as total dry biomass per unit of water input (including both precipitation and irrigation); and WUE_2, as total dry biomass per unit ET_c as calculated from the soil water balance described above. WUE_1 provides a practical estimate for growers of the amount of biomass one can expect given a certain water input. WUE_2 can be used to resolve any additional soil water that might be used to supplement the plant water requirements.

The statistical program JMP Pro 10 (SAS Institute Inc., Cary, NC, USA) was used to assess the differences in survivorship and biomass productivity in year 2 among species and irrigation treatments, and whether the effect of irrigation depended on species, using a two-way analysis of variance (ANOVA). The same statistical program was used to test differences in (i) survivorship, (ii) reproductive offsets, (iii) biomass productivity, and (iv) WUE of *A. americana* among irrigation treatments in year 3 using a one-way ANOVA. The Shapiro–Wilk W-test was used to determine whether the distribution in response variables met the assumption of normality. Tukey's HSD (honest significant differences) was used to resolve significant differences among treatment levels. Linear regression was used to compare the two estimates of WUE, and nonlinear regression was used to assess the response of WUE to increasing water inputs and ET_c.

Results

Initial establishment of *A. americana*, *A. tequilana*, and *A. fourcroydes* was successful, with a mean survival at the plot level of 100%, 97%, and 92%, respectively, by August of 2012. The few plants that were lost during this period were replaced. In January of 2013, the minimum temperature on five consecutive nights ranged from −6 °C to −8 °C, and cold damage was immediately evident in the *A. tequilana* and *A. fourcroydes* even though mortality was not entirely evident until March of 2013. A survey of the field in July of 2013 revealed that 74% of the *A. tequilana* plants were killed by the prolonged cold (Fig. 3), and 88% of the *A. fourcroydes* were severely damaged. There was no mortality or cold damage observed in the plots of *A. americana* after the cold event in January of 2013, and survival of the *A. americana* remained near 100% until 2014. There was greater mortality in *A. tequilana* than in *A. fourcroydes* or *A. americana*, and irrigation treatments did not affect survival in the second year for any species. Statistical significance is not reported for these differences because the distribution of response variables was bimodal (violating the assumption of normality in the statistical model) and could not be corrected through transformation. Ranking data for analysis using nonparametric tests was not possible because survival was 100% for all *A. americana* plots (insufficient range for ranking). Nevertheless, the difference in survivorship among species was clear (Fig. 3).

Although many of the *A. fourcroydes* survived, the cold damage was severe and the aboveground tissue completely died back in many of the plots. After recovery, the aboveground biomass of *A. fourcroydes* in year two was less than 0.3 Mg ha^{-1} yr^{-1}, even in the highest irrigation treatment, and was less than 9% of the biomass production measured for *A. americana* in all irrigation treatments (Fig. 4). While there was a significant main effect of irrigation on biomass ($P < 0.01$), this response differed by species ($P < 0.01$), with *A. americana* biomass increasing more with irrigation levels than biomass of *A. fourcroydes*. *A. americana* biomass was also significantly greater than *A. fourcroydes* biomass in all irrigation treatments ($P < 0.01$). Before statistically resolving the differences between *A. americana* and *A. fourcroydes*, biomass data were log-transformed to achieve a normal distribution. There were so few *A. tequilana* surviving that replicate plots of living plants no longer remained in all treatments (some plots had 0 surviving plants), but the biomass production of the remaining plots was ≤1% of *A. americana* production (Fig. 4). *A. tequilana* and *A. fourcroydes* did not recover nor make any significant gain in biomass production by the end of the third year.

The first evidence of mortality in the *A. americana* plots was recorded in January of 2014, and this was due to an infestation by *Scyphophorus acupunctatus* (agave snout weevil). The plots were first treated with an insecticide (Merit, Bayer CropScience, Monheim, Germany) in February of 2014, and two repeat applications occurred in the summer and fall (each at a rate of 67–87 kg ha^{-1}). The weevil was, however, never fully eradicated from the field. In February of 2015, 77% of the originally established plants (mother plants) in the highest irrigation treatment had been killed by the snout weevil, significantly more than the other irrigation treatments ($P < 0.01$). Survivorship was greatest (80%) in the plots with the lowest annual water inputs (Fig. 5).

Asexual reproduction rates of *A. americana* were greater in the plots with 780 mm annual water inputs in 2014, but the mean number of reproductive offsets (pups) produced for each original plant (mother plant) was similar across treatments in 2015 (Fig. 6). An average of 3 pups per mother plant were produced across

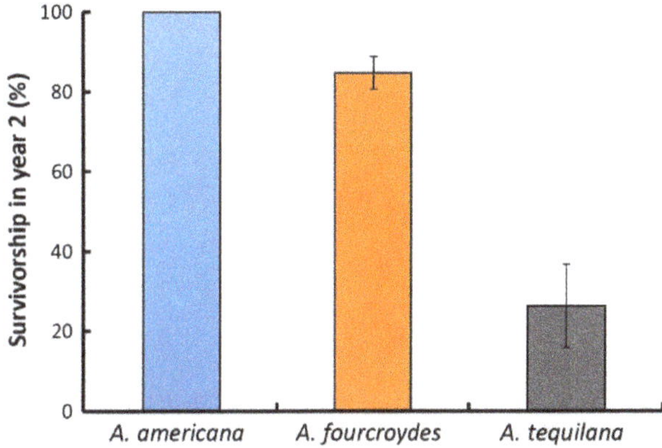

Fig. 3 Percent survivorship of three *Agave* species by July of 2013. Bars represent means of replicate plots across all irrigation levels (n = 16), and error bars represent standard error.

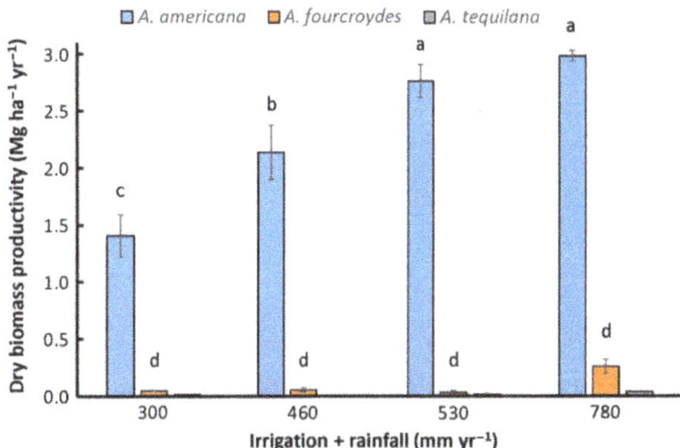

Fig. 4 Annualized productivity of three *Agave* species across the four irrigation treatments by January of 2014. Bars represent treatment means with plots as the replicate (n = 4), and error bars represent standard error. Different letters indicate statistically significant differences ($P < 0.05$).

all plots of *A. americana* in 2014, and a mean of 4 pups per mother plant were produced by 2015. There was not a significant difference among irrigation treatments in the number of surviving pups remaining in the plots ($P = 0.14$), or in the total number of pups produced per mother plant ($P = 0.10$), although the trend in 2015 was toward greater pup production with annual water inputs of 460 mm yr^{-1} and 530 mm yr^{-1}. Reproductive offsets were not observed in the other species.

Biomass of *A. americana* plants increased with water inputs of up to 530 mm in all years (Fig. 7). In 2014, plots with the greatest water inputs (530 mm yr^{-1} and 780 mm yr^{-1}) resulted in the highest annual biomass production (2.8–3.0 Mg ha^{-1} yr^{-1}), significantly greater than the 300-mm or 460-mm treatments ($P < 0.01$). The difference among treatments in biomass of dead plants harvested in 2015 was not significant ($P = 0.77$). There was, however, a trend for greater biomass productivity of healthy mother plants in the plots that received 530 mm yr^{-1} of water when compared to those in the other treatments ($P = 0.06$; Fig. 7). Total aboveground biomass, including biomass of both pups and mother plants, was significantly different across irrigation treatments in 2015 ($P = 0.02$), with the biomass production reaching 9.3 Mg ha^{-1} yr^{-1} in plots with 530 mm of annual water input (Fig. 7). After 3 years, annual production of healthy plants was more than double the annual production after 2 years, although mean production rates of plants killed by the snout weevil were much lower (Fig. 7).

WUE_1, calculated as total aboveground biomass production per unit of total water input, was similar in irrigation treatments that ranged from total inputs of

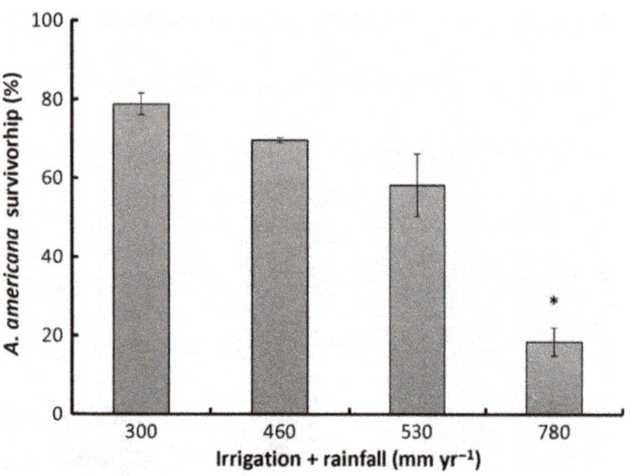

Fig. 5 Percent survivorship of *A. americana* after the agave snout weevil spread through the field in 2014. Survivorship was assessed in February of 2015. Bars represent treatment means with plots as the replicate ($n = 4$) and error bars represent standard error. * indicates statistically significant difference ($P < 0.05$).

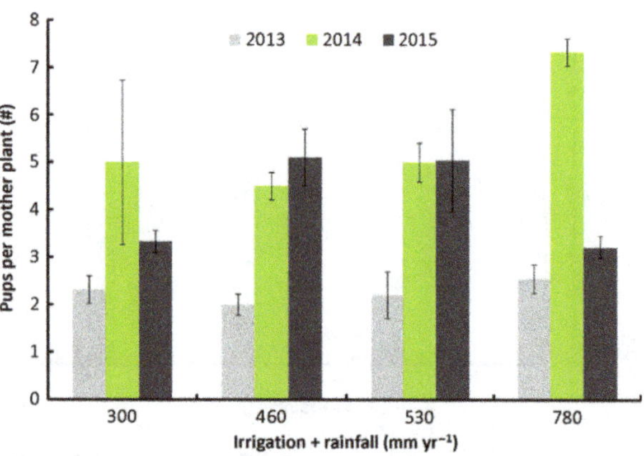

Fig. 6 Number of asexually propagated offsets (pups) observed for each *A. americana* plant that was initially established in the field (mother plants). Bars represent mean of replicate plots ($n = 4$), and error bars represent standard error.

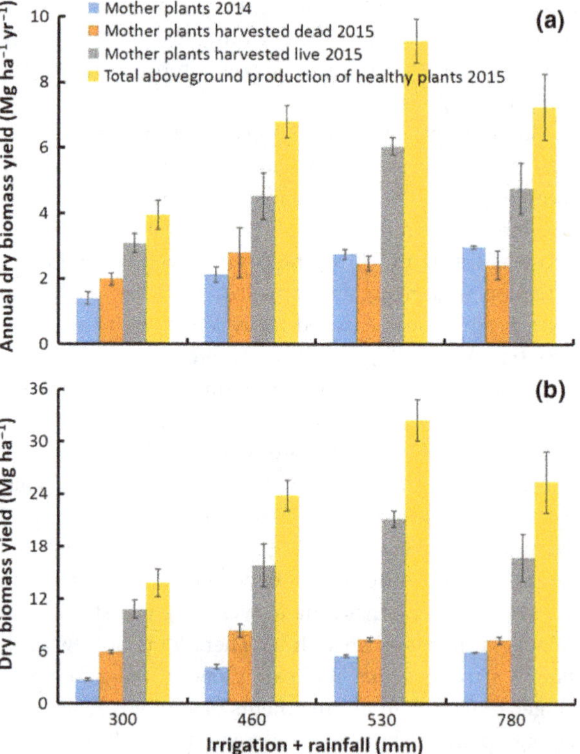

Fig. 7 *Agave americana* oven-dried biomass production annualized (a) and total (b) for 2-year-old mother plants (blue, 2014), dead 3-year-old mother plants (orange, 2015), living 3-year-old mother plants (gray, 2015), and total aboveground biomass of healthy mother and pups (yellow, 2015). Bars represent a mean of individuals from replicate plots ($n = 4$), but the number of dead plants per plot varied ($3 < n < 31$) and error bars represent standard error.

300 mm yr^{-1} to 530 mm yr^{-1} ($P > 0.05$), but WUE_1 declined in the highest irrigation treatment and was significantly lower with inputs of 780 mm yr^{-1} ($P = 0.01$; Fig. 8a). Trends in the response of WUE_2, calculated as total aboveground biomass production per unit of $ET_{c.}$ (Fig. 8b), to water inputs were very similar to the response of WUE_1 because there was little $ET_{c.}$ derived from soil water storage (Table 1). Results from the two methods estimating WUE were closely correlated (Fig. 8d; $R^2 = 0.99$). There was a nonlinear correlation between water inputs and WUE_1 ($R^2 = 0.88$) and between $ET_{c.}$ and WUE_2 ($R^2 = 0.82$), with the greatest WUE resulting from conditions with total water inputs of 460 mm yr^{-1} to 530 mm yr^{-1} (Fig. 8).

The greatest difference in carbon assimilation among treatments, as measured by dusk-to-dawn gas exchange, occurred in June (Fig. 9), which was the time period with the lowest native rainfall (relative to March and November). During the dry summer nights, plants growing in plots with higher water inputs (530 mm and 780 mm) opened stomata up to 3 h earlier than plants growing in the control plots (Fig. 10). Carbon assimilation increased with irrigation treatment levels in June, but differences across treatments were not evident in March and November (Fig. 9). The measurement campaign in November confirmed that there was no midday carbon assimilation.

Discussion

Agave americana was successfully established as an agricultural planting in the arid conditions of the southwestern USA. Although biomass production of *A. americana* responded to irrigation, there was no bene-

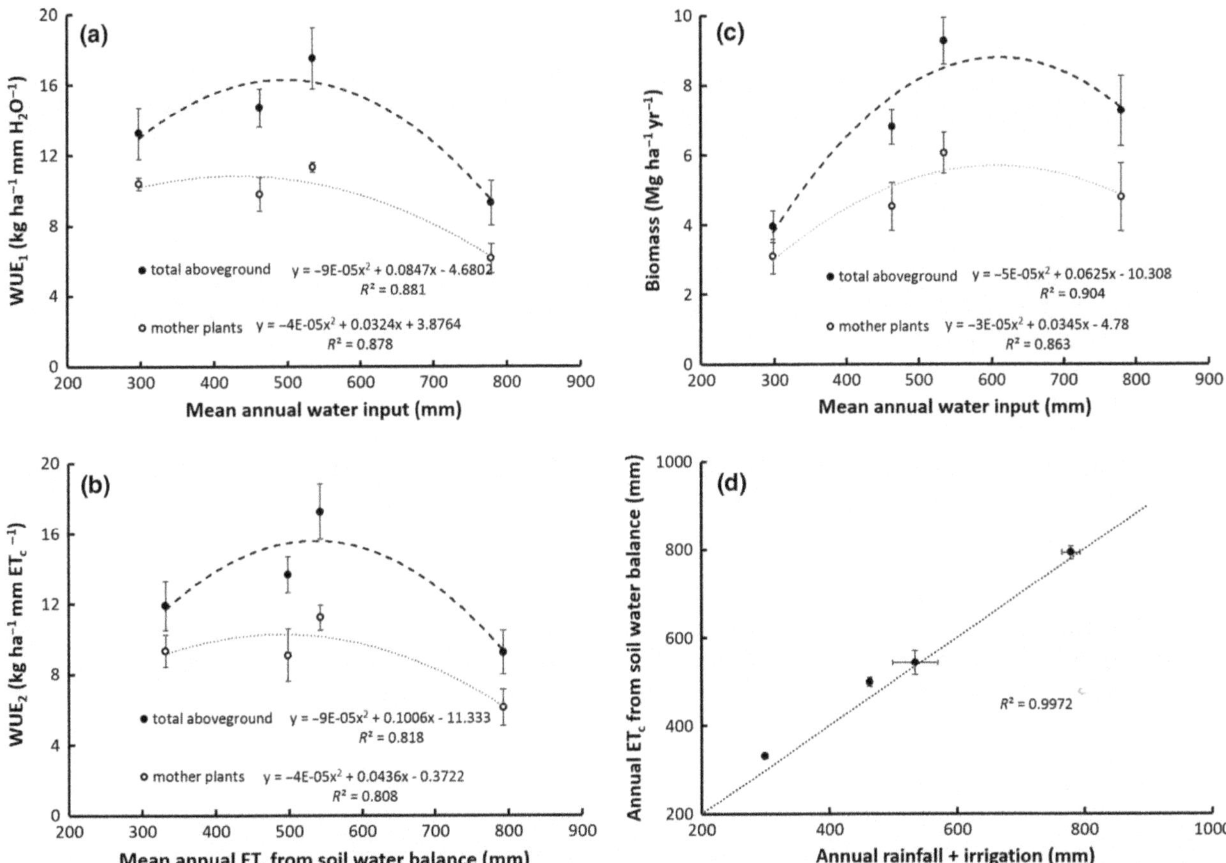

Fig. 8 Relationships between increasing annual water inputs and water use efficiency calculated as the biomass per unit of water inputs, WUE_1 (a); between increasing annual crop evapotranspiration calculated from soil water balances, ET_c, and water use efficiency calculated as biomass per unit of ET_c, WUE_2 (b); and between increasing annual water inputs and oven-dried biomass (c) of *A. americana*. Each point in panels (a), (b), and (c) represents a mean of individuals from replicate plots ($n = 4$), open circles indicate biomass or WUE of mother plants only, and closed circles indicate biomass or WUE of all aboveground plant mass including mother plants and pups. Data here are representing only healthy living plants. Also shown is the close relationship between annual water input (rainfall + precipitation) and annual ET_c (d). Each point in panel (d) represents a mean of replicate irrigation sections in the experimental field ($N = 2$). All error bars indicate standard error.

Table 1 Treatment means* for *Agave americana* soil water balance components† expressed as yearly averages for 2013–2014, where IW is exact irrigation applied without rounding, *R* is measured rainfall, ΔS‡ is the change in soil water storage within a soil profile of 2.0 m, and $ET_{c.}$ is annual crop evapotranspiration

Treatment	IW (mm)	*R* (mm)	ΔS (mm)	ET_c (mm)
580	577	202	14	793
330	333	202	8	543
260	261	202	36	499
100	96	202	34	332

*Treatments are denoted as amount of annual irrigation applied.
†Note deep percolation water penetrating below 2.0 m was assumed negligible for all treatments.
‡ΔS is the change averaged from Jan. 1 to Dec. 31 in 2013 and from Jan. 1 to Dec. 31 in 2014.

fit of irrigating beyond 330 mm yr^{-1} and the optimum water input (including rainfall) for an *A. americana* crop was ~530 mm yr^{-1}. Yields under these conditions ranged from 2.5 to 9.3 Mg ha^{-1} yr^{-1} (dry mass) over 3 years depending on pest infestations and whether harvests included only mother plants or all plant biomass. The lower production estimate is comparable to that of soybean (~2.9 Mg ha^{-} yr^{-1}), a crop grown for biodiesel, and the upper production estimate is comparable to corn grain (~9.4 Mg ha^{-1} yr^{-1}), a feedstock for bioethanol production and the most abundant biofuel crop globally, in prime agricultural regions of the USA based on the national average of these crops over the last 5 years (NASS, 2015).

In the most arid conditions tested here (300 mm of annual water inputs), biomass production of *A. americana* ranged from 2.0 to 4.0 Mg ha^{-1} yr^{-1}, indicating that *A. americana* crops may have commercially viable yields in dry regions with minimal irrigation required for establishment. Approximately 30% of global land is arid or semiarid, most of which is considered poorly suited or unsuitable for agriculture (van Velthuizen *et al.*, 2007), and *Agave* crops offer opportunity for production in these regions. Irrigation for agriculture consumes more water than any other practice and continued high consumption is unsustainable, especially in dry regions (Giovannucci *et al.*, 2012).

In the southwestern USA, agriculture that relies heavily on irrigation is economically unstable. Cotton, for example, one of the most important crops in Arizona, requires an average of 1,046 mm yr^{-1} of irrigation water (in addition to rainfall) and yields only 1.46 Mg ha^{-1} annually (USDA 2012). This high water demand is increasingly problematic; recent estimates suggest that 20–60 Mha of irrigated cropland globally

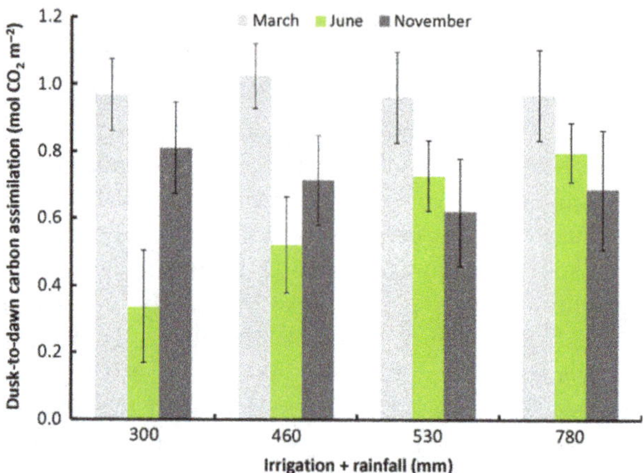

Fig. 9 Net CO_2 uptake of *A. americana* measured at approximately 30-min intervals between the hours of 6 pm and 9 am in March (light gray), June (green), and November (dark gray) of 2014. Each bar represents a mean of individuals (n = 3 for most time points) in each irrigation treatment, and error bars represent the standard error integrated over the full 15 h sampling period. Note that 29 mm of rain occurred prior to the March measurements, soil conditions were dry during the June measurements, and the agave snout weevil was disproportionately present in the higher irrigation treatments in November.

will be forced out of production by the end of the century due to a decline in the availability of freshwater resources (Elliott *et al.*, 2014). *A. americana* is an alternative crop that can be cultivated in dry climates with as little as a tenth of the water inputs required from conventional crops.

Although moderately productive with 300 mm – 460 mm of annual water input, the soil water balance measurements (Table 1) suggest that a greater amount of stored soil water was withdrawn by *A. americana* in these conditions (Fig. 8d). The additional water demand is small (34–36 mm), but could have an impact on soil water recharge rates in dry regions if commercially scaled plantations were established. Effects of agriculture on groundwater recharge are a major challenge for the southwestern USA (*e.g.*, Scanlon *et al.*, 2005). The optimum annual water input therefore seems to be between 460 mm and 530 mm of total water inputs. At inputs ≥530 mm, $ET_{c.}$ is nearly equal to water inputs (Fig. 8d).

The agave snout weevil is a major threat to crops of *A. americana,* and the impact of the snout weevil on *A. americana* survival increased with irrigation levels. The susceptibility of the plants in the high irrigation treatments could be due to the larger initial size of these plants. Horticultural texts indicate that the weevils preferentially burrow into larger plants to reproduce (Irish & Irish, 2000). Alternatively, and because the plant mortality due to the snout weevil was observed in plants that were only 2 years old, the additional moisture in the high irrigation treatments may provide an optimum environment for the agave snout weevil.

The agave snout weevil is a well-known threat to *A. americana* plants, but effective treatments have not been fully developed. Recent studies indicate that extracts from castor oil plants (*Ricinus communis* L) may be effective repellents for adult weevils (Pacheco-Sánchez *et al.*, 2012a,b). Adult weevils burrow into the stem of the *Agave* plant and reproduce there, so the larvae are nourished by the readily available carbohydrates at the leaf bases. The result is that the base of the stem is destroyed before the rest of plant shows symptoms (Fig. 11). The agave snout weevil is widespread in the United States, and should be expected in an *A. americana* crop unless a resistant genotype is identified. There is some anecdotal evidence of resistant genotypes (Irish & Irish, 2000), and it is possible that resistant individuals will be identified as this field trial continues.

If the timing of the snout weevil infestation can be delayed until at least the third or fourth growing season, which seems most likely in conditions with low water inputs, biomass yields can still be competitive with other bioenergy crops. It is noteworthy that the plants are easier to harvest manually after the snout weevil has weakened the stems, but minimizing the spread of this pest would be most desirable. Commercial production in the USA will likely require mechanized harvesting instead of the traditional manual harvesting that is common in Mexico. A recent economic analysis indicated that manual harvests were too costly for economically viable production of *Agave* for biofuel in the USA (Nuñez *et al.*, 2011). Existing mechanized harvesters could be modified to cut the plants at the base as in tequila plantations or to trim leaves annually as in fiber plantation.

This study is the first field experiment to compare *Agave* production under a range of controlled water inputs. The production of *Agave* species measured in this study was lower than previously reported *Agave* biomass in semiarid regions (*e.g.*, Davis *et al.*, 2011), but *A. tequilana* production was similar to modeled projections of *A. tequilana* biomass potential in the United States (Owen *et al.*, 2015). There are no previous reports

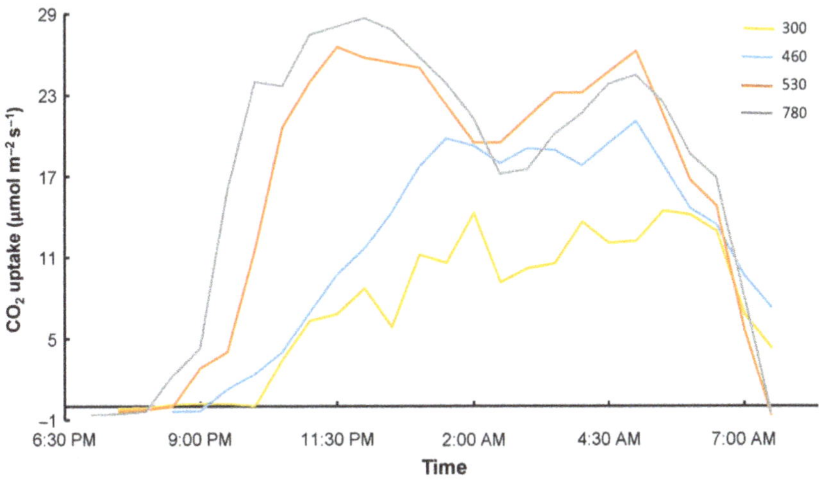

Fig. 10 Dusk-to-dawn CO_2 uptake of *A. americana* measured in June of 2014 with four levels of water inputs (290 mm yr^{-1}, yellow; 410 mm yr^{-1}, blue; 520 mm yr^{-1}, orange; 680 mm yr^{-1}, gray).

Fig. 11 Stem of *A. americana* affected by the agave snout weevil (left) and the stem of a healthy *A. americana* (right).

of *A. americana* production, but yields were lower than theoretical production rates for CAM plants (Nobel, 1991; Borland *et al.*, 2009; Davis *et al.*, 2014), and far below the theoretical predictions of 43 Mg ha^{-1} yr^{-1} for high-yielding CAM plants (Borland *et al.*, 2009). Davis *et al.* (2014) projected that, in arid conditions with only 200 mm annual precipitation, optimum yields of CAM plants could be 8.9 Mg ha^{-1} annually. Although the lowest irrigation treatment in this study, with 300 mm total water input, yielded 2.0–4.0 Mg ha^{-1} yr^{-1}, it is possible that a more mature crop (≥4 years), optimized genetic lines of *A. americana*, or other cold-tolerant *Agave* spp. will yield greater biomass.

Changes in planting density can have substantial impact on biomass production per unit area. The planting density of each plot in this study was ~2500 plants/ha. This is consistent with plant spacing used in sisal production and tequila plantations, and allows enough space for field managers to move between rows. If planting density could be increased to 4000 plants/ha, and managed with specialized equipment or remote technology, then higher yields could be expected. The crop canopy was not closed (there was exposed ground) in all plots measured in this study.

The results of this experiment to date indicate that *A. tequilana* and *A. fourcroydes*, two important commercial species in Mexico, were not productive 3 years after planting in Maricopa, AZ. This region has been predicted to have marginally appropriate growing conditions for *A. tequilana*, while more optimal conditions are predicted in Texas, California, and Florida (Lewis *et al.*, 2015; Owen *et al.*, 2015). Still, the sensitivity of these species to cold make them vulnerable to sporadic extreme climate events, and additional risk would be introduced when planting these species in the southwestern USA. Variable cold tolerances among *Agave* spp. have been observed and described in previous literature (*e.g.*, Nobel & Smith, 1983). The response of the tonoplast to temperature increases during the day has been implicated in controlling phase changes that are essential to CAM (Kluge & Schomburg, 1996), but research to resolve mechanisms of cold tolerance in CAM plants is needed. Carbohydrate concentrations, particularly high molecular weight fructans, are correlated with freeze tolerance in grasses (Dionne *et al.*, 2010), and may explain the difference in cold tolerance between *Agave* species in this study because *A. tequilana* leaves have lower concentrations of water-soluble carbohydrates than *A. americana* leaves (Li *et al.*, 2012; Corbin *et al.*, 2015).

Past life-cycle assessment indicated that there was potential for *Agave* to have equal or greater environmental benefits for displacing fossil fuels when compared to sugarcane (*Saccharum officinarum* L.), corn (*Zea mays* L.), and switchgrass (*Panicum virgatum* L.) (Yan *et al.*, 2011). Yan *et al.* (2011) assumed production rates of approximately 30 Mg ha^{-1} after 6 years (5 Mg ha^{-1} yr^{-1}) in *A. tequilana* stems, the sugar content of which was assumed to be 80% of the dry mass. Corbin *et al.* (2015) recently estimated the sugar content of *A. americana* plant tissue to be 60% of dry mass, suggesting that bioenergy production from *A. americana* in the southwestern USA could have similar life-cycle benefits if the optimum yields (6.1–9.3 Mg ha^{-1} yr^{-1}) observed in this study were achieved. The feasibility of a full commercial production system that uses *A. americana* as feedstock depends on other economic factors specific to the location where it would be cultivated.

Conclusions

Agave americana is a potential alternative bioenergy crop for the southwestern USA, even in arid conditions. *A. tequilana* and *A. fourcroydes* were vulnerable to extreme cold events that can occur in this region, and were far less productive than *A. americana* in this side-by-side trial. The agave snout weevil poses a clear risk to the production of *A. americana*, but it may be manageable in conditions where annual water inputs are ≤530 mm. Genotypic selection for resistance to the agave snout weevil might also reduce this problem in the future. Healthy 3-year-old *A. americana* plots yielded 4.0–9.3 Mg ha^{-1} yr^{-1} total biomass with 300–530 mm yr^{-1} of water input (including 100–330 mm yr^{-1} irrigation, respectively), and plot yields without pup mass included were 3.1–6.1 Mg ha^{-1} yr^{-1}. The production rates of *A. americana* were greater than conventional crops in Arizona (*e.g.*, cotton) with less water input than is typical in this dry agricultural region.

Acknowledgements

We thank Glenda Simer, Richard Simer, Greg Main, and the Maricopa Agricultural Center for managing the experimental field. We also thank Michael Kampwerth and Michael Masters for helping with the initial establishment of the experimental field and appreciate assistance from Jose Ignacio Del Real Laborde in obtaining seedlings for the experiment. This project was funded in part by the Energy Biosciences Institute, and we appreciate valuable support and feedback from Steve Long and Chris Somerville.

References

Borland AM, Griffiths H, Hartwell J, Smith JAC (2009) Exploiting the potential of plants with crassulacean acid metabolism for bioenergy production on marginal lands. *Journal of Experimental Botany*, **60**, 2879–2896.

Corbin K, Byrt CS, Bauer S *et al.* (2015) Prospecting for energy-rich renewable raw materials: *Agave* leaf case study. *PlosOne*, **10**, e0135382. doi: 10.1371/journal.pone.0135382

Cushman J, Davis SC, Yang X, Borland A (2015) Development and use of bioenergy feedstocks for semi-arid and arid lands. *Journal of Experimental Botany*, **66**, 4177–4193.

Davis SC, Frank G, Long S (2011) The global potential for *Agave* as a biofuel feedstock. *GCB Bioenergy*, **3**, 68–78.

Davis SC, LeBauer D, Long S (2014) Light to liquid fuel: theoretical and realized energy conversion efficiency of plants using Crassulacean Acid Metabolism (CAM) in arid conditions. *Journal of Experimental Botany*, **65**, 3471–3478.

Davis SC, Ming R, LeBauer D, Long S (2015) Toward systems-level analysis of agricultural production from crassulacean acid metabolism (CAM): scaling from cell to commercial production. *New Phytologist*, **208**, 66–72.

Dionne J, Rochefort S, Huff DR, Desjardins Y, Bertrand A, Castonguay Y (2010) Variability for freezing tolerance among 42 ecotypes of green-type annual bluegrass. *Crop Science*, **50**, 321–336.

Elliott J, Derying D, Müller C *et al.* (2014) Constraints and potentials of future irrigation water availability on agricultural production under climate change. *PNAS*, **111**, 3239–3244.

Garcia-Moya E, Romero-Manzanares A, Nobel PS (2011) Highlights for agave productivity. *GCB Bioenergy*, **3**, 4–14.

Gee GW, Bauder JW (1986) Particle size analysis. In: *Methods of Soil Analysis, Part 1, Physical and Mineralogical Methods*, Agronomy Monograph No 9, 2nd edn (ed. Klute A), pp. 383–411. American Society of Agronomy, Madison.

Gentry H (1982) *Agaves of Continental North America*. The University of Arizona Press, Tucson, AZ.

Giovannucci D, Scherr SJ, Nierenberg D, Hebebrand C, Shapiro J, Milder J, Wheeler K. (2012) *Food and Agriculture: The Future of Sustainability (March 1, 2012). The Sustainable Development in the 21st Century (SD21) Report for Rio+20.* United Nations, New York. http://dx.doi.org/10.2139/ssrn.2054838

IPCC (2014) *Climate Change 2014: Impacts, Adaptation, and Vulnerability. Part A: Global and Sectoral Aspects. Contribution of Working Group II to the Fifth Assessment Report of the Intergovernmental Panel on Climate Change* (eds Field CB, Barros VR, Dokken DJ, Mach KJ, Mastrandrea MD, Bilir TE, Chatterjee M, Ebi KL, Estrada YO, Genova RC, Girma B, Kissel ES, Levy AN, MacCracken S, Mastrandrea PR, White LL). Cambridge University Press, Cambridge, United Kingdom and New York, NY, USA.

Irish M, Irish G (2000) *Agaves, Yuccas, and Related Plants*. Timber Press, Portland, OR.

Jensen ME, Burman RD, Allen RG (1990) *Evapotranspiration and Irrigation Requirements*. American Society of Civil Engineers Manuals and Reports on Engineering Practices No. 70. ASCE, New York, NY.

Kluge M, Schomburg M (1996) The Tonoplast as a Target of Temperature Effects in Crassulacean Acid Metabolism. In: *Crassulacean Acid Metabolism*, vol. **114** (eds Winter K, Smith JAC), pp. 72–77. Springer-Verlag, Berlin-Heidelberg.

La Barre W (1938) Native American beers. *American Anthropologist*, **40**, 224–234.

Lewis S, Gross S, Visel A, Kelly M, Morrow W (2015) Fuzzy GIS-based multi-criteria evaluation for US *Agave* production as a bioenergy feedstock. *GCB Bioenergy*, **7**, 84–99.

Li H, Foston MB, Kumar R *et al.* (2012) Chemical composition and characterization of cellulose for Agave as a fast-growing, drought-tolerant biofuels feedstock. *RSC Advances*, **2**, 4951–4958.

Mylsamy K, Rajendran I (2010) Investigation on physio-chemical and mechanical properties of raw and alkali-treated *Agave americana* fiber. *Journal of Reinforced Plastics and Composites*, **29**, 2925–2935.

NASS. (2015). National Agricultural Statistics Service, United States Department of Agriculture. Available at: http://www.nass.usda.gov/ (accessed 8 July 2015).

Neales T (1973) The effect of night temperature on CO_2 assimilation, transpiration, and water use efficiency in *Agave americana* L. *Australian Journal of Biological Science*, **26**, 705–714.

Pacheco-Sánchez C, Villa-Ayala P, Montes-Belmont R, Figueroa-Brito R, Jiménez-Pérez A (2012a) Effect of *Ricinus communis* extracts on weight and mortality of *Scyphophorus acupunctatus* (Coleoptera: Curculionidae). *International Journal of Applied Science & Technology*, **2**, 83–94.

Pacheco-Sánchez C, Villa-Ayala P, Montes-Belmont R, Figueroa-Brito R, Jiménez-Pérez A (2012b) Repellency of hydroethanolic extracts of *Ricinus communis* (Euphorbiaceae) to *Scyphophorus acupuntatus* (Coleoptera: Curculionidae) in the laboratory. *Florida Entomologist*, **95**, 706–710.

Post DF, Mack C, Camp PD, Sulliman AS. 1988. Mapping and characterization of the soils on the University of Arizona Maricopa Agricultural Center. *Proc. Hydrology and Water Resources in Arizona and the Southwest*, 49–60. University of Arizona, Tucson, Ariz.

Saxton KE, Rawls WJ (2006) Soil water characteristic estimates by texture and organic matter for hydrologic solutions. *Soil Science Society of America Journal*, **70**, 1569–1578.

Scanlon B, Reedy R, Stonestrom D, Prudic D, Dennehy K (2005) Impact of land use and land cover change on groundwater recharge and quality in the southwestern US. *Global Change Biology*, **11**, 1577–1593.

Scanlon B, Jolly I, Sophocleous M, Zhang L (2007) Global impacts of conversions from natural to agricultural ecosystems on water resources: quantity versus quality. *Water Resources Research*, **43**, W03437. doi:10.1029/2006WR0 05486.

U.S. Department of Agriculture (USDA) (2012) Pima Cotton 2012 Yield Per Harvested Acre by County for Selected States. U.S. Department of Agriculture, National Agricultural Statistics Service. Available at: http://www.nass.usda.gov/Charts_and_Maps/Crops_County/ (accessed 2 April 2014).

van Velthuizen H, Huddleston B, Fischer G *et al.* (2007) *Mapping Biophysical Factors That Influence Agricultural Production and Rural Vulnerability*. Food and Agriculture Organization of the United Nations, Rome. ISBN 1684-8241.

Winter K, Smith JAC (1996) Crassulacean Acid Metabolism: Current Status and Perspectives. In: *Crassulacean Acid Metabolism*, vol. **114** (eds Winter K, Smith JAC), pp. 389–426. Springer-Verlag, Berlin-Heidelberg.

Yan X, Tan DKY, Inderwildi OR, Smith JAC, King DA (2011) Life cycle energy and greenhouse gas analysis for agave-derived bioethanol. *Energy & Environmental Science*, **4**, 3110–3121.

Yang X, Cushman J, Borland A *et al.* (2015) A roadmap for research on crassulacean acid metabolism (CAM) to enhance sustainable food and bioenergy production in a hotter, drier world. *New Phytologist*, **207**, 491–504.

Nobel PS (1996) High Productivity of Certain Agronomic CAM Species. In: *Crassulacean Acid Metabolism*, vol. **114** (eds Winter K, Smith JAC), pp. 255–265. Springer-Verlag, Berlin-Heidelberg.

Nobel PS (2003) *Environmental Biology of Agave and Cacti*. Cambridge University Press, Cambridge, UK and New York, NY, USA.

Nobel PS, Smith SD (1983) High and low temperature tolerances and their relationships to distribution of agaves. *Plant, Cell, and Environment*, **6**, 711–719.

Nuñez HM, Rodríguez LF, Khanna M (2011) Agave for tequila and biofuels: an economic assessment and potential opportunities. *GCB Bioenergy*, **3**, 43–57.

Osmond C (1978) Crassulacean Acid Metabolism: a curiosity in context. *Annual Reviews in Plant Physiology*, **29**, 379–414.

Owen N, Fahy K, Griffiths H (2015) Crassulacean acid metabolism (CAM) offers sustainable bioenergy production and resilience to climate change. *GCB Bioenergy*, in press. doi: 10.1111/gcbb.12272.

Nobel PS (1991) Achievable productivities of certain CAM plants: basis for high values compared with C_3 and C_4. *New Phytologist*, **119**, 183–205.

14

Genotype × environment interaction analysis of North American shrub willow yield trials confirms superior performance of triploid hybrids

ERIC S. FABIO[1], TIMOTHY A. VOLK[2], RAYMOND O. MILLER[3], MICHELLE J. SERAPIGLIA[1,*], HUGH G. GAUCH[4], KEN C. J. VAN REES[5], RYAN D. HANGS[5], BEYHAN Y. AMICHEV[6], YULIA A. KUZOVKINA[7], MICHEL LABRECQUE[8], GREGG A. JOHNSON[9], ROBERT G. EWY[10], GARY J. KLING[11] and LAWRENCE B. SMART[1]

[1]Horticulture Section, School of Integrative Plant Science, New York State Agricultural Experiment Station, Cornell University, Geneva, NY 14456, USA, [2]Department of Forest and Natural Resources Management, State University of New York College of Environmental Science and Forestry, Syracuse, NY 13210, USA, [3]Forest Biomass Innovation Center, Michigan State University, Escanaba, MI 49829, USA, [4]Soil and Crop Sciences Section, School of Integrative Plant Science, Cornell University, Ithaca, NY 14853, USA, [5]Department of Soil Science, University of Saskatchewan, Saskatoon, SK S7N 5A8, Canada, [6]Center for Northern Agroforestry and Afforestation, University of Saskatchewan, Saskatoon, SK S7N 5A8, Canada, [7]Department of Plant Science, University of Connecticut, Storrs, CT 06269, USA, [8]Institut de Recherche en Biologie Végétale, University of Montréal, Montréal, QC H3C 3J7, Canada, [9]Southern Research and Outreach Center, University of Minnesota, Waseca, MN 56093, USA, [10]Department of Biology, State University of New York at Potsdam, Potsdam, NY 13676, USA, [11]Department of Crop Sciences, University of Illinois, Urbana, IL 61801, USA

Abstract

Development of dedicated bioenergy crop production systems will require accurate yield estimates, which will be important for determining many of the associated environmental and economic impacts of their production. Shrub willow (*Salix* spp) is being promoted in areas of the USA and Canada due to its adaption to cool climates and wide genetic diversity available for breeding improvement. Willow breeding in North America is in an early stage, and selection of elite genotypes for commercialization will require testing across broad geographic regions to gain an understanding of how shrub willow interacts with the environment. We analyzed a dataset of first-rotation shrub willow yields of 16 genotypes across 10 trial environments in the USA and Canada for genotype-by-environment interactions using the additive main effects and multiplicative interactions (AMMI) model. Mean genotype yields ranged from 5.22 to 8.58 oven-dry Mg ha^{-1} yr^{-1}. Analysis of the main effect of genotype showed that one round of breeding improved yields by as much as 20% over check cultivars and that triploid hybrids, most notably *Salix viminalis* × *S. miyabeana*, exhibited superior yields. We also found important variability in genotypic response to environments, which suggests specific adaptability could be exploited among 16 genotypes for yield gains. Strong positive correlations were found between environment main effects and AMMI parameters and growing environment temperatures. These findings demonstrate yield improvements are possible in one generation and will be important for developing cultivar recommendations and for future breeding efforts.

Keywords: AMMI, biomass, ploidy, *Salix*, short-rotation coppice, yield

Introduction

If dedicated bioenergy crops are to play a significant role in climate change mitigation strategies, a clear understanding of yield potential on a regional basis must be established. Development and delivery of high-yielding, well-adapted crops is a key underlying assumption in the estimation of bioenergy production capacity in the USA (U. S. Department of Energy, 2011). Crop management and breeding will play crucial roles in meeting the challenges of producing more biomass on limited land in a sustainable manner (Karp & Shield, 2008). Shrub willow (*Salix* spp.) grown in short rotation has shown promise as a viable, regionally based feedstock for marginal land due to its adaptability to cool, moist climates with short growing seasons (Kuzovkina & Quigley, 2005) and large potential for genetic

*Current address: Sustainable Biofuels and Co-Products Research Unit, Eastern Regional Research Center, United States Department of Agriculture-Agricultural Research Service, Wyndmoor, PA, USA

Correspondence: Lawrence B. Smart
e-mail: lbs33@cornell.edu

improvement through breeding (Smart *et al.*, 2005; Smart & Cameron, 2008). This substantial yet mostly underexploited genetic variability among taxa is expected to provide a basis for developing key traits desirable for sustainable bioenergy production, both through traditional breeding and with advanced molecular techniques (Karp *et al.*, 2011).

Willow breeding for biomass production has been most extensively researched in the UK (Lindegaard & Barker, 1997) and Sweden (Larsson, 1998). A thorough account of the global breeding history of shrub willow is provided by Kuzovkina *et al.* (2008). In North America, shrub willow breeding efforts began at the University of Toronto in the early 1980s with collection of native species and exchange of plant material with the UK and European countries (Zsuffa, 1990), but efforts were focused largely on hybridizations between North American native species (Mosseler, 1990). In the USA, acquisition of plant material from collaborators in Canada, as well as China, Japan, New Zealand, Ukraine and Sweden provided the basis of a breeding program focused largely on novel interspecific hybridizations displaying heterosis for biomass yield traits (Kopp *et al.*, 2001; Smart *et al.*, 2005; Smart & Cameron, 2012). Polyploidy is common in *Salix* and novel triploid species hybrids have been linked to improved yields and biomass quality in early selection trials (Smart *et al.*, 2008; Serapiglia *et al.*, 2014, 2015). Cultivar development, however, is a multistage screening process requiring substantial investments in time and resources prior to commercialization (Hanley & Karp, 2014). An accurate evaluation of cultivar yield potentials is ultimately assessed through testing on multiple sites with a diverse range of environmental characteristics.

Biomass yield is an important factor determining the environmental and economic impacts associated with growing shrub willow. Life cycle analyses have shown that yield assumptions can significantly impact net energy ratios and greenhouse gas balances (Heller *et al.*, 2003; Keoleian & Volk, 2005; Caputo *et al.*, 2014), as well as economic returns on investment and production costs (Buchholz & Volk, 2011; Hauk *et al.*, 2014). There have been numerous research trials conducted over the past two decades in regions of the northeastern USA and Central and Eastern Canada that have quantified yields. Some studies have analyzed the differential response of genotype to contrasting environmental conditions (Labrecque & Teodorescu, 2003; Wang & Macfarlane, 2012; Serapiglia *et al.*, 2013; Mosseler *et al.*, 2014a); however, the limited number of test sites and use of diverse cultivars of various levels of genetic improvement makes it difficult to generalize specific genotypic responses to larger growing regions. Others have summarized mean yields from multiple test sites across geographical regions (Kiernan *et al.*, 2003; Volk *et al.*, 2011), but due to unequal representation of genotypes among trials, little insight into the genotypic contribution to yield variability can be gained. Furthermore, efforts to model yields across broad geographic ranges may use general yield estimates from obsolete cultivars or ones that may not be well adapted for a particular region (Walsh *et al.*, 2003; Wang *et al.*, 2015). Plant breeders and agronomists are therefore challenged with assessing genotypic sensitivity to certain edaphic and climatic conditions. This process is also important for assuring continued improvement within a breeding program and for providing a basis for recommending cultivars for broadscale production.

These recommendations become complicated when significant crossovers occur in genotype (GEN) yield rankings, as a response to contrasting environments (ENV). These genotype-by-environment (G×E) interactions are prominent and important phenomena in agriculture, which present both challenges and opportunities for plant breeders and agronomists. Selection and deployment of elite cultivars must be based on results of rigorous testing through coordinated multilocation trials, combined with appropriate statistical analyses for assessing the adaptability of genotypes and predicting performance in untested ENV (Annicchiarico, 2002). A long-standing theme in plant breeding is to focus the search for GEN that exhibit stable yields or broad adaptability across ENV in the targeted growing region. This concept was popularized by Finlay & Wilkinson (1963) and Eberhart & Russell (1966), who developed regression parameters that seek to identify superior yielding GEN that maintain stable performance across ENV. Selection on the basis of stability can help to minimize the complicating effects of G×E interactions, adding efficiency to the selection process by focusing resources on material that has the best promise for widespread optimal performance (Eberhart & Russell, 1966). Selection based on stability should also guard against a potentially detrimental tendency to select GEN based on greater yields in only favorable ENV (Simmonds, 1991; Annicchiarico, 2002). However, when the G×E component is strong and meaningful, there is also a counter argument that contrasting performance among GEN can be capitalized upon, and breeding efforts should focus on specific adaptation (Cooper & Hammer, 1996; Piepho & Möhring, 2005). Thus, G×E interactions are viewed as a useful and informative aspect of cultivar testing, and subdividing a growing region into smaller areas and targeting GEN at those areas can improve overall yields (Gauch & Zobel, 1997). This argument is reinforced by the fact that much of the world's crop production occurs on land that is less favorable than that where the crops were developed (Simmonds, 1991; Gauch & Zobel,

1997). As dedicated bioenergy crops are presumably targeted for marginal lands (Richards *et al.*, 2014; Stoof *et al.*, 2015), perhaps this concept is particularly relevant. The relative merits of these two perspectives, exploiting only broad adaptations or else both broad and specific adaptations, vary from case to case and depend substantially on the relative magnitudes of genotype main effects and G×E interaction effects.

We present an analysis of G×E interactions in first-rotation yields across a network of shrub willow yield trials in North America covering 16 GEN and 10 ENV. The objectives of this study were (1) to identify shrub willow genotypes with broad adaptability in biomass yield across target growing regions in North America, (2) analyze G×E interactions for identifying and characterizing specific adaptation of certain genotypes and (3) identify edaphic and climatic variables that are most closely associated with G×E interaction patterns. This represents the most comprehensive analysis of North American shrub willow yields to date and will serve as a basis for making cultivar-site matching recommendations and will inform future breeding efforts.

Materials and methods

Breeding material and yield trial network

Foundational breeding material used in developing the GEN tested in these yield trials was obtained from the University of Toronto and from accessions of native and naturalized species collected in the northeastern USA and eastern Canada in the 1980s to mid-1990s (Kopp *et al.*, 2001; Smart & Cameron, 2008). Crosses were performed between 1998 and 1999 at SUNY-ESF, and after initial family screening in field trials for biomass yield traits, a group of genetically diverse individuals were deployed in regional yield trials. Between 2005 and 2011, 23 trials were established mainly in the northeastern and Midwestern USA and contained between six and 30 genotypes. To provide an unbiased comparison of genotype yields, we restricted our analysis to 16 cultivars (Table 1) that were all present in each of 10 environments (Table 2). The yield trials were planted between 2006 and 2009 and hosted by institutions located in six US states and two Canadian provinces, and the cultivars have been placed into diversity groups based on pedigree. Trials were established and maintained generally in a consistent manner across sites according to a standardized protocol and methods followed those described in Serapiglia *et al.* (2013) and Volk *et al.* (2011). Conventional tillage was used to prepare the sites in either the fall or spring prior to planting, which generally occurred between May and June. All trials were planted by hand using dormant 25 cm cuttings sourced from nursery beds at the Tully Genetics Field Station of SUNY-ESF in Tully, NY. Trials were laid out in a double-row configuration with 1.52 m between double rows, 0.76 m within the double rows and 0.61 m between plants along the row, for a planting density of 14 400 plants ha^{-1}. Within each yield trial, genotype was the experimental treatment and the experimental units were plots consisting of three double rows, each 13 plants long, with the outer double rows serving as border rows. Genotypes were replicated four times in a randomized complete block design. Preemergence herbicides, generally oxyfluorfen (1.1 kg ai ha^{-1}) and simazine (2.2 kg ai ha^{-1}), were applied prior to budbreak, except at Boisbriand, QC, where no herbicide was used. Periodic

Table 1 Description of the 16 cultivars included in the genotype × environment interactions analysis of first-rotation shrub willow yields

Clone ID	Epithet	Species/pedigree	Mother	Father	Diversity group*	Sex	Ploidy†	Source
99239-015	'Allegany'	*S. koriyanagi* × *S. purpurea*	SH3	95058	6b	F	2X	Bred
9970-036	'Canastota'	*S. miyabeana*	SX61	SX64	5	M	4X	Bred
99202-004	'Fabius'	*S. viminalis* × *S. miyabeana*	SV2	SX67	8	F	3X	Bred
9882-34	'Fish Creek'	*S. purpurea*	94006	94001	6a	M	2X	Bred
99217-015	'Millbrook'	*S. purpurea* × *S. miyabeana*	95026	SX64	9	F	3X	Bred
9980-005	'Oneida'	*S. purpurea* × *S. miyabeana*	94006	SX67	9	M	3X	Bred
99113-012	'Onondaga'	*S. koriyanagi* × *S. purpurea*	SH3	94002	6b	M	2X	Bred
99201-007	'Otisco'	*S. viminalis* × *S. miyabeana*	SV2	SX64	8	F	3X	Bred
99207-018	'Owasco'	*S. viminalis* × *S. miyabeana*	SV7	SX64	8	F	3X	Bred
S25	'S25'	*S. eriocephala*			4	F	2X	Bred
9871-31	'Sherburne'	*S. miyabeana*	SX61	SX67	5	F	4X	Bred
SV1	'SV1'	*S.* × *dasyclados*			1	F	2X	Unknown‡
SX61	'SX61'	*S. miyabeana*			5	F	4X	Natural accession
SX64	'SX64'	*S. miyabeana*			5	M	4X	Natural accession
99207-020	'Truxton'	*S. viminalis* × *S. miyabeana*	SV7	SX64	8	M	3X	Bred
99202-011	'Tully Champion'	*S. viminalis* × *S. miyabeana*	SV2	SX67	8	F	3X	Bred

Lower case letters indicates that these are subgroups of diversity group 6.

*Diversity group refers to Species/Pedigree.

†Ploidy level estimated by flow cytometry (Serapiglia *et al.*, 2015).

‡Collected in Ontario Canada, but possibly the cultivated hybrid *S. viminalis* × (*S. caprea* × *S. cinerea*) (Stott, 1991).

Table 2 Yield trial location characteristics. Precipitation, temperature and solar radiation data are means across four years of first rotation

Location	Code	Year planted	Latitude	Longitude	Elevation (m)	Annual precip. (mm)	Annual GDD (base 10 °C)	Mean Annual $Temp_{min}$ (°C)	Solar radiation (MJ m^{-1} day^{-1})	SOM (%)	Soil pH	Water table depth (cm)
Saskatoon, SK	Sask	2007	52.13	−106.61	510	408	767	−4.1	4329	4.5	7.07	200
Constableville, NY	Cons	2006	43.56	−75.53	513	1457	812	−1.7	4606	8.2	5.66	45
Skandia, MI	Skan	2009	46.36	−87.25	287	822	870	0.1	4909	3.6	6.47	30
Escanaba, MI	Esca	2008	45.77	−87.20	222	714	1016	−0.1	5147	2.8	6.10	200
Brimley, MI	Brim	2009	46.40	−84.47	200	791	1021	1.7	4980	4.0	5.25	15
Boisbriand, QC	Bois	2007	45.63	−73.89	30	1038	1162	1.2	4612	4.0	6.09	–
Middlebury, VT	Midd	2007	44.01	−73.20	114	1138	1419	2.0	4612	6.8	6.70	30
Waseca, MN	Wase	2006	44.06	−93.54	349	843	1459	1.7	4944	5.7	5.40	30
Fredonia, NY	Fred	2008	42.44	−79.29	255	909	1477	4.9	5082	3.6	4.80	30
Storrs, CT	Stor	2009	41.80	−72.23	198	1274	1487	5.4	5171	3.4	6.12	45

GDD, growing degree-days; SOM, soil organic matter.

mechanical or spot chemical weed control was performed as needed. After the first year of growth, all stems were cut back during dormancy to promote coppice regrowth the following spring, at which time 112 kg N ha^{-1} was applied, normally as ammonium sulfate, except for the trial located in Saskatoon, SK, where no fertilizer was applied. Harvests were conducted three years postcoppice (4-year-old root systems) during dormancy, except at Brimley, MI, which was harvested in July at 3.5 years postcoppice due to extreme moisture conditions in the field. Two to four plants on each end of the middle double row of each plot were excluded from harvest measurements to avoid plot-to-plot edge effects, resulting in 18–22 plants available for measurements. All stems from the 18–22 plants from the middle double row of each plot were cut with brush saws or cut and chipped with a mechanical harvester and weighed in a bin with weigh cells in the field. A subsample from each plot consisting of either whole stems or chips was collected, weighed fresh, dried at 65 °C to a constant weight and reweighed to determine moisture content. Moisture content was used to calculate dry matter yield for each plot based on the area occupied by the harvested plants across a 3-year rotation. All yields reported here are expressed as oven-dried Mg ha^{-1} yr^{-1}. Survival data was also recorded on the harvested plants at the time of harvest, except Boisbriand, QC, where no survival data was collected, but survival was >90% for each plot (M. Labrecque, Personal observation) and Savoy, IL where survival was assessed in the middle of the second rotation, but most of the mortality occurred early in the first rotation (G. Kling, personal observation).

Site environmental characteristics

Daily temperature and precipitation data for all 4 years of the harvest cycle were obtained from weather stations nearest to each trial location with the most complete records using publically accessible databases, National Oceanic and Atmospheric Agency, National Centers for Environmental Information (NOAA NCEI, 2015) for US trials and Canadian National Climate Data (Environment Canada, 2015) for CA trials. Daily solar radiation estimates at a 1° by 1° grid scale were obtained from National Aeronautics and Space Adminstration, Prediction of Worldwide Energy Resource (NASA POWER, 2015). Soil samples were generally collected at the time of planting, or occasionally soon after harvest. Due to some differences between soil extraction methods, only pH (1 : 1 soil/water by weight) and % organic matter (typically by loss on ignition) were considered for data analysis.

Dataset refinement

Initially, our dataset was comprised of 640 independent observations where the harvested plot was the experimental unit. While overall mean survival was >90% across the yield trial network, some trials and individual plots experienced greater mortality. Two trials, Escanaba, MI, and Saskatoon, SK, had mean survival values below 80%. Damage from herbicide drift appeared to be the main cause of mortality in Saskatchewan (Amichev *et al.*, 2015), while at Escanaba mortality could not be associated with any particular issue (R. Miller, personal observation). In total, 39 independent observations (experimental units) with >65% mortality at the time of harvest were removed from the original dataset of 640 observations. Data were also inspected for extreme moisture content values, which can impact dry matter yield calculations. Two likely sources of error were 1) premature removal of samples from the drying oven, resulting in excessively low moisture content estimates, and 2) extraneous moisture contamination at harvest during wet conditions, leading to excessively high moisture values. Moisture values that were greater or less than three standard deviations of the mean across all samples were considered outliers and were removed from the dataset. To retain a yield value for the particular observation, the mean of the remaining three replicates was applied to the plot fresh weight of the outlying moisture content. This was performed for 15 samples in total from four of nine trials, which represented 2.3% of the total observations considered in this analysis. For the Boisbriand, QC trial, a single wood sample was collected for each cultivar and the moisture content was applied to all four replicates of that cultivar. The mean moisture content of these samples was reported at 32.7% (SD 2.90%), and it was assumed that all samples had not dried sufficiently, considering the overall mean moisture content across all other trials was 46.9% (SD 4.77%). A correction factor equaling the difference between these two mean values was added to each of the observations originally reported for the Boisbriand trial. This value was 14.2%, which was also very close to the value of 3 standard deviations of the overall mean (14.3%). After accounting for survival and moisture content adjustments, 601 observations were available for analysis from the original 640 from 16 GEN in 10 ENV.

Statistical analysis

We chose to analyze G×E interactions in our first-rotation yield dataset using the additive main effects and multiplicative interactions (AMMI) model. The AMMI model is a combination of analysis of variance (ANOVA) and principal components analysis (PCA), where the G×E interaction, contained in the residual of the additive GEN and ENV main effects model, is subjected to PCA and the interaction sum of squares (SS) are partitioned into a series of interaction principal component (IPC) axes, where IPC1 is the first interaction PC axis, and so on. The AMMI0 model indicates that no IPC axes are included and the model is equivalent to the additive (main effects only) ANOVA. AMMI1 includes the first IPC axis, and so on, while AMMIF is the full model that includes all axes and is equal to the raw data. The maximum number of IPC axes for a given dataset is one less the minimum number of GEN or ENV (Gauch, 1992; p. 85), but typically the majority of the interaction signal is captured in the first few IPC axes, with later IPC axes containing increasing amounts of interaction noise, and decreasing amounts of signal (Gauch, 2012). Therefore, a more parsimonious model containing a lower number of axes is favored and the remaining axes are relegated to a pooled residual. The statistical significance of each axis is often assessed with an *F*-test where the degrees of freedom (DF) for each axis are calculated according to Gollob (1968) (see also Gauch, 1988). More recently, Piepho (1995) demonstrated that the traditional *F*-test can be too liberal and suggested a more conservative F_R-test for determining the significance of each IPC axis. Also, Gauch (1992, p. 147) suggested a simple test for model diagnosis and selection, which involves estimation of the G×E noise SS by multiplying the G×E DF by the mean square error. Consequently, the signal G×E SS can be estimated by subtracting the G×E noise SS from the G×E total SS. Therefore, IPC axes can be assessed by the amount of G×E signal SS that are recovered, instead of the total G×E SS. Selection of higher order models that include greater numbers of axes must be weighed not only in terms of statistical significance, but also in terms of practicality, because higher order AMMI models and especially the noisy AMMIF tend to have a large roster of genotypes that win in at least one environment, and hence, such models produce an unmanageable number of mega-environments (Gauch, 2013). Parsimonious models are often preferred, and AMMI models are most useful when the multiplicative terms have agricultural interpretability (Gauch, 2013). Equation 1 gives the general form of the AMMI model:

$$Y_{ger} = \mu + \alpha_g + \beta_e + \Sigma_n \lambda_n \gamma_{gn} \delta_{en} + \rho_{ge} + \varepsilon_{ger} \qquad (1)$$

where Y_{ger} is the yield of the *g*th genotype in the *e*th environment for the *r*th replicate, μ is the grand mean, α_g is the genotype *g* mean deviation (genotype mean minus grand mean), β_e is the environment *e* mean deviation, λ_n is the singular value for *n*th IPCA axis, γ_{gn} is the genotype g eigenvector value for IPCA axis *n*, δ_{en} is the environment *e* eigenvector value for IPCA axis *n*, ρ_{ge} is the residual, and ε_{ger} is the experimental error.

The AMMI analysis was performed using MATMODEL V3.0, an open source statistical program designed specifically for the analysis of G×E interactions (Gauch & Furnas, 1991; Gauch, 2007). MATMODEL performs the combined ANOVA/PCA and also delineates mega-environments, which allows for the exploration of specific adaptation to particular edaphic or climatic conditions (Gauch & Zobel, 1997).

MATMODEL does not analyze the experimental design; hence, we first analyzed the experimental design using PROC MIXED in SAS® version 9.4 (SAS Institute Inc., 2013), with block nested within ENV, and coded as a random effect. The main effects of GEN and ENV and the G×E interaction term were considered fixed effects. The TYPE3 option in PROC MIXED was invoked to obtain expected SS for the fixed effects. Tukey's studentized range (HSD) *post hoc* test was performed for means separation among GEN. The least squared means from the SAS output were then supplied to MATMODEL to perform the PCA of the interaction and for the mega-environment analysis. The random error variance from the SAS PROC MIXED output was used to calculate the F-tests in the AMMI analysis. Finally, the environment main effects and IPC scores were used in correlation analyses with the various environmental variables using PROC CORR in SAS® version 9.4.

Results

ANOVA and genotype main effects

The grand mean for first-rotation yields for this base case dataset based on the mixed model analysis of variance was 6.96 Mg ha^{-1} yr^{-1}. The main effects of ENV and GEN and the G×E interaction were all highly significant ($P < 0.0001$; Table 3) and accounted for 82, 6 and 12% of treatment SS, respectively. The blocking effect (nested within ENV) accounted for a relatively large amount of random error variance (Table 3).

Although the main effect of genotype accounted for only 6% of the treatment SS, it contains important information about patterns of broad adaptation. The cultivar *Salix viminalis* × *S. miyabeana* 'Fabius', a triploid hybrid in diversity group 8 (DG8) was the overall greatest yielding GEN across the 10 ENV tested, with a mean yield of 8.58 Mg ha^{-1} yr^{-1} (Table 4). Two other *Salix viminalis* × *S. miyabeana* triploid hybrids in DG8, 'Otisco' and 'Tully Champion', ranked 2nd and 4th in overall yield, respectively (Table 4, Fig. 1a). Two of the top five GEN were tetraploid *S. miyabeana* (DG5), including 'Canastota' a progeny selected from a cross between 'SX61' and 'SX64'. Two triploid *S. purpurea* × *S. miyabeana* cultivars from DG9, 'Oneida' and 'Millbrook', were ranked 6th and 7th overall, respectively (Fig. 1a). Triploids showed the greatest relative yields (GEN yield/ENV mean yield averaged over all ENV), with the exception of two cultivars, 'Truxton' and 'Owasco', which performed at or below ENV means (Fig. 1b). The cultivars *S. miyabeana* 'SX61', 'SX64' and *S.* × *dasyclados* 'SV1' were cultivars that showed promise in earlier trials, and subsequently were used as check cultivars in this yield trial network. The top two improved cultivars, 'Fabius' and 'Otisco', performed better than all three check cultivars, and four other improved cultivars performed better than the check clone mean yield of 7.14 Mg ha^{-1} yr^{-1}. The GEN with the lowest mean yields were diploid *S. eriocephala* 'S25' (DG4), a North American native willow species, and two hybrid cultivars of *S. koriyanagi* × *S. purpurea* 'Onondaga' and 'Allegany' (DG6b). ENV mean yields ranged from 2.57 to 11.3 Mg ha^{-1} yr^{-1}, with the greatest yields in eastern USA and Canada, and the lowest yields occurring in the Upper Peninsula of MI, USA, and Saskatoon, SK, CA.

Table 3 Mixed model analysis of variance for first-rotation yields of 16 shrub willow genotypes in 10 environments showing *F*-test results for fixed effects and variance components for random effects

Source	DF	SS	MS	*F* Value	Pr > *F*
Fixed					
ENV	9	5247.57	583.06	28.01	<0.0001
GEN	15	381.03	25.40	9.19	<0.0001
G×E	135	765.39	5.67	2.05	<0.0001
	DF	VC	Pct		
Random					
BLK (ENV)	30	1.27	31.43		
Error	411	2.76	68.57		

DF, degrees of freedom; SS, sum of squares; MS, mean square; ENV, environment; GEN, genotype; BLK, block; VC, variance component; Pct, percent of total variance.

Table 4 Adjusted mean first-rotation shrub willow yields (Mg ha^{-1} yr^{-1}) for 16 genotypes in 10 environments in North America

	Environments*										
Epithet	Midd	Bois	Stor	Wase	Fred	Cons	Esca	Sask	Skan	Brim	MEAN
'Fabius'	13.96	13.17	13.27	10.01	8.19	8.76	8.83	4.38	1.95	3.25	8.58
'Otisco'	11.95	12.39	11.10	7.66	7.54	6.55	9.38	4.46	3.18	2.91	7.71
'SX64'†	11.11	13.61	9.15	10.13	8.01	8.61	5.94	3.34	3.21	3.08	7.62
'Tully Champion'	13.34	13.15	6.65	7.57	8.22	5.61	7.51	5.80	3.75	3.70	7.53
'Canastota'	14.56	13.36	8.78	8.98	7.71	7.75	5.20	3.53	2.29	2.47	7.46
'Oneida'	10.63	8.85	12.31	9.80	6.50	8.89	6.91	4.67	3.25	2.41	7.42
'Millbrook'	10.77	10.04	9.23	9.11	8.00	8.11	7.96	4.13	2.24	2.63	7.22
'SX61'	11.97	10.45	9.44	9.62	6.78	7.05	7.89	2.75	3.05	3.14	7.21
'Fish Creek'	11.18	10.78	9.77	9.07	7.59	8.17	6.30	4.03	2.00	1.74	7.06
'Truxton'	12.32	9.95	9.84	8.96	8.32	6.59	5.71	3.60	3.40	1.74	7.04
'Sherburne'	11.92	10.72	9.82	8.61	6.53	7.02	5.50	3.32	1.89	2.30	6.76
'SV1'	10.92	11.54	6.03	4.11	7.62	7.81	9.02	3.28	2.47	3.23	6.60
'Owasco'	10.13	8.75	7.18	8.85	8.29	6.32	6.44	4.20	3.15	2.56	6.59
'Allegany'	8.31	9.31	6.88	7.51	5.26	6.32	6.94	3.10	2.21	2.13	5.80
'Onondaga'	8.36	7.54	5.54	7.34	5.10	6.84	6.67	3.60	2.06	2.02	5.51
'S25'	9.44	8.52	5.39	6.60	6.52	3.87	3.90	3.81	2.39	1.74	5.22
MEAN	11.30	10.76	8.77	8.37	7.26	7.14	6.88	3.87	2.66	2.57	

*Environment names are truncated to the first four letters, with full environment names given in Table 2.
†Check cultivars are underlined.

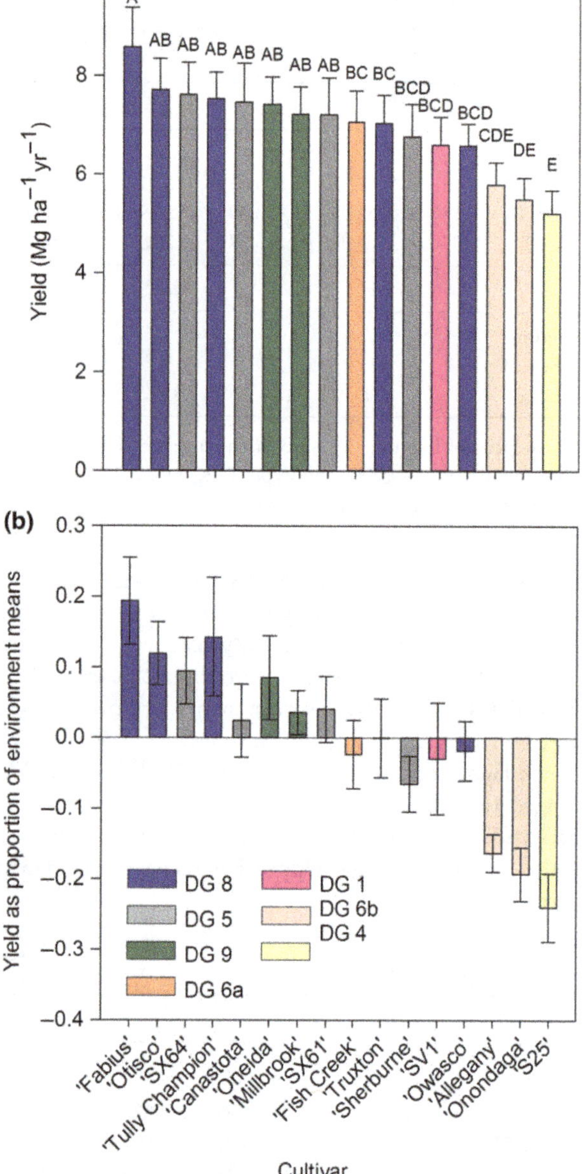

Fig. 1 First-rotation shrub willow mean yields (a) and yields as a proportion of environment means (b) for 16 genotypes in 10 environments in North America. Bars are shaded according to the diversity groups (pedigrees) presented in Table 1. Bars in (a) with different letters indicate significant ($P < 0.05$) differences according to Tukey's studentized range test (HSD). Error bars in (a) are ± one standard error of the mean. Proportions in (b) were calculated as the yield of a genotype divided by the environment mean, averaged across the 10 environments.

AMMI analysis

The AMMI2 model ANOVA revealed that the main effects of GEN and ENV plus the first two IPC axes accounted for 95.5% of the treatment SS with only 68 (of 159) treatment DF (Table 5). Using the error variance obtained from the SAS PROC MIXED output (Table 3), both the F-tests and the more conservative F_R-tests proved IPC1 and IPC2 to be significant ($P < 0.0002$); however, all subsequent axes were not significant and were therefore pooled into a discarded residual. The G×E interaction term is inherently a mixture of true signal in the data (i.e., systematic rank changes amount GEN) and noise. The amount of G×E noise can be estimated by multiplying the G×E DF by the mean square error (Gauch, 2013). Using the mean square error variance obtained from the SAS PROC MIXED output (Table 3), the G×E noise SS for this dataset was estimated to be 373 (46%), while the G×E signal accounts for 438.5 (54%) of the total G×E SS. IPC1 captured 291 SS, or 66.4% of the G×E signal SS. The cumulative SS accounted for by the first two IPC axes is 507, which is greater than the estimated G×E signal SS, suggesting that IPC2 contains a combination of mostly signal and some noise. Therefore, AMMI1 does a reasonable job of capturing the majority of the G×E signal in a parsimonious model. However, in terms of accurately describing our data, AMMI2 might be slightly better, although the more important consideration is that AMMI1 (and AMMI2) is considerably more accurate than the raw data AMMIF as its additional IPC3 and higher components (which are combined in the residual in Table 5) capture a SS of 304, which is mostly noise.

A useful tool for interpreting the AMMI analysis is the AMMI1 biplot, where the additive main effects of GEN and ENV are plotted on the x-axis in units of yield (Mg ha^{-1} yr^{-1}), and the IPC1 scores are plotted on the y-axis, which have units expressed as the square root of the yield (Gauch, 1992; p. 85). By representing both the main effects and the majority of the interaction signal in one projection, the AMMI1 biplot captures 92.3% of the treatment SS, and the relationships between main effects and interaction can be observed simultaneously (Fig. 2). The vertical reference line represents the grand mean yield of the dataset, and the horizontal line is placed at zero for the IPC scores, where points farther away from this line indicate GEN or ENV with larger interactions. Genotypes that occur close to the horizontal line can be regarded as having relatively stable yields across ENV. Genotypes and ENV having the same sign for their IPC1 scores have positive interactions, whereas opposite signs give negative interactions. For instance, 'Fabius' was the highest yielding GEN, and consequently, it is farthest along the right side of the x-axis. It also had one of the largest IPC1 scores, generating a positive interaction in Storrs, CT, but a negative interaction in Skandia, MI. Indeed, the lowest yield for 'Fabius' of 1.95 Mg ha^{-1} yr^{-1} occurred at that location, in comparison with Storrs, CT, where 'Fabius' yielded 13.27 Mg ha^{-1} yr^{-1} (Table 4). The IPC1 scores for 'SV1'

Table 5 Additive main effects and multiplicative interactions (AMMI) analysis of variance for first-rotation yields of 16 shrub willow genotypes in 10 environments showing the first two IPC axes

Source	DF	SS	MS	F value*	Pr > F	F_R Value	Pr > F_R
TRT	159	6774.19	42.60	15.42	<0.0001		
GEN	15	454.50	30.30	10.97	<0.0001		
ENV	9	5508.20	612.02	221.51	<0.0001		
G×E	135	811.49	6.01	2.18	<0.0001		
IPC1	23	290.96	12.65	4.58	<0.0001	2.18	<0.0001
IPC2	21	216.04	10.29	3.72	<0.0001	1.68	<0.0002
Residual	91	304.49	3.35	1.21	0.99	1.21	0.11

DF, degrees of freedom; SS, sum of squares; MS, mean square; TRT, treatment; ENV, environment; GEN, genotype; BLK, block.
*The error mean square from Table 3 was used in calculations of F and F_R values.

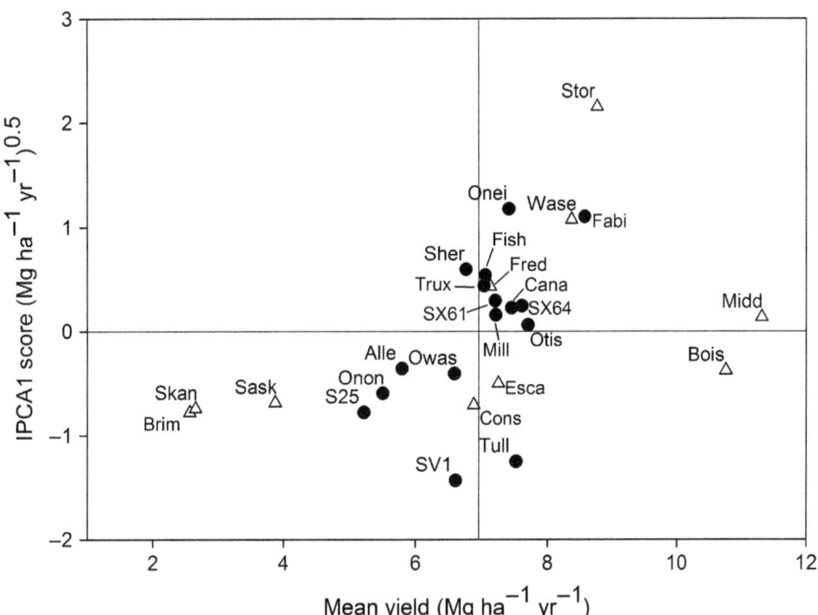

Fig. 2 AMMI1 biplot for the shrub willow yield trial network. Genotype (•) and environment (△) means are on the x-axis, and the IPC1 scores are shown on the y-axis. Genotype and environment names are truncated to the first four letters, and full names can be found in Tables 1 and 2, respectively. The vertical reference line represents the grand mean, while the horizontal line crosses at an IPC score of zero.

and 'Tully Champion' are similar and were the most extreme negative scores out of the 16 GEN, so their interaction patterns are the opposite of 'Fabius'. However, 'Tully Champion' has a substantially greater mean yield compared to 'SV1', and it was the top-yielding cultivar in many low-yielding ENV (Table 4).

When IPC2 is significant (Table 5), the AMMI2 biplot can be used to investigate the interaction pattern in IPC1 and IPC2 together (Fig. 3). The reference lines in Fig. 3 are drawn through zero for both axes. The crossing of these two lines in the middle of the graph indicates no interaction, and therefore, a GEN close to this point would be characterized as having stable yields across ENV, and for this dataset, the triploid 'Otisco' and the tetraploids 'SX61' and 'SX64' are closest to the origin. 'Fabius' and 'Tully Champion are on the opposite ends of IPC1 scores, but have very similar IPC2 scores. In contrast, 'Onondaga' and 'Canastota' were on the opposite extremes of IPC2 scores, with 'Canastota' having a strong positive interaction with Boisbriand, QC in terms of IPC2, as their IPC scores are of the same sign. Unlike Fig. 2, there is no information about main effects in Fig. 3, only interactions, but it is useful for understanding the relationships between AMMI1 and AMMI2.

Mega-environment delineations

Another useful graphical representation of AMMI1 analysis is a plot of the AMMI1 nominal yields (AMMI1

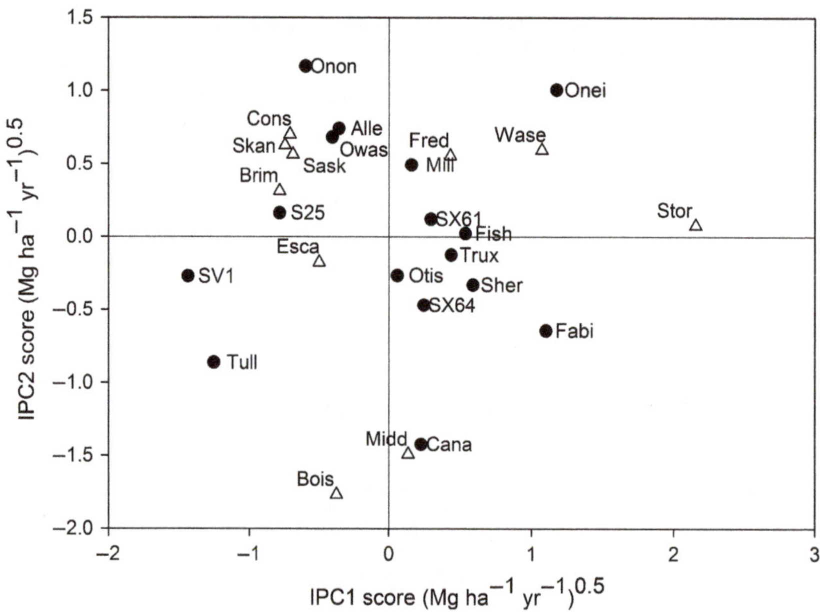

Fig. 3 AMMI2 biplot for the shrub willow yield trial network. Genotype (•) and environment (△) IPC1 scores are shown on the *x*-axis, and IPC2 scores are on the *y*-axis. Genotype and environment names are truncated to the first four letters, and full names can be found in Tables 1 and 2, respectively. The vertical and horizontal reference lines are drawn through zero for both axes.

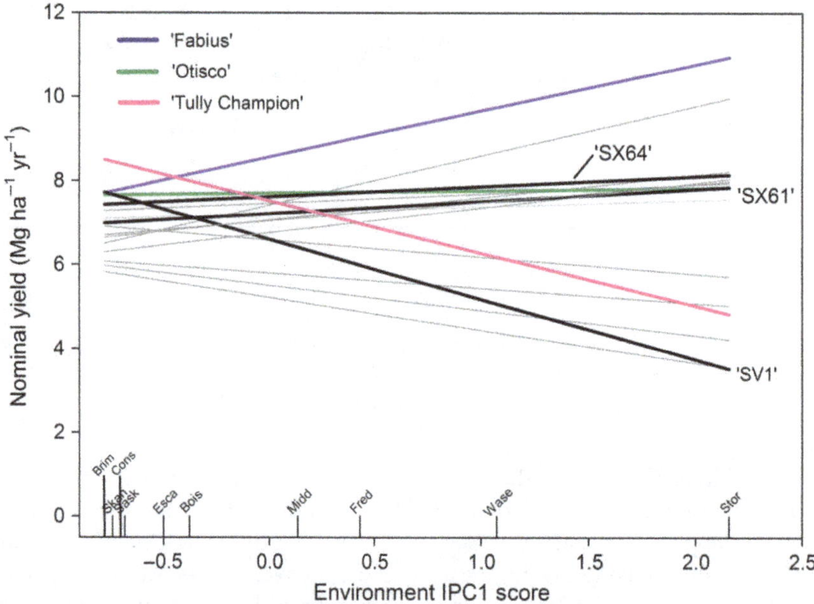

Fig. 4 Shrub willow genotypic responses according to the AMMI1 model. The 10 environment IPC1 scores are represented on the *x*-axis, while the genotype nominal yields (AMMI1 model yields without the environment deviations) are shown on the *y*-axis. All 16 genotypes are represented individually by a straight line. Only relevant genotypes have been labeled. Check cultivars are solid black lines. Environment names are truncated to the first four letters, and full names can be found Table 2.

model yields for each GEN, without the ENV deviation, see equation 1), against the ENV IPC1 scores (Gauch & Zobel, 1997). Figure 4 shows nominal yields for the 16 GEN across the 10 ENV plotted as straight lines, and the vertical positions of the lines relative to one another at a given ENV IPC1 score indicate the yield rankings. This plot is very useful for visualizing GEN crossovers, and for ease of interpretation, only the most relevant GEN have been highlighted and their names provided. For instance, 'Fabius' had the greatest yields in five

ENV ranging in IPC1 scores from −0.38 to 2.16. But for the other five ENV with scores −0.50 and lower, a switch point occurs where 'Tully Champion' outperforms 'Fabius' according to the AMMI1 model predicted yields. This situation illustrates the concept of narrow adaptation, where GEN perform similarly in a group of ENV with similar characteristics.

The degree of environmental sensitivity of a GEN can also be inferred by observing the slopes in Fig. 4. 'Fabius' and 'Tully Champion' have steep, but opposite slopes due to their opposite interactions. 'Otisco' was ranked 2nd in overall yield and was relatively insensitive to ENV, as evidenced by its near zero slope (Fig. 4). Also, similarities among GEN can be observed where slopes are nearly parallel. The line just below 'Fabius' on the right side of Fig. 4 is that of 'Oneida', which is similar to 'Fabius', suggesting that they reacted to ENV similarly. However, the mean yields of 'Fabius' were considerably greater than 'Oneida', making 'Fabius' an obvious superior choice. Similarly, 'SV1' had an interaction pattern that mirrored 'Tully Champion', but again, the predicted AMMI1 yields were lower for 'SV1'.

AMMI can also be used to delineate mega-environments, which are defined as subdivisions of a crop's potential growing range that share similar genotype winners, presumably due to similar biotic and abiotic stresses, but that are not necessarily contiguous (Gauch & Zobel, 1997; Yan *et al.*, 2000). Subdivision of a crop's growing region can help to improve yields by targeting GEN with specific adaptation to ENV where they are likely to perform best (Gauch & Zobel, 1997), especially if there are identifiable environmental or biological patterns associated with the IPC analysis that can be extended to ENV beyond those where crop yields have been tested (Gauch, 2013). MATMODEL uses key switch points in nominal yields among winning GEN across IPC1 scores to delineate mega-environments. For this shrub willow yield dataset, the main switch point for the AMMI1 model is clearly illustrated in Fig. 4, where 'Fabius' and 'Tully Champion' switch between first and second rank at a value of about −0.44 along the ENV IPC1 range. Therefore, AMMI1 defines two winners which happen to divide the 10 ENV equally between them, where 'Fabius' is declared the overall winner, but 'Tully Champion' is superior in the lower yielding ENV.

As the AMMI analysis consists of a family of models, with higher order models incorporating greater numbers of interaction components, higher order models generally result in an increased number of mega-environments and winners (Gauch, 2013). To illustrate this point, Table 6 shows the top five rankings for three AMMI model family members. The ENV are listed by IPC1 scores. AMMI1 divides the growing region into two mega-environments as described above, with only two-first place winners, 'Fabius' and 'Tully Champion'. 'Fabius' takes second place in most ENV where 'Tully Champion' was ranked first. 'Oneida' ranks behind 'Fabius' in more favorable ENV, while 'Otisco' ranks third in many of the less favorable ENV. The main differences between AMMI1 and AMMI2 are that Boisbriand, QC is declared as a separate mega-environment with 'Canastota' the top-ranking GEN, and to some degree, 'Fabius' is relegated to lower ranks in the less favorable ENV. The AMMIF model, which is equivalent to the raw data, is also included for comparisons.

Table 6 Rankings of the top five genotypes for in 10 environments (ENV) based on three AMMI models. AMMI1 includes only the first IPC axis and AMMI2 includes both the first and second IPC axes which were both significant based on the F_R-test. AMMIF represents the raw data

	AMMI1 Rank					AMMI2 Rank					AMMIF Rank				
ENV*	1	2	3	4	5	1	2	3	4	5	1	2	3	4	5
Stor	Fabi†	Onei	Fish	SX64	Sher	Fabi	Onei	Fish	SX64	Sher	Fabi	Onei	Otis	Trux	Sher
Wase	Fabi	Onei	SX64	Otis	Cana	Fabi	Onei	Mill	Fish	Otis	SX64	Fabi	Onei	SX61	Mill
Fred	Fabi	Onei	Otis	SX64	Cana	Fabi	Onei	Otis	Mill	SX64	Onei	Fabi	SX64	Fish	Mill
Midd	Fabi	Otis	SX64	Onei	Cana	Fabi	Cana	*Tull*	SX64	Otis	Cana	Fabi	*Tull*	Trux	SX61
Bois	Fabi	*Tull*	Otis	SX64	Cana	Cana	*Tull*	Fabi	SX64	Otis	SX64	Cana	Fabi	*Tull*	Otis
Esca	*Tull*	Fabi	Otis	SX64	Cana	*Tull*	Fabi	Otis	Cana	SX64	Trux	Owas	*Tull*	Fabi	SX64
Sask	*Tull*	Fabi	Otis	SV1	SX64	*Tull*	Otis	Fabi	SV1	Mill	*Tull*	Onei	Otis	Fabi	Owas
Cons	*Tull*	Fabi	Otis	SV1	SX64	*Tull*	Otis	Mill	SV1	Owas	Otis	SV1	Fabi	Mill	SX61
Skan	*Tull*	Fabi	Otis	SV1	SX64	*Tull*	Otis	SV1	Mill	Fabi	*Tull*	Trux	Onei	SX64	Otis
Brim	*Tull*	SV1	Fabi	Otis	SX64	*Tull*	SV1	Otis	Fabi	SX64	*Tull*	Fabi	SV1	SX61	SX64

*ENV, Environment. Names are truncated to the first four letters. Full names can be found in Table 2.

†Genotype names are truncated to four letters, with full names presented in Table 1. The overall top-yielding cultivar, 'Fabius' is underlined throughout the table, and the second mega-environment winner from AMMI1, 'Tully Champion' is italicized throughout the table.

Should the full data be accepted as most accurate, it would result in a very complex array of mega-environments with seven different cultivars being declared winners in seven narrowly defined subregions. AMMI1 with its simplified mega-environment winner pattern is likely best for making cultivar selections far less complicated. AMMI2 is likely the most accurate choice for modeling overall genotypic performance in this dataset, but the inclusion of one winner in one ENV complicates recommendations and justification for its inclusion would likely need to be verified with more testing.

Edaphic and climatic variables

The AMMI analysis of the G×E interaction operates solely on yield data, with an implicit interpretation that general patterns in yields are a reflection of underlying environmental characteristics of test locations. However, it may be desirable to associate yield patterns with known environmental characteristics, and IPC scores can be used in simple correlations when environmental variables for test locations are available (Van Eeuwijk, 1995). ENV yield was significantly ($P < 0.05$) correlated with mean annual growing degree-days and marginally significantly ($P < 0.10$) associated with longitude and growing season precipitation (Table 7). ENV IPC1 scores were also significantly correlated with growing degree-days, mean annual minimum temperatures and latitude (Table 7), which would covary with each other, but suggest a strong relationship between interaction patterns and temperature. IPC2 scores were positively correlated with elevation (Table 7).

Expanded datasets and analyses

As noted earlier, our yield trial network contained more GEN and test ENV than those included in the above AMMI analysis (601 observations of ~1530 available). However, the MATMODEL program allows for imputation of missing treatment cells in the two-way table of means using the expectation maximization (EM) algorithm (Gauch & Zobel, 1990). To more fully explore the yield database, two additional datasets were assembled that expanded the original complete set of 16 GEN in 10 ENV. The first dataset had an additional 6 GEN that included 10 missing treatments (4.5%) of a possible 220, for a total of 22 GEN in 10 ENV. The second dataset added 6 ENV with 15 missing treatments (5.9%) of 256. It has recently been shown that using the EM method for data imputation with missing data proportions of 5–10% do not significantly reduce the predictive ability of the AMMI model, especially when a small number of IPC axes (one or two) are chosen (Rodrigues *et al.*, 2011; Paderewski & Rodrigues, 2014). The ANOVA tables and mega-environment winners for the two expanded datasets are provided in Table S1. The ANOVA for the two expanded datasets show greater overall variance in the treatments, as expected, but the G×E signal SS were similar at 38.7% for the increased GEN and 36.7% for the increased ENV scenarios (Table S1). The IPC1 axes accounted for 82.4 and 76% of the G×E signal SS for the expanded GEN and expanded ENV scenarios, respectively. Interestingly, the mega-environment analyses remain remarkably similar to those of the base case scenario (Table S2). Even with six additional GEN, 'Fabius' remains the winner in six ENV and 'Tully Champion' is the winner in four of the lowest yielding ENV. When six ENV were added for a total of 16, 'Fabius' wins in seven and 'Tully Champion' in eight (Table S2). 'SV1' wins in one environment, Belleville, NY, but 'Tully Champion' is a close second, and often combining a small mega-environment with a larger one occupied by a close winner is justified (Gauch, 2013). The addition of these two scenarios reinforces the superiority of 'Fabius'

Table 7 Pearson product-moment correlation coefficients between mean yields (Mg ha^{-1} yr^{-1}), IPC scores and edaphic and climatic variables for the 10 environments in the yield trial network

Variable	Yield	IPC1	IPC2
Latitude	−0.534	−0.619*	0.098
Longitude	0.587*	0.300	−0.503
Elevation (m)	−0.496	−0.204	0.785**
Mean annual precipitation (mm)	0.527	0.346	−0.221
Mean April–October precipitation (mm)	0.578*	0.525	−0.195
Mean annul growing degree-days (base 10 °C)	0.650**	0.827**	−0.287
Mean annual minimum temperature (°C)	0.394	0.729**	−0.166
Mean annual solar radiation (MJ m^{-1} day^{-1})	−0.038	0.479	0.238
Soil organic matter (%)	0.261	−0.116	−0.013
Soil pH	0.028	−0.191	−0.320
Depth to water table (cm)	−0.131	−0.300	0.009

*$P < 0.10$; **$P < 0.05$.

and 'Tully Champion' relative to the cultivars tested in the analysis.

Incidentally, MATMODEL can also provide the Finlay–Wilkinson joint regressions analysis (Zobel *et al.*, 1988; Gauch, 2007), which we applied to the original dataset of 16 GEN in 10 ENV. The joint regression analysis only accounted for 23.2% of the G×E SS, compared to 35.9% for AMMI IPC1 (Table S3). Based on GEN slopes, the joint regression analysis of the GEN with the best combination of stable slopes and greater yields would be as 'Otisco' and 'SX64', with the high yields and slopes slightly above 1, and 'Tully Champion' and 'Oneida' with somewhat lower yields and slopes slightly less than 1 (Table S3). These results are somewhat in agreement with AMMI, but selection based on stability would likely rule out the potential gains from including 'Fabius' in optimal locations.

Discussion

Our analysis of this North American shrub willow yield trial network data has demonstrated the importance of G×E interactions in short-rotation woody crop productivity. Our evaluation of first-rotation yields using the AMMI model has accomplished the main objectives of the study, which were to identify GEN with broad adaptation, to define subregions where particular GEN exhibit specific adaptation and to identify environmental variables that help to explain patterns in yield. Despite the proven success of AMMI in accurately diagnosing G×E interactions from a range of agronomic and horticultural crops, we have not seen any other published reports of its use in short-rotation coppice research. AMMI helped to resolve signal from noise, and we selected the more parsimonious model, AMMI1, as the best representation of the data, simplifying the mega-environment analysis with just two winning GEN, 'Fabius' and 'Tully Champion', as opposed to the raw data with seven of the 16 GEN in our dataset winning in at least one environment (Table 6). In further support of the AMMI1 analysis winners, inclusion of additional GEN or additional test locations did not change the mega-environment analysis outcomes for AMMI1 (Table S1).

'Fabius' was the top-yielding cultivar with an overall mean of 8.58 Mg ha^{-1} yr^{-1}, setting a standard for optimal yields and broad adaptability. Also, because of its positive interaction with high-yielding ENV, it is well adapted to favorable growing conditions. 'Tully Champion' had an opposite pattern of interaction and appears to be adapted to less desirable growing conditions. 'Tully Champion' and 'SV1' interacted with the test ENV similarly (Fig. 4), but 'Tully Champion' consistently outperformed 'SV1' make it a much better check cultivar for evaluating future breeding material. Similarly, 'Otisco' and 'SX64' had fairly stable yields, showing little interaction with environment regardless of quality, but 'Otisco' was ranked second in overall mean yield. All three of these superior cultivars ('Fabius', 'Tully Champion' and 'Otisco') are of the triploid pedigree of *S. viminalis* × *S. miyabeana*, and they outperformed the mean of the check cultivars by 5 to 20%, demonstrating that yield gains can be achieved through novel hybridization in shrub willow. This affirms the findings of Serapiglia *et al.* (2014) for early selections of triploids on a single site, but extends it more broadly across a large set of diverse ENV.

It should be noted that 'Fabius' and 'Tully Champion' are siblings from the same cross, which makes their divergent patterns in yield across our sites rather remarkable. These two were selected after an evaluation of the family in a replicated trial at a single location in central NY (Smart *et al.*, 2008). There was considerable variability in growth potential across this family, and clearly the single test location was not adequate to detect differiential response to broad environmental conditions. The cultivar with the most stable yields in our dataset, 'Otisco', shares the same *S. viminalis* mother, 'SV2', with 'Fabius' and 'Tully Champion' (Table 1). 'Owasco' and 'Truxton' are also siblings that belong to the *Salix viminalis* × *S. miyabeana* pedigree (DG8). They share the same father with 'Otisco', 'SX64', but their overall yields were much lower, suggesting superior combining ability of 'SV2' for yield traits. This suggests that interspecific triploid hybridization in *Salix* can confer substantial gains in yield, and future breeding efforts will continue to exploit this potential.

By correlating environment mean yields and IPC scores with edaphic and climatic variables from the test sites, we were able to demonstrate that increased temperatures were positively correlated with both mean yields and IPC1 scores, while the correlation between precipitation and mean yield was marginally significant. We have observed that the triploid hybrids tend to retain a leaf canopy much later in the season compared to other pedigrees. This could be an indication of lower sensitivity to frost or an ability to exploit solar resources at the tail end of the growing season. Labrecque & Teodorescu (2003) examined yields of two willow species at two contrasting sites in southern Québec and suggested that growth in that region may be limited by precipitation, but not likely growing season temperatures, based on reported optimal growing conditions in Sweden. Wang & Macfarlane (2012) analyzed poplar (*Populus* spp.) and shrub willow yields of cultivars sourced from Ontario and New York collections at two locations in Michigan and estimated that the difference in growing degree-days between northern and southern locations could explain 60% of the variation in yield. In the UK, Aylott *et al.* (2008) analyzed yields of three willow GEN of differing pedigrees in relation to edaphic and climatic variables from 49 locations. In general, temporal rainfall patterns were found to be the significant factors affecting yields, but the response varied by GEN, with one being particularly sensitive to soil pH. We were unable to find significant correlations with soil pH or organic matter; however, because of the broad geographic range covered in this analysis, perhaps the greatest amount of variability among the test sites was dominated by climatic factors. The subject warrants additional investigation and testing in locations with higher temperatures and moderate rainfall may reveal a greater geographic production range than previously proposed (Walsh *et al.*, 2003).

G×E interactions have been reported elsewhere in recent studies examining willow yields in multiple locations. In eastern Canada, Mosseler *et al.* (2014a) studied biomass traits of multiple clones from natural accessions of two species, *S. discolor* and *S. eriocephala*, planted on three sites of contrasting fertility. They found no significant rank changes between the two species groups, but large differences in genotypic sensitivities among the GEN within species, suggesting a high degree of intraspecific genetic variability in adaptation that could be exploited for biomass yield improvement. In Denmark, Larsen *et al.* (2014) studied yields of Swedish and UK commercial cultivars across five locations and reported significant G×E interactions, but there was one clear winner, 'Tordis' (*S. viminalis* × *S. schwerinii*), with a mean yield of 6.7 Mg ha^{-1} yr^{-1}. In earlier work involving numerous GEN of three willow species across a large number of sites in Sweden, Rönnberg-Wästljung & Thorsén (1988) used a variant of the Finlay–Wilkinson regression to analyze G×E interactions. As in our analysis, linear regressions did not always adequately explain clonal response to ENV. The authors identified stable GEN using regression coefficients and mean yields, but they did not address specific adaptability. In poplar, Zalesny *et al.* (2009) performed an extensive analysis of 53–79 GEN across four sites and multiple ages in the upper Midwestern USA. They assessed genotypic rankings and segregated GEN based on generalists (consistently high rankings) and specialist (especially variable rankings). More recent attempts to analyze G×E interactions in poplar have incorporated some of the more contemporary statistical techniques used in plant breeding and agronomy. In an analysis of first-year growth of nine GEN in five locations in Spain, Sixto *et al.* (2011) applied the genotype and genotype x environment (GGE) biplot (Yan *et al.*, 2000), which has some similarities with AMMI biplots and is used to visualize interactions (Gauch, 2006a; Gauch *et al.*, 2008). More recently, Sixto *et al.* (2014) analyzed biomass production at the end of a 3-year rotation for nine GEN in four of the locations from their previous study. They applied mixed model variants of stability parameter models including Finlay–Wilkinson and Eberhart–Russell and identified patterns of broad and specific adaptation among GEN and pedigrees. These mixed model approaches often consider ENV as random and have the advantage of handling unbalanced datasets (Smith *et al.*, 2005).

The importance of developing cultivars with specific adaptation can be emphasized when willow cultivation is placed into broader biological and geographical contexts. In the UK and Swedish breeding programs, improved *S. viminalis* and hybrids such as *S. viminalis* × *S. schwerinii* are highly productive and have demonstrated superior yields in numerous trials compared to other pedigrees (Aylott *et al.*, 2008; Lindegaard *et al.*, 2011; Larsen *et al.*, 2014). In the USA, potato leafhopper (*Empoasca fabae* Harris) has proven to be extremely damaging to imported European *S. viminalis* cultivars (Smart & Cameron, 2008). Hybrid crosses of *S. viminalis* with *S. miyabeana* have introduced varying degrees of resistance to potato leafhopper, in addition to improved yields. We have observed more severe damage to 'Tully Champion' compared with 'Fabius' in NY locations (Gouker & Smart, 2015), which may have limited its yield potential on sites of more southern latitude and more eastern longitude. Potentially lower potato leafhopper pressure in northerly ENV may explain the superiority of 'Tully Champion' and 'SV1' at those sites. In Canada, *S. eriocephala* and *S. discolor* have been evaluated as North American native species appropriate for biomass and phytoremediation applications (Mosseler *et al.*, 2014a,b). However, testing in the USA has shown that natural accessions as well as improved cultivars of *S. eriocephala* can be highly susceptible to leaf rust (*Melampsora* spp) (Cameron *et al.*, 2008; Serapiglia *et al.*, 2013). The improved *S. eriocephala* cultivar 'S25' in our analysis had the lowest overall yield, and it is possible that rust infection in combination with deer browse contributed to low performance. In Europe, *Melampsora* rust is a major growth-limiting pathogen for multiple willow species and species hybrids (Åhman, 1998; McCracken & Dawson, 2003; Aylott *et al.*, 2008). *Salix purpurea* and hybrid *S. koriyanagi* × *S. purpurea* cultivars are also susceptible to rust, but based on visual observations in the field and the yields of 'Oneida' and 'Millbrook' in this analysis, the *S. purpurea* × *S. miyabeana* diversity group likely has increased rust resistance. Breeding for rust resistance is an important focus in the USA given the European experience.

Inclusion of some ENV in this analysis caused overall yields to be lower compared to other studies, but collectively they provided strong contrasts for discriminating among GEN. In turn, this analysis likely provides a more realistic assessment of yield potentials on marginal lands, given some of the exceptionally low-yielding ENV. As trials are often performed at experiment stations, where growing conditions are optimized or at least less variable, it may seem obvious that many reported yields are often on the high end and they may not be representative (Simmonds, 1991). This becomes even more relevant in the realm of dedicated bioenergy crops, as the premise is that these are purposed for marginal lands; more specifically, where the economic returns are marginal for production of traditional field crops (Stoof *et al.*, 2015), and thus, there is reduced competition between food and energy production. Evaluations of shrub willow yield potentials in the USA have been largely geographically restricted to regions where institutional knowledge exists, and the true extent of production potential has likely not be adequately tested (Walsh *et al.*, 2003). Given the positive correlations of yield and IPC1 scores with factors relating to increased temperature, perhaps growing willow at lower latitudes will produce greater yields. But genotypic variation in water-use efficiency and drought resistance will be important and may help to inform breeding for improved adaptation to warmer climates (Bonosi *et al.*, 2013).

This study has demonstrated the importance of identifying genotypic adaptability for developing cultivar recommendations and guiding future breeding efforts. Genotypes with high yields but varying sensitivities to environmental conditions have been identified as important check cultivars for testing the next generation of promising genotypes, which is currently underway. More work is needed in exploring the underlying environmental causes for the observed yields, and this will be the focus of a future analysis, which will incorporate an expanded dataset of North American willow yields.

Acknowledgments

Establishment and harvesting of the trials were funded largely by the North Central Regional Sun Grant Center at South Dakota State University through a grant provided by the US Department of Energy Bioenergy Technologies Office under Award number DE-FC36-05GO85041 and the US DOE Regional Feedstock Partnership. Maintenance and harvesting required thousands of person-hours, done mostly by technicians too numerous to acknowledge individually here. E. Fabio was supported by Agriculture and Food Research Initiative Competitive Grant No. 2012-68005-19703 from the USDA NIFA.

References

Åhman I (1998) Rust scorings in a plantation of *Salix viminalis* clones during ten consecutive years. *European Journal of Forest Pathology*, **28**, 251–258.

Amichev BY, Hangs RD, Belanger N, Volk TA, Vujanovic V, Schoenau JJ, Van Rees KCJ (2015) First-rotation yields of 30 short-rotation willow cultivars in central Saskatchewan, Canada. *Bioenergy Research*, **8**, 292–306.

Annicchiarico P (2002) *Genotype x Environment Interactions: Challenges and Opportunities for Plant Breeding and Cultivar Recommendations*. Food & Agriculture Organization, Rome, Italy.

Aylott MJ, Casella E, Tubby I, Street NR, Smith P, Taylor G (2008) Yield and spatial supply of bioenergy poplar and willow short-rotation coppice in the UK. *New Phytologist*, **178**, 358–370.

Bonosi L, Ghelardini L, Weih M (2013) Towards making willows potential bioresources in the South: Northern *Salix* hybrids can cope with warm and dry climate when irrigated. *Biomass and Bioenergy*, **51**, 136–144.

Buchholz T, Volk TA (2011) Improving the profitability of willow crops-identifying opportunities with a crop budget model. *Bioenergy Research*, **4**, 85–95.

Cameron KD, Phillips IS, Kopp RF, Volk TA, Maynard CA, Abrahamson LP, Smart LB (2008) Quantitative genetics of traits indicative of biomass production and heterosis in 34 full-sib F_1 *Salix eriocephala* families. *Bioenergy Research*, **1**, 80–90.

Caputo J, Balogh SB, Volk TA, Johnson L, Puettmann M, Lippke B, O'Neil E (2014) Incorporating uncertainty into a life cycle assessment (LCA) model of short-rotation willow biomass (*Salix* spp.) crops. *Bioenergy Research*, **7**, 48–59.

Cooper M, Hammer GL (1996) Synthesis of strategies for crop improvement. In: *Plant Adaptation and Crop Improvement* (eds Cooper M, Hammer GL), pp. 591–623. CABI, Wallingford, UK.

Eberhart SA, Russell WA (1966) Stability parameters for comparing varieties. *Crop Science*, **6**, 36–40.

Environment Canada 2015. National Climate Data, Available at: http://climate.weather.gc.ca/ (accessed 6 October 2015).

Finlay KW, Wilkinson GN (1963) The analysis of adaptation in a plant-breeding programme. *Crop and Pasture Science*, **14**, 742–754.

Gauch HG (1988) Model selection and validation for yield trials with interaction. *Biometrics*, **44**, 705–715.

Gauch HG (1992) *Statistical Analysis of Regional Yield Trials: AMMI Analysis of Factorial Designs*. Elsevier, Amsterdam, The Netherlands.

Gauch HG (2006a) Statistical analysis of yield trials by AMMI and GGE. *Crop Science*, **46**, 1488–1500.

Gauch HG (2006b) Winning the accuracy game. *American Scientist*, **94**, 133–141.

Gauch HG (2007) *MATMODEL Version 3.0: Open Source Software for AMMI and Related Analyses*. Crop and Soil Sciences, Cornell University, Ithaca, NY, USA.

Gauch HG (2012) *Scientific Method in Brief*. Cambridge University Press, New York, NY.

Gauch HG (2013) A simple protocol for AMMI analysis of yield trials. *Crop Science*, **53**, 1860–1869.

Gauch HG, Furnas RE (1991) Statistical analysis of yield trials with MATMODEL. *Agronomy Journal*, **83**, 916–920.

Gauch HG Jr, Zobel RW (1990) Imputing missing yield trial data. *Theoretical and Applied Genetics*, **79**, 753–761.

Gauch HG, Zobel RW (1997) Identifying mega-environments and targeting genotypes. *Crop Science*, **37**, 311–326.

Gauch HG, Piepho H-P, Annicchiarico P (2008) Statistical analysis of yield trials by AMMI and GGE: further considerations. *Crop Science*, **48**, 866–889.

Gollob HF (1968) A statistical model which combines features of factor analytic and analysis of variance techniques. *Psychometrika*, **33**, 73–115.

Gouker FE, Smart LB 2015. Willow cultivar diversity groups fact sheet. Available at: http://willow.cals.cornell.edu/ (accessed 21 October 2015).

Hanley SJ, Karp A (2014) Genetic strategies for dissecting complex traits in biomass willows (*Salix* spp.). *Tree Physiology*, **34**, 1167–1180.

Hauk S, Knoke T, Wittkopf S (2014) Economic evaluation of short rotation coppice systems for energy from biomass—A review. *Renewable and Sustainable Energy Reviews*, **29**, 435–448.

Heller MC, Keoleian GA, Volk TA (2003) Life cycle assessment of a willow bioenergy cropping system. *Biomass and Bioenergy*, **25**, 147–165.

Karp A, Shield I (2008) Bioenergy from plants and the sustainable yield challenge. *New Phytologist*, **179**, 15–32.

Karp A, Hanley SJ, Trybush SO, Macalpine W, Pei M, Shield I (2011) Genetic improvement of willow for bioenergy and biofuels. *Journal of Integrative Plant Biology*, **53**, 151–165.

Keoleian GA, Volk TA (2005) Renewable energy from willow biomass crops: life cycle energy, environmental and economic performance. *Critical Reviews in Plant Sciences*, **24**, 385–406.

Kiernan BD, Volk TA, Tharakan PJ, Nowak CA, Phillipon SP, Abrahamson LP, White EH (2003) Clone-site testing and selections for scale-up plantings. *Final Report prepared for the United States Department of Energy*, SUNY-ESF, Syracuse, NY, 67 pp.

Kopp RF, Smart LB, Maynard CA, Isebrands JG, Tuskan GA, Abrahamson LP (2001) The development of improved willow clones for eastern North America. *Forestry Chronicle*, **77**, 287–292.

Kuzovkina YA, Quigley MF (2005) Willows beyond wetlands: uses of *Salix* L. species for environmental projects. *Water Air and Soil Pollution*, **162**, 183–204.

Kuzovkina YA, Weih M, Romero MA *et al.* (2008) *Salix*: botany and global horticulture. In: *Horticultural Reviews* (ed. Janick J), pp. 447–489. John Wiley & Sons, Inc., Hoboken, NJ, USA.

Labrecque M, Teodorescu TI (2003) High biomass yield achieved by *Salix* clones in SRIC following two 3-year coppice rotations on abandoned farmland in southern Quebec, Canada. *Biomass and Bioenergy*, **25**, 135–146.

Larsen SU, Jørgensen U, Lærke PE (2014) Willow yield is highly dependent on clone and site. *Bioenergy Research*, **7**, 1280–1292.

Larsson S (1998) Genetic improvement of willow for short-rotation coppice. *Biomass and Bioenergy*, **15**, 23–26.

Lindegaard KN, Barker JHA (1997) Breeding willows for biomass. *Aspects of Applied Biology*, **49**, 155–162.

Lindegaard KN, Carter MM, McCracken A *et al.* (2011) Comparative trials of elite Swedish and UK biomass willow varieties 2001-2010. *Aspects of Applied Biology*, **112**, 57–66.

McCracken AR, Dawson WM (2003) Rust disease (Melampsora epitea) of willow (*Salix* spp.) grown as short rotation coppice (SRC) in inter- and intra-species mixtures. *Annals of Applied Biology*, **143**, 381–393.

Mosseler A (1990) Hybrid performance and species crossability relationships in willows (*Salix*). *Canadian Journal of Botany*, **68**, 2329–2338.

Mosseler A, Major JE, Labrecque M (2014a) Genetic by environment interactions of two North American *Salix* species assessed for coppice yield and components of growth on three sites of varying quality. *Trees*, **28**, 1401–1411.

Mosseler A, Major JE, Labrecque M (2014b) Growth and survival of seven native willow species on highly disturbed coal mine sites in eastern Canada. *Canadian Journal of Forest Research*, **44**, 340–349.

NASA POWER 2015. Prediction of worldwide energy resource climatology resource for agroclimatology. Available at: from http://power.larc.nasa.gov/ (accessed 6 October 2015).

NOAA NCEI 2015. Global historical climatology network. Available at: http://www.ncdc.noaa.gov/data-access (accessed 6 October 2015).

Paderewski J, Rodrigues PC (2014) The usefulness of EM-AMMI to study the influence of missing data pattern and application to Polish post-registration winter wheat data. *Australian Journal of Crop Science*, **8**, 640–645.

Piepho HP (1995) Robustness of statistical tests for multiplicative terms in the additive main effects and multiplicative interaction model for cultivar trials. *Theoretical and Applied Genetics*, **90**, 438–443.

Piepho HP, Möhring J (2005) Best linear unbiased prediction of cultivar effects for subdivided target regions. *Crop Science*, **45**, 1151–1159.

Richards BK, Stoof CR, Cary IJ, Woodbury PB (2014) Reporting on marginal lands for bioenergy feedstock production: a modest proposal. *Bioenergy Research*, **7**, 1060–1062.

Rodrigues PC, Pereira DGS, Mexia JT (2011) A comparison between joint regression analysis and the additive main and multiplicative interaction model: the robustness with increasing amounts of missing data. *Scientia Agricola*, **68**, 679–686.

Rönnberg-Wästljung A, Thorsén J (1988) Inter- and intraspecific variation and genotype × site interaction in *Salix alba* L., *S. Dasyclados* Wimm. and *S. viminalis* L. *Scandinavian Journal of Forest Research*, **3**, 449–463.

SAS Institute Inc. (2013) *(Version 9.4)*. SAS Institute, Inc., Cary, NC, USA.

Serapiglia MJ, Cameron KD, Stipanovic AJ, Abrahamson LP, Volk TA, Smart LB (2013) Yield and woody biomass traits of novel shrub willow hybrids at two contrasting sites. *Bioenergy Research*, **6**, 533–546.

Serapiglia MJ, Gouker FE, Smart LB (2014) Early selection of novel triploid hybrids of shrub willow with improved biomass yield relative to diploids. *BMC Plant Biology*, **14**, 74.

Sixto H, Gil P, Ciria P, Camps F, Sánchez M, Cañellas I, Voltas J (2014) Performance of hybrid poplar clones in short rotation coppice in Mediterranean environments: analysis of genotypic stability. *GCB Bioenergy*, **6**, 661–671.

Smart LB, Cameron KD (2008) Genetic improvement of willow (*Salix* spp.) as a dedicated bioenergy crop. In: *Genetic Improvement of Bioenergy Crops* (ed. Vermerris W), pp. 377–396. Springer New York, New York, NY, USA.

Smart LB, Cameron KD (2012) Shrub willow. In: *Handbook of Bioenergy Crop Plants* (eds Kole C, Joshi CP, Shonnard DR), pp. 687–708. Taylor and Francis Group, Boca Raton, FL.

Smart LB, Volk TA, Lin J *et al.* (2005) Genetic improvement of shrub willow (*Salix* spp.) crops for bioenergy and environmental applications in the United States. *Unasylva*, **56**, 51–55.

Smart LB, Cameron KD, Volk TA, Abrahamson LP (2008) Breeding, selection, and testing of shrub willow as a dedicated energy crop. *NABC Report 19 Agricultural Biofuels: Technology, Sustainability, and Profitability*, National Agricultural Biotechnology Council, Ithaca, NY, 85–92.

Smith AB, Cullis BR, Thompson R (2005) The analysis of crop cultivar breeding and evaluation trials: an overview of current mixed model approaches. *Journal of Agricultural Science*, **143**, 449–462.

Stoof CR, Richards BK, Woodbury PB *et al.* (2015) Untapped potential: opportunities and challenges for sustainable bioenergy production from marginal lands in the Northeast USA. *Bioenergy Research*, **8**, 482–501.

Stott KG (1991) Nomenclature of the promising biomass coppice willows, *Salix* × *sericans* Tausch ex Kern., *Salix dasyclados* Wimm. and *Salix* 'Aquatica Gigantea'. *Botanical Journal of Scotland*, **46**, 137–143.

U. S. Department of Energy (2011) U.S. billion-ton update: biomass supply for a bioenergy and bioproducts industry. (Leads Perlack RD, Stokes BJ), Oak Ridge National Laboratory, Oak Ridge, TN, 227 p.

Van Eeuwijk F (1995) Linear and bilinear models for the analysis of multi-environment trials: I. An inventory of models. *Euphytica*, **84**, 1–7.

Volk TA, Abrahamson LP, Cameron KD *et al.* (2011) Yields of willow biomass crops across a range of sites in North America. *Aspects of Applied Biology*, **112**, 67–74.

Walsh M, De La Torre UgarteD, Shapouri H, Slinsky S (2003) Bioenergy crop production in the United States: potential quantities, land use changes, and economic impacts on the agricultural sector. *Environmental and Resource Economics*, **24**, 313–333.

Wang Z, Macfarlane DW (2012) Evaluating the biomass production of coppiced willow and poplar clones in Michigan, USA, over multiple rotations and different growing conditions. *Biomass and Bioenergy*, **46**, 380–388.

Wang DA, Jaiswal D, Lebauer DS, Wertin TM, Bollero GA, Leakey ADB, Long SP (2015) A physiological and biophysical model of coppice willow (*Salix* spp.) production yields for the contiguous USA in current and future climate scenarios. *Plant, Cell & Environment*, **38**, 1850–1865.

Yan W, Hunt LA, Sheng Q, Szlavnics Z (2000) Cultivar evaluation and mega-environment investigation based on the GGE biplot. *Crop Science*, **40**, 597–605.

Zalesny RS Jr, Hall RB, Zalesny JA, Mcmahon BG, Berguson WE, Stanosz GR (2009) Biomass and genotype × environment interactions of *Populus* energy crops in the Midwestern United States. *Bioenergy Research*, **2**, 106–122.

Zobel RW, Wright MJ, Gauch HG (1988) Statistical analysis of a yield trial. *Agronomy Journal*, **80**, 388–393.

Zsuffa L (1990) Genetic improvement of willows for energy plantations. *Biomass*, **22**, 35–47.

Serapiglia MJ, Gouker FE, Hart JF, Unda F, Mansfield SD, Stipanovic AJ, Smart LB (2015) Ploidy level affects important biomass traits of novel shrub willow (*Salix*) hybrids. *Bioenergy Research*, **8**, 259–269.

Simmonds NW (1991) Selection for local adaptation in a plant breeding programme. *Theoretical and Applied Genetics*, **82**, 363–367.

Sixto H, Salvia J, Barrio M, Ciria MP, Cañellas I (2011) Genetic variation and genotype-environment interactions in short rotation *Populus* plantations in southern Europe. *New Forests*, **42**, 163–177.

15

Water use of a multigenotype poplar short-rotation coppice from tree to stand scale

JASPER BLOEMEN[1], RÉGIS FICHOT[2], JOANNA A. HOREMANS[1], LAURA S. BROECKX[1], MELANIE S. VERLINDEN[1], TERENZIO ZENONE[1] and REINHART CEULEMANS[1]

[1]*Department of Biology, Research Centre of Excellence on Plant and Vegetation Ecology, University of Antwerp, Universiteitsplein 1, Wilrijk B-2610, Belgium,* [2]*Université d'Orléans, INRA, LBLGC, EA 1207, F-45067 Orléans, France*

Abstract

Short-rotation coppice (SRC) has great potential for supplying biomass-based heat and energy, but little is known about SRC's ecological footprint, particularly its impact on the water cycle. To this end, we quantified the water use of a commercial scale poplar (*Populus*) SRC plantation in East Flanders (Belgium) at tree and stand level, focusing primarily on the transpiration component. First, we used the AquaCrop model and eddy covariance flux data to analyse the different components of the stand-level water balance for one entire growing season. Transpiration represented 59% of evapotranspiration (ET) at stand scale over the whole year. Measured ET and modelled ET were lower as compared to the ET of reference grassland, suggesting that the SRC only used a limited amount of water. Secondly, we compared leaf area scaled and sapwood area scaled sap flow (F_s) measurements on individual plants vs. stand scale eddy covariance flux data during a 39-day intensive field campaign in late summer 2011. Daily stem diameter variation (ΔD) was monitored simultaneously with F_s to understand water use strategies for three poplar genotypes. Canopy transpiration based on sapwood area or leaf area scaling was 43.5 and 50.3 mm, respectively, and accounted for 74%, respectively, 86%, of total ecosystem ET measured during the intensive field campaign. Besides differences in growth, the significant intergenotypic differences in daily ΔD (due to stem shrinkage and swelling) suggested different water use strategies among the three genotypes which were confirmed by the sap flow measurements. Future studies on the prediction of SRC water use, or efforts to enhance the biomass yield of SRC genotypes, should consider intergenotypic differences in transpiration water losses at tree level as well as the SRC water balance at stand level.

Keywords: bioenergy, evapotranspiration, poplar, sap flow, short-rotation coppice, stand water balance

Introduction

Short-rotation coppice (SRC) of fast-growing and high-yielding hardwood species as poplar and willow offers an important and environmentally sustainable way of producing heat and electricity from a renewable energy source (Herrick & Brown, 1967; Graham *et al.*, 1992; Gustavsson *et al.*, 1995; Berndes *et al.*, 2003; Kauter *et al.*, 2003; Aylott *et al.*, 2008). Poplar SRC showed high biomass production rates of 10–15 t ha^{-1} yr^{-1} (Heilman *et al.*, 1996; Trnka *et al.*, 2008; Broeckx *et al.*, 2012). However, there have been conflicting observations about the water use of SRC or its impact on the local water cycle. High-yielding SRC has high water requirements (Hall & Allen, 1997; Hall *et al.*, 1998; Allen *et al.*, 1999; Meiresonne *et al.*, 1999; Jassal *et al.*, 2013; Navarro *et al.*, 2014) potentially leading to negative effects on regional water resources (see references in Fischer *et al.*, 2013). A number of – experimental and modelling – studies on evapotranspiration (ET, mm day^{-1}) have argued that the water use of SRC is substantially higher than that of conventional agricultural crops or grasslands (see references in Dimitriou *et al.*, 2009; Petzold *et al.*, 2011). In contrast, other studies have reported that the water use rates of SRC are similar to those from agricultural crops and grasslands (Fischer *et al.*, 2013), that is comparable to or lower than the reference crop evapotranspiration (ET_0) (e.g. Meiresonne *et al.*, 1999; Linderson *et al.*, 2007; Migliavacca *et al.*, 2009; Tricker *et al.*, 2009).

It is expected that the transpiration component of ET (E_c, mm day^{-1}) is large as poplar species have high transpiration rates (Hall & Allen, 1997; Hall *et al.*, 1998; Meiresonne *et al.*, 1999; Kim *et al.*, 2008). Simulated transpiration was 71% of ET for a poplar SRC in the Czech Republic (Fischer *et al.*, 2013) and 66% of ET on a seasonal basis for a willow SRC in southern Sweden (Persson & Lindroth, 1994). Measurements of sap flow

Correspondence: Jasper Bloemen, Institute of Ecology, University of Innsbruck, Sternwartestraße 15, A-6020 Innsbruck, Austria

e-mail: Jasper.bloemen@uibk.ac.at

(F_s, kg h^{-1}) of individual trees scaled to the stand level are frequently used to quantify E_c for mature forest ecosystems (e.g. Oren *et al.*, 1998; Schafer *et al.*, 2002; Unsworth *et al.*, 2004; Bovard *et al.*, 2005; Tang *et al.*, 2006; Oishi *et al.*, 2008). Little is known, however, about the contribution of E_c to ET for SRC as only a limited number of studies combined plant-level measurements with stand-level water balance measurements or estimates of the water use of SRC. In addition, measurements of daily fluctuations in stem diameter (ΔD) can provide complementary information on genotype-specific tree water use as short-term shrinkage and swelling are related to internal water storage dynamics (Zweifel *et al.*, 2000, 2001; Larcher, 2003) and therefore changes in transpiration. As the first reports that stem dimensions change with changes in plant hydration (Fritts, 1961; Kozlowski & Winget, 1964; Impens & Schalck, 1965), short-term high temporal resolution dendrometer measurements have been made on different forest species (see references in Zweifel *et al.*, 2000; De Swaef *et al.*, 2015), including on poplar genotypes grown under controlled conditions (Giovannelli *et al.* 2007). Field measurements of daily ΔD fluctuations, in combination with F_s, might help to understand the dynamics in the contribution of E_c to ET for SRC and to identify unexplored intergenotypic differences in plant water use.

In this study, we monitored the water use of a poplar SRC in Flanders, Belgium, both at stand and at tree level. Our specific research hypotheses were as follows: (i) poplar SRC uses more water than a reference grassland under our specific conditions, (ii) E_c is the largest component of the stand water balance, and (iii) there are important intergenotypic differences in plant water use. For the entire growing season of 2011, we analysed the stand-level water balance of the poplar SRC using the AquaCrop model (Hsiao *et al.*, 2009; Raes *et al.*, 2009; Steduto *et al.*, 2009) complemented with eddy covariance measurements of ET. We further focused on tree-level measurements of plant water use during an intensive field campaign performed during the same growing season. Using detailed tree-level measurements of F_s and ΔD, we quantified the contribution of E_c to ET across three genotypes and we identified intergenotypic differences in plant water use.

Materials and methods

Study site and plant material

Measurements were made in a commercial scale multigenotype SRC plantation, established in Lochristi, province East Flanders, Belgium (51°06′44″N, 3°51′02″E), at an elevation of 6.25 m above sea level. Long-term average annual temperature at the site is 9.5 °C and the average annual precipitation is 726 mm, evenly distributed over the year. The soil has a loamy sand texture (clay content of 11% between 30 and 60 cm depth) with deeper clay-enriched sand layers (~75 cm) and is classified as Anthrosol according to the World Reference Base for Soil Resources (Dondeyne *et al.*, 2015). On 7–10 April 2010, large replicated mono-genotypic blocks were established over a total of 14.5 ha. Cuttings of 12 selected and commercially available poplar (*Populus*) genotypes (see Table 2 in Broeckx *et al.*, 2012) were planted at a density of 8000 plants ha^{-1} (Fig. 1). Hardwood cuttings were planted in a double-row design with alternating distances of 0.75 and 1.50 m between the rows and 1.1 m between the individuals within each row. The site was neither fertilized, nor irrigated. More information on the site, on the management and on soil characteristics is provided by Broeckx *et al.* (2012) and Verlinden *et al.* (2015).

An extendable eddy covariance and meteorological mast was positioned in the north-eastern part of the plantation (Fig. 1) at the beginning of June 2010. Continuous ecosystem flux and microclimate measurements were then initiated (Zona *et al.*, 2013a). The prevailing wind direction was from the south-west (Fig. 1). Tree-level sap flow (F_s) and stem diameter variation (ΔD) measurements were therefore performed within the flux footprint on the upwind side of the mast. These measurements were confined to a subset of the three genotypes closest to the mast (<15 m) characterized by a different parentage, namely Skado (parentage *Populus trichocarpa* T. & G. × *P. maximowiczii* A. Henry), Oudenberg (parentage *P. deltoides* Bartr. ex Marsh. × *P. nigra* L.) and Grimminge (parentage *P. deltoides* Bartr. ex Marsh. × *P. trichocarpa* T. & G. × *P. deltoides* Bartr. ex Marsh.) (Fig. 1). More details on the origin, the selection and the gender of these species are given by Broeckx *et al.* (2012). Stand-level measurements were performed for the entire growing season of 2011, that is during the second growth year of the plantation and before the first coppice of the plantation (performed on 2–3 February 2012). Tree-level measurements were made during an intensive field campaign from 19 August 2011 (day of the year, DOY 231) until 27 September 2011 (DOY 270). All tree-level measurements were made on single stem trees as trees had not yet been coppiced.

Stand-level measurements and modelling

Climate variables. Climate variables were continuously recorded at the site: air temperature (T_{air}) and relative humidity were recorded on the extendable mast at 5.4 m above the ground surface using Vaisala probes (model HMP 45C; Vaisala, Helsinki, Finland); these data were used to calculate vapour pressure deficit (VPD). Incoming photosynthetically active radiation (PAR, 400–700 nm) was recorded at the same height using a quantum sensor (model LI-190; Li-COR, Lincoln, NE, USA). Precipitation was recorded using a tipping bucket rain gauge (model 3665 R; Spectrum Technologies Inc., Plainfield, IL, USA). Soil water content (SWC) was measured diagonally in the 0–10 cm soil layers and horizontally at a specific depth of 40 cm next to the extendable eddy covariance mast using moisture probes (model TDR CS616; Campbell Scientific, Logan, UT, USA). Water table depth was recorded with a

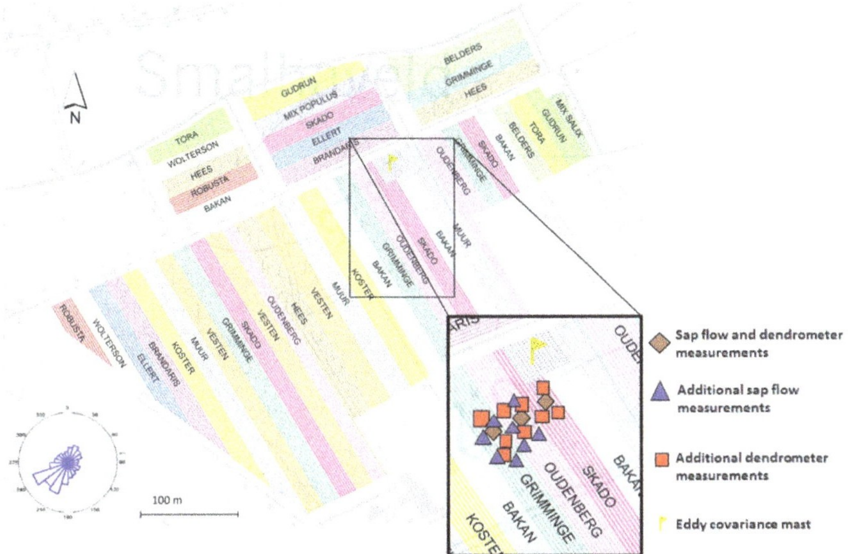

Fig. 1 Map of the study site, indicating the location of the eddy covariance mast and the trees that were equipped with sap flow and/or dendrometer sensors during the 2011 field campaign. The wind rose for the period June 2010–December 2011 is also shown. The differently coloured wide bands indicate the mono-genotypic blocks in the plantation.

pressure transducer (model PDCR 1830; Campbell Scientific) installed in a pipe inserted into the ground to a depth of 1.85 m. Two data loggers (model CR5000 and CR1000; Campbell Scientific) recorded 30-min averages for each environmental variable. If an instrument occasionally failed, the missing environmental variable (T_{air}, relative humidity, or precipitation) was gap-filled using data from nearby standard meteorological stations at 10 and 14 km from the research site. More information on the logging and the gap-filling procedures for the environmental measurements has been published previously (Zona *et al.*, 2013a,b).

Ecosystem evapotranspiration. Ecosystem level fluxes (of carbon, water and energy) were continuously monitored from the eddy covariance mast; in this study, only the water vapour fluxes are considered. High-frequency (10 Hz) measurements of the three-dimensional wind speed components were made using a sonic anemometer (model CSAT3; Campbell Scientific Inc.). Vertical wind velocity was combined with measurements from a closed-path, fast-response gas analyzer (model LI-7000; Li-COR) to measure the latent heat and to calculate evapotranspiration. Additionally, sensible heat fluxes were derived from vertical wind speed and sonic temperature measurements. The sonic anemometer and the inlets of the gas sample lines were positioned at 5.8 m above the ground for the period before 31 August 2011 (DOY 243) and afterwards raised to 6.6 m. Fluxes of latent heat (LE) were calculated using the EdiRe software (R. Clement, University of Edinburgh, UK; www.geos.ed.ac.uk/abs/research/micromet/EdiRe/) from high-frequency data series divided into half-hourly averaging periods. The two-component rotation was applied to set mean lateral and vertical wind velocity components to zero, while the time delay between scalar and vertical wind velocity fluctuations was determined by cross-correlation optimization. A filter rejected data using the following criteria: (i) more than one standard deviation for H_2O vapour and (ii) for quality flags 9 as suggested by Foken & Wichura (1996) and Foken *et al.* (2004). The fluxes of LE were gap-filled using the marginal distribution sampling (MDS) method (implemented in www.bgc-jena.mpg.de/~MDIwork/eddyproc/). This method is adopted by the FLUXNET community as a standardized gap-filling technique. A detailed description of the ecosystem flux measurements at our site has been given by Zona *et al.* (2013a).

Additionally, we calculated reference crop evapotranspiration (ET_0) for the site based on the Penman–Monteith method, as described by Allen *et al.* (1998). ET_0 was integrated either daily or annually and compared with measured and modelled ET to assess the SRC water use relative to a reference crop.

Modelling of the stand water balance. We used the AquaCrop model to determine soil evaporation (E_{soil}) as well as the E_c and ET components of the stand water balance over the growing season. AquaCrop is a crop water productivity and yield response model developed by the FAO that simulates daily biomass production and final crop yield (Table 1, Vanuytrecht *et al.*, 2014). Briefly, E_c is calculated by multiplying ET_0, determined using the Penman–Monteith method, with a crop coefficient (Kc_{Tr}) and a water stress coefficient (K_s). Kc_{Tr} is proportional to the fraction of the soil covered by the canopy (canopy cover, CC) and to $Kc_{Tr,x}$, the maximum crop transpiration coefficient for the specific crop relative to the grass reference surface. The E_{soil} also depends on ET_0 and is further proportional to the fraction of the soil not covered by the canopy (1-CC), to K_{ex} which is the maximum soil evaporation coefficient for a fully wet and unshaded soil surface, and to K_r which is the evaporation reduction coefficient that reduces transpiration when soil water content is low. In this study, we used the default value of 1 for both K_s and K_r as no water stress

Table 1 Definition of parameters, coefficients and variables of climate of stand- and tree-level measurements, of sap flow scaling and of the AquaCrop model

Parameter/variable/coefficient	Symbol	Units	Value	Source
Climate variables				
Air temperature	T_{air}	°C	Variable	
Vapour pressure deficit	VPD	kPa	Variable	
Photosynthetically active radiation	PAR	$\mu mol\ m^{-2}\ s^{-1}$	Variable	
Soil water content	SWC	$m^3\ m^{-3}$	Variable	
Stand-level measurements				
Reference crop evapotranspiration	ET_0	$mm\ day^{-1}$	Variable	Allen *et al.* (1998)
Evapotranspiration	ET	$mm\ day^{-1}$	Variable	
Transpiration component of evapotranspiration	E_c	$mm\ day^{-1}$	Variable	
Tree-level measurements				
Stem diameter fluctuations	ΔD	μm	Variable	
Maximum daily shrinkage	MDS	μm	Variable	
Day- and night-time stem diameter fluctuation over time	$\Delta D/\Delta t$	$\mu m\ h^{-1}$	Variable	
Sap flow scaling variables				
Average sapwood area for all trees equipped with dendrometers	$A_{s\text{-}avg}$	m^2	Variable	
Sapwood area of the sample tree	A_s	m^2	Variable	
Ground surface area per tree	SA	m^2	Table 2	Broeckx *et al.* (2015)
Sapwood area scaled transpiration component of evapotranspiration	$E_{c\text{-}sapwood}$	$mm\ day^{-1}$	Variable	
Leaf area scaled transpiration component of evapotranspiration	$E_{c\text{-}leaf}$	$mm\ day^{-1}$	Variable	
Genotype-specific leaf area index	LAI	$m^2\ m^{-2}$	Table 2	Broeckx *et al.* (2015)
Leaf area of the tree equipped with sap flow sensor	LA	m^2	Table 2	
AquaCrop model				
Soil evaporation	E_{soil}	$mm\ yr^{-1}$	177	Raes *et al.* (2012)
Transpiration component of evapotranspiration	E_c	$mm\ yr^{-1}$	259	Raes *et al.* (2012)
Evapotranspiration	ET	$mm\ yr^{-1}$	437	Raes *et al.* (2012)
Reference crop evapotranspiration	ET_0	$mm\ yr^{-1}$	531	Allen *et al.* (1998)
Maximum soil evaporation coefficient for fully wet and not shaded soil surface	K_{ex}	–	1.1	Raes *et al.*(2012)
Evaporation reduction coefficient	K_r	–	1	Raes *et al.*(2012)
Green canopy cover	CC	%	Variable	Raes *et al.*(2012)
Actual canopy cover adjusted for micro-advective effects	CC_{star}	%	Variable	Raes *et al.*(2012)
Coefficient for maximum crop transpiration for well-watered soil and complete canopy cover	$Kc_{Tr,x}$	–	Variable	Raes *et al.*(2012)
Crop transpiration coefficient	K_{cTr}	–	1.2	Broeckx *et al.*(2015); Zenone *et al.* (2015)
Soil water stress coefficient	K_s	–	1	Raes *et al.*(2012)
Soil surface covered by an individual seedling at 90% emergence	CC_s	cm^2	15	Field data (webcam images)
Number of plants per hectare	Den	$plants\ ha^{-1}$	8000	Broeckx *et al.*(2015)
Initial canopy cover at time = 0	CC_0	%	0.12	Raes *et al.*(2012)
Maximum canopy cover	CC_{max}	$m^2\ m^{-2}$	0.67	Broeckx *et al.*(2015)
Increase in canopy cover	CGC	Fraction day^{-1}	0.058	Broeckx *et al.*(2015)
Decrease in canopy cover	CDC	Fraction day^{-1}	0.075	Broeckx *et al.*(2015)

conditions were observed for the SRC plantation. All parameters relevant to the AquaCrop model, including the ones not mentioned above, are listed and explained in Table 1. Interception evaporation is not considered in the AquaCrop model. The ET is calculated as the sum of E_{soil} and E_c (Vanuytrecht *et al.*, 2014). We performed a sensitivity analysis of the AquaCrop model for our site conditions. The results of this analysis were used to evaluate the effect of relative changes of a number of distributed parameters on the model outputs. Details on the simulated processes have been extensively documented in a set of publications at the model's release (Hsiao *et al.*, 2009; Raes *et al.*, 2009; Steduto *et al.*, 2009) as well as in the FAO irrigation and drainage paper # 66 (Steduto *et al.*, 2012) and in the reference manual (Raes *et al.*, 2012).

Tree-level measurements

Sap flow measurements. The sap flow rate (F_s, kg h^{-1}) of individual trees was measured using the heat balance principle established in previous studies of SRC trees (e.g. Hinckley *et al.*, 1994; Hall *et al.*, 1998; Tricker *et al.*, 2009; Petzold *et al.*, 2011). F_s was monitored continuously during the entire field campaign, using three Dynamax sensors (model SG-EX 25; Dynamax Inc., Houston, TX, USA), one on a tree of each genotype. The sensors were mounted at a height of 50 cm above the base of the stem. The sensors were thermally insulated from the environment with an insulation sleeve and several layers of aluminium foil wrapped around the sensor. F_s was calculated according to the standard procedure for heat balance sensors, described in detail by Sakuratani (1981) and Baker & Van Bavel (1987). In this study, we additionally tested for the heat storage effect (Steppe *et al.*, 2005), but we did not observe early morning spikes in the F_s data. As the poplars were fast growing, we accounted for the increase in stem surface area during the F_s measurements using the increase in stem diameter recorded by high-resolution dendrometers (see further below).

To validate the F_s measurements performed at 50 cm height and to account for variation in F_s among individuals, we also installed F_s sensors (models SG-EX 16 and model SG-EX 19; Dynamax Inc.) on four additional trees per genotype for the last twelve days of the intensive field campaign (15 September 2011–27 September 2011, DOY 258–270). Due to the limited number of sensors, these measurements were only performed on the Oudenberg and Grimminge genotypes and on a smaller stem section higher up the stem at approximately 2.5 m above the stem base. The basal stem diameter of the trees at the start of the additional F_s measurements ranged from 1.63 to 2.16 cm and from 1.51 to 1.84 cm for Oudenberg and Grimminge, respectively. Data from sap flow sensors were collected at 30-s intervals with a data logger (model CR800; Campbell Scientific) and 30-min averages recorded.

The relationship between F_s and VPD was analysed according to Tang *et al.* (2006) and Ewers *et al.* (2001) by fitting the following exponential saturation equation:

$$F_s = a(1 - e^{-b\mathrm{VPD}}) \times 24\frac{h}{d} \qquad (1)$$

with a (kg day^{-1}) and b (kPa^{-1}) corresponding to the fitted coefficients, and 24 h day^{-1} corresponding to a time conversion factor. The relationship between F_s and PAR was analysed using a linear regression. For both analyses, F_s was summed per day and expressed relative to the daytime-averaged VPD (with daytime defined as periods when PAR >5 μmol m^{-2} s^{-1}) or the PAR summed per day.

Stem diameter measurements. Stem diameter fluctuation (ΔD) was continuously measured using automatic point dendrometers (model ZN11-O-WP; Natkon, Hombrechtikon, Switzerland) installed with a ring-shaped carbon frame at a height of 22 cm. Sensors were installed on four trees per genotype (12 sensors in total); one of these trees was also equipped with an F_s sensor as described above. Trees were selected to be representative of the whole range of stem diameters measured during an extensive inventory (n = 1742) performed in February 2011. Data from the dendrometers were collected at 30-s intervals with a data logger (model CR800; Campbell Scientific) and 30-min averages recorded. ΔD was expressed relative to the start of the measurement campaign, by setting the initial stem diameter to zero.

Changes in the stem water status were characterized by calculating the maximum daily shrinkage (MDS) as the difference between the maximum and minimum values of stem diameter during the day (Giovanelli *et al.* 2007). We also determined the day- and night-time increases in stem diameter over time ($\Delta D/\Delta t$, with daytime defined as periods when PAR >5 μmol m^{-2} s^{-1}) to determine different patterns in ΔD of trees of the different genotypes. We limited the combined analysis of F_s, ET and ΔD to the first week of the measurement campaign (DOY 231–236) to clarify the links between the different variables. This period was characterized by a strong variation in VPD, as large dynamics in T_{air} were observed, leading to strong dynamics in F_s, ET and ΔD.

Scaling of sap flow from tree to stand level. Two approaches were applied to scale F_s to canopy E_c, which was then compared to ET. In a first approach, F_s was scaled to E_c by multiplying it by the ratio of the genotype-specific leaf area index (LAI, m^2 m^{-2}) of the whole canopy to the total leaf area (LA, m^2) of the individual tree equipped with a sap flow sensor:

$$E_{c\text{-leaf}} = F_s \frac{\mathrm{LAI}}{\mathrm{LA}} 24\frac{h}{d} \qquad (2)$$

with $E_{c\text{-leaf}}$ (mm day^{-1}) corresponding to the leaf area scaled E_c and 24 h day^{-1} corresponding to a time conversion factor. LA was determined from genotype-specific regressions relating leaf area with leaf length × leaf width (R^2 ≥0.99 for all genotypes) for a minimum sample of 25 leaves spanning the whole leaf size range from trees neighbouring those equipped with sap flow sensors. Harvested leaves were scanned and analysed using ImageJ software (NIH, Bethesda, MD, USA). Leaf length and width of all leaves on the trees equipped with sap flow sensors were measured before (i.e. 10 August 2011, DOY 222) and after the intensive field campaign (i.e. 28 September 2011, DOY 271) to account for the change in LA (Table 2). LAI was monitored for different locations in the study site for three occasions during the period July–September 2011 [on 22 July (DOY 203), on 2 September (DOY 245) and on 23 September (DOY 266)] using cross-calibrated plant canopy analyzers (models LAI-2000 and LAI-2200; Li-COR). We selected data from the measurements performed closest to the F_s sensors, which we assumed to best represent the LAI in the footprint of the eddy covariance measurements (Table 2). More information on the LAI measurements has been published previously (Broeckx *et al.*, 2012).

In a second approach F_s was scaled to E_c by multiplying it by the ratio of the average sapwood area for all trees equipped with dendrometers ($A_{s\text{-avg}}$, m^2) to the sapwood area of the sample tree (A_s, m^2) and per unit of average ground surface area per tree (SA, m^2):

$$E_{c\text{-sapwood}} = F_s \frac{A_{s\text{-avg}}}{A_s \mathrm{SA}} 24\frac{h}{d} \qquad (3)$$

with $E_{c\text{-sapwood}}$ (mm day^{-1}) corresponding to the sapwood area scaled E_c and 24 h day^{-1} corresponding to a time conversion factor. Both A_s and $A_{s\text{-avg}}$ were estimated using the dendrometer data, assuming that for these young trees the entire stem consisted of functional sapwood except for a small fraction of bark tissue. SA was estimated for each genotype based on the

Table 2 Parameters used for scaling sap flow rate to canopy transpiration during the field campaign from 19 August (DOY 231)–27 September (DOY 270) 2011, according to a leaf-based approach [leaf area (LA) or genotype-specific leaf area index (LAI)] and a stem-based approach [ground surface area per tree (SA)] for three poplar genotypes

	LA (m^2)		LAI (m^2 m^{-2})			
Genotype	DOY 222	DOY 271	DOY 203	DOY 245	DOY 266	SA (m^2)
Skado	1.2	1.4	1.9	2.1	2.1	1.2
Oudenberg	1.2	1.5	1.6	1.9	2.1	1.3
Grimminge	1.8	2.0	1.6	2.4	2.2	1.4

DOY, day of the year. See Materials and Methods for additional description.

spacing and the consistent stocking of trees in the mono-genotypic block design of the site (Table 2). Finally, both $E_{c\text{-leaf}}$ and $E_{c\text{-sapwood}}$ were averaged for the three genotypes and summed per day for comparison with the daily sums of ecosystem ET.

Statistical analysis

For the stand-level measurements, we used Pearson correlation and linear regression analysis to determine correlations and regressions between measured and modelled data. For the intensive field campaign, the rate of change in diameter, $\Delta D/\Delta t$, was analysed using a repeated-measures ANOVA model with genotype (n = 3) and night- or daytime period (n = 11) as fixed factors and individual tree (n = 4) treated as random factor. A similar model was used to analyse MDS; however, data were confined to the daytime period (n = 5). The Akaike information criterion correcting for small sample sizes (AICc) was used to determine the covariance structure that best estimated the correlation among individual trees over time. Treatment means were compared using Fisher's least significance difference test. ANOVA analyses were performed using the mixed model procedure (PROC MIXED) of SAS (Statistical Analysis System, Cary NC, USA) with α = 0.05.

Results

Environmental conditions during the measurement campaign

Variable weather conditions were experienced during the measurement campaign (Fig. 2a,b). For instance, VPD (Fig. 2a) varied strongly during the measurement period leading to a range of E_c and ET rates (Fig. 2c). The maximum VPD of 3.6 kPa was observed on 27 June 2011 (DOY 178), which coincided with maximum modelled ET and E_c, and measured ET. Precipitation patterns were dynamic as relatively dry periods alternated with periods of rainfall (Fig. 2b). In response to precipitation events, SWC measured at 0–10 cm depth increased, together with a less pronounced increase in SWC at 40 cm depth and a rising water table (Fig. 2b).

Yearly stand water balance

Both modelled and measured daily ET showed similar dynamics (Pearson's correlation coefficient: 0.861) that were strongly related to changes in VPD (Fig. 2c). Modelled daily E_c started to increase from mid-April onwards up to a maximum of 3.9 mm day^{-1} at 27 June 2011. At the end of the growing season, E_c decreased from late September onwards, as leaf fall started around that period. The average modelled daily E_c for the growing season was 1.3 mm day^{-1}.

Summed over 2011 modelled ET was 437 mm, which was 87 mm higher than measured ET (350 mm) but still 94 mm lower than ET_0 (531 mm, Fig. 2d). Cumulative E_c was 259 mm, representing 59% of ET over the whole year, as derived from the modelled data. When considering the actual growing season (from mid-April to late September), E_c represented 69% of ET. Total modelled E_{soil} was smaller than E_c (177 mm) and accounted for 41% of the total ET. Total cumulative measured precipitation was 669 mm, which was higher than the total ET and ET_0. Run-off at our site was negligible for the stand water balance. The remainder of the precipitation was lost to groundwater leaching. The results from the sensitivity analysis of the model parameters showed that CGC had the largest impact on modelled E_{soil}, E_c and ET followed by $Kc_{Tr,x}$ (Table 3). In contrast, CDC had a limited impact on modelled E_{soil}, E_c and ET. Overall, the deviation in model output observed during the sensitivity analysis ranged from −22.4% to +14.1%. Changes in parameter values had the largest impact on E_c, except for the parameter K_{ex} which only impacted E_{soil}.

Sap flow

The highest F_s rates were observed for Oudenberg (Fig. 3b), as compared to Skado (Fig. 3a) and Grimminge (Fig. 3c), with a maximum F_s rate of 0.3 kg h^{-1} on DOY 247. Skado had the lowest F_s rates with a maximum F_s rate of 0.2 kg h^{-1}; this occurred on the same day as the maximum F_s for Oudenberg. The additional F_s measurements performed for both Oudenberg and Grimminge between 15 September (DOY 258) and 27 September (DOY 270) 2011 confirmed the higher F_s rates of Oudenberg as compared to Grimminge (data not shown). F_s measured with these sensors installed higher up the tree varied in synchrony with the observed patterns in F_s obtained with sensors installed at the stem base. For Oudenberg, F_s at the stem base was within the range of F_s rates measured during the additional campaign. For Grimminge, the daytime F_s measured

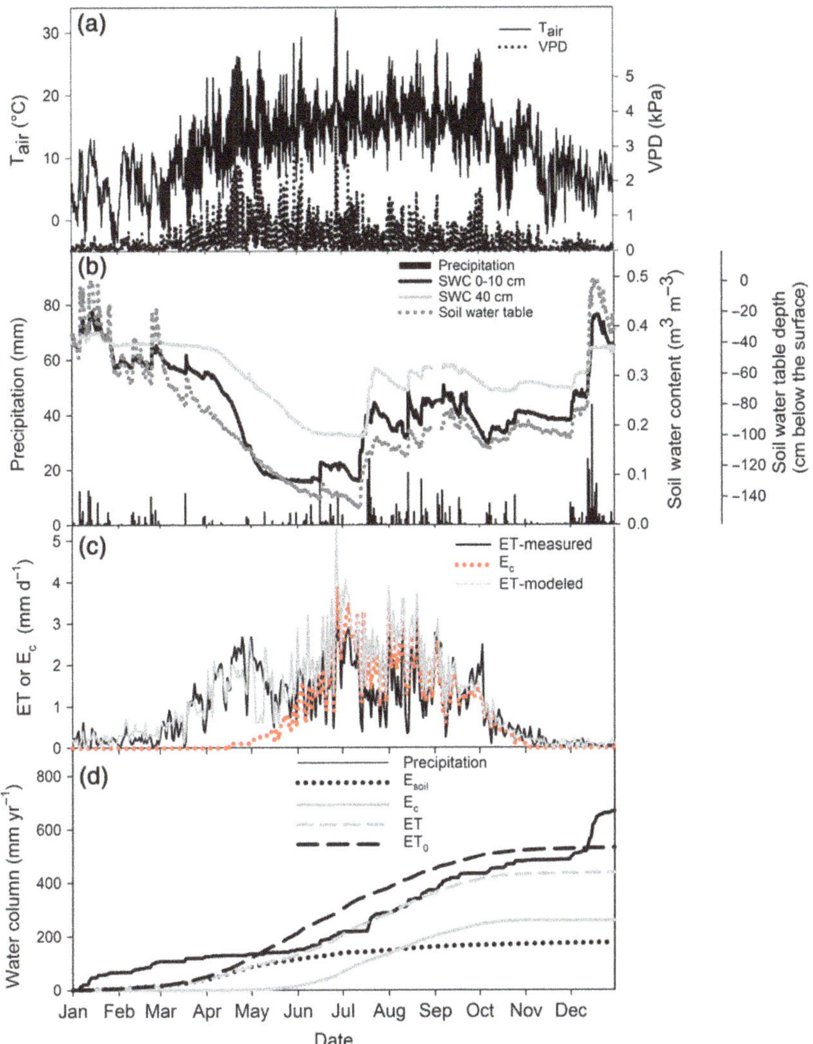

Fig. 2 Time course of meteorological variables of measured and modelled evapotranspiration (ET) and of the cumulative evaporative components during the year 2011. (a) Air temperature (T_{air}; solid line) and vapour pressure deficit (VPD; dotted line); (b) daily summed precipitation (black bars), soil water content (SWC) measured at 0–10 cm (black solid line) and 40 cm (grey solid line) depth, and water table depth (dark grey dotted line); (c) daily measured ET (solid black line), modelled ET (solid grey line) and modelled canopy transpiration (E_c, dotted red line). (d) Cumulative precipitation (black solid line), soil evaporation (E_{soil}, black dotted line), E_c (grey solid line), ET (grey dashed line) and reference evapotranspiration (ET_0, black dashed line).

higher up the stem was on average 0.04 kg h^{-1} lower as compared to the F_s measured at 50 cm.

Daily sums of F_s were significantly correlated with daytime-averaged VPD (Fig. 4a, $P < 0.001$). The maximum F_s rate, estimated by coefficient '*a*' in Eqn (2), was higher for Oudenberg (3.2 kg h^{-1}) than for Grimminge (2.6 kg h^{-1}) and Skado (2.3 kg h^{-1}) (Fig. 4a). A similar genotypic difference was observed for the F_s – PAR regression (Fig. 4b).

Daily stem diameter variation

Growth during the 39-day measurement campaign significantly differed among the three genotypes. Average (±SE) stem diameter increase during the intensive field campaign was significantly higher ($P < 0.01$) for Skado (0.8 ± 0.1 mm) as compared to Oudenberg (0.4 ± 0.1 mm) and Grimminge (0.4 ± 0.1 mm). More interestingly, daily ΔD variations were observed for all three genotypes (Fig. 5) as trees tended to shrink during daytime when F_s was high and swelled during the night when they replenished their water reserves. The Skado tree, equipped with both F_s and dendrometer sensors, did not show a significant shrinkage during the day (except at the onset of F_s during DOY 231, Fig. 5a). The Oudenberg tree, and to a lesser extent the Grimminge tree, significantly shrank during days with high F_s rates (Fig. 5b,c).

Table 3 Sensitivity analysis of the parameters used to model soil evaporation (E_{soil}), the transpiration component of evapotranspiration (E_c) and evapotranspiration (ET) for the multigenotype SRC over the 2011 growing season. The analysis evaluates the effect of changes in a range of site-specific realistic parameter values on the model output. Given are the % deviation at the minimum parameter value (min % deviation), the % of deviation at the maximum parameter value (max % deviation) and the total (total % deviation) deviation of modelled E_{soil}, E_c and ET relative to the base value used in the study

	CC_{max}	CC_0	K_{ex}	$Kc_{Tr,x}$	CGC	CDC
Value used in study	0.67	0.12	1.1	1.2	0.058	0.075
Minimum value	0.57	0.1	1	1	0.048	0.065
Maximum value	0.77	0.4	1.2	1.3	0.068	0.085
Min % deviation E_{soil}	+5.0	+0.6	−6.3	+0.4	+8.8	−0.3
Max % deviation E_{soil}	−4.9	−4.5	+6.1	−0.2	−5.7	+0.3
Min % deviation E_c	−10.8	−1.5	0.0	−16.7	−22.4	+0.9
Max % deviation E_c	+9.8	+11.3	0.0	+8.3	+14.1	−0.7
Min % deviation ET	−4.2	−0.6	−2.6	−9.6	−9.4	+0.4
Max % deviation ET	+3.7	+4.7	+2.5	+4.7	+5.9	−0.3
Total % deviation for E_{soil}	9.9	5.1	12.4	0.7	14.4	0.6
Total % deviation for E_c	20.6	12.8	0.0	25.0	36.5	1.6
Total % deviation for ET	7.9	5.4	5.1	14.3	15.3	0.7

CC_{max}, maximum canopy cover; CC_0, initial canopy cover at time = 0; K_{ex}, maximum soil evaporation coefficient for fully wet and not shaded soil surface; $Kc_{Tr,x}$, crop transpiration coefficient; CGC, increase in canopy cover; CDC, decrease in canopy cover.

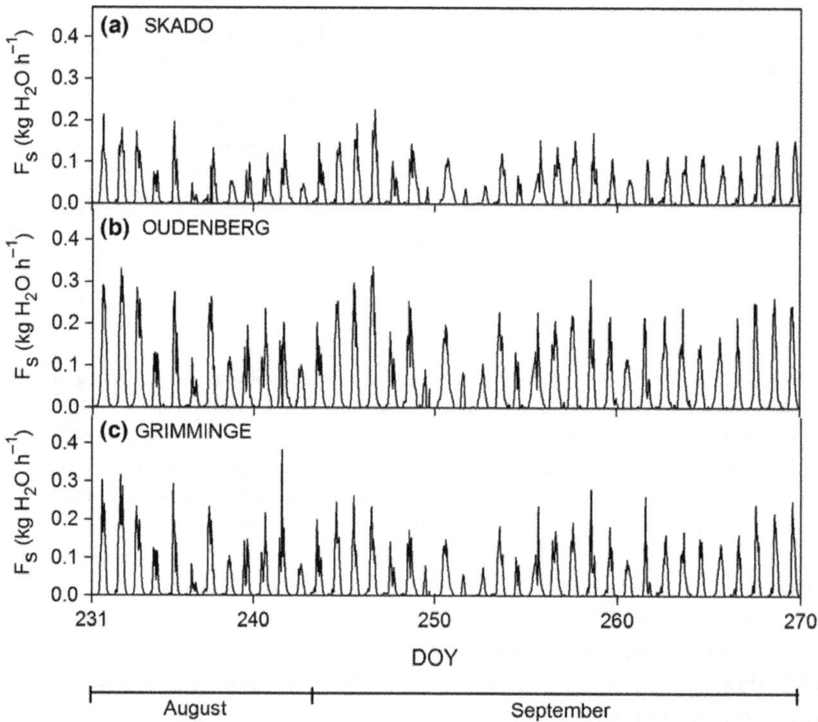

Fig. 3 Time course of sap flow (F_s) monitored using heat balance sensors during the intensive field campaign [19 August (DOY 231)–27 September (DOY 270) 2011] for genotypes (a) Skado, (b) Oudenberg and (c) Grimminge. DOY, day of the year.

Similar patterns in ΔD were observed for the other trees equipped with dendrometers (Fig. 6). Regardless of the genotype, stem diameter growth was observed for all trees of the different genotypes (Fig. 6b–d), but differences in the day- and night-time ΔD were observed among genotypes (Fig. 6e). The change in $\Delta D/\Delta t$ for both day- and night-time confirmed the intergenotypic differences in tree water use as observed with F_s measurements. Oudenberg showed a consistently higher $\Delta D/\Delta t$ during the night as compared to the daytime (Fig. 6e). Genotypes Grimminge and in particular Skado showed a larger variability in the

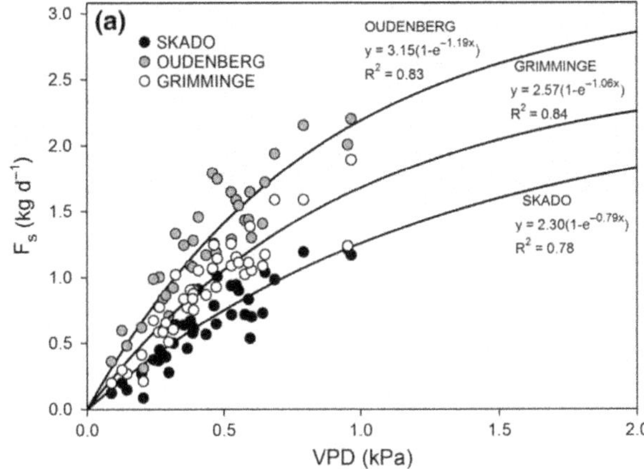

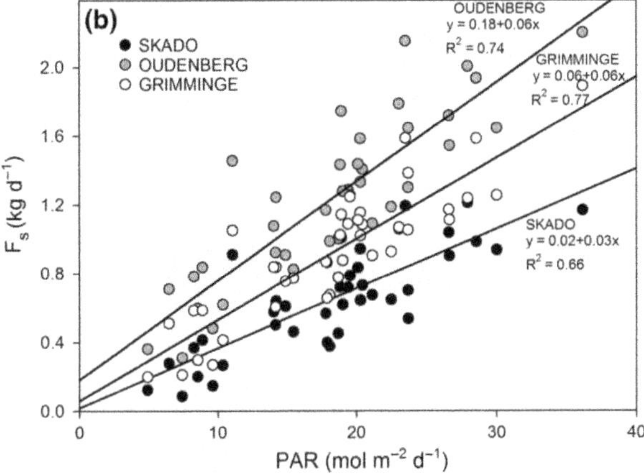

Fig. 4 Daily summed tree sap flow (F_s) measured at stem base for genotypes Skado (black dots), Oudenberg (grey dots) and Grimminge (white dots) both as a function of (a) daytime average VPD and (b) photosynthetically active radiation (PAR). Lines are exponential saturation curves [Eqn.: $y = a(1 - e^{-b\ \mathrm{VPD}})$] and linear curves (Eqn.: $y = ax + b$) in (a) and (b), respectively. VPD used to calculate daytime averages was selected when PAR >5 μmol m^{-2} s^{-1}.

day- and night-time patterns of $\Delta D/\Delta t$, resulting in significant differences in daytime $\Delta D/\Delta t$ among genotypes on DOY 232 ($P = 0.0447$) and DOY 233 ($P = 0.0458$).

Significant differences in MDS ($P < 0.01$) were observed among the three genotypes (Fig. 6f). Oudenberg showed the highest MDS (maximum 10.5 μm), significantly higher than Skado (maximum 4.7 μm) and Grimminge (maximum 5.5 μm) on both DOY 231 ($P < 0.05$) and DOY 233 ($P < 0.05$). Small differences in MDS among the genotypes were observed during the days when lower ET (Fig. 6a) and F_s (Fig. 5) occurred, that is DOY 234 and 235.

Scaling of sap flow from tree to stand level

Scaling of F_s to $E_{\text{c-sapwood}}$ and $E_{\text{c-leaf}}$ resulted in daily average canopy transpiration rates of 1.1 and 1.3 mm day^{-1}, respectively (Fig. 7a). Total $E_{\text{c-sapwood}}$ and $E_{\text{c-leaf}}$ for the measurement campaign were 43.5 and 50.3 mm, respectively. Both values were lower than the total ET from eddy covariance, that is 59.9 mm. For most of the days during the field campaign, ET was higher than both $E_{\text{c-sapwood}}$ and $E_{\text{c-leaf}}$, resulting in a ratio of E_c/ET lower than unity (Fig. 7b). Assuming that ET and E_c were comparable, transpiration accounted for 74% and 86% of total ET, using average $E_{\text{c-sapwood}}$/ET and $E_{\text{c-leaf}}$/ET averaged over the period of the intensive field campaign as an estimate of the contribution of E_c to ET, respectively. Overall, the leaf area-based approach tended to overestimate E_c relative to ET (i.e. E_c/ET >1) more than the sapwood area-based scaling of F_s to E_c. The differences between ET and E_c resulted from E_{soil} and from the transpiration of understory weed vegetation, accounting for 26% and 14% of total ET when estimating E_c using the sapwood and leaf area-based approach, respectively.

Discussion

Stand water balance

It has been argued that SRCs might have a strong impact on the regional water cycle. The stand water balance analysis at our site suggests that the impact of the SRC on the regional water cycle was not negative. First, our site was not water limited, as precipitation was around 53% and 26% higher than annual ET and ET_0, respectively. Secondly, for the year 2011, ET of our poplar SRC was around 18% lower as compared to ET_0, suggesting that the site used less water as compared to a reference grassland. Thirdly, an increase in the ecosystem water use efficiency over the year 2011 (reported earlier by Broeckx *et al.*, 2014b) suggested that the poplars at our site could reduce the transpiration water loss per unit of fixed carbon by regulating stomatal opening. Caution is advised in generalizing these results to other SRCs. While studies on SRC water use in the Czech Republic (Fischer *et al.*, 2013), in Mongolia (Hou *et al.*, 2010) and in the USA (Nagler *et al.*, 2007) showed similar results, almost a same number of studies (see references in Fischer *et al.*, 2013) have shown that SRCs across the globe consume more water as compared to traditional agricultural crops or grasslands. Therefore, site location, local climatic conditions, species considered and the age of the plantation are important factors that determine the actual SRC water use and stand water balance (IEA-Bioenergy, 2011).

In line with results from previous studies (e.g. Persson & Lindroth, 1994; Fischer *et al.*, 2013), both the modelling approach of E_c and the scaling approach of F_s measurements to stand level showed that E_c represented the largest component of ET. The average daily

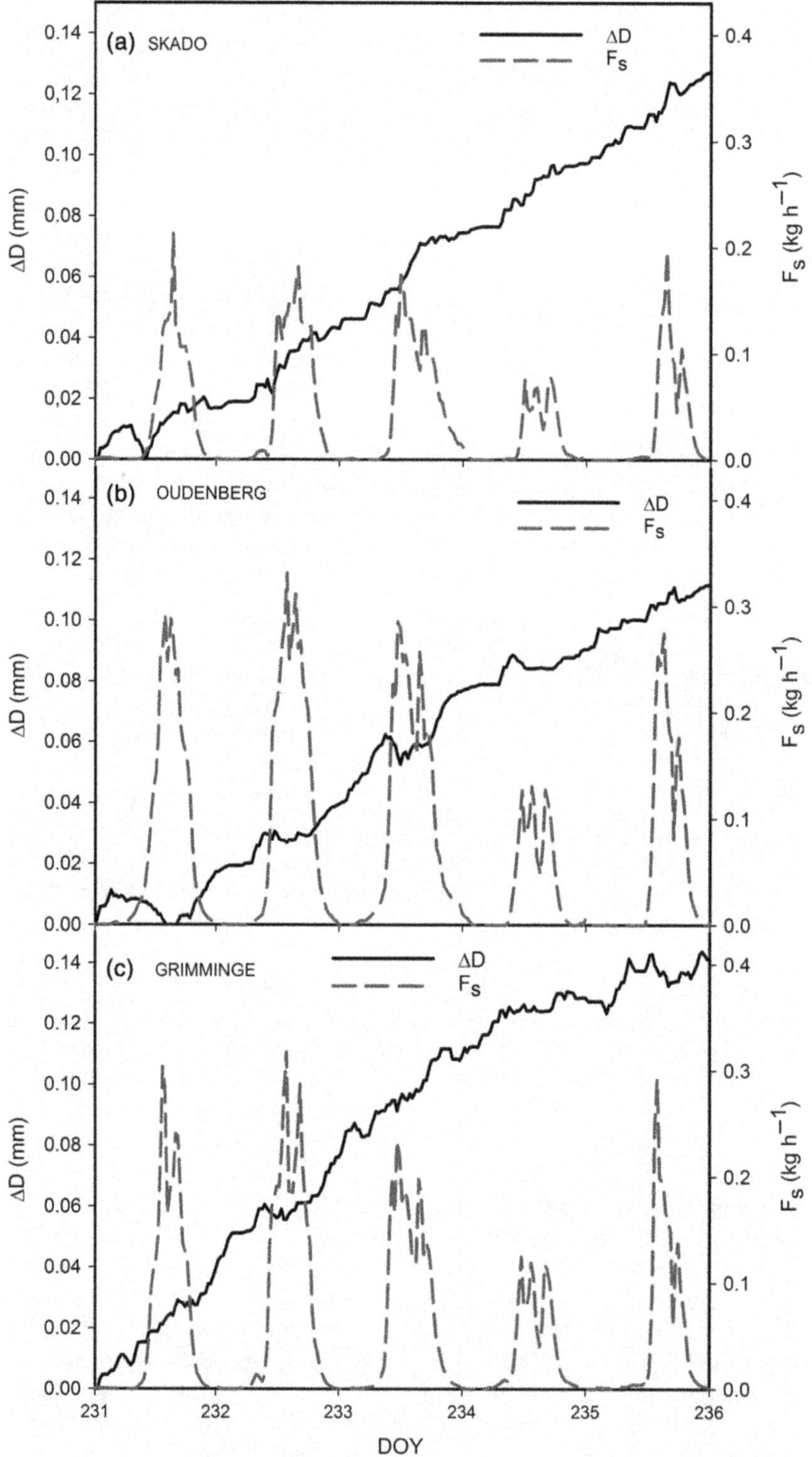

Fig. 5 Time course of short-term stem diameter variations (ΔD; solid line) and of sap flow (F_s; dashed line) both measured at stem base for the genotypes (a) Skado, (b) Oudenberg and (c) Grimminge during the period 19 August (DOY 231)–23 August (DOY 236) 2011. ΔD is expressed relative to the start of the measurement campaign, by setting the initial stem diameter to zero. DOY, day of the year.

E_c values for our site (1.3, 1.1 and 1.3 mm day^{-1}, for modelled, sapwood and leaf area scaled E_c, respectively) were within the lower end of the range of 1–8 mm day^{-1} reported for poplar stands of different genotypes, stand age and geographic locations in temperate climate zones (Meiresonne *et al.*, 1999). For an

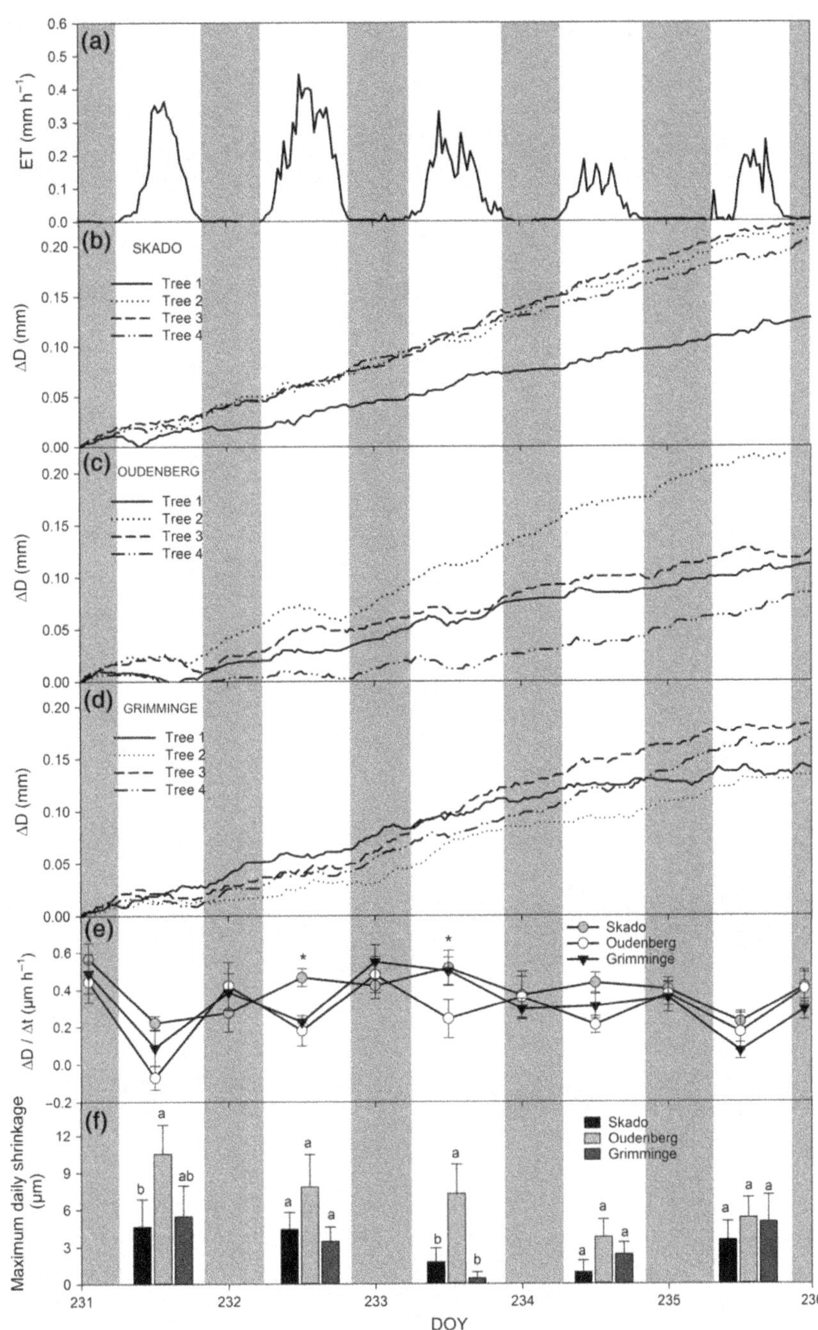

Fig. 6 Time course of evapotranspiration (ET) measured using eddy covariance (a), of stem diameter variations (ΔD) (b–d), of day- and night-time stem diameter variation over time ($\Delta D/\Delta t$; n = 4 per genotype) (e) and of daytime maximum daily shrinkage (MDS) (f). (a) ET; (b–d) ΔD for four trees per genotype; and (e) $\Delta D/\Delta t$ for genotypes Skado (grey dot), Oudenberg (white dot) and Grimminge (black triangle) during the period 19 August (DOY 231)–23 August (DOY 236) 2011. $\Delta D/\Delta t$ was calculated as the difference in diameter at the start and at the end of a day- or night-time period divided by the duration of that period. For data selection, daytime was taken as the period when photosynthetically active radiation (PAR) >5 μmol m^{-2} s^{-1}. Asterisks indicate significant (P < 0.05) differences in $\Delta D/\Delta t$. (f) MDS for genotypes Skado (black bar), Oudenberg (grey bar) and Grimminge (dark grey bar). MDS was calculated as the difference in maximum and minimum stem diameter during the daytime. Different letters indicate significant (P < 0.05) differences in MDS among genotypes. Grey shaded areas represent night-time periods. DOY, day of the year.

irrigated *P. trichocarpa* × *P. deltoides* plantation in the Pacific Northwest of the USA, an average E_c of 4 mm day^{-1} was observed (Kim *et al.*, 2008). Sap flow – measured with the same heat balance principle as in the present study – provided a growing season average E_c of 2.2 mm day^{-1} for a *P. maximowiczii* × *P. nigra*

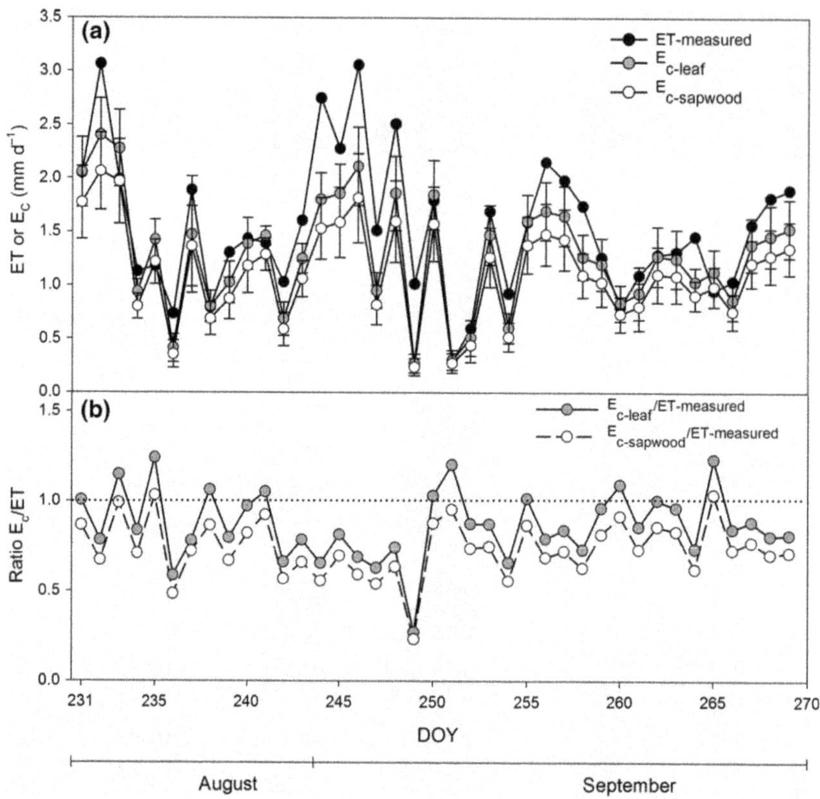

Fig. 7 Time course of (a) daily summed evapotranspiration (ET) and daily summed canopy transpiration (E_c) estimated using sapwood area-based ($E_{c\text{-sapwood}}$) and leaf area-based ($E_{c\text{-leaf}}$) scaling of sap flow, and (b) the ratio of $E_{c\text{-sapwood}}$ or $E_{c\text{-leaf}}$ over ET during the intensive field campaign (19 August (DOY 231)–27 September (DOY 270) 2011). (a) ET (black dots), $E_{c\text{-leaf}}$ (grey dots) and $E_{c\text{-sapwood}}$ (white dots). Error bars represent standard errors (SEs). (b) Ratio of $E_{c\text{-leaf}}$ over ET (grey dots) and $E_{c\text{-sapwood}}$ over ET (white dots). The dotted line indicates a ratio equal to one which implies that $E_{c\text{-sapwood}}$ or $E_{c\text{-leaf}}$ is equal to ET. DOY, day of the year.

plantation in southern Germany (Petzold *et al.*, 2011). Both last mentioned studies on poplar E_c were performed for either single or multishoot stands, while the 2-year-old trees of our site were still single stemmed and had not reached full canopy closure yet (Broeckx *et al.*, 2012). Moreover, the trees considered in previous studies were older and probably had a larger leaf and sapwood conducting area than the trees at our site; this partly explains the higher stand transpiration values found in the literature.

Our E_c/ET estimates based on modelling (69%) and F_s scaling (74% and 86%) were consistent with other published findings for different SRC cultures. A recent review on bioenergy water requirements showed that the average absolute water use of tree-based SRC was 618 mm per year, with 75% of this amount directly transpired by the trees (King *et al.*, 2013). Likewise, a literature survey (Fischer *et al.*, 2013) and simulations for a 'hypothetical' SRC (Grip *et al.*, 1989) showed that transpiration contributed on average to 80% and 71% of the seasonal ET of SRC, respectively. For an experimental pine and switchgrass intercrop forestry system in North Carolina (USA), transpiration modelled over 3 years was on average 62% of ET (Albaugh *et al.*, 2014).

Uncertainties related to estimating stand-level water use

A number of uncertainties may arise when scaling up data from the individual tree to the stand and levels or when modelling stand water balance components. These uncertainties were, however, considered to be minimized in our case for the following reasons.

Uncertainties can be associated with the eddy covariance measurements of ecosystem fluxes (e.g. Baldocchi, 2003). However, for our site, the energy balance closure (based on the assessment of net radiation, latent and sensible heat flux densities, soil heat flux density and energy storage) was 93% in 2011 (Zona *et al.*, 2013a). This value therefore shows the good performance of the eddy covariance system in measuring fluxes at our site during the year 2011 when our measurements were performed. In addition, potential mismatches between spatial footprints of F_s and ET measurements, which depend on the wind direction, were minimized by measuring F_s close to the mast and on the upwind side for the prevailing wind direction.

Uncertainties are also associated with the methods used for scaling F_s. On the one hand, for the sapwood area-based scaling of F_s, we assumed that the whole stem cross section consisted of conducting sap wood; however, this approach was successfully used for scaling F_s to stand level for a *P. nigra* × *P. maximimowczii* SRC in Wisconsin, USA (Zalesny *et al.*, 2006) such that uncertainties were probably limited for this approach. On the other hand, uncertainties were clearly associated with the LA scaling of E_c, leading to a more frequent overestimation of E_c than the sapwood area scaling approach (i.e. E_c/ET frequently higher than 1). These uncertainties likely resulted from both the genotype-specific allometric reconstruction of LA and the heterogeneity in LAI around the mast. Therefore, at our site, a sapwood area scaling approach was preferable for scaling F_s to E_c. Future studies should combine different techniques across different spatial and temporal scales and a range of environmental conditions (Fischer *et al.*, 2013).

In addition to the uncertainties depending on the approach used for scaling F_s, F_s can also vary considerably among individuals of a given genotype, which means that replicates are needed to give an accurate estimate of the mean flux (Oren *et al.*, 1998; Oishi *et al.*, 2008). As F_s measurements in our study were limited to only 39 days, we used the AquaCrop model to estimate E_c for the whole growing season. However, the choice of the model parameter values to determine the yearly stand water balance was also prone to uncertainty. For instance, model parameters used to describe canopy development will have an impact on both the transpiration and the soil evaporation component of the stand water balance. To this end, an additional model sensitivity analysis could be used to evaluate the effect of parameter changes on the model outputs, as performed for the AquaCrop model in our study.

Intergenotypic differences in tree water use

Previously reported measurements at this site, made in the same year 2011 (Broeckx *et al.*, 2014a), revealed differences in stomatal conductance among genotypes to be strongly related to differences in F_s observed during our study. According to Broeckx *et al.* (2014a), Oudenberg showed the highest average stomatal conductance (459 mmol s^{-1} m^{-2}), while Grimminge (319 mmol s^{-1} m^{-2}) and in particular Skado (255 mmol s^{-1} m^{-2}) showed substantially lower average stomatal conductances for the period of the intensive field campaign. Therefore, genetic differences in the control of stomatal opening were an important factor that determined different F_s rates among genotypes. Once stomata are open, VPD is the driving force for F_s, as illustrated by the strong F_s – VPD relationships for all genotypes.

In addition to root water uptake and F_s, stem tissue water storage is an important factor in tree water relations (Zweifel *et al.*, 2000). The pattern of stem diameter variation in response to the replenishment of water storage has been previously observed for poplars under controlled conditions (Giovannelli *et al.*, 2007) and for other species (e.g. Zweifel *et al.*, 2000, 2001; Conejero *et al.*, 2007; Kocher *et al.*, 2012). At our site, the highest shrinking and swelling were observed for the genotype with the highest F_s (Oudenberg), showing that this genotype used an important fraction of its stem water storage to meet its transpiration demand. Skado had lower F_s than the other genotypes, while it was the highest yielding genotype of those considered in our study (Broeckx *et al.*, 2012). This was consistent with previous measurements of leaf gas exchange and intrinsic leaf water use efficiency (Broeckx *et al.*, 2014a) and of carbon isotope discrimination (Verlinden *et al.*, 2015) performed during the same period in 2011. Therefore, at our site, Skado had the highest water use efficiency as compared to Oudenberg and Grimminge. In addition, carbon isotope discrimination techniques already showed that *P. deltoides* and *P. nigra* (both male and female parental species of Oudenberg) were less water use efficient than *P. trichocarpa* (female parent of Skado) at an experimental poplar plantation in central France (Dillen *et al.*, 2008). During a summer drought at a freely draining site in the United Kingdom, genotype Beaupré (*P. trichocarpa* × *P. deltoides*) was able to maintain its transpiration rate for a longer period than genotype Dorschkamp (*P. deltoides* × *P. nigra*) (Hall & Allen, 1997). In response to prolonged drought, stems of SRC poplar genotypes shrank as trees were unable to refill their stem water storage reserves (Giovanelli *et al.*, 2007).

In conclusion, we observed that the SRC poplar of our study, which was not water limited in the year 2011, consumed less water as compared to a reference grassland. Moreover, E_c contributed for 69% to the total growing season ET. The F_s scaling approach for the intensive field campaign yielded similar results as the modelling exercise. At tree level, we observed important intergenotypic differences in both F_s and ΔD showing different water use strategies associated with different growth strategies among the three genotypes. For our site, Skado had the highest water use efficiency as compared to Oudenberg and Grimminge. Besides harvestable yield, tree water use should be considered as a key criterion in SRC management through careful genotype selection. More large-scale experiments combining measurements at leaf, plant and stand level under different rotations are necessary to better understand and quantify the water use of SRC at different scales.

Acknowledgements

This research was funded by the European Commission's Seventh Framework Program (FP7/2007–2013) as European Research Council Advanced Grant (no. 233366, POPFULL) as well as by the Flemish Hercules Foundation as Infrastructure Contract ZW09-06. Further funding was provided by the Flemish Methusalem Programme and by the Research Council of the University of Antwerp within the framework of a bilateral exchange programme between the Universities of Antwerp and Orléans (French–Flemish EGIDE programme Tournesol, project no. 27323PA, 2012–2013). We thank D. Zona for eddy covariance flux data, J. Cools for technical support, K. Mouton for logistic support at the field site and K. Steppe for advice on sap flow sensor set-up.

References

Albaugh JM, Domec J-C, Maier CA, Sucre EB, Leggett ZH, King JS (2014) Gas exchange and stand-level estimates of water use and gross primary productivity in an experimental pine and switchgrass intercrop forestry system on the Lower Coastal Plain of North Carolina, U.S.A. *Agricultural and Forest Meteorology*, **192–193**, 27–40.

Allen RG, Pereira LS, Raes D, Smith M (1998) Crop evapotranspiration – guidelines for computing crop water requirements. In: *FAO Irrigation and Drainage Paper 56*. (ed. FAO). FAO, Rome, Italy.

Allen SJ, Hall RL, Rosier PTW (1999) Transpiration by two poplar varieties grown as coppice for biomass production. *Tree Physiology*, **19**, 493–501.

Aylott MJ, Casella E, Tubby I, Street NR, Smith P, Taylor G (2008) Yield and spatial supply of bioenergy poplar and willow short-rotation coppice in the UK. *New Phytologist*, **178**, 358–370.

Baker JM, Van Bavel CHM (1987) Measurement of mass flow of water in the stems of herbaceous plants. *Plant, Cell & Environment*, **10**, 777–782.

Baldocchi DD (2003) Assessing the eddy covariance technique for evaluating carbon dioxide exchange rates of ecosystems: past, present and future. *Global Change Biology*, **9**, 479–492.

Berndes G, Hoogwijk M, Van Den Broek R (2003) The contribution of biomass in the future global energy supply: a review of 17 studies. *Biomass and Bioenergy*, **25**, 1–28.

Bovard BD, Curtis PS, Vogel CS, Su HB, Schmid HP (2005) Environmental controls on sap flow in a northern hardwood forest. *Tree Physiology*, **25**, 31–38.

Broeckx LS, Verlinden MS, Ceulemans R (2012) Establishment and two-year growth of a bio-energy plantation with fast-growing Populus trees in Flanders (Belgium): effects of genotype and former land use. *Biomass and Bioenergy*, **42**, 151–163.

Broeckx LS, Fichot R, Verlinden MS, Ceulemans R (2014a) Seasonal variations in photosynthesis, intrinsic water-use efficiency and stable isotope composition of poplar leaves in a short-rotation plantation. *Tree Physiology*, **34**, 701–715.

Broeckx LS, Verlinden MS, Berhongaray G, Zona D, Fichot R, Ceulemans R (2014b) The effect of a dry spring on seasonal carbon allocation and vegetation dynamics in a poplar bioenergy plantation. *Global Change Biology Bioenergy*, **6**, 473–487.

Broeckx L, Vanbeveren S, Verlinden M, Ceulemans R (2015) First vs. second rotation of a poplar short rotation coppice: leaf area development, light interception and radiation use efficiency. *iForest – Biogeosciences and Forestry*, **8**, 565–573.

Conejero W, Alarcon JJ, Garcia-Orellana Y, Abrisqueta JM, Torrecillas A (2007) Daily sap flow and maximum daily trunk shrinkage measurements for diagnosing water stress in early maturing peach trees during the post-harvest period. *Tree Physiology*, **27**, 81–88.

De Swaef T, De Schepper V, Vandegehuchte MW, Steppe K (2015) Stem diameter variations as a versatile research tool in ecophysiology. *Tree Physiology*, doi: 10.1093/treephys/tpv1080.

Dillen SY, Marron N, Koch B, Ceulemans R (2008) Genetic variation of stomatal traits and carbon isotope discrimination in two hybrid poplar families (*Populus deltoides* 'S9-2' × *P. nigra* 'Ghoy' and *P. deltoides* 'S9-2' × *P. trichocarpa* 'V24'). *Annals of Botany*, **102**, 399–407.

Dimitriou I, Busch G, Jacobs S, Schmidt-Walter P, Lamersdorf N (2009) A review of the impacts of Short Rotation Coppice cultivation on water issues. *Landbauforschung Volkenrode*, **59**, 197–206.

Dondeyne S, Vanierschot L, Langohr R, Van Ranst E, Deckers J (2015) – De grote bodemgroepen van Vlaanderen: Kenmerken van de "Reference Soil Groups" volgens het internationale classificatiesysteem World Reference Base. KU Leuven & Universiteit Gent in opdracht van Vlaamse overheid, Departement Leefmilieu, Natuur en Energie, Afdeling Land en Bodembescherming, Ondergrond, Natuurlijke Rijkdommen. doi: 10.13140/RG.2.1.2428.3044.

Ewers BE, Oren R, Johnsen KH, Landsberg JJ (2001) Estimating maximum mean canopy stomatal conductance for use in models. *Canadian Journal of Forest Research*, **31**, 198–207.

Fischer M, Trnka M, Kucera J *et al.* (2013) Evapotranspiration of a high-density poplar stand in comparison with a reference grass cover in the Czech–Moravian Highlands. *Agricultural and Forest Meteorology*, **181**, 43–60.

Foken T, Wichura B (1996) Tools for quality assessment of surface-based flux measurements. *Agricultural and Forest Meteorology*, **78**, 83–105.

Foken T, Göckede M, Mauder M, Mahrt L, Amiro B, Munger JW (2004) Postfield data quality control. In: *Handbook of Micrometeorology, A Guide for Surface Flux Measurement and Analysis* (eds Lee X, Massmann W, Beverly L), pp. 181–208. Kluwer Academic Publisher, Dodrecht, The Netherlands.

Fritts DC (1961) An evaluation of three techniques for measuring radial tree growth. *Bulletin of Ecological Society America*, **42**, 54–55.

Giovannelli A, Deslauriers A, Fragnelli G, Scaletti L, Castro G, Rossi S, Crivellaro A (2007) Evaluation of drought response of two poplar clones (*Populus x canadensis Mönch 'I-214' and P. deltoides* Marsh. 'Dvina') through high resolution analysis of stem growth. *Journal of Experimental Botany*, **58**, 2673–2683.

Graham RL, Wright LL, Turhollow AF (1992) The potential for short-rotation woody crops to reduce United-States CO_2 emissions. *Climatic Change*, **22**, 223–238.

Grip H, Halldin S, Lindroth A (1989) Water-use by intensively cultivated willow using estimated stomatal parameter values. *Hydrological Processes*, **3**, 51–63.

Gustavsson L, Borjesson P, Johansson B, Svenningsson P (1995) Reducing CO_2 emissions by substituting biomass for fossil-fuels. *Energy*, **20**, 1097–1113.

Hall RL, Allen SJ (1997) Water use of poplar clones grown as short-rotation coppice at two site in the United Kingdom. *Aspects of Applied Biology*, **49**, 163–172.

Hall RL, Allen SJ, Rosier PTW, Hopkins R (1998) Transpiration from coppiced poplar and willow measured using sap-flow methods. *Agricultural and Forest Meteorology*, **90**, 275–290.

Heilman PE, Hinckley TM, Roberts DA, Ceulemans R (1996) Production physiology. In: *Biology of Populus and its Implications for Management and Conservation* (eds Stettler RF, Bradshaw HD Jr, Heilman PE, Hinckley TM), Chapter 18, pp. 459–490. NRC Research Press, Ottawa, Canada.

Herrick AM, Brown CL (1967) A new concept in cellulose production silage sycamore. *Agricultural Science Review*, **5**, 8–13.

Hinckley TM, Brooks JR, Cermak J, Ceulemans R, Kucera J, Meinzer FC, Roberts DA (1994) Water flux in a hybrid poplar stand. *Tree Physiology*, **14**, 1005–1018.

Hou LG, Xiao HL, Si JH, Xiao SC, Zhou MX, Yang YG (2010) Evapotranspiration and crop coefficient of *Populus euphratica Oliv.* forest during the growing season in the extreme arid region northwest China. *Agricultural Water Management*, **97**, 351–356.

Hsiao TC, Heng L, Steduto P, Rojas-Lara B, Raes D, Fereres E (2009) AquaCrop—the FAO crop model to simulate yield response to water: III. Parameterization and testing for maize. *Agronomy Journal*, **101**, 448–459.

IEA-Bioenergy (2011) Quantifying environmental effects of short rotation coppice (SRC) on biodiversity, soil and water – Task 43. pp. 34.

Impens II, Schalck JM (1965) A very sensitive electric dendrograph for recording radial changes of a tree. *Ecology*, **46**, 183–184.

Jassal RS, Black TA, Arevalo C, Jones H, Bhatti JS, Sidders D (2013) Carbon sequestration and water use of a young hybrid poplar plantation in north-central Alberta. *Biomass and Bioenergy*, **56**, 323–333.

Kauter D, Lewandowski I, Claupein W (2003) Quantity and quality of harvestable biomass from Populus short rotation coppice for solid fuel use – a review of the physiological basis and management influences. *Biomass and Bioenergy*, **24**, 411–427.

Kim HS, Oren R, Hinckley TM (2008) Actual and potential transpiration and carbon assimilation in an irrigated poplar plantation. *Tree Physiology*, **28**, 559–577.

King JS, Ceulemans R, Albaugh JM *et al.* (2013) The challenge of lignocellulosic bioenergy in a water-limited world. *BioScience*, **63**, 102–117.

Kocher P, Horna V, Leuschner C (2012) Environmental control of daily stem growth patterns in five temperate broad-leaved tree species. *Tree Physiology*, **32**, 1021–1032.

Kozlowski TT, Winget CH (1964) Diurnal and seasonal variation in radii of tree stems. *Ecology*, **45**, 149–155.

Larcher W (2003) *Physiological Plant Ecology*. Springer-Verlag, New York, NY, USA.

Linderson ML, Iritz Z, Lindroth A (2007) The effect of water availability on stand-level productivity, transpiration, water use efficiency and radiation use efficiency of field-grown willow clones. *Biomass and Bioenergy*, **31**, 460–468.

Meiresonne L, Nadezhdin N, Cermak J, Van Slycken J, Ceulemans R (1999) Measured sapflow and simulated transporation from a poplar stand in Flanders (Belgium). *Agricultural and Forest Meteorology*, **96**, 165–179.

Migliavacca M, Meroni M, Manca G *et al.* (2009) Seasonal and interannual patterns of carbon and water fluxes of a poplar plantation under peculiar eco-climatic conditions. *Agricultural and Forest Meteorology*, **149**, 1460–1476.

Nagler P, Jetton A, Fleming J *et al.* (2007) Evapotranspiration in a cottonwood (*Populus fremontii*) restoration plantation estimated by sap flow and remote sensing methods. *Agricultural and Forest Meteorology*, **144**, 95–110.

Navarro A, Facciotto G, Campi P, Mastrorilli M (2014) Physiological adaptations of five poplar genotypes grown under SRC in the semi-arid Mediterranean environment. *Trees*, **28**, 983–994.

Oishi AC, Oren R, Stoy PC (2008) Estimating components of forest evapotranspiration: a footprint approach for scaling sap flux measurements. *Agricultural and Forest Meteorology*, **148**, 1719–1732.

Oren R, Phillips N, Katul G, Ewers BE, Pataki DE (1998) Scaling xylem sap flux and soil water balance and calculating variance: a method for partitioning water flux in forests. *Annales des Sciences Forestieres*, **55**, 191–216.

Persson G, Lindroth A (1994) Simulating evaporation from short-rotation forest – variations within and between seasons. *Journal of Hydrology*, **156**, 21–45.

Petzold R, Schwarzel K, Feger KH (2011) Transpiration of a hybrid poplar plantation in Saxony (Germany) in response to climate and soil conditions. *European Journal of Forest Research*, **130**, 695–706.

Raes D, Steduto P, Hsiao TC, Fereres E (2009) AquaCrop – the FAO crop model to simulate yield response to water: II. Main algorithms and software description. *Agronomy Journal*, **101**, 438–447.

Raes D, Steduto P, Hsiao TC, Fereres E (2012) Reference Manual AquaCrop (Version 4.0). AquaCrop. Available at: http://www.fao.org/ (accessed 1 August 2015).

Sakuratani T (1981) A heat balance method for measuring water flux in the stem of intact plants. *Journal of Agricultural Meteorology*, **37**, 9–17.

Schafer KVR, Oren R, Lai CT, Katul GG (2002) Hydrologic balance in an intact temperate forest ecosystem under ambient and elevated atmospheric CO_2 concentration. *Global Change Biology*, **8**, 895–911.

Steduto P, Hsiao TC, Raes D, Fereres E (2009) AquaCrop – the FAO crop model to simulate yield response to water: I. Concepts and underlying principles. *Agronomy Journal*, **101**, 426–437.

Steduto P, Hsiao TC, Fereres E, Raes D (2012) *Crop yield response to water, FAO Irrigation and drainage paper 66*. FAO, Rome, Italy. 500 pp.

Steppe K, Lemeur R, Dierick D (2005) Unravelling the relationship between stem temperature and air temperature to correct for errors in sap-flow calculations using stem heat balance sensors. *Functional Plant Biology*, **32**, 599–609.

Tang JW, Bolstad PV, Ewers BE, Desai AR, Davis KJ, Carey EV (2006) Sap flux-upscaled canopy transpiration, stomatal conductance, and water use efficiency in an old growth forest in the Great Lakes region of the United States. *Journal of Geophysical Research-Biogeosciences*, **111**. doi: 10.1029/2005jg000083.

Tricker PJ, Pecchiari M, Bunn SM, Vaccari FP, Peressotti A, Miglietta F, Taylor G (2009) Water use of a bioenergy plantation increases in a future high CO_2 world. *Biomass and Bioenergy*, **33**, 200–208.

Trnka M, Trnka M, Fialova J, Koutecky V, Fajman M, Zalud Z, Hejduk S (2008) Biomass production and survival rates of selected poplar clones grown under a short-rotation system on arable land. *Plant Soil and Environment*, **54**, 78–88.

Unsworth MH, Phillips N, Link T *et al.* (2004) Components and controls of water flux in an old-growth Douglas-fir-western hemlock ecosystem. *Ecosystems*, **7**, 468–481.

Vanuytrecht E, Raes D, Steduto P *et al.* (2014) AquaCrop: FAO's crop water productivity and yield response model. *Environmental Modelling & Software*, **62**, 351–360.

Verlinden MS, Fichot R, Broeckx LS, Vanholme B, Boerjan W, Ceulemans R (2015) Carbon isotope compositions ($\delta^{13}C$) of leaf, wood and holocellulose differ among genotypes of poplar and between previous land uses in a short-rotation biomass plantation. *Plant, Cell & Environment*, **38**, 144–156.

Zalesny RS Jr, Wiese AH, Bauer EO, Riemenschnieder DE (2006) Sapflow of hybrid popler (Populus nigra L. x P. maximowiczii A. Henry 'NM'6) during phytoremediation of landfill leachate. *Biomass and Bioenergy*, **30**, 784–793.

Zenone T, Zona D, Gelfand I, Gielen B, Camino-Serrano M, Ceulemans R (2015) CO_2 uptake is offset by CH_4 and N_2O emissions in a poplar short-rotation coppice. *Global Change Biology Bioenergy*, doi: 10.1111/gcbb.12269.

Zona D, Janssens IA, Aubinet M, Gioli B, Vicca S, Fichot R, Ceulemans R (2013a) Fluxes of the greenhouse gases (CO_2, CH_4 and N_2O) above a short-rotation poplar plantation after conversion from agricultural land. *Agricultural and Forest Meteorology*, **169**, 100–110.

Zona D, Janssens IA, Gioli B, Jungkunst HF, Serrano MC, Ceulemans R (2013b) N_2O fluxes of a bio-energy poplar plantation during a two years rotation period. *Global Change Biology Bioenergy*, **5**, 536–547.

Zweifel R, Item H, Hasler R (2000) Stem radius changes and their relation to stored water in stems of young Norway spruce trees. *Trees-Structure and Function*, **15**, 50–57.

Zweifel R, Item H, Hasler R (2001) Link between diurnal stem radius changes and tree water relations. *Tree Physiology*, **21**, 869–877.

PERMISSIONS

 Every chapter published in this book has been scrutinized by our experts. Their significance has been extensively debated. The topics covered herein carry significant findings which will fuel the growth of the discipline. They may even be implemented as practical applications or may be referred to as a beginning point for another development.

The contributors of this book come from diverse backgrounds, making this book a truly international effort. This book will bring forth new frontiers with its revolutionizing research information and detailed analysis of the nascent developments around the world.

We would like to thank all the contributing authors for lending their expertise to make the book truly unique. They have played a crucial role in the development of this book. Without their invaluable contributions this book wouldn't have been possible. They have made vital efforts to compile up to date information on the varied aspects of this subject to make this book a valuable addition to the collection of many professionals and students.

This book was conceptualized with the vision of imparting up-to-date information and advanced data in this field. To ensure the same, a matchless editorial board was set up. Every individual on the board went through rigorous rounds of assessment to prove their worth. After which they invested a large part of their time researching and compiling the most relevant data for our readers.

The editorial board has been involved in producing this book since its inception. They have spent rigorous hours researching and exploring the diverse topics which have resulted in the successful publishing of this book. They have passed on their knowledge of decades through this book. To expedite this challenging task, the publisher supported the team at every step. A small team of assistant editors was also appointed to further simplify the editing procedure and attain best results for the readers.

Apart from the editorial board, the designing team has also invested a significant amount of their time in understanding the subject and creating the most relevant covers. They scrutinized every image to scout for the most suitable representation of the subject and create an appropriate cover for the book.

The publishing team has been an ardent support to the editorial, designing and production team. Their endless efforts to recruit the best for this project, has resulted in the accomplishment of this book. They are a veteran in the field of academics and their pool of knowledge is as vast as their experience in printing. Their expertise and guidance has proved useful at every step. Their uncompromising quality standards have made this book an exceptional effort. Their encouragement from time to time has been an inspiration for everyone.

The publisher and the editorial board hope that this book will prove to be a valuable piece of knowledge for researchers, students, practitioners and scholars across the globe.

LIST OF CONTRIBUTORS

Gonzalo Berhongaray, Melanie S. Verlinden, Laura S. Broeckx, Ivan A. Janssens and Reinhart Ceulemans
Department of Biology, Research Centre of Excellence on Plant and Vegetation Ecology, University of Antwerp, Universiteitsplein 1, B-2610 Wilrijk, Belgium

Ji Gao, Erda Lin and Aiping Zhang
Institute of Environment and Sustainable Development in Agriculture, Chinese Academy of Agricultural Sciences, Beijing 100081, China

Shu Kee Lam
Crop and Soil Sciences Section, Faculty of Veterinary and Agricultural Sciences, the University of Melbourne, Melbourne, Vic. 3010, Australia

Leon E. Clarke, James A. Edmonds, Page G. Kyle and Sha Yu
Joint Global Change Research Institute, Pacific Northwest National Laboratory and University of Maryland, College Park, MD 20740, USA

Xuesong Zhang
Joint Global Change Research Institute, Pacific Northwest National Laboratory and University of Maryland, College Park, MD 20740, USA
Great Lakes Bioenergy Research Center, Michigan State University, East Lansing, MI 48824, USA

Allison M. Thomson
Field to Market, The Alliance for Sustainable Agriculture, 777 N Capitol St. NE, Suite 803, Washington, DC 20002, USA

Kejun Jiang
Energy Research Institute (ERI), Beijing 100038, China

Yuyu Zhou
Department of GeologicalandAtmospheric Sciences, Iowa State University, Ames, IA 50011, USA

Sheng Zhou
Institutes of Energy, Environment and Economy, Tsinghua University, Beijing 100084, China

Zoe M. Harris, Maud Viger, Joe R. Jenkins and Gail Taylor
University of Southampton, Southampton, SO17 1BJ, UK

Giorgio Alberti
University of Southampton, Southampton, SO17 1BJ, UK
University of Udine, Via delle Scienze 206, 33100 Udine, Italy

Rebecca Rowe and Niall P. Mcnamara
Centre for EcologyandHydrology, Lancaster Environment Centre, Library Avenue, Bailrigg, Lancaster, UK

Järvi Järveoja, Martin Maddison and Alar Teemusk
Department of Geography, Institute of Ecology and Earth Sciences, University of Tartu, 46 Vanemuise St, Tartu 51014, Estonia

Matthias Peichl
Department of Forest Ecology and Management, Swedish University of Agricultural Sciences Skogsmarksgränd 1, 90183 Umeå, Sweden

Ülo mander
Department of Geography, Institute of Ecology and Earth Sciences, University of Tartu, 46 Vanemuise St, Tartu 51014, Estonia
Hydrosystems and Bioprocesses Research Unit, National Research Institute of Science and Technology for Environment and Agriculture (Irstea), 1 rue Pierre-Gilles de Gennes CS 10030, F92761 Antony Cedex, France

Jacob M. Jungers, James O. Eckberg, Kevin Betts, Donald L. Wyse and Craig C. Sheaffer
Department of Agronomy and Plant Genetics, University of Minnesota, 411 Borlaug Hall, 1991 Upper Buford Circle, Saint Paul, MN 55108, USA

Margaret E. Mangan
Minnesota Department of Agriculture, 625 Robert Street N, Saint Paul, MN 55155, USA

Jérôme Laganière, David Paré and Pierre Y. Bernier
Natural Resources Canada, Canadian Forest Service, Laurentian Forestry Centre, Québec, QC G1V 4C7, Canada

Evelyne Thiffault
Département des sciences du bois et de la forêt, Université Laval, Québec, QC G1K 7P4, Canada

John E. Erickson, Lynn E. Sollenberger and Diane L. Rowland
Agronomy Department, University of Florida, 3105 McCarty Hall B, Gainesville, FL 32611, USA

Xi Liang
Agronomy Department, University of Florida, 3105 McCarty Hall B, Gainesville, FL 32611, USA
Department of Plant, Soil and Entomological Sciences, University of Idaho, 1693 S 2700 W, Aberdeen, ID 83210, USA

Maria L. Silveira
Department of Soil and Water Sciences, University of Florida, 3401 Experimental Station, Ona, FL 33865, USA

Geoffrey P. Morris and Zhenbin Hu
Department of Agronomy, Kansas State University, Manhattan, KS 66506, USA

Paul P. Grabowski
USDA-ARS Dairy Forage Research Center, Madison, WI 53706, USA

Justin O. Borevitz
Research School of Biology, Australian National University, Acton, ACT 2601, Australia

Marie-Anne De Graaff
Department of Biological Sciences, Boise State University, Boise, ID 83725, USA

R. Michael Miller and Julie D. Jastrow
Biosciences Division, Argonne National Laboratory, Argonne, IL 60439, USA

Zhangcai Qin, Qianlai Zhuang and Ximing Cai
Department of Earth, Atmospheric, and Planetary Sciences, Purdue University, West Lafayette, IN 47907, USA
Department of Agronomy, Purdue University, West Lafayette, IN 47907, USA
Ven Te Chow Hydrosystems Laboratory, Department of Civil and Environmental Engineering, University of Illinois at UrbanaChampaign, Urbana, IL 61801, USA

Scott M. Swinton and Sophia Tanner
Department of Agricultural, Food, and Resource Economics, Great Lakes Bioenergy Research Center, Michigan State University, East Lansing, MI, USA

Bradford L. Barham and Daniel F. Mooney
Department of Agricultural and Applied Economics, Great Lakes Bioenergy Research Center, University of Wisconsin-Madison, Madison, WI, USA

Theodoros Skevas
Gulf Coast Research and Education Center, University of Florida, Wimauma, FL, USA

Raj Cibin
Department of Agricultural and Biological Engineering, Purdue University, West Lafayette, IN, USA

Elizabeth M. Trybula
Department of Agricultural and Biological Engineering, Purdue University, West Lafayette, IN, USA
Department of Agronomy, Purdue University, West Lafayette, IN, USA

Jennifer L. Burks, Sylvie M. Brouder and Jeffrey J. Volenec
Department of Agronomy, Purdue University, West Lafayette, IN, USA

Indrajeet Chaubey
Department of Agricultural and Biological Engineering, Purdue University, West Lafayette, IN, USA
Department of Earth, Atmospheric and Planetary Sciences, Purdue University, West Lafayette, IN, USA

Gary Lanigan
Teagasc Environmental Research Centre, Johnstown Castle, Co. Wexford, Ireland

Órlaith níchoncubhair
Teagasc Environmental Research Centre, Johnstown Castle, Co. Wexford, Ireland
UCD School of BiologyandEnvironmental Science, University College Dublin, Dublin 4, Ireland

Bruce Osborne
UCD School of BiologyandEnvironmental Science, University College Dublin, Dublin 4, Ireland
UCD Earth Institute, University College Dublin, Dublin 4, Ireland

John Finnan
Teagasc Crops Research Centre, Oak Park, Carlow, Ireland

Emily R. Kuzmick and Nicholas Niechayev
Voinovich School of Leadership and Public Affairs, Ohio University, Athens, OH 45701, USA

Sarah C. Davis
Voinovich School of Leadership and Public Affairs, Ohio University, Athens, OH 45701, USA
Department of Environmental and Plant Biology, Ohio University, Athens, OH 45701, USA

Douglas J. Hunsaker
USDA-ARS Arid Lands Agricultural Research Center, Maricopa, AZ 85138, USA

Eric S. Fabio, Michelle J. Serapiglia and Lawrence B. Smart
Horticulture Section, School of Integrative Plant Science, New York State Agricultural Experiment Station, Cornell University, Geneva, NY 14456, USA

Timothy A. Volk
Department of Forest and Natural Resources Management, State University of New York College of Environmental Science and Forestry, Syracuse, NY 13210, USA

Raymond O. Miller
Forest Biomass Innovation Center, Michigan State University, Escanaba, MI 49829, USA

Hugh G. Gauch
Soil and Crop Sciences Section, School of Integrative Plant Science, Cornell University, Ithaca, NY 14853, USA

Ken C. J. Van Rees and Ryan d. Hangs
Department of Soil Science, University of Saskatchewan, Saskatoon, SK S7N 5A8, Canada

Beyhan Y. Amichev
Center for Northern Agroforestry and Afforestation, University of Saskatchewan, Saskatoon, SK S7N 5A8, Canada

Yulia A. Kuzovkina
Department of Plant Science, University of Connecticut, Storrs, CT 06269, USA

Michel Labrecque
Institut de Recherche en Biologie Végétale, University of Montréal, Montréal, QC H3C 3J7, Canada

Gregg A. Johnson
Southern Research and Outreach Center, University of Minnesota, Waseca, MN 56093, USA

Robert G. Ewy
Department of Biology, State University of New York at Potsdam, Potsdam, NY 13676, USA

Gary J. Kling
Department of Crop Sciences, University of Illinois, Urbana, IL 61801, USA

Jasper bloemen, Joanna A. Horemans, Laura S. Broeckx, Melanie S. Verlinden, Terenzio Zenone and Reinhart Ceuleman
Department of Biology, Research Centre of Excellence on Plant and Vegetation Ecology, University of Antwerp, Universiteitsplein 1, Wilrijk B-2610, Belgium

Régis fichot
Université d'Orléans, INRA, LBLGC, EA 1207, F-45067 Orléans, France

Index

L

M

N

O

P

R

S

T

V

W

Y

www.ingramcontent.com/pod-product-compliance
Lightning Source LLC
Chambersburg PA
CBHW082044190326
41458CB00010B/3458

* 9 7 8 1 6 4 7 4 0 1 2 0 7 *